Strength in Numbers: The Rising of Academic Statistics Departments in the U.S.

Alan Agresti · Xiao-Li Meng
Editors

Strength in Numbers: The Rising of Academic Statistics Departments in the U.S.

Springer

Editors
Alan Agresti
Department of Statistics
University of Florida
Gainesville, FL
USA

Xiao-Li Meng
Department of Statistics
Harvard University
Cambridge, MA
USA

ISBN 978-1-4614-3648-5 ISBN 978-1-4614-3649-2 (eBook)
DOI 10.1007/978-1-4614-3649-2
Springer New York Heidelberg Dordrecht London

Library of Congress Control Number: 2012942702

Printed on acid-free paper

Springer is part of Springer Science+Business Media (www.springer.com)

To our mentor and friend Stephen Stigler on his 70th birthday.

Contents

Contributors

Alan Agresti Department of Statistics, University of Florida, 102 Griffin-Floyd Hall, Gainesville, FL 32611-8545, USA, e-mail: aa@stat.ufl.edu

Jesse Arnold Department of Statistics, Virginia Tech University, 406 A Hutcheson Hall, Blacksburg, VA 24061-0439, USA, e-mail: jca@vt.edu

James Booth Department of Statistical Science, Cornell University, 301 Malott Hall, Ithaca, NY 14853-3801, USA, e-mail: jim.booth@cornell.edu

John E. Boyer, Jr. Department of Statistics, Kansas State University, 101 Dickens Hall, Manhattan, KS 66506-0802, USA, e-mail: jboyer@ksu.edu

Norm Breslow Department of Biostatistics, University of Washington, Health Sciences Building F, Seattle, WA 98195, USA, e-mail: norm@u.washington.edu

Jianwen Cai Department of Biostatistics, University of North Carolina, 3101 McGavran–Greenberg, Campus Box 7420, Chapel Hill, NC 27599-7420, USA, e-mail: cai@bios.unc.edu

Ming-Hui Chen Department of Statistics, University of Connecticut, 215 Glenbrook Road, U-4120, Storrs, CT 06269-4020, USA, e-mail: ming-hui.chen@uconn.edu

Martha Coventry School of Statistics, University of Minnesota, 313 Ford Hall, 224 Church Street S.E., Minneapolis, MN 55455, USA, e-mail: coven002@umn.edu

Dipak Dey Department of Statistics, University of Connecticut, 215 Glenbrook Road, U-4120, Storrs, CT 06269-4020, USA, e-mail: dipak.dey@uconn.edu

Rebecca Doerge Department of Statistics, Purdue University, 250 N. University Street, West Lafayette, IN 47907-2066, USA, e-mail: doerge@purdue.edu

Norman Draper Department of Statistics, University of Wisconsin, 1220 Medical Sciences Center, 1300 University Avenue, Madison, WI 53706-1685, USA, e-mail: draper@stat.wisc.edu

Nancy Flournoy Department of Statistics, University of Missouri–Columbia, 146 Middlebush Hall, Columbia, MO 65211, USA, e-mail: FlournoyN@Missouri.edu

Melody Gaden Department of Statistics, University of Missouri–Columbia, 146 Middlebush Hall, Columbia, MO 65211, USA, e-mail: GalenM@missouri.edu

Joe L. Gastwirth Department of Statistics, George Washington University, 2140 Pennsylvania Avenue NW, Washington, DC 20052, USA, e-mail: jlgast@gwu.edu

Edward I. George Department of Statistics, The Wharton School, University of Pennsylvania, Philadelphia, PA 19104-6302, USA, e-mail: edgeorge @wharton.upenn.edu

Tom Gerig Department of Statistics, North Carolina State University, 2311 Stinson Drive, Campus Box 8203, Raleigh, NC 27695-8203, USA, e-mail: tom_gerig@ncsu.edu

Dennis Gilliland Department of Statistics, Michigan State University, A415/ A435 Wells Hall, East Lansing, MI 48824, USA, e-mail: gillilan@stt.msu.edu

Leon Gleser University of Pittsburgh, 2717 Cathedral of Learning, Pittsburgh, PA 15260, USA, e-mail: gleser@pitt.edu

Richard Gunst Department of Statistical Science, Southern Methodist University, P.O. Box 750332, Dallas, TX 75275-0332, USA, e-mail: rgunst@mail.smu.edu

Bill Harkness Department of Statistics, Penn State University, 326 Thomas Building, University Park, PA 16802, USA, e-mail: wlh@stat.psu.edu

John Hartigan Department of Statistics, Yale University, P.O. Box 208290, New Haven, CT 06520-8290, USA, e-mail: john.hartigan@yale.edu

Tom Hettmansperger Department of Statistics, Penn State University, 326 Thomas Building, University Park, PA 16802, USA, e-mail: tph@stat.psu.edu

Klaus Hinkelmann Department of Statistics, Virginia Tech University, 406 A Hutcheson Hall, Blacksburg, VA 24061-0439, USA, e-mail: khstat@vt.edu

David Hoaglin 73 Hickory Road, Sudbury, MA 01776, USA, e-mail: dchoaglin@gmail.com

Bob Hogg Department of Statistics, The University of Iowa, 241 Schaeffer Hall, Iowa, IA 52242-1409, USA, e-mail: rvhogg@gametrail.org

Alan Hutson Department of Biostatistics, University at Buffalo School of Public Health and Health Professions, 706 Kimball Tower, 3435 Main St., Buffalo, NY 14214-3000, USA, e-mail: ahutson@buffalo.edu

Richard Johnson Department of Statistics, University of Wisconsin, 1220 Medical Sciences Center, 1300 University Avenue, Madison, WI 53706-1685, USA, e-mail: rich@stat.wisc.edu

Karen Kafadar Department of Statistics, Indiana University Bloomington, 309 North Park Avenue, Bloomington, IN 47408, USA, e-mail: kkafadar@gmail.com

Douglas Kelly Department of Statistics and Operations Research, University of North Carolina, 318 Hanes Hall CB#3260, Chapel Hill, NC 27599-3260, USA, e-mail: dgkelly@email.unc.edu

Kenneth Koehler Department of Statistics & Statistical Laboratory, Iowa State University, Snedecor Hall, Ames, IA 50011-1210, USA, e-mail: kkoehler@iastate.edu

Michael Kosorok Department of Biostatistics, University of North Carolina, 3101 McGavran–Greenberg, Campus Box 7420, Chapel Hill, NC 27599-7420, USA, e-mail: kosorok@unc.edu

Abba M. Krieger Department of Statistics, The Wharton School, University of Pennsylvania, Philadelphia, PA 19104-6302, USA, e-mail: krieger@wharton.upenn.edu

Lynn Kuo Department of Statistics, University of Connecticut, 215 Glenbrook Road, U-4120, Storrs, CT 06269-4020, USA, e-mail: lynn.kuo@uconn.edu

John Lachin III Department of Statistics, George Washington University, 2140 Pennsylvania Avenue NW, Washington, DC 20052, USA, e-mail: jml@bsc.gwu.edu

Nan Laird Department of Biostatistics, Harvard University, 655 Huntington Avenue SPH2, 4th Floor, Boston, MA 02115, USA, e-mail: laird@hsph.harvard.edu

Lisa LaVange Department of Biostatistics, University of North Carolina, 3101 McGavran–Greenberg, Campus Box 7420, Chapel Hill, NC 27599-7420, USA, e-mail: lisa_lavange@unc.edu

Fred Leysieffer Department of Statistics, Florida State University, 214 OSB, 117 N. Woodward Ave, P.O. Box 3064330, Tallahassee, FL 32306-4330, USA, e-mail: fleysief@mailer.fsu.edu

Dennis Lin Department of Statistics, Penn State University, 326 Thomas Building, University Park, PA 16802, USA, e-mail: dkl5@psu.edu

Hosam M. Mahmoud Department of Statistics, George Washington University, 2140 Pennsylvania Avenue NW, Washington, DC 20052, USA, e-mail: hosam@gwu.edu

Glen Meeden School of Statistics, University of Minnesota, 313 Ford Hall, 224 Church Street S.E., Minneapolis, MN 55455, USA, e-mail: glen@stat.umn.edu

William Mendenhall III Deceased

Xiao-Li Meng Department of Statistics, Harvard University Science Center, 1 Oxford Street, Cambridge, MA 02138-2901, USA, e-mail: meng@stat.harvard.edu

Reza Modarres Department of Statistics, George Washington University, 2140 Pennsylvania Avenue NW, Washington, DC 20052, USA, e-mail: reza@gwu.edu

Jane Monaco Department of Biostatistics, University of North Carolina, 3101 McGavran–Greenberg, Campus Box 7420, Chapel Hill, NC 27599-7420, USA, e-mail: monaco@email.unc.edu

Donald F. Morrison Department of Statistics, The Wharton School, University of Pennsylvania, Philadelphia, PA 19104-6302, USA, e-mail: donaldm@wharton.upenn.edu

Nitis Mukhopadhyay Department of Statistics, University of Connecticut, 215 Glenbrook Road, U-4120, Storrs, CT 06269-4020, USA, e-mail: nitis.mukhopadhyay@uconn.edu

Joseph Naus Department of Statistics & Biostatistics, Rutgers University, Hill 506, Busch Campus, 110 Frelinghuysen Rd, Piscataway, NJ 08854-8019, USA, e-mail: naus@rci.rutgers.edu

James Neill Department of Statistics, Kansas State University, 101 Dickens Hall, Manhattan, KS 66506-0802, USA, e-mail: jwneill@ksu.edu

Jim Pitman Department of Statistics, University of California, Berkeley, 367 Evans Hall 3860, Berkeley, CA 94720, USA, e-mail: pitman@stat.berkeley.edu

Carol Redmond University of Pittsburgh, 2717 Cathedral of Learning, Pittsburgh, PA 15260, USA, e-mail: ckr3@pitt.edu

John Rice Department of Statistics, University of California, Berkeley, 367 Evans Hall 3860, Berkeley, CA 94720, USA, e-mail: jariceiii@gmail.com

Howard Rockette University of Pittsburgh, 2717 Cathedral of Learning, Pittsburgh, PA 15260, USA, e-mail: herbst@pitt.edu

Jim Rosenberger Department of Statistics, Penn State University, 326 Thomas Building, University Park, PA 16802, USA, e-mail: JLR@psu.edu

Jayaram Sethuraman Department of Statistics, Florida State University, 214 OSB, 117 N. Woodward Ave, P.O. Box 3064330, Tallahassee, FL 32306-4330, USA, e-mail: sethu@stat.fsu.edu

Bill Schucany Department of Statistical Science, Southern Methodist University, P.O. Box 750332, Dallas, TX 75275-0332, USA, e-mail: schucany@mail.smu.edu

Richard Scheaffer Department of Statistics, University of Florida, 102 Griffin-Floyd Hall, Gainesville, FL 32611-8545, USA, e-mail: scheaffe@stat.ufl.edu

Paul Shaman Department of Statistics, The Wharton School, University of Pennsylvania, Philadelphia, PA 19104-6302, USA, e-mail: shaman@wharton.upenn.edu

Eric P. Smith Department of Statistics, Virginia Tech University, 406 A Hutcheson Hall, Blacksburg, VA 24061-0439, USA, e-mail: epsmith@vt.edu

Margaret Smykla Department of Statistics, Carnegie Mellon University, Baker Hall, Pittsburgh, PA 15213, USA, e-mail: mk74@stat.cmu.edu

Jennifer South Department of Statistics, Texas A&M University, 3143 TAMU, College Station, TX 77843-3143, USA, e-mail: jennifer@stat.tamu.edu

Terry Speed Department of Statistics, University of California, Berkeley, 367 Evans Hall 3860, Berkeley, CA 94720, USA, e-mail: terry@stat.berkeley.edu

Sanvesh Srivastava Department of Statistics, Purdue University, 250 N. University Street, West Lafayette, IN 47907-2066, USA, e-mail: sanvesh@purdue.edu

James Stapleton Department of Statistics, Michigan State University, A415/A435 Wells Hall, East Lansing, MI 48824, USA, e-mail: staplet5@stt.msu.edu

Stephen Stigler Department of Statistics, The University of Chicago, 5734 S. University Avenue, Chicago, IL 60637, USA, e-mail: Stigler@galton.uchicago.edu

Kam-Wah Tsui Department of Statistics, University of Wisconsin, 1220 Medical Sciences Center, 1300 University Avenue, Madison, WI 53706-1685, USA, e-mail: kwtsui@stat.wisc.edu

Geoff Vining Department of Statistics, Virginia Tech University, 406 A Hutcheson Hall, Blacksburg, VA 24061-0439, USA, e-mail: Vining@vt.edu

Martin Wells Department of Statistical Science, Cornell University, 301 Malott Hall, Ithaca, NY 14853-3801, USA, e-mail: mtw1@cornell.edu

Wayne Woodward Department of Statistical Science, Southern Methodist University, P.O. Box 750332, Dallas, TX 75275-0332, USA, e-mail: waynew@mail.smu.edu

Brian Yandell Department of Statistics, University of Wisconsin, 1220 Medical Sciences Center, 1300 University Avenue, Madison, WI 53706-1685, USA, e-mail: byandell@wisc.edu

Marvin Zelen Department of Biostatistics, Harvard University, 655 Huntington Avenue SPH2, 4th Floor, Boston, MA 02115, USA, e-mail: zelen@jimmy.harvard.edu

Dale Zimmerman Department of Statistics, The University of Iowa, 241 Schaeffer Hall, Iowa, IA 52242-1409, USA, e-mail: dzimmer@stat.uiowa.edu

Statistics as an Academic Discipline

Alan Agresti and Xiao-Li Meng

Statistics as an academic discipline, having its own department in most research universities, is now well into its middle age. The recent years have seen a wealth of birthday parties. For example, the *IMS Bulletin* recently noted the seventy-fifth birthday celebration of the Statistical Laboratory at Iowa State and the sixtieth birthday of the Statistics Department at North Carolina State, and *Amstat News* highlighted the sixtieth birthday of Statistics at Virginia Tech, Biostatistics at the University of North Carolina, and Biostatistics at the University of Michigan (and the fortieth of Statistics at Michigan), and the fiftieth birthdays of Statistics at Harvard and Statistics at Wisconsin.

In fact, all but a few of the existing Statistics and Biostatistics departments have been created within the past 50–60 years. In many instances, Statistics had a significant presence on campus before the formation of a department, such as a Statistical Laboratory geared toward collaborative consulting. But in many universities Statistics consisted merely of elementary courses scattered around campus, and there was little, if any, scholarly activity in the field. The creation of a formal department typically resulted in a dramatic change-point in the magnitude and nature of such activity at a university.

University of Florida,
Department of Statistics,
102 Griffin-Floyd Hall,
Gainesville, FL 32611-8545, USA
A. Agresti (✉)
e-mail: aa@stat.ufl.edu

Harvard University,
Department of Statistics,
1 Oxford Street,
Cambridge, MA 02138-2901, USA
X.-L. Meng
e-mail: meng@stat.harvard.edu

A. Agresti and X.-L. Meng (eds.), *Strength in Numbers: The Rising of Academic Statistics Departments in the U.S.*, DOI: 10.1007/978-1-4614-3649-2_1,

Why This Book?

In many departments, the founders and the movers and shakers who fostered the growth of the Statistics or Biostatistics Department have been retiring and gradually leaving us. As an unfortunate consequence, the institutional memory of the early days of many departments is disappearing rather rapidly.

This book project began at a luncheon on September 15, 2009 at Harvard Square in which the two of us discussed the many recent birthday celebrations and our worries that the early days of Statistics and Biostatistics Departments were not well documented. The purpose of our book is to capture some of this information before it is lost forever. The book consists of 39 chapters, one for each department, that describes some of the key aspects of the history of the department, including its founding, its growth, key people in its development, notable students it produced, major research and pedagogical accomplishments, and a summary of where the department stands today and its vision for the future. Although the degrees of emphases on these aspects vary considerably with the chapters, we hope collectively they provide both a holistic and a detailed picture of the evolution of Statistics in the United States as a scientific discipline. Our goal has been to produce a volume that would arouse readers' interests by providing both historic vignettes and current snapshots of many major departments in Statistics and Biostatistics.

When we originally conceived this project, our vision was broader and included departments around the world in academia as well as in industry and government. It soon became apparent that this was overly ambitious. Because of space limitations for this book, we cover only the major U.S. Statistics and Biostatistics departments that were founded by the mid-1960s or at least had the direct roots of their founding well established by that time. If this book is viewed as successful, we hope that there will be at least one other volume to cover the "rest of Statistics," which of course is substantial.

The Evolution of Statistics/Biostatistics Departments

We hope that this book will help to show statisticians how our field has evolved, in terms of its academic development in the U.S. We envision that a young statistician who joins a department documented in this book will find it interesting to learn how that department got to where it is today and about some of the notable achievements along the way. We also hope this book could make a valuable contribution to the field of "disciplinary history," and may even encourage other scientific disciplines to consider editing analogous volumes.

Some other articles that contribute to an overview of how academic Statistics arose in the U.S. are listed in the bibliography below. As David (1998) pointed out, in the first part of the twentieth century statisticians tended to work in isolation within departments such as Mathematics, Economics, Psychology, Education, and Business. Indeed, at the twenty-fifth anniversary of IMS, Craig (1960) stated that "prior to 1920 a scant half-dozen American colleges and universities had, as member of the department, anyone who was seriously interested in the newly developing methods of scientific inference called mathematical statistics. The American Mathematical Society took a very dim view of the whole business and looked upon these mavericks (statisticians) with a suspicion of quackery."

The first departments to focus primarily on Statistics seem to have been the Johns Hopkins Department of Biometry and Vital Statistics, founded in 1918, and the University of Pennsylvania Department of Economic and Social Statistics, founded in 1931. The Johns Hopkins group seems to have been the first department with the word "Statistics" in its title (in the U.S., as Karl Pearson's department at University College, London, started in 1911). David (1998) noted that the formation of the Statistical Laboratory at Iowa State in 1933 may have been the first grouping of statisticians recognized as deserving a unit of their own. Other groups formed before World War II included the George Washington Statistics Department in 1935 (the first in a College of Arts and Sciences) and the Statistical Laboratory at the University of California at Berkeley in 1938. Harshbarger (1976) surveyed the early development of statistics groups at many universities. He credits some whose contributions may have been forgotten over the years. (An example is A. E. Brandt, who was apparently influential in having Gertrude Cox named to head the new department at North Carolina State and who himself later founded a department at the University of Florida).

Of course, Statistics departments were not needed in order for instruction in Statistics to take place. According to an interesting overview by Fitzpatrick (1955), such instruction was available by the 1880s at universities such as Columbia, Johns Hopkins, Michigan, MIT, Pennsylvania, Virginia, and Yale. Stigler (1978) pointed out that a PhD in Mathematical Statistics was awarded at Yale in 1876. In his banquet speech on October 26, 2007, celebrating the fiftieth anniversary of Harvard Statistics, Stigler (2008) traced back at Harvard a course in probability during 1909–1910 and a course in Statistics in 1923. Indeed, by the mid-1920s, elementary statistics courses were available in a relatively large number of departments of Economics, Mathematics, and Education (Glover 1926).

About this time, advanced statistics courses became more common in mathematics departments, such as mathematical statistics courses that followed Carl West's 1918 book, *Introduction to Mathematical Statistics*, or Henry Rietz's 1924 book, *Handbook of Mathematical Statistics*, or his 1927 book, *Mathematical Statistics*. Noether (1989) observed, "At many institutions, the center of statistics instruction had moved from departments of economics to departments of mathematics. As concerned statisticians began to realize, the move often had the unfortunate consequence that mathematicians entrusted with the teaching of introductory statistics courses preferred to concentrate on mathematical, rather than statistical ideas, resulting in the increasingly held belief that statistics was simply a branch of mathematics."

Thus, Statistics departmental activity was slow to develop in the first half of the twentieth century. Hotelling (1940, 1949) gave compelling arguments for why Statistics was well deserving of its own department in the university, pointing out that academia had not been well served by having Statistics taught in various departments by teachers inadequately prepared in statistical theory. In short, "The task of leading the blind must not be turned over to the blind." In addition, he argued against Statistics being organized under the jurisdiction of the Department of Mathematics, pointing out that "The teacher of statistics must have a genuine sympathy and understanding for applications, and these are not well possessed by a great many pure mathematicians, at least in the opinion of some of those concerned with the applications. For so long as such an opinion is maintained, for example by psychologists and economists, these specialists will be suspicious that courses in statistics given by a department consisting largely of pure mathematicians are unsuitable for their purposes. The result is likely to be a sabotaging of attempts at centralization, the different departments reverting to the old and ultimately objectionable system of teaching their own separate courses in statistical methods."

Post World War II, the chapters in this book show that many renowned statisticians besides Hotelling were also influential in departmental development. Notable examples are Jerzy Neyman at Berkeley, Fred Mosteller at Harvard, and George Box at Wisconsin. Yet, at the same time, these chapters mention people not nearly as well known today yet who had their own significant influence. A prime example is W. Allen Wallis, who was responsible for founding the departments at Chicago and Rochester and indirectly responsible at Stanford and Harvard.

Hogg (1989) gave an entertaining overview of many of the personalities involved in the formation of many departments and Odell (1989) gave a related overview focusing on the Southwest, with both articles mentioning many groups formed too recently to be included in our book. The quarter century following World War II saw rapid expansion of higher education in the U.S., and the field of Statistics was no exception. Minton's discussion of the Hogg and Odell articles mentioned that by the 1970 *Amstat News* annual listing of colleges and universities offering degrees in Statistics, there were 99 departments or programs with "Statistics" in the title.

A View of the Past and a Vision for the Future

As you read this book, we think you will be pleased and perhaps a bit surprised to hear how many "success stories" there have been in academic Statistics in the U.S. Yes, there is the occasional failure, Princeton and Buffalo being most notable, but even there we see signs of regeneration and hope for the future.

But, as Statistics as an academic discipline grows rapidly, some growing pains are also inevitable. For example, as our subject expands through a variety of new types of applications, what is the core theory material that should be taught to all statisticians? Some chapters mention how recent economic realities have forced departments to downsize substantially in numbers of faculty. This makes many "growing-pains" issues, such as balancing teaching and research, lack of staff and infrastructural support, etc., even more challenging (Meng 2004). Many administrators undoubtedly still view Statistics merely as a course that many students must take to satisfy some requirement in their chosen field. A very encouraging sign is the development of links that are happening between many departments and other disciplines around the university. In some departments, such links may be crucial to the department's survival. Also, some departments still do not have much of an undergraduate program, apart from service courses, and many (e.g., Moore 2001) have argued in recent years that our place in academia would be more secure if we offered much more than the first course.

Indeed, there has been a rather substantially increased demand for intermediate courses for graduate students from other disciplines, whose research now call for more sophisticated statistical methods than those they learned in introductory courses. Given the fact that at many colleges and universities, even finding enough qualified teachers for the introductory level courses is an ongoing struggle (Meng 2009), how to provide quality intermediate-level courses is a substantial challenge that our profession now faces. We certainly hope that a future volume of this kind, say 30–50 years from now, will consider our current period as another golden age of our discipline because we turned these "growing pains" into "growing gains," making our field larger, stronger, and more mature.

Proposed Departmental and ASA Documentation of Our History

We imagined that in preparing their articles, many authors would have or would discover interesting and useful materials archived from the past, for instance from past departmental anniversary celebrations. We have encouraged and continue to encourage departments to post such material and additional documentation such as annual reports and pictures at a link from their departmental home page that is especially devoted to history.

In conjunction with this book, we plan to ask the American Statistical Association (ASA) to set up a location at their website devoted to the history of Statistics departments. At that site, links would be created to the various departmental history sites. This material could then be periodically updated by each department to help create a historical record for each department. More broadly, we envision that a new section of ASA devoted to The History of Statistics could be created to stimulate the organization of Internet records and other archives at departmental links as well as to present sessions with historical content at national meetings. This can help to supplement the historical material that is currently available, such as the Iowa State "Archives of Statistics" set up through the American Statistical Association, which contains 195 boxes of material including photographs, lectures and symposia, collections from famous statisticians, and other artifacts. See www.lib.iastate.edu/spcl/manuscripts/MS349.html.

Acknowledgments and Dedication

Histories are made by people and written by people, but almost never the same group of people, and the passage of time often scatters the material and challenges our memories. Our space for this book has also been very limited, as we could afford to give each department only about 5,000 words, with exceptions for a few departments with substantial history and an unusually major impact in our field (e.g., UC Berkeley). Some readers may feel we should have given more space yet to the major-impact departments at the expense of other departments. We felt, however, that each department has made its own contributions that were worthy of a full chapter at this stage of their history. And we hope that the authors of the chapters for the major-impact departments will consider writing extended, in-depth histories that can appear in future journals and books.

Because of the space constraints, it was a great challenge for the authors of each chapter not only to collect material but also to decide what to include. We therefore would like to sincerely thank all the chapter contributors for very generously devoting their time and energies to this project, and to offer our apologies to a few for whom our request for preparing such a chapter has apparently imposed more burdens than from their own deans. We also offer our apologies to those readers who might find inaccurate recollections or gross omissions. Both the chapter authors and we as editors have tried to avoid either occurrence, but unfortunately separating signal and noise is never something that can be achieved with 100% success, especially when a signal to one person can be noise to another, or vice versa. We can only hope for fewer errors in future editions or volumes of this kind. Also, we sincerely hope that all departments in this book as well as others founded since the mid-1960s will, in the future, prepare extended histories and post them at their departmental websites.

We also are very grateful to the many colleagues who made suggestions about these chapters. The referees we approached nearly unanimously agreed to help us. Their names are listed separately at the end of the book. At the beginning and throughout the project, eminent statisticians such as Sir David Cox, Ingram Olkin, and Stephen Stigler gave us helpful advice about the entire project. We especially thank Ingram for his tireless effort in helping us to review a large number of chapters, providing all of us the benefit of his long memories and vast experiences. Many thanks to John Kimmel, who as then Executive Editor at Springer, kindly encouraged and supported this project. Thanks also to Marc Strauss and Hannah Bracken for their support during the later stages. We thank Alan's wife, Jacki Levine, for the suggested title for the book, Kat Worden for her help as our editorial assistant, and Dale Rinkel for her assistance as well.

Finally, both of us have benefitted immensely over the years from the friendship and mentorship of Stephen Stigler. Steve was Alan's major professor when he was a graduate student at the University of Wisconsin in 1968–1972. Steve was also the department chair who recruited and then served as a much valued mentor of Xiao-Li during Xiao-Li's tenure from 1991–2001 at the University of Chicago. Indeed our entire profession owes so much to this eminent historian/statistician for bringing vividly to life the fascinating history of our profession. It is therefore with great pleasure that we dedicate the book to Steve, on the occasion of his seventieth birthday.

References

Craig AT (1960) Our silver anniversary. Ann Math Stat 31:835–837

David HA (1998) Statistics in U.S. universities in 1933 and the establishment of the statistical laboratory at Iowa State. Stat Sci 13:66–74

Fitzpatrick PJ (1955) The early teaching of statistics in American colleges and universities. Am Stat 9:12–18

Glover JW (1926) Statistical teaching in American colleges and universities. J Am Stat Assoc 21:419–424

Harshbarger B (1976) History of the early developments of modern statistics in America (1920–1944). In: Owen DB (ed) On the history of statistics and probability. Marcel Dekker, New York, pp 132–145

Hogg RV (1989) On statistical education in the United States since 1920. In: Proceedings of the American Statistical Association sesquicentennial invited paper sessions. American Statistical Association, Alexandria, pp 286–292

Hotelling H (1940) The teaching of statistics. Ann Math Stat 11:457–472

Hotelling H (1949) The place of statistics in the university. In: Neyman J (ed) Proceedings of the Berkeley symposium on mathematical statistics and probability. University of California Press, Berkeley, pp 21–40

Meng X-L (2004) Statistics departments: time for the second divorce? Amstat News, pp 2–3, Nov

Meng X-L (2009) Desired and feared—what do we do now and over the next 50 years? Am Stat 63:202–210 (rejoinder 64, 26–29; 362–363)

Moore DS (2001) Undergraduate programs and the future of academic statistics. Am Stat 55:1–6

Noether GE (1989) The teaching of statistics in the USA: the early years. In: Proceedings of the American Statistical Association sesquicentennial invited paper sessions. American Statistical Association, Alexandria, pp 279–285

Odell PL (1989) A historical summary of graduate statistical education in the Southwest. In: Proceedings of the American Statistical Association sesquicentennial invited paper sessions. American Statistical Association, Alexandria, pp 293–307 (and discussion by Minton PD on pp 308–311)

Stigler SM (1978) Mathematical statistics in the early states. Ann Stat 6:239–265

Stigler SM (2008) Statistics at Harvard? A toast to the Harvard Statistics Department. Am Stat 62:246–250

Carnegie Mellon University Statistics Department

Margaret L. Smykla and with the assistance of the department faculty

The Department of Statistics at Carnegie Mellon University—then Carnegie Institute of Technology–was created in 1966. Prior to that time, Carnegie Tech's Department of Mathematics served as the training ground for an outstanding mix of young statistical talent, such as undergraduate and masters degree students Frederick Mosteller (1930s); David L. Wallace (1940s); Mel Hinich (1950s); and Henry Block, Stephen Lagakos, Gary Simon, and Miron Straf (1960s). Among the statistically oriented instructors who began their distinguished careers in the Mathematics Department were Abraham Charnes, Carlton Lemke, Lester E. Dubins, Donald P. Gaver, Jr., and M. M. Rao.

In 1957, Morris H. DeGroot joined the Department of Mathematics as it was moving toward pure mathematics, a shift that did not bode well for the future hiring of statisticians. The only other statistician in the department was former Institute of Mathematical Statistics (IMS) President Edwin G. Olds who, despite being overshadowed in numbers by mathematicians for decades, was able to effectively make the case for statistics.

Olds' influence on budding statisticians at Carnegie Tech can be traced back to the mid-1930s. In *The Pleasures of Statistics: The Autobiography of Frederick Mosteller*, Mosteller recalled having to compute the probability of casting a total of nine and ten using three ordinary dice in a physical-measurements course. When as a sophomore he asked how to solve the problem using a large number of dice, he was directed to Olds.

Carnegie Mellon University,
Department of Statistics,
Baker Hall,
Pittsburgh, PA 15213, USA
M. L. Smykla (✉)
e-mail: mk74@stat.cmu.edu

A. Agresti and X.-L. Meng (eds.), *Strength in Numbers: The Rising of Academic Statistics Departments in the U.S.*, DOI: 10.1007/978-1-4614-3649-2_2,

> He [Olds] began showing me slowly and carefully how to do the three-dice problem…The method was that of generating functions, and it magically, and I do mean magically, counted how many ways there were to get each total with the dice. Although I had loved mathematics all along, this was the first time I ever felt that I'd been working with a peashooter when I could have had a cannon…
>
> Soon I was a mathematics rather than a physics major.

Hooked on probability and statistics, Mosteller would go on to earn a PhD in Mathematics from Princeton University. In 1957, en route to becoming one of the premier statisticians of the twentieth century, Mosteller founded the Department of Statistics at Harvard University almost a decade before Carnegie Tech had one (as fate would have it, one of his early PhD students at Harvard was Stephen Fienberg, who later became head of Carnegie Mellon's Department of Statistics and dean of its College of Humanities and Social Sciences (H&SS)). In 1964, Mosteller and fellow Carnegie Tech and Princeton alumnus Wallace authored their classic book, *Inference and Disputed Authorship: The Federalist.* When it was featured on the cover of *Time* magazine for its use of statistical methodology to solve the disputed authorship of the Federalist Papers, Carnegie Tech's Mathematics Department basked in the reflected limelight.

By the mid-1960s, DeGroot's interests had been diverging from his colleagues for years, leading him to yearn for a group of colleagues with whom to share his passion for research in decision making and other statistical interests. His friend and lifelong collaborator, Richard M. Cyert, then dean of the Graduate School of Industrial Administration (GSIA; now the Tepper School of Business) agreed. Together they developed a plan for a Department of Statistics with DeGroot as its head and Gaver as the other senior faculty member. A year later, Carnegie Tech became Carnegie-Mellon University. (In 1986, the hyphen was removed and the school assumed its present name of Carnegie Mellon University.)

In its early years the department had only a few, albeit outstanding, faculty members, such as Norman Starr, Paul Shaman, Stanley Sclove, Michael Woodroofe, and Madabhushi Raghavachari. The department, while not part of any college, reported to Cyert, first as dean and then as Carnegie Mellon president. Notable graduate students during this time included Prem Goel (former chairman of the Department of Statistics at Ohio State University), the late David Hildebrand (former chairman of the Department of Statistics at the University of Pennsylvania), and Benjamin Kedem (Professor of Mathematics at the University of Maryland).

In 1969, Jay Kadane joined the department. Three years later, DeGroot stepped down as head when his wife became ill and Kadane, who was 6-years out of graduate school, succeeded him. In addition to DeGroot and Kadane, the faculty at the time consisted of John Lehoczky, William W. Davis, Paul Shaman, Al Kinderman, and Gerry Ramage. There were a handful of graduate students, no undergraduates, and two staff members.

The department's offices in the early years were on the eighth floor of the concrete-block-walled, dungeon-like Science Hall, now Wean Hall, with odd aromas provided by the nearby chemistry laboratories. While Carnegie Mellon had

Commencement 1969: (left to right): Profs. DeGroot and Raghavachari; grad student Goel; and unidentified friend

Commencement 1971: (left to right) Profs. Kinderman, Ramage, and Lehoczky

some computers which used racks of IBM cards, Statistics had none. The department also still lacked a college to call home, as it was reporting both to the dean of GSIA and to the dean of Engineering. Blazing its own path, the department decided to focus on Bayesian Statistics, applications of Statistics, and, later,

Statistical Computing, when none of these was widely recognized as crucial to a growing department. In keeping with the latter focus, Statistics rented its first terminal: an interactive terminal hard-wired into the university computer system.

An early important proposal by Kadane that was adopted by departmental consent was to evaluate applied research on the basis of whether it was a contribution to the applied area and not necessarily a contribution to Statistics. Another part of that strategy was to forego operating a consulting center for drop-in statistical advice. Instead, faculty were encouraged to participate in joint work with researchers in other fields. As a result, the department managed to develop collaborative relationships with many scholars across campus for decades to come.

George Duncan joined the department in 1974, followed by William Eddy and Diane Lambert 2 years later. As almost none of the faculty had spouses or children at the time, there was a lot of after-hours socializing, such as Wednesday evening cocktails and hors d'oeuvres in the Faculty Club in Skibo (the campus student union named after Andrew Carnegie's ancestral home in Scotland). Everyone also gathered at noon each day in their shared hallway in Science Hall to walk for lunch to Skibo.

In what would become a hallmark of the department to the present day—journal and magazine editorships—DeGroot followed up his 1971–1975 editorial work as book review editor of the *Journal of the American Statistical Association* (*JASA*) by serving as the publication's theory and methods editor from 1976 to 1978. Kadane was his deputy editor for the latter. Duncan took over that *JASA* editorship from DeGroot in 1979.

Selected Articles from the Early Years

Brown, G. F., Kadane, J. B., and Ramage, J. G. (1974). The asymptotic bias and mean-squared error of double K-class estimators when the disturbances are small. *International Economic Review*, October, 15, 667–679.

Cyert, R. M. and DeGroot, M. H. (1971). Interfirm learning and the kinked demand curve. *Journal of Economic Theory*, 3, 272–287.

DeGroot, M. H. and Starr, N. (1969). Optimal two-stage stratified sampling. *Annals of Mathematical Statistics*, 40, 575–582.

Gaver, D. P. (1968). Headstart strategies for combating congestion. *Transportation Science,* 2 (2), 172–181.

Kadane, J. B. and Lehoczky, J. P. (1976). Random juror selection from multiple lists. *Operations Research*, 24, 207–219.

Kinderman, A. J. (1974). On the distribution of the deviations from the mean. *Indian Journal of Statistics*, 36, B, 2, 209–211.

Sclove, S. L. (1968). Improved estimators for coefficients in linear regression. *Journal of the American Statistical Association*, 63, 596–606.

In 1978, Wen-chen Chen came aboard; about the same time, the department offices moved from the eighth to the seventh floor of Science Hall. Mark Schervish arrived in 1979—an important year for collaborations. During a graduate seminar taught by Kadane, Schervish met University of Pittsburgh philosophy faculty member Teddy Seidenfeld (who later became a Carnegie Mellon joint Philosophy/Statistics faculty member). Their interactions would eventually evolve into weekly research meetings and numerous papers by the trio, both of which continue today.

Meanwhile, across the Atlantic at the First Valencia International Meeting on Bayesian Statistics in Valencia, Spain, DeGroot and Fienberg were beginning their collaboration on comparing probability forecasters. At this meeting, DeGroot was able to finally convince Fienberg, whom he and Kadane were aggressively courting, to join the department a year later in 1980. The final negotiating item involved finding opportunities in Pittsburgh for Fienberg and his son to play hockey.

The dawn of the 1980s was a time of change and development, with faculty numbering about 10 and graduate students roughly 15. Among them was Luke Tierney, in his first faculty post, and PhD student Kathryn Chaloner. The couple met at Carnegie Mellon and later married. Today, both are professors and department heads at the University of Iowa—she in the Department of Biostatistics, and he in the Department of Statistics and Actuarial Science.

In 1980, statistics finally landed a home when it joined H&SS, where it remains a core department today.

In 1981, Robert Kass was hired, and Fienberg became head just as he was starting his 6-year stint as chair of the Committee on National Statistics (CNSTAT) at the National Academy of Sciences. As department head, he oversaw submission of a proposal to the first Scientific Computing Research Equipment in the Mathematical Sciences (SCREMS) program at the National Science Foundation, for which the department received funding to purchase its first mini-computer for $250,000. This was the start of the department's leadership in statistical computing beyond Carnegie Mellon, a position it retains to the present day.

Collaboration continued to flourish, as DeGroot, Fienberg, and Kadane began work on the use of statistics in legal settings, which would become the edited volume, *Statistics and the Law*. At a downtown Pittsburgh hotel, the department hosted the international Computer Science and Statistics 13th Symposium on the Interface, chaired by Eddy. This was to be the first of many conferences and workshops hosted by the department and its faculty.

Amidst a swirl of activity, tragedy struck. On July 3, 1981, while visiting family in Taiwan with his wife and 1-year-old son, 31-year-old Wen-chen Chen was found dead. Chen had undergone extensive questioning by Taiwanese authorities regarding his alleged political activities in the U.S. With DeGroot as the driving force in bringing attention to what the Taiwanese government was dismissing, he and renowned U.S. pathologist and Pittsburgh resident Cyril Wecht traveled to Taiwan. While Wecht concluded through autopsy findings that the promising young faculty member was murdered, there was no further action by the Taiwanese government.

The department continued to grow and, with the university-wide tentacles of a sprouting bureaucracy compounding regulations and paperwork, in 1982 Statistics hired its first business manager. Ruey Tsay came aboard that year also, while a year later Joel Greenhouse arrived to begin a post-doctoral position supported by a National Institute of Mental Health (NIMH) grant for training in Psychiatric Statistics with Lehoczky and DeGroot as principal investigators.

In keeping with the tradition of departmental editorships of journals and magazines, DeGroot was spearheading the creation of *Statistical Science*, dedicated to communication among statisticians regardless of specialty, while Kadane was in the middle of a 3-year stint as applications and coordinating editor of *JASA*. On the honors front, Fienberg was the 1982 recipient of the Committee of Presidents of Statistical Societies Presidents' Award (COPSS) as the outstanding statistician aged 40 or younger—the first of three Carnegie Mellon Statistics faculty to receive this honor.

Selected Books Written by Faculty

Anderson, M. J. and Fienberg, S. E. (1999). *Who Counts? The Politics of Census-Taking in Contemporary America*. Russell Sage Foundation.

DeGroot, M. H. (1970). *Optimal Statistical Decisions*. McGraw-Hill.

DeGroot, M. H. (1975). *Probability and Statistics*. Addison-Wesley.

DeGroot, M. H., Fienberg, S. E., and Kadane, J. B., editors (1986). *Statistics and the Law*. Wiley.

Devlin, B., Fienberg, S. E., Resnick, D., and Roeder, K., editors (1997). *Intelligence, Genes, & Success: Scientists Respond to "The Bell Curve"*. Springer-Verlag.

Kadane, J. B. (2011). *Principles of Uncertainty*. Chapman and Hall (free on the web).

Kadane, J. B., Schervish, M., and Seidenfeld, T., editors (1999). *Rethinking the Foundations of Statistics*. Cambridge University Press.

Kadane, J. B. and Schum, D. A. (1996). *A Probabilistic Analysis of the Sacco and Vanzetti Evidence*. Wiley-Interscience.

Kass, R. E. and Vos, P. (1997). *Geometrical Foundations of Asymptotic Inference*. Wiley.

Mockus, J., Eddy, W. F., Mockus, A., Mockus, L., and Reklaitis, G. (1997). *Bayesian Heuristic Approach to Discrete and Global Optimization* (Algorithms, Visualization, Software, and Applications). Kluwer Academic Publishers.

Mosteller, F., Fienberg, S. E., and Rourke, R. E. K. (1983). *Beginning Statistics With Data Analysis*. Addison-Wesley.

Schervish, M. J. (1995). *Theory of Statistics*, Springer-Verlag.

Wasserman, L. W. (2003). *All of Statistics: A Concise Course in Statistical Inference*. Springer-Verlag.

By the early 1980s, the department enjoyed a worldwide reputation in Bayesian Statistics and decision making, and was developing strength in statistical computation. At the same time, Carnegie Mellon was stressing these same areas, as well as cross-disciplinary research, in its strategic planning. The university's recognition of the importance of statistics in such research would result, a decade later, in the department's playing a key role in several major cross-disciplinary efforts on campus, such as the Center for the Neural Basis of Cognition (CNBC), and the Center for Automated Learning and Discovery (CALD), now the Machine Learning (ML) Department.

In 1984, the department moved from Wean Hall to two newly renovated wings of Baker Hall, almost doubling its space in faculty, staff, and student offices, and acquiring two conference rooms, a lounge, and more. To ensure frequent interaction, graduate student offices were interspersed among faculty offices. A spring evening was the setting for faculty, staff, and university officials to unite in celebration as DeGroot was recognized for his decades of contributions and achievements with Carnegie Mellon's highest honor: a University Professorship. Later in the year, Fienberg stepped down as head, and Lehockzy assumed the post.

Early 1980s: a house party with some faculty and students

In 1985, Fienberg, by now Maurice Falk Professor of Statistics and Social Science, and Eddy began developing the idea for a quarterly magazine for anyone with an interest in the analysis of data—*CHANCE, New Directions for Statistics and Computing*. It would reach fruition 3 years later, and eventually become a publication of the American Statistical Association for the next 20 years and counting. By mid-decade, the department housed about 15 faculty and 30 graduate students.

Selected Articles by faculty

Eddy, W. F. and Schervish, M. J. (1995). How many comparisons does Quicksort use? *Journal of Algorithms*, 19, 402–431.

Fienberg, S. E. (1994). Conflicts between the needs for access to statistical information and demands for confidentiality. *Journal of Official Statistics*, 10, 115–132.

Perone Pacifico, M., Genovese, C. R., Verdinelli, I., and Wasserman, L. (2004). False discovery control for random fields. *Journal of the American Statistical Association*, 99, 1002–1014.

Greenhouse, J. B., Kaizar, E. E., Kelleher, K., Seltman, H., and Gardner, W. (2008). Generalizing from clinical trial data: A case study. The risk of suicidality among pediatric antidepressant users. *Statistics in Medicine*, 27(11): 1801–1813.

Junker, B. W. and Sijtsma, K. (2000). Latent and manifest monotonicity in item response models. *Applied Psychological Measurement*, 24, 65–81.

Schervish, M. J., Seidenfeld, T., and Kadane, J. B. (2002). Measuring incoherence. *Sankhya*, 64, Part 3(i), 561–587.

Kruk, L., Lehoczky, J., Shreve, S., and Yeung, S.-N. (2003). Earliest-deadline-first for a multiple-input heavy-traffic real-time queue. *Annals of Applied Probability*, 13, 55–99.

Roeder, K., Bacanu, S. A., Wasserman, L., and Devlin, B. (2006). Using linkage genome scans to improve power of association in genome scans. *American Journal of Human Genetics,* 78, 243–252.

Tierney, L. and Kadane, J. B. (1986). Accurate approximations for posterior moments and marginal densities. *Journal of the American Statistical Association*, 81, 82–86.

Academic statistics departments were afforded a unique opportunity to acquire their own equipment as a result of the awarding of the first SCREMS funds in 1982, followed by Department of Defense funds under its University Research Instrumentation Program (DURIP). The Statistics Department moved from general availability of terminals in 1981, to a VAX 11/750 in 1983, to several workstations in 1985—all maintained by Eddy. By 1986, the department's equipment included one VAX 11/750 superminicomputer with 912 MB of disk storage, one VAX-station 500 color workstation, one CSPI Mini-Map array processor, five GIGI color microcomputers, and more. All machines were interconnected by a 10 Mb Ethernet using DECnet and TCP/IP software.

The mid-1980s marked the graduation of the department's first undergraduate statistics majors. Meanwhile, the labors of the late-1970s through the mid-1980s PhD graduates were bearing fruit in government, industry, and academia. At the National Institutes of Health, Todd Sahlroot and Dean Follmann engaged in research on clinical trial and statistical modeling of biomedical data. Kathy Blackmond Laskey, at the Decision Sciences Consortium, analyzed data from the national pesticide survey of well water for the Environmental Protection Agency, while Sheryl Kelsey, as co-director of the Epidemiology Data Center at the University of Pittsburgh, oversaw research studies sponsored by the National Institutes of Health and other agencies. Diane Saphire and Elizabeth Stasny were beginning their academic careers at, respectively, Trinity University and Ohio State University.

Late 1980s: Jay Kadane; 1990 Ph.D. students, left to right, Giovanni Parmigiani (Harvard School of Public Health), and Mario Peruggia (Ohio State University)

In 1987, Greenhouse was the recipient of the university's prestigious William H. and Frances S. Ryan Award for Meritorious Teaching. Fienberg was appointed dean of H&SS and, a year later, Larry Wasserman was hired. Ngai Hang Chan followed shortly afterward. Eddy was developing a journal to extend the use of computational and graphical methods in statistical and data analysis that would become the *Journal of Computational and Graphical Statistics*, a joint publication of the ASA, IMS, and the Interface Foundation of North America.

By 1989, the department's computing equipment included nine terminals, five printers, 11 personal computers, and 32 workstations. To maintain the growing inventory, the department hired its first full-time systems manager. That same

Morris H. DeGroot, 1931–1989

year, faculty member Michael Meyer developed StatLib, a system for distributing statistical software, data sets, and information by electronic mail, FTP, and WWW, and which is hosted by the Statistics Department to this day.

Sadness gripped the statistical community with the death of Morrie DeGroot on November 2, 1989. As former student and collaborator Prem Goel remarked at the memorial service, mirroring the department's sentiment:

> Morrie has not gone anywhere, that he has merely changed his body which was worn out, that memories of all the good times with Morrie are still with me, and that his ideas and advice will keep on showing me the right path.

Two years later the department formally honored its founder through its ongoing, biannual Morris H. DeGroot Lecture held in conjunction with the first of a long series of Case Studies in Bayesian Statistics workshops co-organized by Kass on applications of Bayesian statistics to problems in science and technology.

The department also transformed a conference room into the Morris H. DeGroot Memorial Library containing the bequeathed books and periodicals of its namesake. Over the years, the collection expanded to include hundreds of faculty contributions and some new purchases.

By the start of the next decade the major thrusts of the department had evolved into Bayesian Statistics, Computational Statistics, Industrial Statistics and Stochastic Modeling, and Biostatistics and Psychiatric Statistics. In keeping with the latter, Brian Junker joined the department as a postdoctoral fellow in the Program in Psychiatric Statistics, jointly operated by the department and Western Psychiatric Institute and Clinic. Fienberg returned in 1993 following a 2-year stint as academic vice-president at York University in Toronto. With the arrival of Kathryn Roeder and Christopher Genovese a year later, personnel rose to 14 tenure-track, one research scientist, and three full-time-equivalent visiting faculty.

By 1995, StatLib had grown to a collection of about 150 Mbytes, and the StatLib server was servicing about 60,000 monthly transactions. In the classroom, curricular and pedagogical innovations introduced in new H&SS core courses, "Statistical Reasoning" and "Introduction to Statistical Methods," were partly credited with the rise in the number of Statistics majors and minors, and with the overall interest in the discipline: by spring 1996, there were over 900 undergraduates enrolled in department courses.

Journal/Magazine Editorships by Faculty

Founding Editors (in approximate chronological order)

Statistical Science (DeGroot); *Journal of Computational and Graphical Statistics* (Eddy); *CHANCE, New Directions for Statistics and Computing* (co-founders Eddy, Fienberg); *Bayesian Analysis* (Kass); *Electronic Journal of Statistics* (Wasserman); *Annals of Applied Statistics* (co-founding editor Fienberg); *Journal of Privacy and Confidentiality* (co-founder Fienberg); *Annual Review of Statistics and its Application* (Fienberg).

Editors

Journal of the American Statistical Association (DeGroot, Duncan, Eddy, Fienberg, Kadane, Schervish); *Journal of Official Statistics* (Fienberg); *Proceedings of the National Academy of Sciences* (Fienberg); *Psychometrika* (Junker); *Statistics in Medicine* (Greenhouse); *IMS Lecture Notes-Monograph Series* (Greenhouse); *Statistical Science* (Fienberg, Kass); *Annals of Statistics* (Kadane); *CHANCE, New Directions for Statistics and Computing* (Schervish).

The development of the graduate program continued with roughly 36 students. The number of special graduate education programs by now included the joint PhD with the H. John Heinz III School of Public Policy and Management (now the H. John Heinz III College) and the joint PhD in Statistics and Machine Learning. The program expanded further with the Masters of Science in Computational Finance (MSCF), a joint venture of four departments, including Statistics, for future investment professionals. Besides local classroom instruction, a second group of students in New York City take the program in a suite of classrooms using distance learning technology.

In 1995, Kass was named head, replacing Lehoczky after 11 years. On the research front, brain science provided fodder for ever-widening statistical collaboration. A new tool to measure the change in blood flow related to neural activity in the brain—functional Magnetic Resonance Imaging (fMRI)—offered statistical challenges for Eddy, Genovese, and new hire Nicole Lazar. Kass' interest in statistical modeling of neural activity culminated in a faculty appointment—and executive committee membership—in Carnegie Mellon's new CNBC, established to foster interdisciplinary research on the neural mechanisms of brain function. (A joint CNBC-Statistics PhD program was approved in 2011.)

To keep pace with the increasing number of Statistics undergraduate and graduate degree recipients, outreach initiatives were instituted, like departmental newsletters and the annual alumni breakfasts at the Joint Statistical Meetings. Meanwhile, the influx of young faculty not only provided a balance with senior faculty, but heralded a change in family profiles. The after-hours socializing of the prior decade gave way to afternoon baby showers and child care schmoozing. New faculty also meant new honors. Roeder was the 1997 recipient of the COPSS Award, with Wasserman accepting the award 2 years later. Genovese received an NSF CAREER Award for young researchers. For senior faculty, Lehoczky's being named the Thomas Lord Professor of Statistics raised the number of chairs to four: Kadane was the Leonard J. Savage Professor of Statistics and Social Sciences; Fienberg, recently elected president of the IMS and the International Society for Bayesian Analysis, was named a University Professor; and Seidenfeld became the first recipient of the Herbert A. Simon Chair in Philosophy.

As the world prepared for the new millenium, the department set its sights on NSF funding that was awarded jointly with Carnegie Mellon's Department of Mathematical Sciences through a Vertical Integration of Research and Education grant, or VIGRE. The goal of the VIGRE program, as articulated by NSF, is to "increase the number of well-prepared U.S. citizens, nationals, and permanent residents who pursue careers in the mathematical sciences". A few years after the joint award, Statistics was singly awarded VIGRE funding. Through that assistance, the Statistics Department enhanced its undergraduate, graduate, and postdoctoral programs. With the graduate and postdoctoral programs, mentored cross- disciplinary work, communication, and teaching were emphasized.

Newly instituted teaching teas and research teas encouraged interaction among regular faculty and VIGRE trainees at all levels. (By 2006, VIGRE participants numbered 3 faculty principal investigators, 15 faculty mentors and project leads, 6 VIGRE-supported post docs/visiting assistant professors, 11 VIGRE-supported graduate students, and 28 VIGRE- supported undergraduates.) The grant also gave birth to the department's 8-year Summer Undergraduate Research Experience (SURE), directed by Greenhouse and Howard Seltman, M.D., who received his PhD from the department in 1999. Students—primarily from historically black colleges and universities—interested in public health careers designed supervised research projects such as "Factors Associated with the Perceptions of Obesity and Depression Among Adolescents in the U.S.," and "The Hidden Cost of Caregiving: Emotional Stress." Students presented their work at a poster session to the campus community at the conclusion of the 8-week SURE program.

Maintaining the focus on undergraduate education, by 2000 the department offered a variety of programs, with the most popular being: (1) the BS and minor in Statistics in H&SS; (2) the statistics and operations research concentrations of the BS in Mathematical Sciences in the Mellon College of Science; and (3) the statistics and data mining concentration of the BS in Information Systems in H&SS. The college itself was expanding, as a 20,000-square-feet addition provided new research areas, multimedia classrooms, conference rooms, meeting space, and more. When H&SS lost its dean to another university, Lehoczky was named interim dean. Faculty honors during this period included Kadane being named a University Professor, and Fienberg's election to the National Academy of Sciences.

Cross-disciplinary work among faculty continued to span a variety of scientific, business, and policy domains, including archeology, astronomy, biostatistics, data mining, education, environment, finance, government, marketing, neuroimaging, neurophysiology, and social science/public policy. In 2001, the faculty published 67 articles and commentaries, and 6 books. A year later, an advisory board report called the department "world class in the caliber and rigor of its theoretical and methodological research, in its contributions to advancing the discipline of statistics, and in its education programs. Moreover, it is preeminently so in interdisciplinary research."

Cross-disciplinary Research Interests of Faculty

Areas in which Carnegie Mellon faculty have made significant research contributions include biology (Eddy, Genovese, Kadane, Roeder, Seltman, Wasserman); cognition and education (Junker); finance (Kadane, Lehoczky, Schervish); forensic statistics and the law (DeGroot, Eddy, Fienberg, Kadane, Lehoczky, Schervish); government statistics (Eddy, Fienberg, Kadane); machine learning (Eddy, Fienberg, Kass, Lee, Wasserman); medicine/health (Fienberg, Greenhouse, Seltman); neuroscience (Eddy, Genovese, Kass, Shalizi, Ventura); online learning (Meyer); physics (Genovese, Jin, Kadane, Schafer, Shalizi, Verdinelli, Wasserman); social science (Fienberg, Greenhouse, Kadane, Rinaldo, Shalizi).

Statistics faculty have held joint appointments in the departments of biological sciences, machine learning, mathematics, philosophy, social, and decision sciences, as well as in the Tepper School of Business and the Heinz College (public policy and management), and in numerous interdisciplinary centers such as CNBC, Cylab, and the Center for Computational Biology.

Oded Meyer, appointed in 1999, was named the department's first regular teaching faculty member. At the same time, reformulated classes like Advanced Data Analysis II, in which students prepare semester-long projects to present at Carnegie Mellon's competitive annual "Meeting of the Minds" undergraduate research symposium, continued to attract newcomers to the field.

The growth of the graduate and undergraduate programs, and the research activities, also added pressure on the small amount of space the department occupied, and still does. Appropriate office space for new faculty hires can be problematic. Some graduate student offices are in other buildings, both on and off campus, and important computing equipment is located in rented remote locations. While the reconfiguration of former faculty offices into multi-desk offices has proven a viable option for fostering a sense of community for students, contiguous space remains the department's most pressing issue.

In that vein, Eddy opened, in 2002, a 2500-square-feet "branch office" three blocks from campus with a staff engaged in imaging research—optical video imaging, microscopic histological imaging, and laser fluorescence imaging. Faculty honors during this period included Eddy becoming chair of CNSTAT, Junker's appointment as editor of *Psychometrika* and Kass's as editor of the new electronic *Bayesian Analysis*, and Seidenfeld's being named a University Professor. Eddy and Fienberg were designated lifetime National Associates by the National Academy of Sciences.

On July 1, 2004, Schervish became head, replacing Kass who had served 9 years in the post. In addition to the 17 regular faculty, there were seven visitors, ten instructors, and two faculty with external appointments. On the academic front, the department launched the new BS in Economics and Statistics degree program, and began restructuring its graduate program so students could engage in research earlier in their careers with fewer required courses. Fienberg was elected to the Royal Society of Canada in 2004 and to the American Academy of Arts and Sciences in 2007, joining Seidenfeld who was elected several years earlier. On the social front, departmental celebrations kicked into high gear over the next 5 years with events in honor of Kadane attaining emeritus status (first in the department's history), Fienberg's 65th birthday, Eddy being named the first John C. Warner Professor of Statistics, and Lehoczky's 40 years at Carnegie Mellon.

By 2008, the number of undergraduate majors totalled about 100; not coincidentally, that same year Meyer won the Elliott Dunlap Smith Award for excellence in teaching in the college. On the graduate side, the new Masters in Statistical Practice (MSP) program premiered a year later: a 1-year, two-semester professional masters degree program emphasizing statistical practice, methods, data analysis, and workplace skills, and directed by Greenhouse.

The department concluded one decade and began another with a resounding mix of junior and senior faculty recognition: Jiashun Jin was winner of the 2009 Tweedie New Researcher Award from the IMS, and Rebecca Nugent received the 2009 Chikio Hayashi Award from the International Federation of Classification Societies to a promising young researcher. Kadane was elected to the American Academy of Arts and Sciences in 2010, while Eddy completed his second three-year term as chairman of CNSTAT, becoming the only person to have chaired both statistics committees at the National Academies, having previously chaired the Committee on Applied and Theoretical Statistics. In 2011, Lehoczky joined the ranks of University Professors, while Ann Lee was awarded the Estella Loomis McCandless Professorship as a junior faculty member who has shown great promise in her field.

As of late 2010, 145 undergraduates were pursuing majors in the department, and seven were pursuing minors. There were 882 students enrolled in undergraduate Statistics courses.

Fall 2011 is record-setting on a number of fronts: There are 65 graduate students including 18 MSP students—our most ever! And we now have the largest group of undergraduate Statistics majors in the country. Faculty, who comprise professors, special faculty, professors in residence, postdoctoral research associates, and more, total 33—one of our biggest numbers—with the newest members being assistant professor Ryan Tibshirani, and Heinz College associate professor and Anna Loomis McCandless Professor Amelia Haviland, a 2003 department alumnus with a joint PhD with Heinz; she also has an appointment in the Statistics Department.

Finally, in Fall 2011, the department's college became the Marianna Brown Dietrich College of Humanities and Social Sciences in honor of the mother of William S. Dietrich II, who provided a gift of a $265 million fund to support Carnegie Mellon. It is the eighth largest gift ever by an individual to a private higher education institution in the U.S.

"The naming of a college is a momentous occasion at any university," said Carnegie Mellon President Jared L. Cohon. He went on: "Let me underscore now that for Carnegie Mellon, the naming of its Humanities and Social Sciences College is especially significant. It represents a statement of the progress that H&SS has made over the last several years, especially under the leadership of Dean John Lehoczky. It's also a statement about Carnegie Mellon's commitment to the future of humanities and social sciences at the university," he said.

Recent MS and PhD Graduates

The solid grounding MS and PhD students receive here in Theoretical, Applied, and Computational Statistics makes them especially valued in scientific, industrial, and government settings where issues are complex and data are often difficult to collect or incomplete, and where there are no easy answers. Recent graduates and their current employers include: University of Washington (Adrian Dobra, Elena Erosheva); Morgan Stanley New York (Mihaela Serban); RAND (J. R. Lockwood, Louis Mariano); Duke University (Dalene Stangle, Al Erkanli, Daniel Manrique); University of North Carolina (Justin Gross); Battelle Institute (David Friedenberg); Committee on National Statistics (Daniel Cork); Carnegie Mellon University (Alessandro Rinaldo, Howard Seltman); University of Chicago (Daniel Heinz); University of California, Berkeley (Cari Kaufman); University of Buffalo (Jeffrey Miecznikowski); Los Alamos National Laboratory (Kary Myers); Bristol-Myers Squibbs (Can Cai); National Cancer Institute (Michelle Dunn, Eric Huang); U.S. Food and Drug Administration (Xiting Yang); Georgia Institute of Technology (Nicoleta Serban); University of Pittsburgh (Marnie Bertolet); Genentech, Inc. (Diana Luca); Jump Tracing, LLC (Libo Xie); Swarthmore College (Lynne Steuerle Schofield).

What has Made Statistics at Carnegie Mellon Special?

Several features of the research and other activities of the department have made it distinctive at Carnegie Mellon, and a leader among its peers at other universities around the world.

Joel Greenhouse and some MSP students taking a break to enjoy the outdoors

Orientation 2011 Picnic welcoming 26 new graduate students

First, from its founding, the department has been home to a focus on Bayesian inference, and this focus flourished as faculty developed computational approaches such as Laplace's method and in the 1990s with the emergence of Markov chain Monte Carlo methods. This intellectual focus fit nicely with the department's

A recent picture of the "old-timers" here prior to 1983: (front, left to right:) Jay Kadane, Bill Eddy, John Lehoczky; (back, left to right:) Rob Kass, Margie Smykla (staff), Mark Schervish, Steve Fienberg, and Joel Greenhouse

emphasis on Statistical Computing and its involvement in the creation of the Machine Learning Department in the School of Computer Science.

Second, the department's emphasis on collaborative research established it within Carnegie Mellon as an essential locus of research and teaching linked to almost all units of the university. Statistics graduate students and faculty engaged in interdisciplinary research before it became fashionable in most other departments around the United States, and faculty and students alike publish widely in scientific journals outside the field of statistics; they are often known as much for their substantive contributions in other areas as for their publications in statistics journals. Many of the department's students are enrolled in joint programs, further solidifying this interdisciplinary focus.

Finally, the department and the university have taken pride in the professional leadership roles played by its faculty, from editorships of journals and book series, to leadership in professional societies and at the National Academy of Sciences.

Ed Olds might not recognize the department were he alive today, but he would surely appreciate its commitment to teaching, to its undergraduate and graduate students, and to the professional leadership role which he initiated over half a century ago. The introduction in 2010 of an electronic application system for the PhD Program (MS and PhD) and the MSP Program garnered record applications to both. The department's many and varied successes, including its swelling numbers of undergraduate majors, remain cornerstones crucial to the department's strategic plan, which is the same as at the department's founding 45 years ago: training problem solvers to be in a position to choose amongst a variety of professional opportunities in the U.S. and beyond.

Columbia University Statistics

Tian Zheng and Zhiliang Ying

Statistical Activities at Columbia Before 1946

Statistical activities at Columbia predate the formation of the department. Faculty members from other disciplines had carried out research and instruction of statistics, especially in the Faculty of Political Sciences (Anderson 1955). Harold Hotelling's arrival in 1931, a major turning point, propelled Columbia to a position of world leadership in Statistics. Hotelling's primary research activities were in Mathematical, Economics, and Statistics but his interests had increasingly shifted toward Statistics. According to Paul Samuelson (Samuelson 1960), "It was at Columbia, in the decade before World War II, that Hotelling became the Mecca towards whom the best young students of economics and mathematical statistics turned. Hotelling's increasing preoccupation with mathematical statistics was that discipline's gain but a loss to the literature of economics." His many contributions include canonical correlation, principal components and Hotelling's T^2.

In 1938, after attending a conference of the Cowles Commission for Research in Economics, Abraham Wald, concerned by the situation in Europe, decided to stay in the United States. With Hotelling's assistance, Wald came to Columbia on

A substantial part of this chapter was developed based on Professor T. W. Anderson's speech during our department's sixtieth year anniversary reunion. We thank Professor Anderson for his help (as a founding member of our department) during the preparation of this chapter, our current department faculty members, Professors Ingram Olkin, and and Tze Leung Lai for their comments and suggestions on earlier versions of this chapter.

Columbia University, Room 1005 SSW,
MC 4690 1255 Amsterdam Avenue, New York, NY 10027, USA
T. Zheng (✉) · Z. Ying
e-mail: tzheng@stat.columbia.edu

Z. Ying
e-mail: zying@stat.columbia.edu

A. Agresti and X.-L. Meng (eds.), *Strength in Numbers: The Rising of Academic Statistics Departments in the U.S.*, DOI: 10.1007/978-1-4614-3649-2_3,

a fellowship from the Carnegie Corporation (Wolfowitz 1952; Anderson 1955) and began to study Modern Statistics with Hotelling, becoming a regular faculty member in the Economics department in 1941. In addition to their prolific research in Statistics, Hotelling and Wald also carried out broad and advanced instruction and training in Statistics at Columbia, which was "ahead of that offered at any other American university."Anderson (1955) As a result of their efforts, in 1942, a Joint Committee on Instruction was appointed to supervise the Ph.D. in Mathematical Statistics, which consisted of Frederic Mills (Economics), Paul Lazersfeld (Sociology), Helen Walker (Teachers College), and people from the Psychology department and Columbia College. These activities in research, instruction and training in Statistics paved the road for the foundation of a new department.

On July 1st, 1942, soon after the United States entered World War II, the Statistical Research Group (SRG), financed by the Applied Mathematics Panel of the National Defense Research Committee, was formed at Columbia University to support the emerging statistical needs of the war effort (Rees 1980). Over the 39 months of its duration, a total of 18 principal statisticians (mostly visitors from other universities) served in the SRG with W. Allen Wallis as director. This was "the most extraordinary group of statisticians ever organized taking into account both number and quality" (Wallis 1980). The SRG played an important role in the history of Statistics in the United States. As also documented in the *University of Chicago* chapter, members of the SRG produced a number of important contributions to Modern Statistics, the most important of which is sequential analysis (see Wallis (1980) for his full account of the origins of sequential analysis). The SRG also prepared a number of statisticians for their future career as leaders in Statistics and other fields, as evidenced by the fact that eight members became IMS presidents, four became ASA presidents, two received the Nobel Prize in Economics, and two became university presidents.

Founding of a Statistics Department, Early Years and Influential Contributions

Instruction and research in Statistics continued at Columbia during and after the war, due to the efforts of Hotelling and Wald (Anderson 1955). In 1946, Hotelling moved to North Carolina to help found the Statistics department at UNC. He also invited Wald to join him. Afraid of losing two prominent scholars in Economics and Statistics, Columbia decided to create a Statistics department in the Faculty of Political Science. To differentiate its purpose from the widely held perception that Statistics was primarily about data collection and analysis by means of conventional methods, the new department was named Mathematical Statistics,[1] referring

[1] As more and more departments of Statistics were formed in the country, in November, 1982, Columbia decided to change its name to the simpler "Department of Statistics".

to a scientific discipline that focused more on developing new methodology using sophisticated Mathematics.

The new department had three founding members: Wald, Jacob Wolfowitz, and Theodore W. Anderson. During the first year, Jerzy Neyman and Joseph Leo Doob were visiting professors; Wald unsuccessfully invited them to join Columbia. With Howard Levene joining the faculty in 1947, Henry Scheffé in 1948, and Kai Lai Chung in 1949 (as a research associate), the department started to take form, small yet vigorous. These were exciting years. The faculty members were productive and contributed a number of seminal papers, including:

- Wald, A. and J. Wolfowitz (1948). "Optimum Character of the Sequential Probability Ratio Test." Annals of Mathematical Statistics **19**(3): 326–339.
- Anderson, T. W. and H. Rubin (1949). "Estimation of the Parameters of a Single Equation in a Complete System of Stochastic Equations." Annals of Mathematical Statistics **20**(1): 46–63.
- Levene, H. (1949). "On a Matching Problem Arising in Genetics." Annals of Mathematical Statistics **20**(1): 91–94.
- Wald, A. (1949). "Note on the Consistency of the Maximum Likelihood Estimate." Annals of Mathematical Statistics **20**(4): 595–601.
- Anderson, T. W. (1951). "Estimating Linear Restrictions on Regression Coefficients for Multivariate Normal Distributions." Annals of Mathematical Statistics **22**(3): 327–351.
- Anderson, T. W. and D. A. Darling (1952). "Asymptotic Theory of Certain Goodness of Fit Criteria Based on Stochastic Processes." Annals of Mathematical Statistics **23**(2): 193–212.
- Kiefer, J. and J. Wolfowitz (1952). "Stochastic Estimation of the Maximum of a Regression Function." Annals of Mathematical Statistics **23**(3): 462–466.
- Scheffé, H. (1952). "An Analysis of Variance for Paired Comparisons." Journal of the American Statistical Association **47**(259): 381–400.
- Levene, H. (1953). "Genetic Equilibrium When More Than One Ecological Niche Is Available." American Naturalist **87**(836): 331–333.
- Scheffé, H. (1953). "A Method for Judging All Contrasts in the Analysis of Variance." Biometrika **40**(1–2): 87–104.

Because the department was small, a scheme was developed to invite faculty to teach. In addition to Neyman and Doob, the department also hosted other visitors, including R. C. Bose, Donald Darling, P. L. Hsu, Salem Khamis, Michele Loève, Erich Lehmann, E. J. G. Pitman S. N. Roy, John Curtis, Joe Gani, William Feller, Albert Bowker, Harald Cramér, Wassily Hoeffding, Gerald Lieberman, Leo Goodman, Theodore E. Harris, Klaus Krickeberg, Sigetti Moriguti, Harold Ruben, Milton Sobel, Bernard Friedman, Warren Hirsch, W. J. Youden, Jerome Cornfield, and Somish Das Gupta. Many of them taught courses. At that time, there were few textbooks in Probability and Statistics. According to Anderson, "a number of basic books in our field of statistics in some sense came out of the teaching activities of the department". When Doob was a visiting professor, his course on probability theory contained a lot of new material and ideas that were incorporated into his book

Fig. 1 Founding members of Statistics Department at Columbia: Hotelling, Wald, Wolfowitz, and Anderson. Even though Hotelling left before the department was officially established, we included him here as he was so instrumental for the developments at Columbia that led to the creation of the department

Stochastic Processes. Erich Lehmann was working on his book on *Testing Statistical Hypotheses*. Loève's book on Probability Theory was partially developed at Columbia. Anderson's book, *An Introduction to Multivariate Statistical Analysis*, came out of his lectures at Columbia. Henry Scheffé wrote his book, *The Analysis of Variance*, while at Columbia. Kai Lai Chung wrote one of his books largely there.

In November 1950, Wald took a trip with his wife to visit universities and research centers in India. On December 13, 1950, their flight from northern India to southern India crashed in the Nilgiris district, killing all abroad (Wolfowitz 1952). Despite his career being tragically cut short, during his years at Columbia (1939–1950), Wald published nearly 70 papers, two books, and made several important contributions to Statistics. Among them, the most widely known (besides those mentioned above) are:

- Wald, A. (1939). "Contributions to the theory of statistical estimation and testing hypotheses." Annals of Mathematical Statistics **10**: 299–326.
- Wald, A. (1940). "The fitting of straight lines if both variables are subject to error." Annals of Mathematical Statistics **11**: 284–300.
- Mann, H. B. and A. Wald (1943). "On the Statistical Treatment of Linear Stochastic Difference Equations." Econometrica **11**(3–4): 173–220.
- Wald, A. (1943). "Tests of statistical hypotheses concerning several parameters when the number of observations is large." Transactions of the American Mathematical Society **54**(1–3): 426–482.
- Wald, A. (1945). "Sequential Tests of Statistical Hypotheses." Annals of Mathematical Statistics **16**(2): 117–186.
- Wald, A.: *Sequential Analysis* (1947), John Wiley and Sons.
- Wald, A.: *Statistical Decision Functions* (1950), John Wiley and Sons.

Ups and Downs

Wald's untimely death was not only a monumental loss but also a major blow to the young department, as keeping Wald at Columbia had been the primary reason for the birth of the department. For more than 40 years after Wald's death, the Statistics department remained small and survived "a series of crises," according to a conversation with T. W Anderson in Statistical Science (DeGroot 1986)

After Wald's death, Anderson (appointed as acting executive officer before Wald's trip), Scheffé and Levene kept the department going. The university quickly organized a committee to review the young Statistics department. In particular the committee considered the future development of the department (Columbia University Archives), including "expansion or contraction in personnel." Key questions included: (1) "should the department continue as a separate department?" and (2) after the death of Wald, should one or more appointments be made and to whom? In their March 1951 report, the committee concluded that Columbia's Statistics department remained "one of the few strong groups" but had been "severely impaired by Wald's death." It recommended Columbia keep Mathematical Statistics as a separate department and make an appointment in Theoretical Probability and Statistics to maintain international leadership in Statistics. Wolfowitz was also recommended to chair the department (referred to as executive officer at that time).

Statistics was expanding rapidly in the 1950s, probably more so than now, with many major universities setting up departments or groups. Wolfowitz was recruited away from the small Columbia department by Cornell in 1951. Fortunately, however, the department was able to successfully recruit Howard Raiffa. Aryeh Dvoretzky, who worked with Wald and Wolfowitz, also spent a fraction of his time at Columbia. Within a year, the department was reviewed a second time, resulting in the formation of a special committee for the academic year of 1952–1953, to fill several possible permanent and visiting positions (Columbia University Archives). Then, Scheffé received an offer from Berkeley in March 1953 and decided to accept it. He officially resigned on April 9, 1953. During the 1952–1953 academic year, Herbert Robbins of the University of North Carolina at Chapel Hill was at the Institute for Advanced Study and the department saw an opportunity. With efforts from Anderson, Levene, Raiffa and Herbert Solomon (then at Teachers College), Robbins accepted an offer from Columbia University, joining the department in the fall of 1953. During the critical times between November 1950 and the summer of 1953, the department held together strong due to the devotion of her very few members: Anderson, Scheffé, Levene, Raiffa, and Solomon, who supported the many daily activities (teaching, supervision, committee, planning, and recruiting) that were key to the survival of a small yet vital department.

The department stabilized after Robbins' arrival. From 1953 to the mid 1960s, the department hired a number of assistant professors: Jerome Sacks, Emanuel Parzen, Gopinath Kallianpur, Ronald Pyke, Lajos Takács, Donald Ylvisaker,

Simeon Berman, Cyrus Derman, Ronald Schaufele, Innis Abrahamson, Morris Skibinsky, Max Woodbury, Leon Gleser, David Siegmund, Alvin Baranchik, and John Rolph. The department also made special appointments to people in the New York City area. Chief among them were Colin Mallows (At & T Bell Labs) and Y. S. Chow (IBM Watson Center).

A similar crisis occurred in 1967–1968, when Robbins received multiple offers and took a 2-year leave to visit several places, and Anderson decided to join Stanford. But the return of Robbins and David Siegmund[2] and the hiring of Y. S. Chow resulted in another period of stability and productivity, lasting to the mid 80 s. During this period, a number of assistant professors were hired, including Burton Singer, Raoul Lepage, Tze Leung Lai, Jack Cuzick, Bruce Levin, Richard Bradley, Ioannis Karatzas, and Steven Lalley. In 1981, John Van Ryzin joined the department (jointly with the division of Biostatistics).

Another crisis developed in the late 1980s, starting with the retirements of Levene (1984) and Robbins (1985). Then came the departures of Singer (1985) to Yale and Lai (1986) to Stanford, followed by the death of Van Ryzin (1987). Karatzas played the key role in ensuring the survival of the department at this time. The department also received help and support from alumni, former members, and friends at other institutes and in the industry.

On February 27, 1992, Howard Levene made a $1 million pledge for an endowed professorship in Statistics. It was his hope that this gift will "insure the survival of this small but vital department." Mentioned in the public announcement, Columbia made the decision to name this professorship after Levene in appreciation of his lifelong devotion to teaching and research at Columbia. After spending 1 year visiting the department in 1991–1992, Paul Meier officially joined the department in 1992 as the new chair and became the first holder of the Howard Levene Professorship. Heyde arrived at the same time and added further strength in applied probability. Since then, the department has been stable and growing.

60 Years of Evolution: Space, Name, Undergraduate and Graduate Teaching

Due to its close relation with Economics, the Department of Mathematical Statistics was initially located within the Faculty of Political Science, occupying several offices in Fayerweather Hall, where Economics was housed. The department ran into space issues (scattered and inadequate space) in the late 60 s and was gradually moved to the Mathematics building between 1968 and 1971. Due to another wave of rapid growth after 2000, in 2004, the department, still part of the

[2] Siegmund was assistant professor at Columbia from 1966 to 1967 before joining Stanford in 1967.

Fig. 2 Photographic memories from the 1970s and 1980s. *Left* Robbins and Lai in the departmental office. *Right* photos of Siegmund and Robbins from the 1970s

Fig. 3 The "ever-lasting" Howard Levene

Graduate School of Arts and Sciences, moved to its current location in Columbia's new building for the School of Social Work.

For a brief time, the department of Mathematical Statistics was also home for Computer Science at Columbia under the leadership of Jonathan L. Gross until the formation of a new Computer Science department in the Engineering school. During that time, the department's name temporarily changed to "Mathematical

Fig. 4 Holiday party during Meier's visit in 1991 or 1992. From left to right in the two panels: de la Peña, Heyde, Kalicharan, Civtanic, Spivak, Meier, Lo, and Levene

Statistics and Computer Sciences." The first Ph.D. in Computer Science was awarded in May, 1977.

The department was initially formed as a strictly graduate department, primarily to avoid competing with other departments for the teaching of undergraduate Statistics. However, around 1950, it began teaching elementary Statistics to students enrolled in the School of General Studies, which is for students beyond their usual college age seeking part or full time undergraduate education. Starting in the mid 1960s, undergraduate teaching became a primary function of the department. In a letter dated Oct 1966, Levene reported that the department was offering undergraduate service courses to a total of 330 students. Another report in fall 1977 stated that the department was offering introductory Statistics with and without calculus, which were attended by a total of 132 students. Although the discussion of majors can be found in earlier departmental reports, the official record at Columbia states that the major in Statistics was established in 1982. Over the years, a concentration (minor) and multiple joint majors—Economics-Statistics, Mathematics-Statistics, Political Science-Statistics—were created to accommodate different interests from the undergraduate population. As part of a nationwide trend, the undergraduate major in Statistics at Columbia has become an increasingly popular choice, with more than 100 majors/minors last year.

The graduate teaching of statistics started even before the department was formed. As mentioned earlier, around the time the SRG was launched, a special committee (the Joint Committee on Instruction) on the Ph.D. program in mathematical statistics was also formed. The committee included Frederic Mills (Economics), Paul Lazersfeld (Sociology), Helen Walker (Teachers College) and people from the Psychology department and Columbia College. Kenneth Arrow was in the program but switched to Economics after getting his masters degree. Abe Girshick and William Madow were also in that program. By the time the department was formed, there were many graduate students participating under the GI Bill, and some of them held jobs in New York City. To accommodate these students, courses were offered in late afternoons and evenings. As a center in the New York area, many students outside Columbia University came to audit courses offered by the department. For example, Herman Chernoff,[3] and Franco Modigliani who later received the Nobel Prize in Economics, took Anderson's time series course. When Doob was teaching a course on stochastic processes, many students from NYU took it. In recent years, with the expansion of the Ph.D. and master's programs, total enrollments in graduate courses now exceeds 1000.

Contributions to Statistics

Graduates

The first Ph.D. graduate of the department was Howard Levene in 1947, who then became an "ever-lasting" member of the department. The first female Ph.D. graduate was Rosedith Sitgreaves in 1952, who worked with T. W. Anderson on the behavior of classification procedures. Sitgreaves remained in academia until her retirement in 1981 and held professorships at Stanford, Columbia Teachers College, and California State University. Among early doctoral graduates (before 1970) in Statistics from Columbia, Bowker (1949), Girshick (1947), and Stein (1947) played important roles in the formation and build-up of the Stanford University statistics department. In particular, Bowker became its founding chair in 1948, while officially still a Ph.D. student at Columbia. He later also served as Chancellor of the City University of New York (1963–1971) and Chancellor of the University of California at Berkeley (1971–1980). Other students in the first 25 years of the department's history included[4] :

- 1946–1955: Ralph Brookner (Wald), Howard Levene (Wolfowitz), Charles Stein (Wald), Gobind Ram Seth (Wolfowitz), Edward Paulson (Wald), Abe Girshick (Wald), Albert Bowker (Hotelling), Gottfried Noether (Wolfowitz), Stanley Issacson (Scheffé), Henry Teicher (Wolfowitz), Robert E. Bechhofer (Anderson), Milton Sobel (Wald and Wolfowitz), Jack Kiefer (Wolfowitz), Louis Cote (Chung), Jack Laderman (Wald), Rosedith Sitgreaves (Anderson), Lionel Weiss (Anderson), Allan Birnbaum (Raiffa), Cyrus Derman (Robbins), Richard Savage (Levene), Zivia Wurtele (Wald and Robbins), and William Kruskal (Levene).

[3] Chernoff was a student at Brown. Because Brown did not have anyone to be an adviser in statistics, Chernoff worked with Wald but was never matriculated in Columbia.

[4] A complete list of Ph.D. in Statistics at Columbia can be found at the department's website.

- 1956–1965: Sylvain Ehrenfeld (Raiffa), M. Vernon Johns (Robbins), Leon Herbach (Raiffa), Leonard Cohen (Raiffa), Peter Frank (Derman), Judah Rosenblatt (Anderson), Ruth Zwerling Gold (Anderson), Theodore K. Matthes (Anderson), Leo J. Tick (Anderson), Simeon M. Berman (Maruyama), Peter E. Ney (Chow), Lakshmi Venkataraman (Tabács), Ester Samuel (Robbins), Theophilos Cacoullos (Anderson), Gideon E. Schwarz (Anderson), Lloyd Rosenberg (Anderson), John P. Comer (Colind), Lee R. Abramson (Anderson), Joseph L. Gastwirth (Anderson), Arthur Cohen (Anderson), Edward Pollack (Levene), Richard G. Krutchkoff (Robbins), Ora Engelberd (Takács), James Pickands, III (Berman), Jerry E. Bramblett (Robbins), Norman Starr (Robbins)
- 1966–1971: Paul Burke (Takács), David Siegmund (Robbins), Saul Shapiro (Takács), Richard B. Will (Schaufele), Paul Shaman (Anderson), Stanle L. Sclove (Baranchik), Chandan K. Mustafi (Abrahamson), Arthur J. Nadas (Siegmund), Michael Kantor (Baranchik), Joseph L. Fleiss (Anderson), Ronald S. Dick (Levene), Michael J. Wichura (Levene), Harold B. Sackrowitz (Baranchik), Pi-Erh Lin (Baranchik), George P. H. Styan (Anderson), Joseph Breuer (Levene), Stephen Strauss (Chow), Ellen S. Hertz (Teicher), Robert A. Rutledge (Levene), Richard M. Stanley (Siegmund), George P. McCabe (Robbins), Serge L. Wind (Robbins), Neville E. O'Reilly (Robbins), Neil W. Henry (Anderson), Rasul A. Khan (Robbins), Joan E. Miller (Gross), Paul Tobias (Singer), Tze Leung Lai (Siegmund), Naomi B. Robbins (Siegmund).

Contributions (Including Textbooks and Papers) by Faculty During Their Time at Columbia

The department started with a strong research program, with people pioneering in areas such as multivariate analysis, sequential analysis, statistical decision theory, and econometrics. Hotelling's seminal contributions and Anderson's book are the classics of multivariate analysis. Wald's two books laid the foundation for sequential analysis and statistical decision theory. Columbia faculty members continued the work of Wald in sequential analysis with their many important publications, including Anderson's sequential boundary, power-one tests of Robbins, and fixed-width confidence intervals of Chow and Robbins. Major advances in statistical decision theory were made by Howard Raiffa as well as by Robbins (empirical Bayes theory), Chow, Robbins and Siegmund (on sequential optimal decision) and Karatzas and Lai (on optimal stochastic control). Anderson had worked on statistical problems arising from Economics, as did Burt Singer (his paper with Heckman in Econometrica was cited over two thousand times).

In addition to the two books already listed under Wald's publication list, faculty members also wrote other influential books. Among them are

- Anderson, T. W. (1957). An Introduction to Multivariate Statistics.
- Chow, Y. S., Robbins, H. and Siegmund, D. (1971). Great Expectations: The Theory of Optimal Stopping. Houghton-Mifflin.
- Chow, Y. S. and Teicher, H. (1978). Probability Theory: Independence, Interchangeability, Martingales. Springer.
- de la Peña, V. and Giné, E. (1999). Decoupling: From Dependence to Independence. Springer.
- de la Peña, V., Lai, T. L. and Shao, Q. (2010). Self-Normalized Processes: Limit Theory and Statistical Applications. Springer.
- Gelman, A., Carlin, B., Stern, H. and Rubin, D. (2004). Bayesian Data Analysis. Chapman & Hall.
- Gelman, A. and Hill, J. (2007). Data Analysis Using Regression and Multilevel/Hierarchical Models. Cambridge.
- Heyde, C. C. (1997) Quasi-Likelihood and Its Application. Springer.
- Karatzas, I. and Shreve, S. E. (1988). Brownian Motion and Stochastic Calculus. Springer.
- Karatzas, I. and Shreve, S. E. (1998). Methods of Mathematical Finance. Springer.
- Jacod, J. and Protter, P. (2012). Discretization of Processes. Springer.

Current Status of the Department

The Department has grown considerably in recent years. Enrollments in undergraduate and graduate classes have grown dramatically as have demands for statistical collaborations from around the campus. The Department now has two joint faculty appointments (with Political Science and with Sociology) and expects to make more appointments of this nature in the coming years. Space has again become a critical problem for the department and we hope the University's future expansion into the Manhattanville campus will provide a long-term solution. The Department has secured a position as one of the key science departments at Columbia and looks forward to making vital contributions at the forefront of statistical research as the discipline evolves in the coming decades (Figs. 1, 2, 3, 4, 5).

List of current faculty members (starting year)

- Richard Davis (2007)
- Victor de la Peña (1988)
- Regina Dolgoarshinnykh (2003)
- Yang Feng (2010)
- Andrew Gelman (1996)
- Gerardo Hernandez del-Valle (2005)
- Michael Hogan (2005)
- Martin Lindquist (2002)
- Jingchen Liu (2008)

Fig. 5 Departmental summer BBQ, May 2003. *Left* Paul and Louise Meier and Mildred Sklar. *Right* Faculty members exhibited deep interest in the open problem of how to [properly] start the grill

- Shaw-Hwa Lo (1989)
- David Madigan (2007)
- Liam Paninski (2005)
- Philip Protter (2011)
- Daniel Rabinowitz (1993)
- Bodhi Sen (2008)
- Michael Sobel (2010)
- Victoria Stodden (2010)
- Jae Kyung Woo (2011)
- Frank Wood (2009)
- Zhiliang Ying (2000)
- Hongzhong Zhang (2010)
- Tian Zheng (2002)

References

Anderson, T. W. (1955). The Department of Mathematical Statistics. A history of the faculty of political science, Columbia University. R. G. Hoxie. New York, Columbia University Press: 250–255.

DeGroot, M. H. (1986). "A Conversation with T. W. Anderson." Statistical Science **1**(1): 97–105.

Samuelson, P. A. (1960). "Harold Hotelling as mathematical economist." The American Statistician **14**(3): 21–25.

Wallis, W. A. (1980). "The Statistical Research Group, 1942–1945—Reply." Journal of the American Statistical Association **75**(370): 334–335.

Wolfowitz, J. (1952). "Wald, Abraham 1902–1950." Annals of Mathematical Statistics **23**(1): 1–13.

Biometrics and Statistical Science at Cornell

James G. Booth and Martin T. Wells

Statistics as a discipline at Cornell has a complex history. Even today, visitors and job applicants find the distributed model, with several groups of statisticians distributed in various departments, colleges, and buildings across campus, quite confusing. This document initially focuses on one of these groups founded as a unit within the Department of Plant Breeding, in the College of Agriculture, in 1947.[1] Key figures that influenced the discipline in other Cornell statistics groups are discussed later, and we conclude with an overview of the recently formed Department of Statistical Science.

The Biometrics Unit 1947–1988

In this history of the first 40 years of the Cornell University Biometrics Unit, we discuss the founding and policy makers, the people, the teaching, the research, the

[1] Much of the material is taken from a technical report written by Walter T. Federer (1989) after his retirement in 1986. The full technical report is available from Cornell's Department of Biological Statistics and Computational Biology upon request.

Cornell University,
Biological Statistics and Computational Biology,
Comstock Hall, Ithaca, NY 14853, USA
J. G. Booth (✉)
e-mail: jb383@cornell.edu

Cornell University,
Social Statistics, Comstock Hall,
Ithaca, NY 14853, USA
M. T. Wells
e-mail: mtw1@cornell.edu

A. Agresti and X.-L. Meng (eds.), *Strength in Numbers: The Rising of Academic Statistics Departments in the U.S.*, DOI: 10.1007/978-1-4614-3649-2_4,

consulting, the computing, and the future outlook for the Biometrics Unit as it was viewed after the first 40 years.

The Biometrics Unit was established in 1947 with the first personnel being hired in 1948. One faculty position, one technical assistant (Biometrician), and one secretarial position were established initially. In addition, monies were made available for purchase of a number of Monroe, Friden, and Marchant desk calculating machines for class and office use.

Since every academic group was required to be part of a department, the committee for the Unit and the College of Agriculture Administration decided that the Biometrics Unit should be attached to the Department of Plant Breeding, which was the department of the Biometrics Unit Policy Committee Chairman, Prof. Harry H. Love. The faculty in Plant Breeding did very complex field experiments that required sophisticated designs and analysis strategies (e.g., strip block experiments) and could only conduct one experiment per year. Thus, deriving the maximal amount of information out of each experiment was essential.

For many years prior to 1948, an introductory statistical methods course had been taught by Prof. Love and later by J. R. Livermore in the Department of Plant Breeding. The committee wished to update the course and to extend it to two semesters. The committee decided that statistical methods at the content and level of George W. Snedecor's book, *Statistical Methods, 4th Edition* in 1946, and material similar to that which later appeared in the book, *Experimental Designs*, by William G. Cochran and Gertrude M. Cox in 1949, should be available for students. The committee also decided that a person trained as a statistician should be giving the courses. Up until this time an animal breeder, a plant breeder, an economist or other subject-matter specialist who knew some statistical methodology had offered the courses in statistics in the College of Agriculture. This was a turning point, since most statistics courses in the College and at Cornell since that time have been offered by individuals trained as statisticians. There also was a Statistics and Research Methods course taught in the department, first by Professor Love and later by Professors Sandford S. Atwood and Carl C. Lowe. In 1966, the name Department of Plant Breeding was changed to the Department of Plant Breeding and Biometry on the initiative of the plant breeders in the department.

People

As already mentioned, when the Biometrics Unit was established, three positions and funds for computing equipment were allocated by the College of Agriculture in 1947. In filling the faculty position, Prof. Love, Chairman of the Statistics Committee of the College of Agriculture and Head of the Department of Plant Breeding, began the search for possible candidates for the position. They searched for a statistician interested and versed in agricultural and biological problems. Biologists with statistical expertise were among the candidates considered. For example, Dr. Max R. Zelle, who was the number three person in the U.S. Atomic Energy Program

Fig. 1 Pioneers in the development of Statistics at Cornell: Walter T. Federer (left), Douglas S. Robson (middle), and Shayle R. Searle (right)

and who came to Cornell in 1948 as a microbiologist–geneticist in the Department of Dairy Science, was one of the candidates under consideration.

In late 1947, Dr. Love traveled to Iowa State College to interview Walter T. Federer (Fig. 1), who was later interviewed at Cornell in February 1948. He was hired as Professor of Biological Statistics and came to Cornell in August 1948. Dr. Federer had an MS in Plant Breeding and a PhD in Statistics and so seemed an ideal candidate. At the time Federer was hired, the college administration stated that a second statistician would be hired within a year. It was the administration's idea that hiring Charles R. Henderson as an animal breeder in 1948 fulfilled the promise. Drs. Federer and Love were under the impression that the position would be in the Biometrics Unit. As a footnote, Henderson always had an open invitation to associate with the Biometrics Unit but he chose not to; instead, he was on the Policy Committee where he was a very effective and very strong supporter of the unit. The second professional person hired in the unit was a biometrician and the second person to fill this position was Douglas S. Robson. The second faculty appointment was Robert G. D. Steel in 1952 as an Associate Professor. The Policy Committee felt that additional statistics courses and statistical consulting were needed. Professor S. S. Atwood, who succeeded Harry Love as Chair of Plant Breeding, was able to obtain the support of 15 out of the 18 Department Chairs in the College of Agriculture for this position over any in their own departments. The third faculty position was obtained in 1955 when Douglas S. Robson was hired as an Assistant Professor. This third position had the very strong backing of the Director of Research, C. E. F. Guterman, who appreciated the statistical consulting for college research projects and the statistical research of the unit.

Professor Shayle R. Searle came to Cornell University in 1962 as a statistical computing specialist for the Cornell Computing Center with the promise that there would be a tenure-track faculty position. Toward the end of his 3-year appointment, Searle found that there was no tenure-track position available despite the promise. This was resolved by the then Cornell University Provost S. S. Atwood, Dean C. E. Palm, and Director W. Keight Kennedy of the College of Agriculture. It was decided that he would be placed permanently in the Biometrics Unit but that his would be a new position rather than filling one of the unit's established positions. Since Searle was involved with computing, this would add another dimension to the unit's activities. The same year that Searle came to Cornell, Keewhan Choi was appointed as an Assistant professor. Leslie N. Balaam, Kali S. Banerjee, B. Leo Raktoe, and Ying Y. Wang were visiting professors (with State and NIH support) who also helped with the teaching and consulting activities of the unit in the 1963–1965 period. Doug Robson was awarded a 10-year NIH Career Development Award, 1964–1974, freeing up a position which was filled by N. Scott Urquhart in 1965. A fifth position in the unit was created when Foster B. Cady was appointed in 1971, specifically to take over the teaching of introductory statistics and consulting. In 1973, a sixth position was created for Robson. A seventh position was created when Carlos Castillo-Chavez joined the unit in June 1988. Since he was a bio-mathematician rather than a statistician, the full implication of biometry was being realized and a new dimension was added to the unit with his appointment. Below is a summary of the people holding the various positions and their replacements over the years.

1. Federer (1948–1986), Altman (1987–2001), Hooker (2006-present)
2. Steel (1952–1960), Choi (1962–1966), Solomon (1968–1981), McCulloch (1982–2000), Strawderman (2000-present)
3. Robson (1952–1963), Urquhart (1965–1971), Davidson (1971–1973), Wood (1973–1977), Schwager (1978–2011)
4. Searle (1962–1996), Contraras (1995–2002)
5. Robson (1973–1987), Churchill (1990–1996)
6. Cady (1971–1979), Casella (1981–2000), Booth (2004-present)
7. Castillo-Chavez (1988–2004)
8. Holmes (1996–1999)

The numerous visiting Faculty and Fellows associated with the Biometrics Unit were a stimulating and motivating force. New ideas and concepts brought by Visitors and Fellows had a strong influence on the history of the unit. Such individuals as Arnold J. King, George W. Snedecor, Frank Yates, David J. Finney, Oscar Kempthorne, Samuel Karlin, and David R. Cox were stimuli that were greatly appreciated. People such as W. Scott Overton, Leslie N. Balaam, B. Leo Raktoe, D. Raghavarao, Esther Seiden, Frederick Pukelsheim, and others were frequent visitors and contributors to the unit's program over the years. The statistics groups at Cornell University were fortunate in obtaining two Andrew D. White Professors-at-Large for Statistics. Samuel Karlin held this position in the 1970s and Sir David R. Cox held this position in the 1980s.

The unit's first graduate student, Douglas S. Robson, together with Robert Bowles, was the recipient of the first undergraduate degree in Statistics in the United States from Iowa State University in 1949. Robson was supported on the biometrician position which was later divided into two State assistantships in 1954. When several contracts and grants and especially the NIH Training Grant were discontinued in 1976, four additional State Assistantships were allocated to the unit. By the late 1950s and early 1960s, eight to ten graduate students were in the program. This grew to 21 graduate students majoring in Biometry or Statistics by 1988. In the early days of the program, graduate students were required to minor in mathematics. Most students selected Jack C. Kiefer as their minor committee member. Professor Kiefer was very exacting and many graduate students had a trying and difficult time getting up to the standard in mathematics. Professor Jacob Wolfowitz seldom took minors or majors in Statistics.

Perhaps, the unit's most famous graduate major is Douglas S. Robson, the first student. Some that became internationally proclaimed as leaders in their field are B. L. Raktoe, A. Hedayat, J. Eccleston, J. Skalski, and C. E. McCulloch. One of several graduates from the minor program in Statistics and Biometry who became internationally famous was the unit's own S. R. Searle. Other famous graduates include Henry Regier in Fishery Biology and David A. Harville in Statistics and Statistical Genetics.

The undergraduate major program in Biometry and Statistics began in 1961–1962. While the number of students in the major was initially quite small, it grew to the 50–60 range by 1988 (and now exceeds 80).

Teaching

As indicated previously, the Policy Committee for the Biometrics Unit desired to have a statistical methods course at the level of G. W. Snedecor's *Statistical Methods*. The course would be for graduate students who needed statistical procedures for their thesis research. It was also desired that more than a perfunctory level of statistical design, both survey and experiment, be included. A two-semester, three-credit-hour, sequence was devised to meet these goals. However, the original course material, covering Snedecor as well as what later appeared in Cochran and Cox in 1949, was considered too strenuous for students, and the courses were considered to be under-credited. Hence, in 1953–1954, two, two-semester, sequences were introduced. One covered the material in Snedecor, and associated with this sequence were two supplementary courses which covered the algebraic aspects of various statistical procedures. The second sequence was on experiment and treatment design and analysis. Beginning in 1976 the nature of the course was changed from a two-semester sequence to two individual courses, one on experiment design (the arrangement of treatments in an experiment) and one on treatment design (the selection of treatments for an experiment). The reason was to accommodate the needs of graduate students from diverse fields.

The next set of courses developed by members of the Biometrics Unit included an introductory methods course for undergraduates, a probability and inference sequence and a course on matrix algebra. The Mathematics Department questioned (about every 10 years) the reason for teaching a mathematics course such as matrix algebra in the Biometrics Unit. There were also occasional objections to having probability taught in the unit.

An experiment with teaching the second semester of the graduate methods course was tried under the leadership of Constance L. Wood. The course was expanded and divided into six one-credit modules in 1977. There were two modules on analyses of designed experiments, two on regression, one on non-parametric and distribution-free procedures, and one on sampling and estimation for biological populations. Since a large number of the same students took the first three modules for credit, the material in them was eventually combined into a four-credit-hour course which is still taught today.

Using increased enrollments in courses and student responses as measures of success, five individuals should be singled out for teaching the graduate methods sequence. These are N. S. Urquhart, W. H. Swallow, D. L. Solomon, C. E. McCulloch, and G. Casella. This course sequence was a nemesis for several individuals in that many complaints about their teaching were received. Over time, Statistical Methods I and II evolved to a place where they were well accepted and taught. They were considered to be two of the most important introductory graduate methodology courses offered at Cornell.

Research

Long-term NIH training and research grants and consistent NSF funding allowed the department to double the number of graduate students it supported and to maintain an extensive program of visiting Fellows and scientists.

During the 40-year period following the establishment of the Biometrics Unit, 1041 technical reports, 546 published papers, and 132 theses were written. Of the theses, 21 were bachelors, 55 were masters, and 56 were PhD theses. Thirty-two papers were published in the Annals of Statistics, 45 in Biometrics, 38 in animal science journals, 32 in medical science journals, and 62 in biological science journals. Some of the fields of research can be broadly categorized as statistical design, regression and linear models, sampling and estimation for biological populations, statistical genetics, statistical modeling and statistical methodology. In addition to being prolific publishers in the top statistics journals, the Biometrics Unit faculty wrote numerous books during this period including some classic texts: Federer (1955), *Experimental Design—Theory and Application*, Searle (1971), *Linear Models*, Searle (1982), *Matrix Algebra Useful for Statistics*, and Casella and Berger (1989), *Statistical Inference*.

Consulting

Statistical consulting with graduate students and faculty from all areas of Cornell University was a sizeable part of the unit's activities from the beginning. Originally in 1948–1949, an open door policy for statistical consulting was in operation. Since the entire day was often taken up for both Federer and Robson, a policy of making appointments was instituted. This reduced the consulting load to one-third or one-half the previous amount. Evidently, when no appointment was necessary, individuals did not fully think out their questions. Over the years, the statistical consulting load reached manageable proportions owing to constant adherence by faculty to committing a fixed number of hours per week to this activity.

Since graduate majors in Statistics and Biometry indicated that consulting was one of the most valuable experiences they received during their graduate training, a formalized system was devised in 1960–1961 whereby students were required to do statistical consulting as part of their training. In 1979–1980 statistical consulting was put on a more formal basis under the direction of Professor Robson. Putting his 30 years of experience to work, credit was given under the course name of Statistical Consulting, a course which still exists today.

Doug Robson was considered to be a super consultant. One Department Head said in a meeting of a large CALS committee, "Every faculty member and graduate student in CALS should have the opportunity of consulting with Robson. His excellent concepts, ideas, and solutions are beyond belief and truly magnificent!" Such a statement demonstrates the impact that one statistical consultant can have. C. E. McCulloch said, "Robson is the best statistical consultant I have ever encountered." Doug was not a meeting person. At one meeting when the dean was present, one faculty member joked "Doug, what are you doing here? I thought you ran a feed store." To which the Dean replied, "He does. It's called the College of Agriculture."

The Ensuing Years in Biometry

Professors Federer, Robson, and Searle retired in 1986, 1987, and 1996, respectively. As the unit personnel embarked on their second 40-year stint, they seemed to be in a very good position to maintain the eminence of the Biometrics Unit. A highly competent faculty was in place and they were creative and energetic. In the decade from 1988 to 1998 the unit continued to flourish and develop its reputation as a core group within in the statistics community at Cornell. New faculty joined the unit with the additional of Gary Churchill in 1990, Martha Contraras in 1995 and Susan Holmes in 1996. That year the unit separated from Plant Breeding and became the Department of Biometry, one of the departments affiliated with the newly formed Department of Statistical Science. By this point in time, Biometry had developed significant strengths in biomathematics, statistical genetics, and

Bayesian statistics, as well as retaining its traditional strengths in the more classical areas of experimental design and linear models.

Around 2000, an exodus of key faculty members occurred (Gary Churchill, Charles McCulloch, George Casella, Naomi Altman) and a period of administrative instability ensued. At the same time, the Cornell Genomics Initiative (now the New Life Sciences Initiative) was gaining momentum as a broad-based effort to increase Cornell's strengths in genomics and computational biology. Under the direction of Dean Susan Henry, the Department of Biometry was dissolved in 2001 and replaced with the Department of Biological Statistics and Computational Biology (BSCB) as a key component of the New Life Sciences Initiative. Martin Wells (from the Department of Social Statistics in ILR) became chair and the majority of the current faculty members were hired over the next several years. Professor Wells stepped down as chair at the end of June 2006 but retains his affiliation with BSCB through a special joint appointment. James Booth is the current chair of BSCB.

Some History of Cornell's Other Statistics Groups

Before the Second World War, the discipline of statistics played only a minor role at Cornell. Exceptions were that Walter Wilcox held the first professorship in statistics in the United States and Professor Edmund Ezra Day, who was to become our fifth University President, was president of the American Statistical Association a generation earlier. But by the late 1940s, Cornell had an extraordinary group of probabilists, rivaled in the world perhaps only by Moscow (Ehrenberg et al. 1994).

With the outbreak of the Second World War, strong groups of statisticians developed in Princeton and New York. Among the luminary at Columbia were Harold Hotelling, Milton Friedman, Henry Scheffe, Abraham Wald, and Jacob Wolfowitz. In the Statistical Research Group at Princeton were William Cochran, Theodore Anderson, Fred Mosteller, Fred Stephan, and John Tukey. The developments of mathematical and applied statistics at Cornell occurred shortly after the Second World War out of these groups, with major appointments in Mathematics, Industrial Engineering, and Industrial and Labor Relations.

Phillip McCarthy was recruited from Princeton to Cornell by Fred Stephan in 1948. Sample surveys had become a major tool as social science researchers sought to put their work on a firm empirical and theoretical base. However, at that time the current designs for most studies had major weaknesses, and modern sampling methods were needed. McCarthy was hired on a Social Science Council Research grant to clarify design and related issues and subsequently developed some of the seminal foundational ideas in survey sampling. McCarthy then took a position on the faculty of the new School of Industrial and Labor Relations (ILR) at Cornell where he remained until 1988 (Ehrenberg et al. 1994). In 1949, Isadore Blumen was recruited into the ILR social statistics group. McCarthy and Blumen

Fig. 2 Pioneers in the development of Statistics at Cornell: Jacob Wolfowitz (left) and Jack Kiefer (right)

played a major role in developing the Cornell Statistics Center, the first cross campus statistics organization, and did the arduous work directing it without any compensation.

The untimely death of Abraham Wald in an airplane crash in India in 1950 was a deep personal and professional loss for many in the Columbia statistics community. At the invitation of brother-in-law Isadore Blumen, Jacob Wolfowitz spent the summer of 1951 at Sheldrake on Cayuga Lake. During the visit, Wolfowitz developed associations with members of the Cornell mathematics faculty that led to an offer of a professorship (Blumen et al. 1981). At Cornell, Wolfowitz's great love of discussing ideas and problems resulted in collaboration with many mathematicians, probabilists, and students. Most outstanding of these students was Jack Kiefer who joined the Cornell mathematics faculty in 1952. Besides extending the frontiers opened up by Wald in the areas of decision theory and sequential analysis, Kiefer and Wolfowitz (Fig.2) also either pioneered or made important early contributions in new fields such as stochastic approximation, many server queuing systems, and the design of experiments. Kiefer and Wolfowitz were central figures in a whirl of mathematical statistics activity at Cornell involving other theoretical statisticians and probabilists, such as Kai-Lai Chung, Gil Hunt, Mark Kac, and Frank Spitzer. In 1959, Roger Farrell joined the Mathematics Department at Cornell thereby adding further depth to Cornell's intellectual dominance in statistical decision theory and optimal experimental design. Numerous visitors in the 1950s including Julius Blum, Aryeh Dvoretzky, William Feller (who had previously been on the Mathematics Department faculty), Esther Seiden, Leonid Hurwicz, Milton Sobel, and Henry Teicher added to the great richness of the intellectual Cornell environment (Blumen et al. 1981).

Wolfowitz was dismayed by what he felt was the betrayal of fundamental values and of academic norms by many students and faculty during the turmoil of the 1960s. Wolfowitz's conservative leanings put him at odds with many around him and he was concerned by the Cornell administration's unwillingness to assure him that he could continue beyond normal retirement at 65. In 1970, Wolfowitz resigned from Cornell and took a position in mathematics at the University of Illinois. On the other side of the aisle Kiefer was always willing to fight for liberal causes. Kiefer held a position at Cornell until 1979 when he retired and joined the faculty at the University of California at Berkeley. After the departure of Kiefer and Wolfowitz, the intellectual dominance in statistical decision theory in the Department of Mathematics was carried on by Larry Brown (a Kiefer student), Roger Farrell, and Juinn (Gene) Hwang.

While Kiefer and Wolfowitz joined the mathematics faculty in the early 1950s, two of Wolfowitz's students from Columbia University, Robert Bechhofer and Lionel Weiss, joined Industrial and Systems Engineering. Bechhofer and Weiss were brought to what became the School of Operations Research and Industrial and Systems Engineering as key components of a vision to shape Industrial and Systems Engineering into a more sophisticated mathematically and statistically discipline, better suited to the rapidly evolving demands of industry for decision-making tools (Bland and Resnick 2000). Bechhofer and Weiss mentored generations of graduate students in theoretical and applied statistical research and provided dedicated and extremely effective courses in modern statistical methodologies to decades of Cornell students; the pair's devotion to the School of Operations Research and to Cornell was unsurpassed.

Although all of the original pioneers moved on or retired, several members of the existing statistics faculty have had long careers at Cornell that overlapped with many of the founding generation, including David Ruppert, Bruce Turnbull, and Sid Resnick in Operations Research, Paul Velleman and Martin Wells in Social Statistics, and Gene Hwang in Mathematics.

Cornell Statistics has trained many PhD students many of whom have gone on to be leaders in the profession. Graduates from Cornell include Doug Robson, Jerry Sacks, Shayle Searle, Praesert Na Nagora, Samuel Kotz, Larry Brown, Sol Kaufman, Richard Schwartz, David Moore, Gary Lorden, Robert F Engle, Sam Hedayat, Robert Jacobsen, John Eccleston, Ching-Shui Cheng, Sally Sievers, Gloria Zerdy, Bruce Turnbull, Avi Mandelbaum, Cavell Brownie, Jim Berger, Ken Pollock, William Swallow, Rich Gelber, David Harville, James Rosenberger, Yehuda Vardi, William Notz, Richard Smith, Chuck McCulloch, Luke Tierney, Iain Johnstone, Mosuk Chow, Constantine Gatsonis, Chris Jennison, Walt Piegorsch, Mark Low, Karen Bandeen-Roche, Mary Dowling, Russell Barton, Ziding Feng, Jon Maata, Calvin Berry, Lance Waller, Naisyin Wang, Hong Zhao, Fushing Hsieh, Victoria Chen, Henry Lynn, Isaac Bekele, Yu-Ling Tseng, Andy Tsao, David Lansky, David Umbach, Kun He, Yana Kane-Esrig, Ed Frongillo, Costas Goutis, Jim Hobert, Kwee Yeo, David Lando, Horng-Shing Lu, Jean Opsomer, Weijing Wang, John Walker, Weizhen Wang, Rick Cleary, Wenxin Jiang, Lynn Eberly, Adam Ding, Tony Cai, Andrew Schulman, Pam Ohman, Craig Borkowf, Adrienne Freeman-Gallant, Sam Wu, Steve Gulyas, Haiqun Lin, Steve Hogan, John Staudenmayer, Ciprian Crainiceanu, Dabao Zhang, Yong Sung Joo, Yan Yu, Daniel Fink, Trevor Park, Rich Lavine, Matt Briggs, Russ Zaretzki, Lisa Madsen,

Chong Wang, Matt Tom, Jing Qui, Min Zhang, Harrison Zhou, Cindy Long, Will Anderson, Bala Rajaratnam, Gongfu Zhou, Ben Shaby, Xin Zhou, Haizhi Wu, Kathryn Barger,Vadim Zipunnikov, Serena Chan, Bret Hanlon, Liz Schifano, David Clement, Kirsten Eilertson, Nikolay Bliznyuk, Matthais Kormaksson, Amy Li, Haim Bar, and Raj Narayanan.

The Department of Statistical Sciences Today

The traditional structure at Cornell for Statistics has been decentralized, with groups established in the Colleges of Engineering, Arts, Industrial and Labor Relations, Agricultural and Life Sciences as well as a consulting service in Human Ecology. For years, this structure worked although there was continual dissatisfaction by statisticians about lack of administrative coherence and autonomy. The advantage of the Cornell traditional structure is that it forces statisticians to be close to their applications areas and prevents an insularity that may result from being concentrated in one department. However, the Cornell organizational structure needed to become more effective.

In light of faculty losses at that time (Math: Larry Brown, Roger Farrell, Persi Diaconis, Richard Liu; Biometry: George Casella, Chuck McCulloch, Naomi Altman, Gary Churchill, Martha Contreras, Susan Holmes, Shayle Searle; ORIE: Lionel Weiss, Tom Santner, Bob Bechhofer, Elizabeth Slate; and Social Statistics: Isadore Blumen, Ali Hadi), the Cornell administration agreed that it was important to devote special attention to enhancing research strength in core areas of statistics. In June 2001, Provost Biddy Martin asked a committee of deans to reconsider the organization and administration of the Department of Statistical Sciences (DSS), in light of the revised mission and focus of the Biometry department in CALS (now the Department of Biological Statistics and Computational Biology). The charge was to devise ways to enhance the capacity of the DSS to serve as a coordinating mechanism and a catalyst across multiple colleges, i.e., to move it from a "virtual" to a "real" department in significant ways.

Initially CALS, and then ILR School, became the lead-college in administering the department and organizing its activities. This start was primarily a collaborative effort between CALS and ILR, but other colleges were invited to participate in accord with their teaching and research interests. In 2005, DSS became a department in Cornell's new Faculty of Computing and Information Science (CIS). The CIS partnership opened potential new avenues for exploring future work with the Departments of Computer Science and Information Science. It seemed that CIS was the most logical intellectual home for DSS.

Today, Statistical Science at Cornell engages in two related activities in a much more unified manner: research and teaching of core statistics including emerging areas such as computational statistics, Bayesian methods, functional data analysis and semiparametric modeling, and spatial data; and outreach to other disciplines including: data mining, machine learning, bioinformatics and genomics, financial engineering and risk management, health, control and measurements of the Internet, environmental science, demand forecasting, empirical legal studies, resource management, and systems engineering.

The Graduate Field of Statistics offers two programs of study: the MS/PhD in Statistics and the Masters of Professional Studies (MPS) in Applied Statistics. The PhD program is intended to prepare students for a career in research and teaching at the university level, and requires writing and defending a dissertation. The MS portion of the MS/PhD program is primarily intended as the first part of the PhD program and typically involves 2 years of graduate-level coursework in statistics and probability; upon successful completion, students are eligible to transition into the PhD program. The MPS in Applied Statistics is intended for students who are interested in professional careers in business, industry, government, or scientific research. Cornell's MPS program provides rigorous training in modern data analytical skills that are sought after in almost any field and our graduates have had excellent career opportunities.

Finally, a new undergraduate degree program in Statistical Science in the College of Arts and Sciences now allows students in that college to major in Statistics, and complements the longstanding Biometry and Statistics major in CALS.

References

Bland R, Resnick S (2000) Lionel Weiss memorial statement, Cornell University. http://ecommons.cornell.edu/bitstream/1813/18704/2/Weiss_Lionel_2000.pdf

Blumen I, Kesten H, Farrell RH (1981) Jacob Wolfowitz memorial statement, Cornell University. http://ecommons.cornell.edu/bitstream/1813/19116/2/Wolfowitz_Jacob_1981.pdf

Casella G, Berger R (1989) Statistical inference. Brooks/Cole, Pacific Grove

Ehrenberg R, Velleman P, Blumen I (1994) Philip McCarthy memorial statement. Cornell University. http://ecommons.cornell.edu/bitstream/1813/18159/2/McCarthy_Philip_J_1994.pdf

Federer WT (1955) Experimental design. The Macmillan Company, New York

Federer WT (1989) The Biometrics Unit: the first 40 years, 1948–1988. Biometrics Unit technical report BU-1000-M, Cornell University

Searle SR (1971) Linear models. Wiley, New York

Searle SR (1982) Matrix algebra useful for statistics. Wiley, New York

The Florida State University Statistics Department

Jayaram Sethuraman and Frederick W. Leysieffer

The Department of Statistics at The Florida State University was established in 1959. It was at a time when a major transformation in Florida's higher education system was occurring at The Florida State University. The university had evolved from Florida State College for Women just 12 years before. The university enrolled just 8,179 students that fall. There was a continuing air of excitement on campus. The state and the administration had made a commitment and had the resources to build strengths in FSU's faculty, particularly in the sciences. Hiring was an ongoing and vigorous activity. New departments were being created. That is when the Department of Statistics first opened its doors.

Ralph Bradley arrived from Virginia Polytechnic Institute to become the department's first head. He held this position for 19 years and was instrumental in growing it from infancy to maturity. Following Ralph Bradley who served as department head from 1959 to 1978, there has been a succession of chairs contributing to its continued development: Myles Hollander (1978–1981, 1999–2005), Fred Leysieffer (1981–1987, 1990–1993), Jayaram Sethuraman (1987–1990), Duane Meeter (1993–1996), Ian McKeague (1996–1999), and currently Dan McGee (2005–2011). Xufeng Niu is slated to be the next chair which began on August 2011.

Florida State University,
Department of Statistics,
214 OSB 117 N. Woodward Ave,
P.O. Box 3064330
Tallahassee, FL 32306-4330, USA
J. Sethuraman (✉) · F. W. Leysieffer
e-mail: sethu@stat.fsu.edu

F. W. Leysieffer
e-mail: fleysieffer@fsu.edu

A. Agresti and X.-L. Meng (eds.), *Strength in Numbers: The Rising of Academic Statistics Departments in the U.S.*, DOI: 10.1007/978-1-4614-3649-2_5,

Ralph Bradley

Published Histories of the Department and Celebrations

Ralph Bradley chronicled the developments in the department in an article, describing it through its first 15 years (Bradley 1975).

The department celebrated its thirtieth anniversary in 1989 and published the invited lectures at that event in a special issue of the *Communications in Statistics*. Fred Leysieffer's account of the 15 years following 1974 is published in that volume.

A wealth of information regarding the professional development of the department through 1990, together with extensive details on faculty, graduate students, distinguished visitors, and significant accomplishments and recognitions, is available in these two references; readers can easily access them from the department web pages www.stat.fsu.edu/history/bradley.pdf and www.stat.fsu.edu/history/leysieffer.pdf.

Faculty, Students and Staff of the Department of Statistics in 1978

The fortieth and fiftieth anniversaries of the department were celebrated in 1999 and 2009 under the chairmanships of Myles Hollander and Dan McGee.

The Early Years

Ralph made two initial senior hires that were to shape the department's future. Frank Wilcoxon joined the department in 1960 and Richard Savage in 1963. Their well-known accomplishments inspired graduate students and served, significantly, to interest prospective, junior, and senior faculty members to come to The Florida State University.

Frank Wilcoxon was hired on a part-time basis in the fall of 1960. He is remembered as enjoying his interactions with students and junior faculty members and being a thoroughly committed environmentalist. He loved taking friends out for trips in his kayak and would frequently be spotted in Tallahassee astride his motorcycle.

Frank Wilcoxon, 1964

Frank passed away in 1965 at the age of 75. The departmental graduate students purchased a plaque designating the departmental reading room as The Frank Wilcoxon Room. His working notes are preserved in that room.

Richard Savage is remembered as giving freely of his time and effort to fashion the graduate curriculum. He read and commented upon every paper and technical report written in the department. His comments were not necessarily laudatory but always constructive and helpful. He set high academic standards for the department, ones that would serve as benchmarks for years to come.

Richard and JoAnn Savage

Early on, the department organized its first holiday lunch. The entire department participated and it was a great feast. It was highlighted by the many international students, each vying to bring out the best specialties representative of his or her country. Later, Doug Zahn served as master chef, with his signature slow baked ham and pecan pies as centerpieces. That tradition continues today.

Ralph encouraged graduate students to become leaders. The students fielded athletic teams. Students organized the department's spring picnics, a tradition that also continues today. (Faculty were responsible for the fall picnic and still are.) In later years many took leadership positions in the organizations they were part of. In the academic world, many became department chairs or deans.

During these years, the department was fortunate to have the resources to hire a series of excellent, well-known, senior faculty members who in turn would attract talented junior faculty and graduate students. As they are familiar to people in the profession of Statistics, their professional accomplishments are not recounted here.

Myles Hollander was the leader in nonparametrics. His forceful delivery of lectures attracted many students to pursue research in nonparametrics or to apply it in their work. His three books have become classics in this field.

Jayaram (Sethu) Sethuraman (still active in his post-retirement) came in to be the leader of the probability program. He brought his profound understanding of mathematics, mathematical statistics, and probability theory along with the high standards he set for his students. Graduate students, who took his classes, will forever remember how challenging his lectures were. All members of the department were grateful beneficiaries of his knowledge, willingness, and patience in helping them to achieve a better understanding of the technical aspects of their work.

Robert Serfling worked on asymptotic methods and wrote his famous book on asymptotic theorems of statistics when he was here.

Frank Proschan, a very caring faculty member, was so very generous with his ideas for new directions in research. His Saturday morning seminar, the "Reliability Club," met weekly with graduate students in attendance, reading papers and discussing new directions. Frank's presence brought a constant flow of visitors to the Reliability Center.

Frank Proschan

Debabrata (Dev) Basu's time in the department came with many visitors, some for extended stays, who journeyed to Tallahassee to explore the foundations of Statistics with him. His noon-time bridge playing became a legendary, daily event, complete with a wealth of postmortem discussions.

George Marsaglia's arrival signaled a new direction for the department in statistical computing. His search for improved random number generators defined his research. George had a passion for solving crossword puzzles, and a favorite pastime for graduate students during lunch time was a session with him solving the latest New York Times puzzle.

The department hosted short-term and long-term visitors. Jaroslov Hajek spent a year and wrote his famous paper on the asymptotic sufficiency of rank statistics in the Bahadur sense during this period. Dennis Lindley gave an inspiring series of lectures on Bayes methods and converted many a faculty to become Bayesians. The visit of Oscar Kempthorne is memorable for the debates with our faculty member Dev Basu on tests of significance and Bayesian methods. Richard Barlow and Al Marshall teamed up with Frank Proschan to work on Reliability. David Blackwell inspired J. (Sethu) Sethuraman on Dirichlet Processes and he wrote his famous paper on its constructive definition. The department buzzed with research activity with summer visitors who came to collaborate with faculty, prominently with Frank Proschan.

Three of our faculty members became presidents of the American Statistical Association, either while at the university or while holding later appointments elsewhere. They are Ralph Bradley, at Florida State University, Richard Savage, at Yale University, and Lynne Billard, at the University of Georgia.

Douglas Zahn was awarded the ASA's W. J. Dixon Award for Excellence in Statistical Consulting in 2009.

In 1990, the faculty numbered 16, and it included leaders such as Dev Basu, Hani Doss, Myles Hollander, Fred Huffer, George Marsaglia, Frank Proschan,

Fred Leysieffer, Duane Meeter, and Jayaram Sethuraman, in probability and statistics theory, statistical inference, Bayes analysis, nonparametrics, large deviations, reliability theory, and random number generation. The accomplishments of the faculty are reflected in their published work. Some selected examples follow.

Books by and about Departmental Faculty Members

Barlow RE, Proschan F (1975) Statistical theory of reliability and life testing. Holt Rinehart and Winston, New York

Boen JR, Zahn DA (1982) The human side of statistical consulting. Lifetime Learning Publ Belmont, CA

Brown BW Jr, Hollander M (2008) Statistics: a biomedical introduction. Wiley, New York

Dasgupta A (2011) Selected works of Debabrata Basu. Springer, New York

Hollander M, Proschan F (1984) The statistical exorcist. Marcel Dekker, New York

Hollander M, Wolfe DA (1999) Nonparametric statistical methods, 2nd eds. Wiley, New York

Selected Significant Faculty Publications

Athreya KB, Doss H, Sethuraman J (1996) On the convergence of the Markov chain simulation method. Ann Statist 24:69–100

Basu D (1980) Randomization analysis of experimental data: the Fisher randomization test. Jour Amer Statist Assoc 75:575–582

Basu D (1977) On the elimination of nuisance parameters. Jour Amer Statist Assoc 72:355–366

Chaganty NR, Sethuraman J (1993) Strong large deviation and local limit theorems. Ann Probab 21:1671–1690

Hajek J (1974) Asymptotic sufficiency of the vector of ranks in the Bahadur sense. Ann Math Statist 2:75–83

Korwar RM, Hollander M (1973) Contributions to the theory of Dirichlet processes. Ann Probab 1:705–711

Lynch J, Sethuraman J (1987) Large deviations for processes with independent increments. Ann Probab 15:610–627

Peña E, Strawderman R, Hollander M (2001) Nonparametric estimation with recurrent event data. Jour Amer Statist Assoc 96:1299–1315

Proschan F, Sethuraman J (1977) Schur functions in statistics — I; The preservation theorem. Ann Statist 5:256–262

Sethuraman J (1994) A constructive definition of Dirichlet priors. Statist Sinica 4:639–650
Zahn DA, Isenberg D (1983) Non-statistical aspects of statistical consulting. The American Statistician, 37:287–302

In the intervening years, some of the original faculty members left for other opportunities and some have taken retirement. Only Fred Huffer and Xufeng Niu remain from the 1990 group though Jayaram Sethuraman and Myles Hollander are retired and active in the department and Fred Leysieffer holds a post-retirement, part-time position in the central administration of the university.

Participants in the 40th anniversary of the Department in 1999

Economic conditions have necessitated a reduction in faculty size. Currently, we have 14 full time faculty in the department including two in endowed professorships. Of the 14, two have mainly teaching duties. The university now enrolls 40,000 students and it has a College of Medicine.

The Department Now

Were the early founders of the department to return now, they would not recognize it. This is true not just because of the change in faculty members but in the expertise in the diversity of disciplines they have brought with them. The subject of statistics has evolved and its scope has widened over the years. With new recruitment, spearheaded by its chair, Dan McGee, the department has become more diverse in its interests.

Current faculty have diverse specializations and are leaders in several modern statistics areas. The department has major emphases in the theory and application of statistical methods which apply to biostatistics and computational methods and also the traditional area of time series and applied probability. More specialized topics include computer vision, shape analysis, bioinformatics, machine learning, Bayes and semiparametric methods in survival analysis, protein shape analysis, biological data mining, and nonparametric methods on manifolds.

Xufeng Niu and Dan McGee

Current faculty members, as of fall 2011, are listed below, with the year they joined the department:

Adrian G. Barbu, Assistant Professor, 2007, PhD Computer Science, UCLA. Research Interests: Artificial production markets, Active random fields, Marginal space learning, Graph partition by Svendsen-Wang cuts.

Radha Rani Bose, Associate in Statistics, 2003, MS Applied Statistics and MS Mathematics, FSU. Professional interests: The teaching of statistics.

Eric Chicken, Associate Professor, 2002, PhD Statistics, Purdue University. Research Interests: Statistical process control, Nonparametric regression and density estimations, statistical estimation via wavelets, Water flow models.

Fred Huffer, Professor, 1982, PhD Statistics, Stanford University. Research Interests: Applied probability.

Daniel L. McGee, Professor, 2002, PhD Biostatistics, Johns Hopkins University. Research Interests: Biostatistics

Xufeng Niu, Professor and Chair, 1991, PhD Statistics, University of Chicago. Research Interests: Time series analysis, Linear and nonlinear models, spatial statistics, Bayesian survival analysis.

Victor Patrangenaru, Professor, 2006, PhD Mathematics Haifa University, PhD Statistics, Indiana University. Nonparametric data analysis on manifolds, Data analysis on sample spaces with a manifold stratification, analysis of non-categorical data, to include directional data, 3D machine vision data from stereo pairs, 3D size-and-shape in medical imaging data.

Steven Ramsier, Research Associate 2001, PhD Mathematical Sciences, Clemson University. Professional Interests: The teaching of statistics and statistical consulting.

Yiyuan She, Assistant Professor, 2008, PhD Statistics, Stanford University. Research Interests: Asymptotic theory, model selection and computational issues for high dimension regressor, Statistical computing, and Bioinformatics.

Debajyoti Sinha, Ron and Carolyn Hobbs Endowed Professor, 2007, PhD Statistics, University of Rochester. Research Interests: Survival analysis, Bayesian biostatistics, Modeling cancer prevention data, Cure rate survival data, Semiparametric empirical Bayes.

Elizabeth Slate, Duncan McLean and Pearl Levine Fairweather Endowed Professor, 2011, PhD Statistics, Carnegie Mellon University. Research Interests: Longitudinal data, correlated data, Bayesian methods, Statistical methods in oral research, Discovering and modeling disease biomarkers.

Anuj Srivastava, Professor, 1997, D.Sc. Electrical Engineering, Washington University. Research Interests: Statistical shape analysis, Shape detection and classification in cluttered point clouds, bioinformatics.

Wei Wu, Assistant Professor, 2006, PhD Applied Mathematics, Brown University. Research Interests: Statistical modeling and decoding of population neuronal activity in motor cortex, Statistical analysis and inference in the function space of spike trains, Spike train metrics.

Jinfeng Zhang, Assistant Professor, 2007, PhD Bioinformatics, University of Illinois at Chicago. Research Interests: Statistical methods for computational biology with focus on protein shape analysis, protein structure prediction, and biological data mining.

University Recognition of Faculty

The university appreciates the excellence of the departmental faculty. Each year, The Florida State University recognizes one member from its entire faculty body as a Robert O. Lawton Distinguished Professor. This is the highest annual honor that it can bestow upon a current faculty member. To date, five members of the Department of Statistics have been so recognized. They are Ralph Bradley (1970–1971), Richard Savage (1973–1974), Frank Proschan (1984–1985), Jayaram Sethuraman (1993–1994), and Myles Hollander (1998–1999). It is unusual for a single department to have had so many of its faculty so honored.

Teaching has been a priority with the administration at the university. Six departmental members were so recognized for their excellence in teaching, receiving university teaching awards. They are Douglas Zahn, Kaisheng Song, Steven Ramsier, Duane Meeter, Ian McKeague, and Radha Bose. Graduate Mentor Awards went to Anuj Srivastava and Daniel McGee.

Outreach

Faculty have participated in bringing statistical awareness in and around Tallahassee. With external funding, J. (Sethu) Sethuraman has been mentoring some 87 high school students during summers from 1981 for 30 years. In recognition, he was awarded the Florida State University President's Continuing Education Award in 2002.

Professors Eric Tollar and Jayaram Sethuraman mentoring students in the outreach program.

Academic Programs

The PhD Program

The PhD program in Statistics has always been the mainstay of the department. When it was established, it was the first statistics PhD program to be offered at a state university in Florida. In the past, students would specialize their studies in two areas of concentration, one of which could be outside the department. Students could choose concentrations in mathematical statistics, applied statistics, probability theory, reliability theory, and operations research, and an external area of study such as population studies or mathematics.

Currently, two tracks toward the PhD are available, one in statistics and the other in biostatistics. Within either track, programs of study are tailored individually to a student's career objectives and may include an interdisciplinary field of study as well. The recent introduction of a PhD in Biostatistics has generated an increase in the number of graduate students. Currently, we have 73 PhD students. The number of PhD graduations in the last 2 years is 10 and 13.

The Masters Program

Three options are available in the Master of Science program; applied statistics for students interested in careers in statistical consulting at the MS level, a mathematical statistics option with more emphasis on probability theory, and statistical inference, which serves as preparation for PhD studies in the department, and a biostatistics option preparing students for further graduate work or employment opportunities in biostatistics.

The Undergraduate Program

The department has always offered an undergraduate degree program in Statistics. Academically strong students have the option of enrolling in a BS/MS program, which allows the students to complete requirements for both degrees over an accelerated 5-year time span. The number of undergraduate students has increased greatly after the recent introduction of SAS certification courses for their benefit. Currently, there are 53 undergraduates majoring in Statistics.

Our Graduates

The department takes pride in the accomplishments of its graduates. Our graduates have been primarily employed in academics, pharmaceutical and industrial firms, and government. Some of them have branched off to open private statistical consulting companies or started businesses outside the field of statistics. While it is impossible to chronicle the successes of all of our graduates, one can, by way of example, note a few of their successes.

The American Statistical Association has recognized the following graduates as fellows. They are listed here along with the year of their designation.

Bruce J. McDonald	1975	Gregory Campbell	1998
William J. Blot	1979	Ibrahim A. Ahmad	2000
Ronald H. Randles	1979	Dennis D. Boos	2001
Jesse C. Arnold	1981	Edsel A. Peña	2002
Richard L. Scheaffer	1983	Ram C. Tiwari	2003
Daniel L. Solomon	1984	Michael A. Proschan	2005
A. Blanton Godfrey	1985	Bradley A. Jones	2008
Ronald E. Wyzga	1990	Iris M. Shimizu	2008
Larry H. Crow	1991	Hulin Wu	2008
Robert L. Taylor	1991	Michael T. Longnecker	2009
Jagbir Singh	1997	Gang Li	2010

Gang Li, Ronald Randles, and Robert Taylor have been recognized as fellows of the Institute of Mathematical Statistics.

A number of our graduates are or have been chairs of academic departments in universities. They include:

Ibrahim Ahmad (University of Northern Illinois, University of Central Florida, and Oklahoma State University)
Jesse Arnold (Virginia Polytechnic Institute and State University)
Kuang-Fu Cheng (National Central University, Taiwan)
Robert Hannum (University of Denver)
Choudary Hanamura (University of Rhode Island)
William Grams (Embry Riddle Aeronautical University)
Victor Kane (Kennesaw State University)
James Lynch (University of South Carolina)
Bruce McDonald (San Francisco State University)
Gillian Mimmack (University College of the Fraser Valley, Canada)
Ronald Randles (University of Florida)
Richard Scheaffer (University of Florida)
Michael Schell (Moffitt Cancer Center)
Daniel Solomon (North Carolina State University)
Robert Taylor (University of Georgia and Clemson University)
Ram Tiwari (University of North Carolina, Charlotte)
Connie Wood (University of Kentucky)

Blanton Godfrey serves as Dean of the College of Textiles at North Carolina State University and Daniel Solomon is Dean of the College of Physical and Mathematical Sciences also at North Carolina State University. Douglas Jones serves as Associate Dean for Academic Affairs at the College of Business at Rutgers University.

Richard Scheaffer has contributed significantly to the teaching of statistics through his leadership of the Quantitative Literacy Program.

We can recognize a number of our graduates among the graduates who pursued careers outside of academic life, again by way of example. Robert Sielken established the consulting firm, Sielken and Associates Consulting, Inc. in Bryon, TX. Thomas Redman established the consulting firm, Navsink Consulting Group in Little Silver, NJ after having established and led the Data Quality Laboratory at Bell Laboratories. William Blot is CEO of the International Epidemiology Foundation in Rockville, MD. Charles Holland established QualPro in Knoxville, TN. Robert Clickner is Associate Director of Westat in Rockville, MD. Victor Kane was plant manager for one of the Ford Motor Corporation's largest manufacturing plants, the Sharonville Transmission Plant in Cincinnati, OH. Ruth Ann Killion served as the Chief of the Planning, Research, and Evaluation Division for the 2000 Census.

Two graduates have established endowed chairs for the department, Ron Hobbs and David Fairweather.

Two graduates have been honored locally by the FSU circle of Omicron Delta Kappa, a student national leadership honorary society, which, each year, celebrates three of the university's graduates as Grads Made Good during the homecoming activities. Blanton Godfrey and Larry Crow were so honored for their achievements.

References

Bradley RA (1975) The Florida State University Department of Statistics—the first fifteen years. Intern Stat Rev 43:355–363. www.stat.fsu.edu/history/bradley.pdf

Leysieffer FW (1990) The second fifteen years: a continuation of the department of statistics at the Florida State University. Commun Stat Theory Methods 19:3961–3983. www.stat.fsu.edu/history/leysieffer.pdf

Conversations with Faculty Members

Hollander M (2001) A conversation with Ralph A. Bradley. Stat Sci 16:75–100

Hollander M (2008) A conversation with Jayaram Sethuraman. Stat Sci 23:272–285

Hollander M, Marshall AW (1995) A conversation with Frank Proschan, Stat Sci 10:118–133

Samaniego FJ (2008) A conversation with Myles Hollander. Stat Sci 23:420–438

Sampson AR, Spencer B (1999) A conversation with I. Richard Savage. Stat Sci 14:126–148

The Department of Statistics at The George Washington University

Hosam M. Mahmoud, Joseph L. Gastwirth, John Lachin III and Reza Modarres

Formed in 1935, the Department of Statistics at The George Washington University is the oldest statistics department within a school of liberal arts and sciences, and one of the oldest departments of statistics in the United States. The evolution of the Department of Statistics at The George Washington University (the department, henceforth) along with accomplishments of its faculty and students is described here. To commemorate its 75th anniversary, the department held a 1-day symposium featuring many distinguished speakers. The mission of the department is to provide quality teaching at the undergraduate and graduate levels with a faculty and PhD student body contributing to the frontiers of research in statistical theory, methodology, and practice. In the last quarter century or so the department has been offering service courses such as introduction to statistics to business and social sciences students. Currently, the department is servicing about 2,100 students annually. It offers BS, MS, and PhD degrees in Statistics. Jointly with the School of Public Health, it offers an MS and a PhD in Biostatistics and participates in the MS and PhD programs in Epidemiology.

George Washington University,
Department of Statistics,
22nd Street NW, Rome Hall, 5th Floor, 801,
Washington, DC 20052, USA
H. M. Mahmoud · J. L. Gastwirth · J. Lachin III · R. Modarres (✉)
e-mail: reza@gwu.edu

A. Agresti and X.-L. Meng (eds.), *Strength in Numbers: The Rising of Academic Statistics Departments in the U.S.*, DOI: 10.1007/978-1-4614-3649-2_6,

History of Chairmanship

Chair	Years
Frank Weida	1935–1963
Harold Bright	1963–1965
Solomon Kullback	1965–1972
Herbert Solomon	1972–1974
Jerome Cornfield	1974–1977
Samuel Greenhouse	1977–1980, 1986
Hubert Lilliefors	1979–1983
Arthur Kirsch	1983–1986, 1990–1994
Robert Smythe	1986–1990, 1994–1998
Hosam Mahmud	1998–2001
Tapan Nayak	2001–2007
Reza Modarres	2007–2012

The First Quarter Century (1935–1960): The Foundation

In 1935, Frank Weida convinced the university to form a department of statistics, even though he was the sole faculty member and chair. Weida came to The George Washington University as a professor of mathematics and offered statistics courses in that department until his proposal to create a department of statistics was approved. While in the Mathematics Department, Weida already had a very successful PhD student, Solomon Kullback. During the subsequent years, the department grew both in the number of students and faculty and in its role in the university. Many undergraduate programs, e.g,. business, economics, and psychology, require statistics courses as part of the basic training in their discipline. The department's MS and PhD programs have also graduated many students, who have been employed in the government, industry, and universities.

Kullback and Important Academic Descendants

Worthy of special mention is Solomon Kullback, who served in the department from 1939 until 1972. Winner of the 1962 Wilks Award, the importance of his research is widely recognized. Hundreds of papers refer to the Kulback–Leibler discrepancy measure. He also wrote a seminal book entitled *Information Theory and Statistics* (Wiley, 1959). His role in building the PhD program was phenomenal. Under his supervision, notable PhD graduates obtained their degree: The late Hubert Lilliefors (known for the Lilliefors test of normality), the late Samuel

Greenhouse (a founding statistician at the National Institutes of Health, NIH), and Robert Shumway (an expert in time series and an author of a leading text in the area). The George Washington University recognized the paramount significance of Kullback in its history with a memorial conference in 1996. Twenty-eight papers on the Kullback–Leibler information measure and related topics were presented at the conference.

In the early days, Kullback taught courses on probability theory, multivariate analysis, information theory, time series analysis, and characteristic functions, while simultaneously pursuing a distinguished career in the National Security Agency. It is purported that his style was that of "story-telling," always a beginning, a plot, and an end. Among many other remarkable achievements, Kullback played an important role in breaking the Japanese code during the Second World War. Kullback was the second in command at the National Security Agency with the rank of Major General, when some members on his staff defected to the Soviet Union. He was forced to resign, and that is when he became a full-time faculty member. Before that, he was a part-time instructor at The George Washington University teaching in the evening program. The department has a long tradition of offering courses in the evening, to cater to the needs of statisticians in the government and local agencies in Washington, D.C.

The Years Following Kullback

During the tenures of Bright (second Department Chair) and Kullback, the PhD program grew and many successful researchers were guided by Kullback. Several, including Hubert Lilliefors, Robert Shumway, and Terrence Ireland became faculty members and eventually Full professor. Their dissertations further developed the information-theoretic approach to statistics that Kullback originated and summarized in his book.

Within the college of arts and sciences, the department assumed responsibility for all statistics courses. Arthur Kirsch developed many classes, especially for students in Psychology. He also served on many PhD committees for their students, whose dissertations involved statistical analyses. Everett Johnson had a similar role for students in economics and business.

The Second Quarter Century (1960–1985)

In 1972, Solomon Kullback retired and Herbert Solomon was hired from Stanford to be chairman. He helped to strengthen the area of biostatistics by encouraging Jerome Cornfield to come and found the Biostatistics Center. Cornfield was the Fisher Lecturer for the 1973 Joint Statistical Meetings, and served as President of the American Statistical Association in 1974. In 1972, Herbert Solomon recruited

Joseph Gastwirth to join the department and develop an MS level class for the Economics Department as well as guide PhD students in the department. Essentially, he replaced Everett Johnson who retired in January 1973. After Herbert Solomon decided to return to Stanford, Cornfield took over as chairman for 2 years. During this time, Samuel Greenhouse, like Cornfield a statistician with a distinguished research career at the NIH, who had taught part-time in the department, joined the department as a professor. He succeeded Cornfield as chairman.

A few of the important contributions of professors Cornfield and Greenhouse to Biostatistics should be mentioned. Both were at the core of the NIH research team that was responsible for the development of many statistical methods. Cornfield served on the committee that conclusively linked smoking to lung cancer, demonstrating that the relative risk, R, of lung cancer in smokers was between 5 and 10. In particular, he rebutted Sir Ronald Fisher's conjecture that there might be an omitted variable, such as a genetic trait that caused both by showing that such a variable would need to raise one's risk of lung cancer by *at least* R and would need to be *at least* R times more prevalent in smokers. It was not scientifically plausible that such a variable existed. Cornfield's inequality stimulated the development of methods for assessing the sensitivity of the main statistical inference drawn from an observational study to unmeasured predictors. Along with Bradford Hill in the UK Cornfield had an important role in establishing a randomized clinical trial as the "gold standard" for medical study. His understanding of the issues raised by multiple comparisons was crucial for the proper interpretation of several important studies. When a drug under investigation does not have an "overall" effect, often the data are subdivided into strata, which are analyzed separately in a search for a positive treatment effect. Cornfield emphasized the fact that the subgroups are usually not defined a priori and that most of the "subgroup effects" vanish when the proper adjustment for multiple comparisons is made. As part of a Diabetes Study team his reanalysis of a drug, tolbutamide, which showed a 2.5-fold increased risk of cardiovascular mortality but was not superior to other treatments, prevented the widespread use of the drug. Although other statisticians suggested that the randomization might not have been adequate so the patients receiving tolbutamide would have a greater risk of cardiovascular disease or that the effect might not be real because it was not observed in all 12 clinics, Cornfield noted that the reason one obtains patients from many clinics is that it is difficult to obtain a sufficient number for a sound statistical analysis. Then he stratified the patients by the number of risk factors for cardiovascular disease they had. Standardizing the data for the number of risk factors showed that patients on the drug still had twice the risk of cardiovascular problems than patients receiving a placebo or insulin. Notice that the Cornfield inequality implies that another risk factor, which was not included, would need to have doubled the risk and been twice as prevalent in the patients receiving tolbutamide. In a randomized study, this is not likely.

At the NIH Greenhouse contributed to both the theory and application of longitudinal studies in addition to the analysis of data on medical screening tests, case–control studies, and logistic regression. Early in his career at NIH

Greenhouse worked with Dunn and Mantel to develop the statistical foundations for the evaluation of screening and diagnostic tests. In their 1950 article Greenhouse and Mantel studied methods for determining whether a test had promise. If the distribution of the assay values in healthy individuals (controls) is $G(x)$ and its distribution among diseased (cases) is $F(x)$, they considered the null hypothesis as $G^{-1}(0.95) < F^{-1}(0.10)$, i.e., a good assay should have sensitivity of at least 0.90 using a cutoff point that assured a specificity of at least 0.95. They developed both parametric and nonparametric methods. In the mid-1950s Greenhouse, Mantel, and Goldin developed methods of bioassay to study the effects of various doses of antileukaemic agents in mice that allowed for more than one drug. His 1959 article with Geisser on correcting the F-statistic used in the ANOVA of a one-factor repeated measurement design for violation of the "sphericity" assumption that requires that there be no interaction between the factor and the treatment has been cited hundreds of times and is included in the standard software. He had a major role in an 11-year longitudinal study of biological and behavioral aspects of aging that culminated in four Public Health Service reports, which helped the development of modern gerontology. In a 1973 article with Seigel, Greenhouse showed that the logistic model can be used in case–control studies. After coming to The George Washington University, Greenhouse collaborated with Cornfield on developing methods for the analysis of sequential trials that did not pair each patient with a control. The standard approach to matched pair data discards the concordant pairs, as it was originally thought that this did not lead to a loss of efficiency. They showed that this holds only when the alternative, an odds ratio greater than 1, approaches the null value but for odds ratios noticeably greater than 1.0, the difference can be substantial. He also collaborated with Gastwirth on adapting biostatistical methods for the analysis of data arising in the legal setting.

Before joining the department, Gastwirth was known for the development of nonparametric and robust methods. In particular, his 1966 article on robust methods and the 1967 article on the large-sample distribution of linear combinations of order statistics, with Chernoff and Johns, are still cited today. The concepts of efficiency robustness laid out in 1966 have been successfully used to create robust methods for the analysis of survival data and genetic associations when there are several scientifically plausible distributions or genetic models that could generate the data. His 1972 article on robust methods for estimating the Lorenz curve and Gini index from grouped data demonstrated how one can obtain accurate estimates of these important measures of economic inequality using upper and lower bounds that can be obtained from grouped data. These three articles continue to have an impact in the field. At The George Washington University, he and Mitchell Gail developed a goodness-of-fit test to see whether data followed the exponential distribution using the Gini index, which has good power properties for a large family of alternative distributions. In the 1970s, Gastwirth began an active research program in legal statistics that has led to methods for showing that a "chain letter" or pyramid scheme is misleading potential participants, combining binomial data sets arising in hiring discrimination cases, adapting Cornfield's inequality for assessing whether an omitted variable can "explain" a significant race or gender effect, using the Peters-Belson regression approach, which estimates a disparity by comparing the actual salary of a minority member to that a majority member would receive. Graubard, Rao, and Gastwirth showed that the P-B approach can be used in studying health inequalities. In collaboration with Hikawa and Bura, nonparametric regression methods have been incorporated in the P-B approach. Recently, Gastwirth and Pan have utilized survival analysis to estimate the lost wages and benefits that an individual who has been discriminated against should receive. An important problem arises in situations where there are more members of the protected group than positions that were available during the period of discrimination. Courts then use a "lost chance" criterion and their forthcoming article demonstrates how the Cox model can be used in this context. Gastwirth is the author of *Statistical Reasoning in Law and Public Policy* and editor of *Statistical Science in the Courtroom.*

Robert Smythe, a student of K. L. Chung, was hired in 1981 to teach and advise PhD students in probability theory and its applications and Tapan Nayak, a student of C. R. Rao, was hired in 1983 to strengthen the department's program in statistical inference.

Biostatistics and the Biostatistics Center

Biostatistics has long been a strength of the department. In the late 1940s, following the War, several bright individuals without formal statistical training were hired to government positions as statisticians at the newly formed NIH, including Jerome Cornfield and Samuel Greenhouse. Greenhouse, among others, then pursued graduate degrees in statistics at The George Washington University. Greenhouse was a 1959 PhD student of Kullback, whose dissertation on extensions of the linear discriminant function comprises Chap. 13 of Kullback's landmark book. While Greenhouse worked at the NIH, he taught as an adjunct professor for many years and then assumed a full professorship in 1974 upon his retirement from the NIH.

In 1972, then-Provost and former Department Chair, Harold Bright, and then-Chair Herbert Solomon, recruited Jerome Cornfield to join the faculty as professor, having recently retired from the NIH. Cornfield received a grant from NIH to establish a research center, which became the Biostatistics Center, a research facility of the department to conduct large-scale, multi-center clinical trials, and epidemiologic studies. The center was located in Bethesda, near NIH and Cornfield divided his time between the center and the campus. Shortly thereafter, Cornfield served as Department Chair after the departure of Solomon.

During the 1970s, Cornfield recruited former colleagues from the NIH to join the research faculty of the center—Nathan Mantel and Max Halperin, among others. After Cornfield's premature death, Halperin became center director and recruited Tavia Gordon from the NIH as well.

Cornfield and his center co-director, Larry Shaw, also recruited a generation of younger statisticians—Janet Wittes, Sarah Fowler (a student of Lilliefors), Murray Selwyn (later a PhD student of Gastwirth), John Lachin, and Susan Ellenberg (later a PhD student of Greenhouse), among others. Lachin followed Halperin as center director and was in turn followed by Fowler. Many other PhD students worked at the Biostatistics Center during their studies and stayed on as faculty, including Oliver Bautista, William Rosenberger, Elizabeth Thom, and Naji Younes.

The medical studies at the Biostatistics Center provided the foundation for methodological research in biostatistics and the introduction of advanced biostatistics courses in the department. L. J. Wei took a sabbatical leave from the National Cancer Institute in 1980–1981 where he worked with David Byar and Mitchell Gail, who also received his PhD from the department and has served as an adjunct faculty. L. J. Wei was then recruited to join the department in 1981. Soon L. J. Wei, and later Peter Thall and Robert Smythe, shared time at the center. More recently, Yinglei Lai and Qing Pan of the department are actively involved with research projects at the center. With joint appointments in the center, the department recruited tenured faculty Gordon Lan and Zhaohai Li, and more recently Michael Larsen who was recruited as an associate professor with a faculty affiliation to the Biostatistics Center on a selective excellence award from the university. All have pursued joint research with the center faculty.

For many years Greenhouse taught a legendary course in multivariate analysis. Cornfield introduced a two-semester course in Biostatistics, later taught by Greenhouse and others. Lachin introduced courses in data analysis and survival analysis. Lan, Li, and others taught advanced topics including sequential analysis and longitudinal analysis.

In 1995, with an established graduate curriculum in biostatistics, Lachin and then-Chair Smythe, together with the Public Health Program, established an MS and a PhD program in Biostatistics. Lachin directed the program until 2005, and was then followed by Zhaohai Li and now by Efstathia Bura.

Expanding the Horizons into Computing and Computational Statistics

By 1978, the department realized the potential of computing as a new direction and an important methodology in modern science. There was an increasing demand for computer science courses within the school of liberal arts and sciences. While the university has a department of computer science in the Engineering School, the department of statistics began to offer courses such as introduction to computing and introduction to data structures. These were intended for an audience of students in liberal arts and sciences, who do not necessarily prepare for a degree in engineering. These courses were service courses accepted toward the quantitative component of meaningful initiation in the arts and sciences, and had a statistical flavor. Faculty hiring took place in 1977, and the first faculty member to serve in this capacity was

Raymond Thomas. Hubert Lilliefors also taught computing classes. In 1982 the department officially changed its name to the *Department of Statistics and Computer and Information Systems*. Hosam Mahmoud (in the area of probabilistic analysis of algorithms), was hired in 1983, and two other hires followed (Ralph Bunker and Richard Epstein). Blaza Toman (in the area of experimental design) was also hired in this period and became the first woman to be awarded permanent tenure.

The Last Quarter Century (1985–2010)

By 1990, the impetus of the computer-oriented teaching of the 1980s as well as new directions in science that globally integrated statistics and computing encouraged the department to tighten its connection to computing, and three hires in the area of computational statistics were made: David Grier (hired in 1989), who is also interested in the history of this area, Catherine Hurley (hired in 1990), whose research is in graphical statistical methods and Reza Modarres (hired in 1991), who specializes in statistical computing and computer-intensive methods. The department officially changed its name to the *Department of Statistics and Statistical Computing*.

To stay current with new directions in the field, the department made additional hires in the 1990s. Sudip Bose (in Bayesian statistics) and Efstathia Bura (in regression, with a specialty in dimension reduction) were hired in 1990 and 1996. Contract positions appeared on the scene of American universities, and in view of the tightening conditions of the U.S. economy, the department followed suit. Within the last 12 years Subrata Kundu (nonparametric statistics) served for about 11 years, with the last eight (continuing into the present) being continuous. Also, Srinivasan Balaji (stochastic differential equations, stochastic processes and now into random structures and algorithms as well) is on a continual contract since 2002. In 2004, Nozer Singpurwalla (Distinguished professor) changed his affiliation. He had been formerly with two departments in the Engineering School, and in 2004 he moved to the department.

The most recent hires are Ying Lei Lai (2003), Qing Pan (2007), Jonathan Stroud (2008), and Michael Larsen (2009). Lai (in biostatistics) received several grants from the NIH. He is now serving as Associate professor. Pan is also in biostatistics, serving in an Assistant professor position. She received the Best Student Paper Award from the Canadian Statistical Society in 2009. Stroud (an expert in computational Bayesian statistics, with interest in financial applications and atmospheric science) came from Wharton School at the University of Pennsylvania and is now serving as Associate professor. Larsen's role is noted in the subsection titled Biostatistics and the Biostatistics Center.

Several developments and important events took place in the last quarter century; we name a few prominent ones. In this quarter century, noted statisticians were affiliated with the department in various capacities. They include Fritz Scheuren (President of the American Statistical Association 2005), who was a popular undergraduate teacher and the instigator of a graduate certificate in Survey Design and Data Analysis. Noted statisticians affiliated with the department also include the late Samuel Kotz (author of many influential probability and statistics textbooks), who was a research affiliate with no teaching duties.

The service and contribution of Joseph Gastwirth were recognized by a 1-day symposium in his honor (organized and chaired by Zhaohi Li). Selected papers of the conference appear in *Law, Probability and Risk* and in *Statistics and Its Interface*.

Since 2003, a series of annual symposia entitled *Frontiers of Statistics, Mathematics and Computation* was organized by Jagdish Chandra (research professor at the department). The series was hosted on campus and supported by the Department of Decision Sciences at The George Washington University.

Some Distinguished Graduates

The department has produced a number of fine graduates, who proceeded to serve elsewhere in prominent positions. Among these are:

(1) Mitchell H. Gail (1975): Chief of the Biostatistics Branch of the Division of Cancer, Epidemiology and Genetics and elected member of the Institute of Medicine of the National Academy of Sciences and former President of the American Statistical Association.
(2) Nancy J. Kirkendall (1974): Director of Statistics and Methods, Energy Information Administration, Vice President, ASA, 2007 Herriot Award for Innovation in Federal Statistics.
(3) Steven W. Lagakos (1972): Walcott professor of Biostatistics, Chair, Department of Biostatistics, Director, Center for Biostatistics in Aids Research, Harvard University and elected member of the Institute of Medicine of the National Academy of Sciences.
(4) Wallace E. Larimore (1972): President, Adaptics, Inc., Washington, D.C., 1994 Statistics in Chemistry Award, ASA.
(5) Thomas W. Nolan (1985): Co-founder of Associates in Progress, Senior Fellow, Institute for Healthcare Improvement, co-author or author of six books on managing quality healthcare, Deming Medal, 2000–2001, American Society for Quality, Washington, D.C.
(6) William Rosenberger (1992): Assumed a research position at the Biostatistics center of The George Washington University, and academic positions at University of Maryland, Baltimore County, and George Mason University. At present, he is a Full professor and Chair of the Department of Statistics at George Mason University. He has co-authored several books.
(7) Michael D. Sinclair (1994): Obtained his PhD while employed at the U.S. Census Bureau. Later he became the Chief Statistician for the Office of Federal Contract Compliance in the Department of Labor, the Acting Head of the Bureau of Justice Statistics and now is at the National Opinion Research Center.

(8) Nancy L. Spruill (1980): Has had a distinguished career in the defense department guiding major acquisition programs, such as fighter and attack aircraft, naval ships and submarines, etc. Since 2000, she has been Director, Acquisition Resources and Analysis, in the Office of the Under Secretary of Defense for Acquisition, Technology and Logistics.
(9) Jeremy Wu (1982): National Ombudsman, Office of Economic Impact and Diversity, Department of Energy. Program Manager, Local Employment Dynamics, Longitudinal Employment Household Dynamics, Center for Economic Statistics, U.S. Census Bureau.

Honors and Awards Received by Faculty

The accomplishments of many faculty members and graduates have been recognized by the major statistical associations, such as ASA, IMS, and ISI. Special awards include:

Wilks Award, 1962 (ASA, Solomon Kullback)
Guggenheim Award, 1985 (Guggenheim Foundation, Joseph Gastwirth)
ASA Founders Award 1993 (ASA, Samuel Greenhouse)
Wilks Award, 1984 (Department of Defense, Nozer Singpurwalla)
Fisher Award, 2009 (ASA, L. J. Wei)
Fulbright, 2008 (U.S. Government, Efstathia Bura)

Photos

A recent photo of the Faculty (Spring, 2010)

From left to right: Hosam Mahmoud, Subrata Kundu, Yinglei Lai, Qing Pan, Srinivasan Balaji, Reza Modarres, Joseph Gastwirth, Jonathan Stroud, Tapan Nayak, Michael Larsen, Sudip Bose, Zhaohai Li, Nozer Singpurwalla.

The department in 1994

From left to right; standing: Joseph Gastwirth, Sudip Bose, Reza Modarres, Hubert Lilliefors, Hosam Mahmoud, Gordon Lan, Tapan Nayak; sitting: Bicky Pham (secretary), Dalila Estrilla (secretary), Robert Smythe, Blaza Toman.

Solomon Kullback (on the left) at the reception of his acceptance of Wilks Award in 1962

Former Biostatistics Center Directors

Jerome Cornfield

Samuel Greenhouse

John Lachin

Harvard University Department of Biostatistics

Nan Laird and Marvin Zelen

The Early Years: 1922–1961

In 1922 the Harvard School of Public Health (HSPH) was founded in the Harvard Longwood Medical Area of Boston. The Department of Biostatistics (then called Vital Statistics) was one of four core departments in the new school. Edwin B. Wilson was appointed as the Chair in Vital Statistics (Fig. 1). Wilson came to HSPH from MIT where he had been Chair of the Physics Department. Trained as a physicist, Wilson's interests in science and its applications were very broad. He was also interested in the fundamentals of mathematics as applied to problems in public health. He published papers on mathematical epidemiology, discussed confidence intervals for a binomial probability in a 1927 publication, was president of the American Statistical Association, and was the first managing editor of the Proceedings of the National Academy of Sciences for which he served for 50 years.

Surviving correspondence from the 1930s demonstrates the intellectual interchange between E. B. Wilson and Professor Ronald A. Fisher as the latter was developing some of the foundational concepts of statistical inference. Wilson was a perceptive critic of Fisher's work, and Fisher's appreciative letters to Wilson bore this out. In 1931, Wilson recruited newly- minted Smith College mathematics major Jane Worcester to be his "computer." They coauthored approximately two

Harvard University,
Department of Biostatistics,
655 Huntington Avenue SPH2, 4th Floor,
Boston, MA 02115, USA
N. Laird (✉) · M. Zelen
e-mail: laird@hsph.harvard.edu

M. Zelen
e-mail: zelen@jimmy.harvard.edu

A. Agresti and X.-L. Meng (eds.), *Strength in Numbers: The Rising of Academic Statistics Departments in the U.S.*, DOI: 10.1007/978-1-4614-3649-2_7,

Fig. 1 Edwin B. Wilson, circa 1940

dozen theoretical papers during their 15-year working relationship; Wilson retired from the chairmanship in 1945.

Hugo Muench succeeded Wilson as chair of the department (now renamed Biostatistics) in 1946 and served for 15 years. Muench was a physician, trained in biostatistics at the Johns Hopkins University (JHU). In 1947, Worcester successfully defended her dissertation and earned the Dr.PH degree in Biostatistics and Epidemiology—the first woman to earn this degree at the Harvard School of Public Health. Muench sought and received a faculty appointment for Worcester and mentored her to a later promotion to associate professor. During these years, Margaret (Marge) Drolette joined the department as a teaching assistant after completing an outstanding undergraduate record in mathematics at Radcliffe College. Drolette subsequently earned her PhD in Statistics in 1965 from the Harvard Faculty of Arts and Sciences (FAS) and ultimately became Professor of Biostatistics. Robert Reed came to the Biostatistics Department as a second-generation biostatistician after completing his bachelor's degree at Johns Hopkins in Applied Statistics and receiving a PhD in Sociology from the University of Chicago. Reed's father was the pioneering biostatistician Lowell J. Reed. The senior Reed was Professor of Biostatistics and later Dean of the School of Public Health at Johns Hopkins and President of JHU (Fig. 2).

1961–1977: Years of Consolidation

Robert Reed succeeded Muench as Chair of the Department of Biostatistics in 1961. During this period, the department consisted of two full professors (Reed and Worcester), one lecturer (Drolette), two emeritus professors (Wilson and

Fig. 2 Hugo Muench, Robert Reed, and Jane Worcester circa 1950

Muench), a handful of graduate students, and several research assistants and computer programmers. Reed served as chair until 1973, and Jane Worcester was appointed the first female chair of the department. The department had a school-wide reputation of being student friendly, especially to those in the MPH program, which numbered about 110 students per class, most of whom were physicians. The department's *raison d'etre* was principally as a service department for HSPH. However, the department also had a robust Master of Science program and, together with the Department of Statistics in the Harvard FAS, offered a Doctor of Science program as well as a Doctor of Public Health degree.

It is impossible to overstate the importance of teaching quality to the mission of the Department of Biostatistics during the Reed-Worcester-Drolette era. The reputation of the department was centered on the universal accessibility of its courses to all students at HSPH, and course evaluations by students documented the success of the department in its reputation for quality instruction. Year after year, a Biostatistics Department faculty member was named "outstanding teacher of the year" for the School of Public Health. Often this recognition was given to Marge Drolette. This tradition of outstanding teaching has continued throughout the department's history to the present time, with numerous faculty receiving prestigious teaching awards from the School.

During this period, additional faculty were added to meet the growing demands for Biostatisticians in teaching and research. New faculty included Yvonne Bishop, Ray Neff (an alumnus of the department who went on to head the Health Sciences Computing Facility at HSPH), and Nan Laird, who initially joined the department to teach in the executive programs of the Department of Health Policy and Management.

1977–1980: The Mosteller Years: A New Beginning

In 1970, Derek Bok became the youngest president in the history of Harvard University. After setting his first priority as the reform of undergraduate education, Bok turned his attention to the Medical Area with the appointment in 1972 of Howard Hiatt as Dean of the School of Public Health. Hiatt had great respect for the role of statistics in medical decision making, and was eager to expand the teaching and research of Biostatistics in the medical area. He began to consult regularly with Frederick Mosteller in the Statistics Department in the Faculty of Arts and Sciences (FAS) about strategies for improving statistical science at the School and finding a new chair for the department. Together they organized an Interdisciplinary Faculty Seminar on Health and Medicine that was widely attended by faculty and graduate students from FAS, the Kennedy School, and the Schools of Medicine and Public Health.

In the spring of 1977, Mosteller accepted the position of Chair of the Biostatistics Department. Mosteller was a familiar and influential figure not only at Harvard, but in the national and international statistical community as well, serving as President of the American Society for the Advancement of Science. Prior to accepting the post in Biostatistics, he had chaired the Statistics Department, served as faculty member in the Department of Psychology and Social Relations in FAS, and as a faculty member of the Committee on Public Policy of Harvard's School of Government (Fig. 3).

Mosteller immediately set about creating a consolidating Biostatistics faculty scattered in other departments and created an academic home for the many Biostatisticians in the Harvard medical area. He also began to recruit new faculty. In the summer of 1977 the department appointed Marvin Zelen Professor of Statistical Science. Zelen also founded the Department of Biostatistics and Epidemiology at the Sydney Farber Cancer Institute (later renamed the Dana-Farber Cancer Institute). He succeeded in bringing many members of his Statistical Laboratory at the State University of New York at Buffalo (SUNYAB) to HSPH, where they were appointed as junior faculty. This group included Colin Begg, Richard Gelber, David Schoenfeld, and Kenneth Stanley. The following year, Marcello Pagano and Stephen Lagakos came from SUNYAB to join the HSPH faculty. Although the main focus of research activities for these faculty members was cancer clinical trials and collaborating/consulting at the Dana-Farber Cancer Institute (DFCI), Zelen viewed the Biostatistics Department at HSPH as most

Fig. 3 Nan Laird, Frederick Mosteller, and Marvin Zelen, late 1990s

appropriate for their faculty appointments and teaching responsibilities. Bringing these faculty to HSPH was a major coup for the department. Tripling the size of the department overnight required no small measure of political savvy on Mosteller's part. In addition to soothing ruffled feathers among the other department chairs, he had to prove the benefits to the School would more than justify the financial investment. Mosteller's strategy was to approach department heads separately, and ask how they would accomplish this for their own department, then set out to follow their advice. Mosteller wrote, "It turned out that, when department and committee chairs understood what we were trying to do and how it could be financed, the benefit for the School seemed obvious, though the story took a great deal of explaining and ploughed through hills of skepticism." (Mosteller, Frederick, *The Pleasures of Statistics*).

In addition to recruiting the faculty group based at the Dana-Farber, Mosteller also recruited Christine Waternaux, James Ware, John Orav, and Thomas Louis, all as faculty based at HSPH. When Mosteller resigned as Chair in 1981 to become Chair of the Department of Health Policy and Management at HSPH, the department had 23 primary faculty and 37 graduate students.

One of Mosteller's priorities was to expand the teaching and graduate program. He set up an extensive set of Department Committees to implement this expansion. Zelen successfully secured a large training grant from the NIH's National Cancer Institute (NCI). Ware, who had been recruited from the NIH to collaborate with the School's environmental health program, was awarded a second training grant from the NIH's National Institute of Environmental Health Sciences (NIEHS). It was an exciting time for the department. Faculty meetings were held regularly and often, and all faculty were fully engaged and committed to building an outstanding department.

New course proposals were submitted to the School's Curriculum Committee almost monthly. At one point a question arose in a School-wide faculty meeting as to when the Department of Biostatistics would finally get its curriculum fixed and stop changing its courses? Mosteller answered very firmly, "Never".

1980–1990: Marvin Zelen's Decade of Leadership

Marvin Zelen became acting chair during the 1980–1981 academic year when Mosteller went on sabbatical leave. He was appointed Chair the following year. During Zelen's chairmanship, the department continued to expand. The department had 27 masters and doctoral students in 1979–1980 and this number increased to 60 by 1990. Faculty increased from 28 to 34. Among the faculty, four professors (Laird, Lagakos, Pagano, and Ware) received tenure at HSPH and Bernard (Bernie) Rosner received tenure in the Medical School, and secondarily in this department. Forty-five doctorates were awarded in this 10-year period. The department ranking, as determined by the National Research Council (as well as other compilations), was always amongst the top biostatistics/statistics departments in the U.S.

The department continued to expand its teaching function in the School of Public Health, especially for students in subject matter departments. A high proportion of students elected to take more advanced courses in the department after taking the mandatory biostatistics course. During 1985–1990 period the Biostatistics' student credit hours averaged 21% of the entire School's credit hours. High enrollment in the department's classes was due to the growing awareness of the importance of Biostatistics in Public Health and the easy availability of computing. It also reflected the high quality of teaching in the department.

During this time period, the rapidly growing Biostatistics department at the Dana-Farber served as the largest focus for the HSPH-based department's collaborative research program. The critical mass of biostatistical scientists at the DFCI motivated research on the many new methodological problems encountered in cancer research. HSPH faculty appointed to the DFCI Department of Biostatistics during this period included: James Anderson, Dianne Finkelstein, Rebecca Gelman, Robert Gray, David Harrington, Myrto Lefkopoulou, Cyrus Mehta, Louise Ryan, and Anastasios Tsiatis.

Faculty based at the DFCI were heavily involved in national and international multicenter cancer clinical trials. At one time there were eight cancer clinical trial cooperative groups having their statistical centers at the DFCI. This constituted about one-third of the National Cancer Institute's multicenter clinical trials program. These included the Statistical Centers for the Eastern Cooperative Oncology Group (ECOG), led by Zelen and Harrington, Cancer and Acute Leukemia Group B (CALGB), led by James Anderson, and the International Breast Cancer Study Group (IBCSG), led by Richard Gelber. These activities made the department the world center for carrying out multicenter cancer clinical trials. Hundreds of cancer trials were carried out through these eight Cooperative Clinical Trial Groups. Many of the ideas and methods, considered routine today for carrying out multi-center trials, were developed during that time—especially in the ECOG.

In addition to the large biostatistical group at the DFCI, smaller statistical groups were being seeded within the Harvard-affiliated hospitals. The Massachusetts General Hospital (MGH) appointed Schoenfeld and Finkelstein to organize a biostatistical collaborative and consulting group, and Waternaux headed a biostatistics group at McLean Hospital.

Under the leadership of James Ware, the department played a major role in the Harvard Six Cities Study of air pollution and health. The School's Department of Environmental Health had initiated this project in the 1970s with the collaboration of Yvonne Bishop. The Six Cities Study was one of the most influential, innovative, and longest running studies concerning the health effects of air pollution in the U.S. For the first time in this country, public health researchers directly correlated environmental exposure to toxins with health impacts; their findings served as key inputs to the Environmental Protection Agency in setting standards for air pollution. Many of the department's faculty collaborated in the project, including Laird and Louis, as well as newly-recruited Victor DeGruttola and later Andrea Rotnitzky. Work on this project was instrumental in developing innovative methods for the statistical analysis of longitudinal data, and made the department a leading center for developing methods for longitudinal data analysis.

During the Zelen chairmanship, the well-known environmental study, often referred to as the "Woburn Study" was carried out by Lagakos and Zelen. The study was named for its setting in a small town of about 35,000 residents located a few miles north of Boston. The study was prompted by the observation of a childhood leukemia cluster in East Woburn. Residents received a blend of water from eight wells, two of which were found to be contaminated with trichloroethylene (TCE), an industrial solvent. Lagakos and Zelen were able to demonstrate an association between exposure to well water and the leukemia cases as well as an array of birth defects. The Woburn Study generated a huge amount of publicity, resulting in several books, popular articles, a PBS Nova program in which Lagakos and Zelen were interviewed, congressional testimony by Zelen and the motion picture "A Civil Action". An important feature of the Woburn Study was the organization of 250 volunteers to collect data using random telephone dialing. This was the founding of what is now referred to today as "citizen epidemiology".

1990–2009: Laird, Lagakos, and Ryan: The Tradition Continues

Nan Laird was appointed Department Chair in 1990, Stephen Lagakos in 1999, and Louise Ryan in 2007. Traditions of departmental growth and excellence continued during this 20-year period, marked by new directions in research,

Fig. 4 Department Senior Faculty (front row: Lee-Jen Wei, Nan Laird, Marvin Zelen; middle row: David Harrington, James Ware, Marcello Pagano; back row: Stephen Lagakos, Anastasios Tsiatis, Bernard Rosner) on Retreat, 1990

building stronger connections with the HSPH Department of Epidemiology, and the establishment of industry partnerships. In 1990, the department had just completed more than a decade of unprecedented growth in faculty, students, and space. Dean Harvey Feinberg established two endowed chairs for department members: the Henry Pickering Walcott Professorship for the department chair, and the Frederick Mosteller Professorship, which was awarded to Ware when he became the HSPH Academic Dean, a position he held until 2009. The department's academic program was firmly in place, but research programs were still expanding and the department had grown in size to nearly 80 doctoral and master's students and over 30 faculty (Fig. 4).

In the wider world of public health, HIV/AIDS had emerged as a major threat to populations throughout the world. In 1989, the department's focus on clinical trials expanded greatly when Lagakos (Fig. 5) was awarded an NIH contract to lead the Statistical Center for the AIDS Clinical Trial Group (ACTG), a multicenter cooperative clinical trial group formed to carry out clinical trials in HIV/AIDS. The ACTG was the NIH/NIAID main program for carrying out HIV/AIDS clinical trials. The award was large and led to the establishment of an independent center named the Center for Biostatistics in AIDS Research (CBAR). CBAR served as the department's focus on AIDS. In addition to being the home of the Statistical Center, CBAR generated seminars, faculty/student support, and enhanced the

Fig. 5 Steve Lagakos and Louise Ryan, circa 2000

department's methodological research on AIDS. It was important in attracting a new training grant concentrating on AIDS/HIV led by Pagano.

The ACTG Statistical Center award provided opportunities to recruit new faculty. Among these new faculty were Michael Hughes, Kenneth Stanley (returning to HSPH from the WHO), Paige Williams, Rebecca Betensky, and Lee-Jen Wei. Additional faculty were recruited in subsequent years, including Tianxi Cai and Judith Lok. In addition, the project required numerous biostatistical scientists ranging from master's level statisticians to senior and Principal Research Scientists. The project was fortunate in that DeGruttola had earlier been interested in the scientific problems of AIDS and had become one of the most knowledgeable faculty at the School on the scientific aspects of the disease. He and Lagakos provided the initial intellectual leadership for CBAR, which became one of the research hubs at HSPH for HIV/AIDS. Lagakos and DeGruttola also formed strong research ties with the work undertaken by Max Essex in the HSPH Department of Immunology and Infectious Diseases.

Another transforming event was the beginning of the Human Genome Project in 1990 and the expansion of investigations for genes underlying disease. The department's research program in Statistical Genetics began in the 1990s when Laird initiated a Working Group seminar series. She received funding for methodological research in Statistical Genetics from the NIMH which led to the recruitment of Christoph Lange. They developed both innovative research methods and software packages which are widely used in genetic analysis throughout the world.

Over the years, several faculty working in the area of Statistical Genetics, including Peter Kraft, Liming Liang, and Alkes Price, were appointed jointly in Epidemiology and Biostatistics. The success of the Statistical Genetics Program has attracted many visitors, postdoctoral fellows, and students. It has generated collaborative research with colleagues throughout the Harvard medical community on a large spectrum of health problems, including cancer, asthma, COPD, bipolar disorder, nicotine addiction, behavioral inhibition, Alzheimer's disease, sleep disorders, malaria, and birth defects.

The Human Genome Project brought many new technological advances in molecular biology and generated new fields of investigation in genomics; e.g., high-dimensional data analysis, bioinformatics, and computational biology. The department made many new appointments in this area beginning in the early 2000s. The newly renamed Department of Biostatistics and Computational Biology at the Dana-Farber Cancer Institute, chaired by David Harrington, led the way for the School. Among those appointed were Wing Wong and John Quackenbush at the senior level, and Robert Gentleman, Xiaole Shirley Liu, Cheng Li, Guocheng Yuan, and Armin Schwartzman at the junior levels. Although both Wing Wong and Robert Gentleman ultimately left the department, major open source software systems were developed during this period with Robert Gentleman starting the development of Bioconductor and Cheng Li continuing the development of the d-Chip system. These two software systems are used worldwide for the analysis of genomic data.

During Louise Ryan's chairmanship, the School-wide interdisciplinary Program in Quantitative Genomics (PQG) was established under the leadership of Xihong Lin, John Quackenbush, and David Hunter. To strengthen this program, computational biologists Winston Hide and Curtis Huttenhower were recruited into the department. The goal of the PQG is to improve health through the study of genetics, behavior, environment, and medicine and encourage interdisciplinary training in quantitative genomics. Today the PQG is directed by Lin, Kraft, Quackenbush, and Hide, and incorporates the Harvard Bioinformatics Core.

The tradition of working closely with researchers in the Department of Environmental Health continued during this period, with the formation of the Program in Environmental Statistics in the Department. Several faculty were recruited to work in this area, including Matthew Wand, Brent Coull, Chris Paciorek, and Xihong Lin. The program in cancer research and clinical trials centered at DFCI continued to flourish with the appointment of several new faculty, including Kyungmann Kim, Stuart Lipsitz, Yi Li, Donna Neuberg, Joe Ibrahim, and Armin Schwartzman. The Biostatistics Group at the MGH continues to flourish. Interaction with the Department of Population Sciences was started with the joint appointment of Grace Wyshak, who has studied many issues in women health.

Ties with Epidemiology were greatly strengthened during this period with several additional secondary faculty appointments. Among these were: James Robins, Donna Spiegelman, Eric Tchetgen Tchetgen, and Tyler VanderWeele. A new program of research in causal inference, initiated by Robins, has attracted faculty, postdocs, and students from both departments, and has made HSPH a noted center for research in this area. Faculty working on causal inference include Robins, Rotnitzky, Lok, Tchetgen Tchetgen, and VanderWeele.

Many faculty, whose primary homes are in the Harvard Medical School, received secondary appointments in the department as they took on substantial training or teaching roles at HSPH. They include Bernard (Bernie) Rosner, Kimberlee Gauvreau, Robert Glynn, and Sharon-Lise Normand. John Orav, first appointed in the 1980s, continues to play a leadership role in the Clinical Effectiveness Program, a Master's Degree Program designed for physicians.

The department was awarded several new NIH training grants during this period. The first was awarded from the NIMH, joint with Epidemiology, headed by Laird and Waternaux. Subsequently, Garrett Fitzmaurice was recruited to work in Psychiatric Statistics and currently heads this effort. As previously mentioned, when the department expanded its research focus to HIV/AIDS, a training grant from NIAID was awarded to the department. Betensky developed a training program funded by NINDS for students and postdoctoral fellows working in Neurostatistics and Neuroepidemiology. Most recently, the department has developed a program funded by the NIGMS to train students working in quantitative genomics, computational biology, and genetic epidemiology. This training grant was initially headed by Wei, and is currently directed by Lin and Quackenbush. In addition, Lagakos established an endowed scholarship program, named in honor of his immigrant father Vasilios Stavros Lagakos, to train international doctoral students from developing countries who intend to return home to continue their careers.

In 1991, the department was approached by statisticians at Schering-Plough to form a new type of Academic–Industry Partnership. Its goal was to create an interactive relationship between HSPH and Schering-Plough biostatisticians for the promotion of methodological research and its application to the scientific challenges facing the pharmaceutical industry. The partnership sponsors an annual joint scientific workshop at HSPH, with an average attendance of 150–200 attendees from academia, government, and industry. A sampling of workshop themes includes: Global Clinical Trials, Vaccines and Control of Disease, Individualized Medical Treatments, Strategies in Drug Safety and Monitoring, Emerging Strategies in the Design and Monitoring of Clinical Trials, Interim Analysis, Adaptive Design, and Bayesian Methods in Clinical Trials. Due to the success of the Schering-Plough partnership, the department soon developed similar relationships with Pfizer, Wyeth, and the Genetics Institute. In 2003, the American Statistical Association awarded the Harvard-Schering Plough Partnership the Statistical Partnerships in Academe, Industry, and Governments Prize, "For an outstanding statistical partnership representing a collaboration between academia and industry of 11 years, which has resulted in the annual Harvard/Schering-Plough Workshop, a unique forum for discussing emerging topics in drug development; interdisciplinary visits between the two institutions to exchange and generate research ideas; a summer intern program at Schering-Plough; and the funding of student training and faculty research at Harvard university. This award is recognition of excellence within the SPAIG concept." When Merck and Schering Plough merged in 2010, the tradition continued with a renamed Harvard-Merck Partnership.

In 2000, the department launched an academic partnership with Kitasato University in Tokyo, Japan. Kitasato had recently formed the first graduate degree program in Biostatistics in Japan, and the newly formed Department of Biostatistics was (and continues to be) chaired by Masahiro Takeuchi, a graduate of the department's doctoral program. The partnership involves department faculty both in the development of Kitasato's doctoral program and in research on statistical issues relating to drug development in Japan and Asia. The Kitasato–Harvard Workshop, held each year in Tokyo, attracts between 500 and 700 attendees from major universities, hospitals, government health departments, and pharmaceutical organizations, mainly from Japan, and focuses on drug development problems in Japan and the Pacific Rim countries.

Three annual lecture series were established during this time period to bring distinguished speakers into the department. The Myrto Lefkopoulou lectureship award was established in 1992 to honor the memory of a beloved colleague who died of cancer at the age of 34. It is given to honor an individual within 15 years of receiving the doctorate who has distinguished him/herself in biostatistics. The Marvin Zelen Leadership Award was established in 1997, on Zelen's 70th birthday, to honor individuals who have provided leadership in biostatistical programs in academia, industry, or government. Finally, the Alumni Award was established in 2004 to honor the department's graduates who have gone on to make notable contributions to biostatistics through their careers in academia, government, or industry.

Under the leadership of Ryan, and now Betensky, the department has made great efforts to increase the diversity of quantitative scientists working in biomedical research. The department has been very successful in attracting minority students to its doctoral program, in part through improving "pipeline" efforts with its Summer Program in Quantitative Sciences, instituted in 1994. In 2009, the department received an award from the American Mathematical Society, which noted its successful efforts in creating a replicable model for bringing more underrepresented minority undergraduates into the pipeline leading to an advanced degree in the mathematical sciences

In 2003, the Department of Biostatistics began to award the PhD under an agreement between HSPH and the Harvard Faculty of Arts and Sciences (FAS). Previously, the Department of Biostatistics had offered the doctor of science (SD) degree. One motivation for the switch was the greater visibility of the PhD over the SD.

2009–Present: The Department Today

The field of Biostatistics is constantly invigorated through developments in biomedical science that require new quantitative approaches, and advances in computing that expand the range of possible methods. When Louise Ryan returned to Australia in 2009 to take a position at the Commonwealth Scientific and Industrial Research Organization, Victor DeGruttola (Fig. 6) was appointed chair of the

Fig. 6 Victor DeGruttola, 2009

department. DeGruttola (Fig.6) has responded to new opportunities by increasing faculty in a variety of quantitative sub-disciplines while maintaining a core in mathematical statistics. He fosters ongoing discussion about the training of statisticians to provide a basis for the lifelong learning that will be necessary to address future challenges. This focus leads not only to development of an academic base for genomics and bioinformatics, but also to provide integrated instruction in Biostatistics and Epidemiology, expanding in new directions while maintaining current strengths. Thus, the department remains poised to develop and expand into emerging new areas of public health.

Sadly, Steve Lagakos was killed in a tragic automobile accident in 2009, cutting short his distinguished career. Steve was respected as a scholar and leader and is greatly missed by colleagues, students, and staffs.

Today, the department continues to grow and thrive. Two new senior faculty members joined the department in the fall of 2009, Francesca Dominici and Giovanni Parmigiani. Parmigiani succeeded Harrington as Chair of the Department of Biostatistics and Computational Biology at the DFCI, while Dominici's appointment greatly strengthened the environmental statistics program. Ware returned to the department after serving as academic dean for 19 years. In the fall of 2011, Franziska Michor, Christopher Barr, and Sebastien Haneuse were appointed to the junior faculty ranks.

The department continues its tradition of excellence in methodological research which has important substantive applications in public health. The faculty based at the DFCI and others in the department are engaged both in applied and methodological work in all aspects of cancer. Senior faculty engaged in this work include Gray, Harrington, Lin, Parmigiani, Wei, and Zelen. The Environmental Statistics Program continues to flourish under the leadership of Professors Coull, Dominici, and Lin. Professors Lin, Quackenbush, Wei, and Laird are engaged in the continuing development of programs in genomics and statistical genetics. Hughes succeeded Lagakos as director of CBAR. DeGruttola, Pagano, and Wei continue to work on important problems in clinical trials, prevention, and surveillance for HIV/AIDS. Gray is leading the Statistical Center of the Eastern Cooperative Oncology Group, succeeding Harrington. Ware and Betensky lead the department's work in Catalyst, the Harvard Clinical and Translational Science Center (CTSC). Wei continues to lead the department's involvement with industry partnerships.

Major methodological areas of research include the design and analysis of clinical trials, survival analysis, sequential methods, statistical genetics, longitudinal analysis, semi-parametric methods, causal inference, measurement error, Bayesian methods, surveillance, screening for the early detection of diseases, bioinformatics, computational biology, personalized medicine, signal and image processing, and computationally intensive statistical methods. The number of publications by department members in the 10-year period ending in 2010 is 1880 (Reference: Google Scholar). As this is being written, the department has 14 master's students, 66 doctoral students, 67 research fellows and associates, 33 research scientists, and 58 faculty at all levels, both primary and secondary.

The department's history is evidence that its research agenda is driven by current public health problems. By long standing tradition, an environment has been created in which research and practice can thrive together. Research in biostatistics theory without input from applications is likely to be sterile and applications without a solid basis of theory can be misleading and inefficient. The curriculum and research agenda continues to change to meet new challenges. The goals in the future are to maintain the department's leadership role in biostatistical science and to train the leaders of tomorrow.

Acknowledgments We wish to acknowledge Ray Neff and Dan Segal, former graduates of our department, for providing input about the early history of the department, and our current faculty colleagues. We especially acknowledge Dianne Finkelstein for her comments.

55 Years of Harvard Statistics: Stories, Snapshots, and Statistics

Xiao-Li Meng

Late evening of Valentine's Day, 1957. The driver of a VW Bug, wearing his trademark leather hat with ear flaps, waited patiently at Boston's Logan Airport. Finally a pair of newlyweds stepped out of a much delayed flight from Toronto. They were not expecting a Valentine's Day party, but rather an Ivy League interview process. However, the VW Bug took the young man and his bride directly to a house on Pierce Street in Belmont, about 15 miles west of the airport. Greeting him was not a set of challenging questions, but a very lively celebratory party already in progress.

The driver was none other than Frederick Mosteller, the founding father of our department. The young man was Arthur Dempster, who was interviewed the next day for an assistant professorship. He, however, did not join the department until 1958 because, as Mosteller wrote (Annual Report, 1957–1958), "Dr. Dempster thought a year of applied work would do him good, and he went instead to the Bell Telephone Laboratories." Having one's statistical research rooted in and motivated by real-life applications became a hallmark of Harvard Statistics' outlook (or "departmentality") from very early on precisely because of the conviction and the practice of our founding generation, and those who came later.

Mosteller and his wife Virginia hosted the party because merely 2 days earlier, Harvard's Faculty of Arts and Sciences (FAS) had voted to establish a Department of Statistics at FAS. As Mosteller opened his first annual report: "Lincoln's birthday, 1957, marked the establishment of the Department of Statistics by a Faculty vote, after debate that lasted about eight minutes. But Harvard has not been as hasty as this might imply".

Indeed, the lucky number "eight" would be appropriate if we replace "minutes" by "years".

Science Center, 1 Oxford Street, Cambridge, MA 02138-2901, USA
X.-L. Meng (✉)
e-mail: meng@stat.harvard.edu

A. Agresti and X.-L. Meng (eds.), *Strength in Numbers: The Rising of Academic Statistics Departments in the U.S.*, DOI: 10.1007/978-1-4614-3649-2_8,

The Early Trimesters

Mosteller joined Harvard during 1946–1947, the year when the Department of Social Relations was formed. Mosteller was recruited to this new department from Princeton as a fresh PhD in Mathematics.[1] During 1948–1949, Mosteller taught *Mathematics 190, Mathematical Statistics*, for the first time. It was also during that year that Harold A. Freeman, a Professor in the Department of Economics at MIT, wrote to Mosteller, proposing a joint Harvard-MIT degree in Statistics.

Freeman, like Mosteller, was a member of the Statistical Research Group, based at Columbia University,[2] where he collaborated with Abraham Wald in developing applications of Wald's groundbreaking 1943 work on sequential analysis. He also served as a consultant to the US Army and a number of other governmental agencies, and became a vice president of the American Statistical Association during 1949–1950.

This background of Freeman provides a glimpse into his motivation for a joint statistics degree, which was not given in his letter dated December 9, 1948. But his list of possible topics and courses outlined his vision: a well-balanced program integrating theory and applications. In particular, he wrote:

> We could quite nicely cover the necessary mathematics, probability, mathematical statistics, and the higher levels of stochastic processes, modern computing techniques, and we would deal with certain applications particularly in fields as electronic communication, physics, econometrics, psychology, neurology, and education.

This description is so strikingly contemporary that a future amateur historian could well mistake it as a proposal from MIT at the turn of the twenty-first century, especially considering that MIT still does not have a statistics program more than 60 years after Freeman's proposal!

Freeman's proposal seems to have been an important impetus to Mosteller's first proposal for establishing a Committee of Statistics at Harvard. Among Mosteller's many meticulously kept and chronologically ordered records, Freeman's letter and Mosteller's proposal were kept back-to-back, immediately after a ten-page document of "Early History of The Department of Statistics" (hereafter, the Document). This document listed no author, but even the most casual reasoning in Mosteller and Wallace's "*Inference and Disputed Authorship*" would lead to the conclusion that Mosteller was the author, likely aided by his life-time collaborator and assistant, Cleo Youtz (1909–2005), who was also the department historian, among many invaluable contributions she made to the department.

The document stated that "About this time Professor Mosteller wrote a proposal for a Committee of Statistics." Whereas the two-page proposal shows no date (but

---

[1] Much information about Mosteller's career at and contributions to Harvard (and beyond) can be found in "*A Statistical Model*" (Eds: Fienberg, Hoaglin, Kruskal, and Tanur, 1990, Springer), and in the edited autobiography "*The Pleasure of Statistics*." (Eds: Fienberg, Hoaglin, and Tanur, 2006, Springer).

[2] See Stephen Stigler's "University of Chicago Department of Statistics" in this volume.

with a penciled "Date?" on it), we can be reasonably certain that the gestation period for the establishment of Harvard Statistics, which officially started its life on July 1, 1957, is just a few months shy of 8 years. (See Stigler[3] (2008) for the pre-Mosteller's era.) The first trimester, however, seems to be an almost unnoticed one, as the document described: "During the year of 1950–1953, no notable progress was made toward the establishment of a statistics department or a committee on statistics." In his autobiography, Mosteller recalled that his proposal was declined in April 1950 by then Provost Paul Buck, who considered the field of statistics was much too narrow for Harvard (see footnote 1).

The documented evidence of the influence of Freeman's proposal came from a 1954 Report by a Visiting Committee and a 1956 Report by a Subcommittee on Statistics (of the Committee on Applied Mathematics), which was chaired by Mosteller. The 1954 Report resulted from Ford Foundation grants to five universities, one of which was Harvard, to review their behavioral sciences. That report suggested "exploring the possibility of developing statistics jointly with M.I.T." The Subcommittee Report, dated January 27, 1956 (hereafter the Report), went further, making eight recommendations (from A to H), three of which are most relevant to the current discussion. They are:

> D. that ways be found to bring two full-time statisticians to Harvard, one of senior status,
> G. that though committees on statistics have universally failed, a committee on statistics initially be established with the view that a successful program would later lead to a departmental structure,
> H. close association be maintained with Massachusetts Institute of Technology so that advantage may be taken of new developments (they also are reviewing their statistics program).

The report included a section detailing a discussion with Freeman, mostly on nine MIT courses on statistics and information theory, and their relevance for the recommendations.

The second trimester therefore showed more vital signs. However, that vitality seemed to be short-lived. In responding to the subcommittee's recommendations, then Dean of FAS, McGeorge Bundy, wrote that while he had no reason to quarrel with the subcommittee's reasoning,

> I do have a remark that I do not see how this can be done without new resources, unless one or another of our existing departments is prepared to allocate a place to statistics. I do hope, therefore, that the Committee on Applied Mathematics may explore the question of the degree of interest with existing departments have in making such a contribution.

It does not take much experience with academic politics to see that this is merely a diplomatic way to say "NO." Virtually every department fights for resources, then, now, and will in the foreseeable future. Suggesting to one's colleagues that they should consider donating departmental resources to help build another department is almost surely a sign that either the chair is tired of her/his

---

[3] Stigler, S. (2008) Statistics at Harvard? A toast to the Harvard Statistics Department. *American Statistician,* **62**, 246–250.

administrative duties or s/he has more administrative aspiration than the chair's chair can accommodate.

A miscarriage therefore seemed to be looming.

Prepared for Complications

Evidently, this chapter would not have existed, literally and figuratively, if Mosteller was completely unprepared for Dean Bundy's response. The details in the report, however, suggested otherwise. I will ask the reader's indulgence for quoting this report extensively, because it provides a rather holistic picture of statistics as an academic field in the United States during the critical "big bang" period, when most of the current major departments came into existence, as documented in this volume.

For example, in its two-page section on "Administrative Arrangement," the report stated:

> In the country as a whole there have been five main ways of handling statistics: (1) as a separate department (in at least one school an entire institute has been formed), (2) as a semi-independent organization attached to another department, usually mathematics, (3) no separate arrangement, the statistics staff is usually part of the mathematics faculty, (4) some sort of committee structure exists to administer programs in case the need for them arises, (5) no special provision for statistics is made, staff belong to their own departments.

The report foresaw the trend over the past half-century, that is, moving from later categories to earlier ones, when it stated that "The writer knows of no American universities that have contributed an interesting number of statisticians to the profession in recent times with setup (5). The same remark applies to setup (4)." The report went on to refute the occasional citations that the committee format had worked at University of Chicago, because "It is a department in all but name—has its own budget, space, makes its own appointments, and grants its own degrees."

Having argued that neither (5) nor (4) was a healthy setup for statistics, the report cited Princeton as a successful example of (3), and cited the main reason being "After the close of the war, John Tukey became interested in the teaching of statistics." This emphasis on the direct link between teaching and success has had a lasting impact on the development of Harvard Statistics, which has produced an impressively large list of influential graduates, despite the fact that it was and still is the smallest department among its peers. Using the same criteria, the report cited the University of Michigan as another possible example of past success but that "They are struggling now to rebuild." Similarly, for the arrangement (2), the report acknowledged Berkeley had some success with that format, but emphasized that "The trend, however, is to the department, and Berkeley is moving toward that solution."

The report then provided a list of universities that the report stated would be generally agreed to be major institutions in Statistics in the United States "from the point of view of training." It included nine institutions, all but one of which are still currently major "statistics institutions" by Mosteller's criteria: Berkeley, Chicago, Columbia, Iowa State (College), Michigan, North Carolina Chapel Hill, North Carolina State (College), Princeton, and Stanford. The report then emphasized that "All but Princeton and Michigan had the equivalent of a department, and both have a Laboratory."

With such supporting evidence, the report naturally concluded that:

> The historical evidence therefore is clear, no organization has led to nothing, committees have uniformly failed, with mathematics departments there have been success, but even these are moving toward separate adjuncts. The more uniform success has come from the department and the semi-independent unit.

At this point, a reader might be as puzzled as I was initially. Why did the report argue so clearly that "committees have uniformly failed," but then went on to make the aforementioned Recommendation G? What was the rationale to repeatedly remind a university administration that the proposal it was asked to invest resources in was something that had never succeeded?

Could it be reverse psychology to stroke the ego of Harvard's Administration, "Wow, no one can do it, so it ought be up to Harvard to show how!"?

A Brilliant Delivery

Well, not impossible—Mosteller was known to be a great poker player. But a more plausible explanation is that Mosteller had a few-steps-ahead chess strategy in place. The temporary stepping-stone nature of Recommendation G is quite clear from its wording, a point was made ever clearer later in the report:

> Departments are not constructed overnight, nor are adjuncts. Furthermore, Harvard has rather more success with Committees than do most institutions. Presumably this stems from the unstinting, indeed unremitting, committee work training given by Harvard to all faculty members from the day of appointment. It should also be noted that it is very easy to kill a proposed endeavor by inflexibility about the initial administrative arrangement of the non-existent organization.

What a witty piece! It not only lightly got the report out of the seemingly awkward position of recommending something it spent pages arguing against by playing a pun on the word "Committee," but also offered truly insightful wisdom on negotiation. That is, asking for the right wine glass only when one is certain that wine will be served!

Of course wisdom and wit alone can only carry the process so far; to accomplish the grand task Mosteller undertook, he also needed a weapon. And he did have one. Perhaps not a pure coincidence, the overall director of the aforementioned 1954 study funded by the Ford Foundation was Allen Wallis of The University of Chicago, where Mosteller spent his sabbatical during 1954–1955. Wallis was recruited by Chicago in 1946 to start a program in statistics, and by then the program was in full swing except for the formal title as a "Department." It also had interest in recruiting Mosteller as early as 1949 (see footnote 2). Indeed, as Mosteller recalled in his autobiography: "By fall I had received a most attractive offer to chair an expanding department of statistics at another university, and Virginia and I concluded that if Harvard did not want a department, I should move, much as we had enjoyed Harvard." The offer came from none other than The University of Chicago in the fall of 1956, as the document revealed:

> On September 14, 1956, Professor Mosteller met with Dean Bundy to discuss the possibility that Professor Mosteller would leave Harvard to accept the chairmanship at the Committee of Statistics at the University of Chicago.

The document then reported that there were many telephone conversations and conferences in the following several months between Mosteller and the Dean and others, including a visit on October 18–22 by Mosteller to Chicago, and a visit to Harvard on November 4 by Wallis and James Lorie. (Lorie was an associate dean at Chicago's Business School, where Wallis became its Dean in 1957, and Lorie was well known for his ability to recruit.[4])

The heavy recruitment pressure seemed to pay off, not for Chicago, but for Harvard. On December 7, 1956, Dean Bundy formally approved the establishment of the Department of Statistics, with funds for three appointments, one senior and two juniors.

Mosteller's 1956 Report asked for a committee with two positions. He ended up getting a department with three positions. Offering Mosteller the chairmanship for a Committee on Statistics as a counteroffer would be ineffective, because the report already argued clearly that Chicago's Committee was effectively a department, and that a "Committee on Statistics" had essentially zero chance to succeed! Apparently the only game-changing move available to Dean Bundy was what Mosteller's 1956 Report was really after: a fully independent Department of Statistics.

We perhaps will never know how difficult this negotiation process was for Mosteller, but we are all grateful that he put himself on the line, literally, in order to put Harvard on the Statistics landscape. This was a profound victory for Harvard Statistics and beyond, especially considering MIT still needs to find its own Mosteller more than 60 years after Freeman's proposal.

Education Programs: Early Accomplishments and Recent Efforts

Among the 1956 Report's remaining five recommendations, three were primarily on graduate education:

[4] "Lorie developed Chicago Approach to management education" *Chicago Chronicle*, October 6, 2005.

C. the establishment of an advanced full course in mathematical statistics,
E. that if D is accomplished, a doctoral program in statistics be established, and
F. that additional courses appropriate to such a degree program be offered, where possible maintaining the prerequisite levels so that the courses could serve the general University as well as statistical students.

Whereas establishing a viable PhD program was the priority, Recommendation F went beyond merely training statisticians by calling for service to the entire university. This department's tradition started on day one. In its first year 1957–1958, among the nine courses offered, Mosteller himself taught two with the highest enrollments; *Statistics 122: Statistics in Social Sciences* had the highest enrollment of 134, serving "Economics, Social Relations, and Psychology."

William Cochran, recruited in 1957 from Johns Hopkins to fill the senior position of the three authorized appointments, taught three courses respectively in regression analysis, experiment design, and analysis of counts, reflecting well Cochran's expertise and the department's long tradition on teaching statistical principles for both data collection and data analysis.

John Pratt, recruited in 1957 from Chicago to fill one of the two junior positions, also taught three courses. His *Statistics 142: Non-parametric Inference* is an example of the impact of faculty research interests on curriculum. Pratt moved to Harvard's Business School when he was promoted to Associate Professor during 1960–1961, the same year Dempster was promoted to Associate Professor. (Back then promotions required availability of allocated slots; happily, the slot at Harvard's Business School helped to retain both Dempster and Pratt at Harvard.) After Pratt's move, Statistics 142 (or its graduate-level counterpart 242) was taught sporadically, with the last listing of Statistics 242 found in the 1975–1976 course catalogue.

The last of the nine courses was *Statistics 250, Theory of Statistical Decisions*, taught by Howard Raiffa. Raiffa was recruited by Harvard's Business School from Columbia also in 1957, but with a quarter of his time in Statistics. Raiffa is a well-known pioneer in the field of decision analysis, and hence it was fitting that not only did he bring a decision analysis course to Harvard, but also he (and his wife) did a formal decision analysis before bringing himself to Harvard. As Raiffa recalled[5] "Harvard was the clear winner. Of course, there were some dimensions where Columbia was better, so it wasn't a dominating solution, but the formalization helped us really decide that it wasn't a close call at all." How nice to see that our founding generation practiced what they preached, benefited from it, and benefited all of us with their sound decisions!

The year of 1958–1959 also saw the first PhD graduate (Joseph Arthur Greenwood) and first Masters degree recipient (Alan Barnard Howson). Since then the department has produced over 170 PhDs, over 400 AMs, and over 100 ABs; see Fig. 1 for time trends. Our alumni's many lasting contributions will undoubtedly

[5] Fienberg, S. (2008) "The Early Statistical Years: 1947–1967—A Conversation with Howard Raiffa." *Statistical Science*, **23**, 136–149.

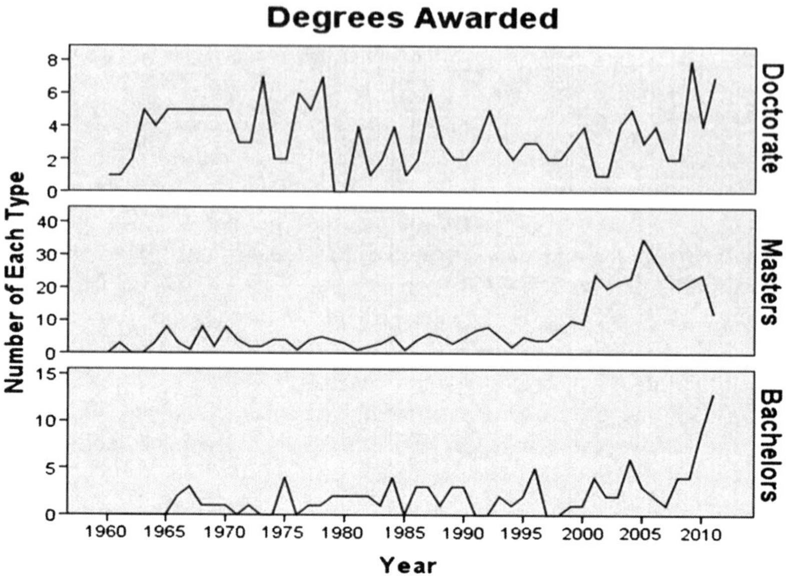

Fig. 1 Time trends in degrees awarded

take an entire book to document. But as an example of leadership in the academic world, the department has produced over three dozen (current or former) full professors at statistics, biostatistics, and other departments and schools (e.g., Education, Business) at many major universities, as well as passionate educators in leading liberal arts colleges; see the full list at the alumni tab at stat.harvard.edu.

Today the department welcomes about 8–10 PhD and 15–20 Masters students each year, and offers about 30 courses annually, including a set of courses aimed at graduate students' professional development beyond the traditional curriculum. These include *Statistics 303: The Art and Practice of Teaching Statistics*, a one-year required course for all first year PhDs; *Statistics 399: Problem Solving in Statistics*, an all-faculty-participating course aimed at helping (mainly 2-year PhD) students to prepare for qualifying exams; and *Statistics 366: Research Cultivation and Culmination*, a course designed for those who have passed the qualifying exams to inquire the skills of developing a research idea into successful publication.

In 2008, the department's overall effort was recognized by a GSAS (Graduate School of Arts and Sciences) Dean's Prize for Innovations in Graduate Education. At students' suggestions, \$1,500 of the \$25,000 award was used to purchase perhaps the world's first endowed chair for graduate students: a full-body massage chair. The students provided very compelling reasons to win the dean's approval for such a purchase, including it being an effective recruitment tool for persuading prospective students because it symbolizes well the department's caring for its students and its general friendly atmosphere.

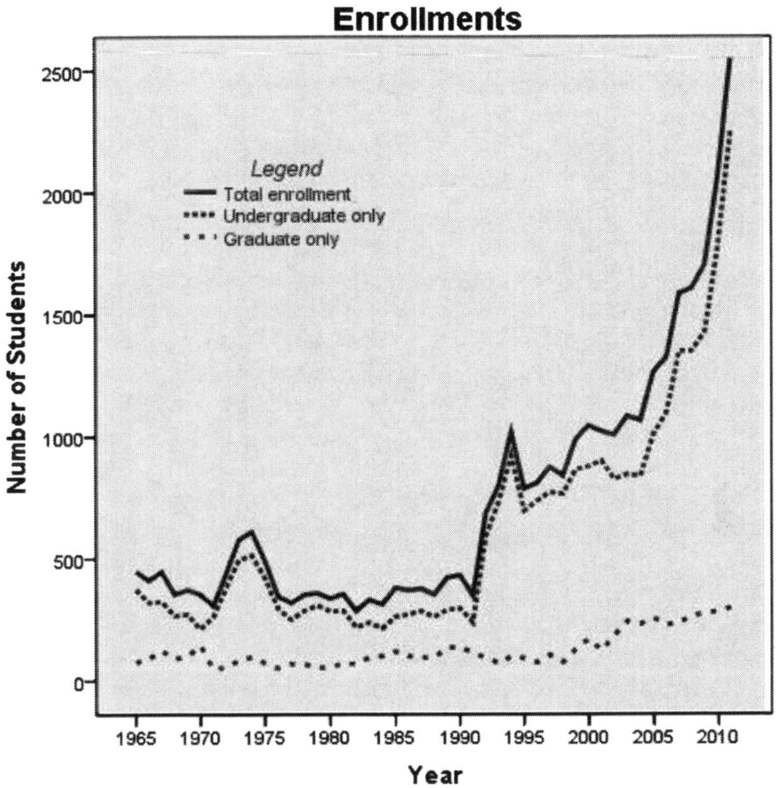

Fig. 2 Enrollment trends

The students were of course right. Much of the recent effort has been devoted to enhancing our intellectually challenging but socially caring environment. The latest installment perhaps is best highlighted by the newly established *Statistics 365*, where students and faculty gather every Tuesday evening to indulge themselves with much food for body (Chinese catering from Yenching Restaurant from Harvard Square) and food for thought (e.g., *Reading R. A. Fisher* in the fall of 2011 and *Reading D. R. Cox* planned for the fall of 2012).

The remaining two recommendations of the 1956 Report perhaps would surprise some readers because they were about *not* taking action:

A. no action with respect to elementary courses given now,
B. that no attempt be made at this time to create a bachelor's degree program in statistics.

Recommendation A displays again Mosteller's leadership skill for building a new program: to win the maximum possible support by minimizing the threats to existing programs. Besides, nothing sinks a new program more swiftly than overloading it before having an adequate navigation staff and plan. Recommendation B was made based on similar considerations, but the report left the door open for the future: "Perhaps when a course structure and staff are available, the question should be reconsidered."

And the future arrived in another 8 years. On February 11, 1964, one day before Lincoln's birthday, FAS faculty voted to approve the establishment of an undergraduate concentration in Statistics. However, effort was made as early as 1959. An article dated November 9, 1959 in *Harvard Crimson* ("*Statistics Dept. Nears Plan for Concentration*") reported the findings of a two-man committee (Cochran and Pratt), indicating that the plan would "go into effect in the fall." Unfortunately, similar to the fate of Mosteller's first proposal to establish a statistics entity at Harvard, there was no report on progress whatsoever in the next 3 years. However, unlike the endeavor for establishing the department, this time the force for breakthrough came within Harvard, and it was from a somewhat unexpected source.

It was from John Monro, Dean of Harvard College from 1958 to 1967, who was known for making unexpected moves, including giving up his Harvard deanship in 1967 for the then unaccredited Miles College in Alabama because of his passion for providing education to the unprivileged.[6] A memo from Mosteller to the statistics faculty, dated October 14, 1963, started with "Dean Monro expressed enthusiasm for the possibility of undergraduate offering in statistics," because "He feels that statistics has the advantage of a mathematical flavor that many students like and some concreteness which many of these same students find lacking in their more advanced mathematical work and it can offer an applied slant without being unduly narrow." (The memo also indicated that Dean Monro was the elder brother of Sutton Monro of the well-known "Robbins and Monro stochastic approximation".)

Dean Monro's enthusiasm proved to be most effective. The concentration started officially during the academic year 1964–1965, with six enrolled; the first two AB degrees were awarded in the following year. But more importantly, Monro's vision that statistics programs should not be restricted to a few "mathematical elites" also proved to be most effective. As Figs. 1 and 2 allude to, our undergraduate concentration remained very small until about 2005, when it began to increase with an exponential rate of 60 % per year, currently reaching 72. Not coincidently, this was about the time the department took a broader view in structuring our undergraduate curriculum, by shifting its traditional focus as an initial preparatory program for future statisticians to a much broader statistical education opportunity for future scientists, policy makers, educators, etc.

Of course excellence in undergraduate education takes tremendous collective effort, and currently we are very fortunate to have a dream team.[7] Our co-Directors of Undergraduate Studies, David Harrington and Joseph Blitzstein, have led much of the recent effort in communicating with our colleagues around the campus and with students from all backgrounds to fruitfully build a program that reflects well

[6] *The New York Times*, April 03, 2002, "John U. Monro, 89, Dies; Left Harvard to Follow Ideals".

[7] See Meng (2011) "Team Dreams and Dream Teams" *Amstat News*, November 1, 2011.

and serves amply the vastly increased interests in and demand for Statistics. Our department has also established several signature courses. Joseph Blitzstein, our first Professor of Practice, is a household name among Harvard students because of his award winning *Statistics 110: Introduction to Probability* (available at Harvard iTunes U). Michael Parzen, our first senior lecturer, coming to the department in 2010 with 17 years of teaching MBA students, attracts over 1,300 students in 2011–2012 with his *Statistics 104* (*Introduction to Quantitative Methods for Economics*) and *Statistics 107* (*Introduction to Business and Financial Statistics*).

Parzen succeeded another phenomenal teacher, Kenneth Stanley, who retired in 2010 and who represented the goodwill from our sister department, the Department of Biostatistics, which has helped us tremendously over the years by providing some of its most effective faculty instructors and educators (such as Harrington) to help us to accomplish our educational mission. Stanley's teaching was so effective that one student wrote: "Taking a course from Professor Stanley is like taking a course in Christianity, and Jesus himself is teaching."

Our first Part-Time Professor of Practice, Stephen Blyth, offers yet another signature course that reflects well both the versatility and the vitality of Statistics: *Statistics 123: Applied Quantitative Finance on Wall Street*. Being an alumnus himself, a veteran of Wall Street and now a managing director at the Harvard Management Company, Blyth offers the students a uniquely practical and rigorous program that benefits them inside and outside the classroom. In return, the students honored him as a Class of 2011 Favorite Professor, the first time he taught the course. The same honor was bestowed upon Stanley by the Class of 2009, to Parzen by the Class of 2012, and to Blitzstein by the Class of 2009 through the Class of 2012, four consecutive years!

The recognitions, of course, do not stop at those who are at the forefront of the educational line. Kevin Rader, our first preceptor, has worked tirelessly behind the scene to provide much needed teaching support, from training facilitators for the study network (an organized study group) to serving as Head Teaching Fellow. For his effort, he was awarded a David Pickard Teaching Award during the inaugural event for the David Pickard Endowment Fund and Lecture.

The David Pickard Fund, which was established in 2010 by generous contributions of alumni, under the leadership of Victor Solo (a faculty member at Harvard Statistics during 1982–1985), reflects fully the lasting impact of excellence in education and mentoring. The opening paragraph of the Fund Document summarizes this fact succinctly:

> The gifts of colleagues, students, and friends of David Pickard establish the David K. Pickard Memorial Endowment Fund. This fund celebrates the memory of David K. Pickard, who served as a junior faculty member in the Harvard Statistics Department from 1977 to 1985. Professor Pickard was known for his outstanding teaching, having won two major Harvard-wide teaching awards: the Phi Beta Kappa Prize in 1982 and the Levenson Prize in 1984. He also won the Hoopes Prize two years in a row for supervising and nominating a senior thesis. Professor Pickard had a strong influence on the Statistics PhD students at Harvard in that period. Sadly, Professor Pickard died of a brain tumor in August 1986 in Kingston Ontario, where he had moved after leaving Harvard.

Stories About and From Our Dozen Senior Faculty in 55 Years

The great teaching and mentoring of students is a perfect segue into the stories about and from our most distinguished faculty through the 55-year history, because our founding father Mosteller was well known as an effective educator and caring mentor. In particular, his course on NBC's *Continental Classroom* in 1960–1961 had over a million viewers across 170 stations. As for his caring mentorship, Stephen Fienberg, an advisee of Mosteller and who himself has received many honors including the election to the National Academy of Sciences, recounts:

> Many of us who worked with Fred benefited greatly from his comment on drafts of papers and book manuscripts. He was always able to make suggestions and comments that would improve our efforts. This was especially the case for *Discrete Multivariate Analysis*, where Fred worked over every chapter multiple times although he refused to be a co-author with Yvonne Bishop, Paul Holland, and me, even though the book would not have existed without his inspiration and hard work.

Mosteller chaired the department for a total of 15 years (1957–1969, 1973, and 1975–1977). He retired from classroom teaching in 1987, but remained very active in research and writing until 2003, when he left the department and moved to West Virginia to live with his daughter, passing away on July 23, 2006. He wrote much in his life, and much has been written about his life. He published over 350 papers, and more than 50 books. A complete list of his (known) publications can be found in "*The Pleasure of Statistics.*"

Like Mosteller, much has been written about Cochran, who served as acting chair in the fall of 1960 and in 1962–1963, and retired in 1976. To this date, he is still among the very few statisticians who are widely acknowledged to have made fundamental contributions in all three areas of data collection: experiment design, sample survey, and observational studies. Such accomplishments undoubtedly benefited from his involvement in astonishingly wide-ranging and large-scale studies that have had a profound impact on humanity. These include studies on World War II bombings, radiation in Hiroshima, the Kinsey Report on human sexual behavior, the Salk polio vaccine, equality in education, and, by far, his foremost contribution, the Surgeon General's Report on Smoking and Health (1964).

In addition to his applied work, he was also well known for his very influential textbooks such as *Experimental Designs* (1950, with G. M. Cox) and *Sampling Techniques* (1977), and for many theoretical and methodological contributions such as Cochran's Theorem, Cochran's Q test, and the Cochran–Mantel–Haenszel test. A good starting point for reading about his life and many contributions is the report[8] on celebrating the centenary of Cochran held by our department on

[8] "Cochran at 100" *Harvard Gazatte*, Nov. 18, 2009. news.harvard.edu/gazatte/story/2009/11/cochran-at-100.

November 14, 2009, and his obituary in *Annals of Statistics* (1982, 1–10)—Cochran passed away on March 29, 1980, on Cape Cod.

Fortunately, there is a founding faculty member who is with us, engaging deeply in some most penetrating foundational research. Art Dempster, who served as the department chair for a total of 11 years (1969–1975, 1977–1979, and 1982–1985), is widely acknowledged as one of the deepest thinkers in statistics in the last half-century. Dempster is most cited for his extremely influential paper on the EM algorithm (*JRSSB* 1977), co-authored with Nan Laird and Donald Rubin, which is one of the top ten most-cited articles in Statistics ever. Incidentally, Nan Laird, like Rubin, was a student of the department (an advisee of Dempster), and then went on to become a faculty member (since 1975) and chair (from 1990 to 1999) of our sister department, the Department of Biostatistics. She is a great example of how alumni have helped the department, having taught departmental courses, served on its search committees, and co-organized our 50th anniversary celebration, among many other activities over nearly four decades.

Among Dempster's theoretical contributions, he is most well known for the Dempster-Shafer (DS) theory. Also known as the theory of belief functions, the DS paradigm goes beyond the Bayesian inference paradigm. A key feature of the DS theory is that ignorance, that is, "don't know" is a distinctive state with its own assigned ("belief") value after specifying the values for "yes" and "no." This development appears to stem from Dempster's emphasis on logic of statistical inferences,[9] leading to a new paradigm.

Herman Chernoff, another leading theoretician of the last half-century, is our other Professor Emeritus (since 1996). Chernoff is known for many contributions bearing his name,[10] such as Chernoff bound, Chernoff information, Chernoff's distribution, and Chernoff face. Chernoff is another example of our "departmentality" because (see footnote 10) "a lot of my ideas came out of the work I did on a contract funded by ONR, much of which involved applications." The following fascinating story from Chernoff shares some of his experience on an NRC committee to examine an FBI report that claimed to prove that President Kennedy was killed by a shot from the Grassy Knoll at Dealey Plaza:

> Although the proof depended on an elaborate statistical match between the echoes in the Plaza as recorded on the tapes of the police radio recordings, I was the only statistician on the committee. The evidence was less than convincing to the committee, but we lacked something strong with which to counter the theory until a rock and roll musician (who should be deaf) who solved our problem. The police had used two radio channels for communication. The echo analysis had been done on the tapes of one. The imputed shot

[9] See Dempster, A. P. (1998) "Logicist Statistics. I. Models and modeling" *Statistical Science*, **13**, 248–276, and Dempster, A. P. (2008) "Logicist Statistics II. Inference" in *Classic Works of the Dempster-Shafer Theory of Belief Functions* (*Studies in Fuzziness and Soft Computing*), **219**, 761–786, Springer.

[10] See Bather, J. (1996) "A Conversation with Herman Chernoff" *Statistical Science*, **4**, 335–350, and Zacks, S (2005) "Herman Chernoff: An Appreciation" *Journal of Statistical Planning and Inference* **130**, 3–12.

> had been on a part of the tape where there was some very noisy talk. The musician had listened to both tapes, one in each ear and claimed that this talk was the same as "Hold everything secure until..." which was clear on the other tape, and had been after the fatal shot. No one had been able to decipher the noisy message before, but sound analysis showed that the musician was correct and this was the basis for rejecting the claim about the fatal bullet.

Evidently, this was not the only time that Chernoff had to deal with a difficult situation. After 22 years on the Stanford faculty, he moved to MIT in 1974, partly in hopes of building a statistics program. But that turned out to be an impossible task, leading to Chernoff's move to Harvard in 1985.

But Harvard was not without its own problems. As Chernoff recalled in his 1996 conversation with Bather (see footnote 10) that although he decided to join Harvard after sending what his wife labeled as a "suicidal" letter to MIT president (telling him how MIT should learn from Purdue University about building a statistics program), "I must admit that Harvard also has a problem. In my opinion it is much too small a department to maintain the robustness necessary to survive." Indeed, a similar sentiment was felt by Peter Huber, most widely known for his very influential work in robust statistics.[11] Huber joined Harvard Statistics in 1978 from ETH, chaired the department from 1979 to 1982, but ultimately decided to leave Harvard and joined MIT in 1988. In recollecting his concerns, he wrote to me:

> The long-range organizational challenge was to transform us into a top-rated statistics department. We certainly would not want to teach everything, but we would need to offer some neglected basics. I insisted on a course on measure theoretic probability—if only to inoculate future PhD level statisticians against mathematical charlatanry. A modest increase of the departmental faculty would go a long way. My attempt to persuade the Dean failed, just as previous attempts by Fred and Art had failed.

Although Huber's effort to expand the department did not bear fruit, his effort to enhance the interaction with other departments in FAS did, and is well remembered by his students. David Donoho, another alumnus who has received numerous honors including the election to the National Academy of Sciences and a MacArthur Fellowship, wrote the following account about his PhD advisor:

> During 1979–1982, Huber finished his book on robust statistics. He also obtained a large grant from the Office of Naval Research to work in graphical methods for data analysis, and set up a team to develop the software package PRIM-H for visualization of high dimensional data on a high-tech graphics unit of the day, the Evans and Sutherland Picture System 2. The graphics research led to plenty of interactions with other Harvard faculty. For example, astronomers Margaret Geller and John Huchra's work discovering walls and voids in galaxy catalogs was spurred by Margaret's seeing PRIM-H in action on the Smithsonian redshift catalog. I think that seeing their data in PRIM-H inspired them to approach their data analysis in a new way.

---

[11] See Buja, A. and Kunsch, H. (2008) "A Conversation with Peter Huber". *Statistical Science*, **23,** 120–135.

Interaction with other faculty indeed has been a part of our "departmentality." The arrival of Donald Rubin in 1984 further enhanced this tradition. Rubin spent a decade at Educational Testing Service before becoming a Professor of Statistics and of Education at University of Chicago in 1982, and at Harvard he has worked closely with a number of leading psychologists, economists, etc. Motivated by a large array of real-life problems, Rubin has made many methodologically influential and practically impactful contributions, for which he has received numerous honors, most recently the election to the National Academy of Sciences. The edited volume[12] in honor of his 60th birthday documents well his contributions up to early 2000. A book review[13] of this volume perhaps best summarizes Rubin's influence and impact:

> ... Donald Rubin, who may well be the most influential contributor to applied statistics in the past three decades. Consider, for example, the EM algorithm, multiple imputation, the potential outcomes framework for causation, propensity scores, and matching methods. Rubin has also been instrumental in championing an "if it works" approach to empirical Bayesian statistics, which has gone far in creating a détente between frequentists and Bayesians, at least insofar as the application of statistics to real data is concerned.

Rubin also chaired the department for a total of 13 years (1985–1994, 2000–2004), and served as a great mentor to many graduate and undergraduate students, winning the aforementioned Hoopes Prize four times (1999, 2002, 2003, and 2010).

But after the founding generation, the formal interaction with other departments did not start until the arrival of Carl Morris in 1990, who held a 50–50 split joint appointment between Statistics and Health Care Policy at the Medical School. Morris arrived with such a joint appointment because of his well-known work in both theory and methodologies (e.g., hierarchical and empirical Bayes, and natural exponential families) and substantial experience in real-life applications, including his earlier experimental design work in the RAND Health Insurance Experiment. In recalling his experience with this joint appointment and how it impacts his current teaching, Morris wrote:

> I joined Harvard's department because it always has been exemplary in the interface of application and theory, and that tradition continues. My five year split appointment with Harvard's Health Care Policy Department provided an ideal opportunity to develop applications and theory together, including developing hierarchical models for profiling medical units and for meta-analyses. When teaching in Stat, these values involve choosing theoretical topics that matter for applications.

During his chairmanship in 1994–2000, Morris took a critical leadership role to sustain our undergraduate introductory courses, at a time when the department was severely understaffed and hence there was a serious consideration that it might

---

[12] Gelman, A. and Meng, X.-L. (2004) *Applied Bayesian Modeling and Causal Inference from Incomplete-Data Perspectives*. Wiley and Sons.

[13] Smith, H. (2007) Book Review of Gelman and Meng (2004), *Sociological Methods & Research*, **36**, 140–143.

Photo 1 Founding generation, taken on May 22, 1959, at 2 Divinity Street, Cambridge. *Front left* to *right*: Pratt, Cochran, Mosteller. *Back left* to *right*: Raiffa and Dempster

have to focus on PhD training only. Indeed, in 1999–2000, the department only had five ladder faculty, with Dempster, Morris, and Rubin as the three senior faculty.

Luckily, just about that time, Rubin was heavily recruited by another university. Like Mosteller, Rubin used the opportunity skillfully to expand the department, resulting in a clustered hiring that ultimately brought in Jun Liu and Wing Wong as full professors in 2000, and Sam Kou (as assistant professor) and myself in 2001. To continue the interface tradition, Wong's appointment was jointly between Biostatistics and Statistics, with a larger percentage in Biostatistics. This reflected Wong's moving his research focus to computational biology, where he has made path-breaking contributions, just as he did with his earlier work on statistical theory (e.g., the theory of partial likelihood) and computation (e.g., the Data Augmentation Algorithm). Although Wong only stayed at Harvard until 2004, he recounts the importance of these 4 years in his research transition:

> I was Professor of Statistics and Professor of Biostatistics from 2000 to 2004, a period of explosive growth in genomics that was then my research focus. I got my first NIH grant in 2000 to build up an interdisciplinary bioinformatics lab, and (with Cheng Li) developed a popular method for analyzing microarrays. It was not an easy switch from pure statistical research to interdisciplinary genomics research, but the transition was eased by the rich

> Harvard environment, where we had access to world-leading collaborators in almost any biomedical topics.

Like Wong, Liu has made fundamental contributions in many areas of computational biology and bioinformatics, and is also widely known for his work in Markov chain Monte Carlo and related areas. But unlike Wong, who moved to Stanford from Harvard, Liu moved—or more precisely, returned—in the opposite direction, as he recalls:

> I first came to Harvard Statistics Department as Assistant Professor in 1991, at which time there were 6 faculty members including me. Because I got my PhD degree in Statistics from University of Chicago in only three years, I knew very little real statistics at that time. So my first two years at Harvard were really time for my re-education. I sat in almost all major courses taught by Don, Carl, and Art, and Herman. Persi Diaconis was a professor at the Harvard math department at that time, and I also took some courses from him. It was an amazing time, and I learnt a great deal in two years, ranging from causal inference to missing data analysis, hierarchical models, and group theory.

Liu's experiences of learning a great deal certainly echo my own, being a PhD student of the department myself (1986–1990), under the supervision of Rubin. For example, for my first 30-min post-qualifying presentation in the department, I xeroxed my 30-page post-qualifying paper onto 30 transparencies. I even practiced, and convinced myself that I could *read* one page per minute. But before I finished reading the first slide, Chernoff asked a question. I mumbled something and tried to press on. Chernoff stopped me again: "Xiao-Li, you are not answering my question."

In my chair's role since 2004, I have had the great fortune to work with my colleagues and experts from Harvard's Bok Center for Teaching and Learning to improve our students' communication skills. Whenever people asked me where my passion for such training came from, Chernoff's "stop sign" always appears as the initial impetus. Much of what we do as statisticians is to help answer complex questions posed by nature and by mankind, and in such a process effective communication is of critical importance, especially as statisticians are increasingly called upon to take on leadership roles in scientific inquiries.

We are therefore fortunate to have Sam Kou on our faculty, an exemplary model of future leaders of our field. Since his graduation from Stanford in 2001, Kou has made fundamental contributions to methodological, computational, mathematical, and interdisciplinary statistics through his work on Bayesian and Monte Carlo methods, nonparametric methods, stochastic inference in biophysics, and stochastic modeling in finance and economics. The speed of Kou's promotion in 2008 to full professor is a great testimony to his being a "complete package": a great researcher, an effective teacher, and a dedicated citizen. It was the fastest one: it took only 11 days (normally 3–6 months) from the day his dossier was submitted to the day I received the news that that his promotion was approved by President Drew Faust. It is also the "slowest" one: the last successful internal promotion to full professor was Dempster in 1963–1964.

Photo 2 Current faculty, taken on January 17, 2012, at 1 Oxford Street, Cambridge (Junior search meeting over catering from Yenching). *Front left* to *right*: Jun Liu (Professor), Xiao-Li Meng (Professor and Chair), Carl Morris (Professor), Stephen Blyth (Professor of Practice (part-time)), and Joseph Blitzstein (Professor of Practice, co-Director of Undergraduate Study). *Back left* to *right*: Michael Parzen (Senior Lecturer), Samuel Kou (Professor, co-Director of Graduate Study), Tirthankar DasGupta (Assistant Professor, co-Director of Graduate Student), Donald Rubin (Professor), and Edo Airoldi (Assistant Professor). *Not pictured*: David Harrington (Professor of Biostatistics, co-Director of Undergraduate Study), and Natesh Pillai (Assistant Professor)

We yet have another extremely fast senior appointment in the department to speak to the excellence of our faculty. Joseph Blitzstein, a phenomenal teacher and innovative educator, fondly recalls his Harvard experiences so far:

> Just after graduating in 2006, I came to Harvard as an Assistant Professor, and after an extremely busy, interesting, and engaging 5 years, I was promoted to Professor of the Practice in 2011. In my work here, I have followed a principle of reconciliation, aiming to interweave research and pedagogy, theory and application, probability and statistics, and Bayesian and frequentist approaches. As Co-Directors of Undergraduate Study, Dave Harrington and I have helped the concentration (major) grow from 8 to 72 in the last 6 years, promoting a vision of Statistics as a field which is intellectually exciting, aesthetically beautiful, and intensely useful.

Indeed, the intellectual excitement, esthetic beauty, and vast applicability are what make Statistics as a scientific discipline flourish, from being considered much too narrow for Harvard half a century ago, to a field featured in *New York Times* with titles such as "For Today's Graduate, Just One Word: Statistics" (August 5, 2009) and "What Are the Odds That Stats Would Be This Popular?" (January 26, 2012). As a small reflection of the dramatically increased presence of Statistics on the scientific stage and beyond, our department has been authorized to conduct a clustered hiring with the potential of recruiting five faculty members in the next few years. As a matter of fact, as I am closing this chapter today (2/2/2012), I met with our first candidate, and I will meet four more between now and the Valentine's Day. It took the last 55 years to recruit a dozen senior faculty members for the department, but I certainly believe the reader will not need to wait another

55 years to hear the stories about and from the next dozen senior faculty members in Harvard's Statistics Department.

Thank you, Mosteller.

Acknowledgments I thank all my current and ex-colleagues and fellow alumni for helping me to write about our history, and more importantly for making the history. This chapter would not exist without the tremendous assistance from our staff members (Betsey Cogswell, Steven Finch, Dale Rinkel, Maureen Stanton, and Ellen Weene) and Krupa Patel from the Office of Senior Vice Provost Judith Singer (also an advisee of Mosteller). I also thank Alan Agresti, Ingram Olkin, and Stephen Stigler for historical insights and constructive comments. All errors and omissions are mine.

Iowa State University Statistics Department

Kenneth Koehler

Early developments in statistics at Iowa State University hinged on the willingness of a young mathematics professor to collaborate with scientists on issues involving applied mathematics and statistics, and to interact with an agricultural entrepreneur with a developing interest in statistics and computational methods. Increasing demand for statistical support resulted in the establishment of the Statistical Laboratory and the Statistical Section in the Agricultural Experiment Station which fostered development of statistical and computational methodology for agricultural experimentation and surveys, genetics research, and applications to the engineering sciences. It also encouraged the development of graduate and undergraduate degree programs in statistics and the eventual establishment of a Department of Statistics. This evolved into a centralized and highly collaborative program in statistics that now involves 36 faculty, about 140 graduate students, and over 100 undergraduate statistics majors.

Early Years (1913–1933)

The seeds of statistics at Iowa State were planted in 1913 with the appointment of George Snedecor as an Assistant Professor of Mathematics. A southerner, Snedecor held a BS in Mathematics and Physics from the University of Alabama and an AM degree in Physics from the University of Michigan. Upon his arrival at Iowa State College, Snedecor made it clear that he was willing to work with other

Iowa State University,
Department of Statistics and Statistical Laboratory,
Snedecor Hall, Ames, IA 50011-1210, USA
K. Koehler (✉)
e-mail: kkoehler@iastate.edu

A. Agresti and X.-L. Meng (eds.), *Strength in Numbers: The Rising of Academic Statistics Departments in the U.S.*, DOI: 10.1007/978-1-4614-3649-2_9,

faculty on applied mathematical problems, and he soon developed an interest in applications of statistics. Snedecor taught courses in statistics through the Department of Mathematics as early as 1915. The first course, "Mathematical Theory of Statistics", was changed to "Statistical Methods of Interpreting Experimental Data" two years later. Another course, "Biometric Methods of Interpreting Agricultural Data" also appeared in the 1915 catalog.

A seminal event occurred in the spring of 1924 when Henry A. Wallace, then assistant editor of *Wallace's Farmer*, met with 20 faculty researchers on Saturday afternoons to study multiple regression and the use of computing machines. During frequent visits to Washington, D.C. in the early 1920s when his father, Henry C. Wallace, was Secretary of Agriculture, Henry A. Wallace became interested in statistical methods and labor-saving punched card machines that were used at the Federal Bureau of Agricultural Economics. He had several discussions with Snedecor in 1923 concerning applications of correlation and regression analysis in plant breeding research. This led to the series of ten Saturday afternoon seminars in the spring of 1924. Wallace and Snedecor were assisted by Charles F. Sarle, who was stationed in Des Moines at the Federal Division of Crop and Livestock Estimates and engaged in graduate courses in statistics from the Drake University Mathematics Department. Wallace and Sarle were able to borrow card sorting and tabulating equipment from an insurance company in Des Moines and haul it to Ames when needed for the Saturday seminars. A direct result of this series of Saturday meetings was the 1925 Iowa State College bulletin co-authored by Wallace and Snedecor, *Correlation and Machine Calculation* that attained worldwide circulation. At the age of 43, it was Snedecor's first publication in Statistics! Wallace, who founded the Pioneer Hi-Bred Seed Corn Company in 1926 and eventually become U.S. Secretary of Agriculture, 1933–1940, and Vice President, 1940–1944, remained proud of the boost he had been able to give statistics at Iowa State.

As a result of Wallace's seminars, demand for professional help in statistics grew rapidly and in 1927 the Mathematics Statistical Service (MSS) was formed, headed by Snedecor with the assistance of A. E. Brandt and students hired as machine operators. Brandt was most heavily involved with the computing aspects of analyses. An IBM Punched Card Tabulating Machine, the most powerful computing device on campus, was situated in his office area and he often worked late into the night providing assistance in operating that machine. Brandt was promoted to Assistant Professor in 1930 and he earned a PhD degree in Genetics in 1932. He continued to work for the MSS and published many papers on methods for performing statistical computations using punched card processing machines over the ensuing decade.

Gertrude Cox received her BS degree in Mathematics in 1929. She had taken Snedecor's courses in statistics and after graduation decided to work toward the MS degree under his direction. She worked part time in the MSS while completing her graduate studies, mainly assisting researchers with preparation of data for machine processing. In 1931, she received the first MS degree in Statistics from Iowa State College through the Department of Mathematics. After graduation she left to do graduate work in California, but Snedecor persuaded her to return to Ames as a statistical consultant. In the same year, Mary Clem, who would lead the clerical and computational staff for many years, transferred to the MSS after working as a clerk for the Economics Department and learned how to operate calculators and card punch equipment. Thus, the MSS staff in 1931 numbered four (Snedecor, Brandt, Cox, and Clem).

In 1931, Snedecor published an enlarged second edition of *Correlation and Machine Calculation*. The first edition made no reference to the work of R. A. Fisher, but the revision reveals Snedecor's recognition of the importance of Fisher's 1925 *Statistical Methods for Research Workers*. Snedecor was instrumental in bringing Fisher to Iowa State for 6 weeks in the summer of 1931. This visit was made possible by an enlightened Iowa State program that invited one distinguished scientist each summer whose work was of interest to several departments. Fisher lectured on his *Statistical Methods for Research Workers* and *The Genetical Theory of Natural Selection* and discussed experimental work with local faculty. Fisher's visit was a great success and was repeated in 1936 when he received from Iowa State his first honorary doctorate. The Department of Statistics still has the Millionaire calculating machine, with electric motor, that Fisher used during his visits.

The Statistical Laboratory (1933–1947)

To provide a more formal organization of statistics at Iowa State, the MSS was transformed into the Statistical Laboratory in 1933, with Snedecor as director. Its functions were described under five headings: Research, Statistical Counsel, Teaching, Computational Service, and Calculating Machines. Administratively, the Statistical Laboratory was directly under the office of the President. Snedecor was given control of all computing equipment on campus (approximately 150 units). His duties included situating the equipment across the campus to maximize effectiveness, collecting a usage fee to recover cost of machine maintenance, and approving requisitions for new computing equipment. Gertrude Cox joined Snedecor and Brandt as faculty members, with teaching supplemented by members of the Mathematics Department and Clem supervising equipment operators. The addition of Gerhard Tintner (Economics and Mathematics) in 1937 added strength in mathematical statistics. In 1938, C. P. Winsor joined the Statistical Laboratory faculty and Snedecor also brought in W. G. Cochran from Rothamsted as a visiting professor. Cochran agreed to return as professor in the following year (Photo 1). In addition to Fisher, other visitors to the laboratory included John Wishart, who came from Cambridge in 1934 as a summer lecturer on covariance problems, Frank Yates, who came from Rothamsted in 1937–1938 as a resident lecturer in design of replicated experiments, and Jerzy Neyman, who visited in the spring of 1938 to consult on sampling problems.

Photo 1 Statistical Laboratory Faculty 1940 (from left to right William G. Cochran, Gertrude Cox, George W. Snedecor, Charles P. Winsor)

During this time, Snedecor was working on his famous book, *Statistical Methods Applied to Experiments in Agriculture and Biology*. Early drafts of chapters appeared at the laboratory in 1935. In late summer 1936, Snedecor turned over his administrative and consulting duties to Brandt and Cox in order to stay home to work on the book. He walked to the office in any kind of weather, however, when he wanted Miss Cox to locate another set of data to use in his book. Copies of the first edition appeared in September 1937. The book was heavily influenced by Fisher's *Statistical Methods for Research Workers*, but Snedecor wrote in a more easily accessible style and provided a variety of helpful exercises. It filled an immediate need and, not counting printings abroad and translations into nine languages, sold over 200,000 copies. It went through eight editions, eventually co-authored by Cochran and re-titled simply as *Statistical Methods*. It introduced thousands of researchers around the world to statistical inference and the application of basic statistical principles in study design. For many years it was one of the most highly cited publications in the *Science Citation Index*, and it is still in print (Photo 2).

The Statistics Section of the Iowa Agricultural Experiment Station was organized in 1935, with Snedecor as Section Head. The Station provided partial salary support for most of the members of the Statistical Laboratory. This consolidated oversight of statistical consulting and computational services, centralized teaching of courses in statistics, and fostered collaboration in the development of statistical methodologies across varied disciplines. This model with a strong centralized program in statistics and vibrant collaboration across varied scientific disciplines is actively maintained today.

Photo 2 George W. Snedecor (left) and Arnold J. King (right) discuss survey issues with P. C. Mahalanobis (center) during his visit in 1946

A major development in survey methodology and the growth of the Statistical Laboratory took place in 1938, with the establishment of a cooperative agreement between Iowa State College and the U.S. Department of Agriculture for joint research in agricultural statistics and related statistical theory. This agreement fostered research in two major areas: (1) survey methodology for area sampling that led to the development of the Master Sample of Agriculture, and (2) methods of forecasting and estimating crop yields. As a result, Laboratory staff and facilities were significantly increased, including the addition of seven USDA collaborators that included Arnold J. King and Raymond J. Jessen.

The use of random sampling was not yet a well-accepted practice at this time (Fuller, 1984). To estimate crop and livestock production, the USDA largely relied on questionnaires delivered to farmers by rural postal carriers. Under the 1938 Agreement, the USDA and the Statistical Laboratory cooperated in research on efficient methods of sampling farms to assess agricultural production and practises, in addition to developing experimental designs for field plot experiments, developing statistical methodology for the analysis of time series, and incorporating meteorological information in crop forecasting. Early work included Cochran's development of regression estimation and Jessen's development of area sampling procedures. At this time the USDA Bureau of Agricultural Economics, having a need for a data collection procedure that could provide information for simultaneous studies, decided that the Statistical Laboratory should design a national area sample of about 5,000 farms. A. J. King played a key role in conceiving and promoting this initiative that would become known as the Master Sample of Agriculture. The label "Master Sample" came to represent the materials and methods used in the creation of the original sampling frame rather than the sample itself. The Master Sample was distinguished from earlier large-scale area samples by the use of relatively small, efficient, sampling units derived from statistical principles that provided for unbiased estimates of population totals. By 1944 the Bureau of the Census became involved with the Master Sample as a means of collecting information for the 1945 Agricultural Census, and it was eventually expanded to a much larger sample of about 300,000 of the approximately six million farms in the United States at that time. The Bureau of the Census provided significant contributions of money, manpower, and technical assistance. The project was largely administered by A. J. King, with Jessen in charge of the statistical aspects. The 1944–1945 Annual Report of the Statistical Laboratory indicates that "150–200 clerks and supervisors" were working on the Master Sample. This operation was housed in the Armory on the college campus and in rented space in neighboring towns. The 1945 Census of Agriculture was the only use of the complete set of 67,000 area units of the original Master Sample, but the frame and materials were used into the 1970s to draw numerous agricultural samples. The Statistical Laboratory cooperated with the Bureau of the Census in expanding the Master Sample into a labor force survey that became the Current Population Survey. In 1957 the Laboratory developed another national area sample that evolved into current practises for assessing land use in the United States.

Brandt left Iowa State in 1938 to accept a newly created position of Senior Mathematical Statistical Analyst in the U.S. Soil Conservation Service. By this time the Statistical Laboratory was well established and Mary Clem was supervising a large group of machine operators engaged in scientific data processing. Gertrude Cox was finally promoted to Research Assistant Professor in 1939. In the following year she moved, as a full Professor, to North Carolina State College to help establish a statistics program, and she was appointed as the Head of the new Department of Experimental Statistics in January, 1941. Her numerous successes as a statistician, administrator, and promoter of statistics are well known.

By 1938, course offerings in statistics were quite extensive, and W. G. Cochran provided the initiative to establish a PhD program in Statistics within the Department of Mathematics. Six PhDs were awarded prior to the establishment of the Department of Statistics in 1947: Holly C. Fryer, 1940; Richard L. Anderson, 1941; Jerome Ching Li, 1943; T. A. Bancroft, 1943; Samuel L. Crump, 1947; David B. Duncan, 1947. Bernice Brown was the second student to earn an MS degree in Statistics (1932), but most information on MS degrees awarded in Statistics before the establishment of the Department of Statistics has been lost.

The Department of Statistics: The First 25 Years (1947–1972)

George Snedecor who had set the tone for statistics at Iowa State, had shown initiative in inviting Fisher to consult and give summer courses, and had made a national impact on the statistical scene with his book. A detailed chronology of Snedecor's contributions is given in a Festschrift edited by Bancroft: *Statistical Papers in Honor of George W. Snedecor* (1972). An article by Cox and Homeyer (1975) provides more personal views of Snedecor from his students and colleagues. Wishing to avoid additional administrative duties, Snedecor showed no interest in establishing a Department of Statistics and was quite content to let the Department of Mathematics administer the statistics teaching program. After his mandatory retirement from administrative duties in 1947, a Department of Statistics was formed. Jessen was persuaded to become Acting Head of the Department and Acting Director of the Statistical Laboratory, and King administered collaborative work with the USDA and other agencies. During World War II, Cochran worked with the Statistical Research Group at Princeton (headed by Samuel S. Wilks) and he convinced Alexander Mood, who took a doctorate at Princeton, to join the faculty at Iowa State. Mood took charge of teaching and theory. Cochran moved to Raleigh in 1946 to head the graduate program in Experimental Statistics, but the arrival of Oscar Kempthorne 1947 continued the Rothamsted connection. At this time there was a faculty of ten including: G. W. Brown, S. L. Crump, P. G. Homeyer, L. Hurwicz, R. J. Jessen, O. Kempthorne, A. J. King, and A. M. Mood. A full set of courses was offered for BS, MS, and PhD degrees, and an extensive set of service courses was offered for students majoring in other fields in which statistical methods were important research tools.

In 1949, T. A. Bancroft, who was serving as the Director of the Statistical Laboratory of Alabama Polytechnic Institute, was persuaded to return to Iowa State as an associate professor. A year later he became the first permanent Head of the Department of Statistics and Director of the Statistical Laboratory, a position he held for the next 22 years. Bancroft was a dynamic leader and visionary who led a fruitful period of expansion. During his 22-year tenure as Department Head, the faculty grew from 13 to 32. To foster interdisciplinary collaboration, shared faculty appointments were established with the Departments of Mathematics, Psychology, Sociology, Political Science, Industrial Engineering, and the newly formed Computation Center. In addition, about a dozen faculty received partial salary support to provide statistical expertise to researchers associated with the Agricultural Experiment Station, and other faculty received partial salary support through the Statistical Laboratory to provide statistical support for researchers in the social and behavioral sciences, mathematics, computer science, economics, engineering and the physical sciences. These faculty members also contributed to teaching statistics to undergraduate and graduate students in their shared disci-

Photo 3 1949 punch card processing equipment in the Statistical Laboratory Computing Service (C. C. Mosier and Mary Clem, standing)

plines. This centralized the university's Statistics instruction in the Department of Statistics and promoted an interdisciplinary culture that continues to this day.

During this time the number of graduate students grew from 39 to 86 and course offerings were expanded. Outstanding early PhD recipients included W. T. Federer (1948), F. A. Graybill (1952), D. Thompson (1952), D. V. Huntsberger (1954), Helen Bosivich (1955), M. B. Wilk (1955), J. N. K. Rao (1961), R. R. Hocking (1962), W. J. Hemmerle (1963), K. H. Hinkelmann (1963), and W. J. Kennedy (1964). After graduation, many gained experience as research associates in the Statistical Laboratory before moving on to distinguished careers.

From its inception, the Statistical Laboratory made computing equipment available to the entire campus (Photo 3). A difficult problem that was to have a major impact on the development of digital computing arose in 1936 when John Atanasoff, a professor from the Physics Department, asked Brandt for help in computing approximations to the solution to complex systems of partial differential equations. The two men soon determined that the IBM tabulating machine in the Statistical Laboratory was not up to the task, but they had ideas about how they might modify the machine to better serve their needs. That endeavor ended when the IBM office manager in Des Moines would not allow the modification. Atanasoff and Brandt wrote a paper on their work, but Atanasoff continued to seek a solution. Over the next 4 years he designed a new computational device and with Clifford Berry, his graduate student, he built a working model in 1941 that is

Photo 4 Paul Homeyer assisting students in the computational lab for a 1950 Statistics class. The photograph shows old and new Monroe calculators

recognized as the first electronic digital computer. The beginning of World War II ended this effort when Atanasoff and Berry departed for defense-related jobs.

The Statistical Laboratory continued to provide campus-wide computation service for analysis of research data using simple IBM business machines (key punches, verifiers, sorters, and desk calculators) until 1957 when an IBM 650 was installed in the Statistical Laboratory (Photos 4, 5). This was the first such computer facility in Iowa or the surrounding region. Mary Clem was responsible for scheduling time on the computer and H. O. Hartley, who Bancroft recruited from University College, London, in 1953, provided guidance on its use for scientific computations. Users were charged for computer time to recover the cost of machine rental. At the same time, work began on the construction of a digital computer, the Iowa State Cyclone, which was completed in 1959 and greatly expanded in 1961. In 1962 the decision was made to acquire a larger IBM 7074-1401 and form a Computation Center, not connected with the Statistical Laboratory, to manage university computing support. The Department of Computer Science was also established at that time. Several members of the Statistical Laboratory computing staff, including Mary Clem, moved to the Computation Center. The remaining computing staff in the Statistical Laboratory was organized into the Numerical-Analysis Programming Group (later to be called the Statistical Computing Section). This group provided data processing services, programming, and consulting on the use of statistically orientated computational tools to campus researchers. Hartley was in charge until he left for Texas A&M in 1963, passing on leadership to his former student, William Hemmerle. A year later, William Kennedy took charge and very effectively led the section for the next 26 years.

Photo 5 Form left to right George Zyskind, Oscar Kempthorne, Dewey Harris, around 1964

This was also an active period in the publication of texts. Cochran and Cox (1950) published *Experimental Designs*, largely based on their collaboration when they were both at Iowa State. Mood (1950) wrote his *Introduction to Mathematical Statistics* at Iowa State before leaving for industry. Anderson and Bancroft (1952) authored *Statistical Theory in Research*. Kempthorne published *The Design and Analysis of Experiments* in 1952 and *An Introduction to Genetic Statistics* in 1957. Tintner (1952) wrote *Econometrics*. At the undergraduate level, David Huntsberger (1961) published *Elements of Statistical Inference*. Hartley co-authored with E. S. Pearson *Biometrika Tables for Statisticians* (1954) and with J. A. Greenwood *Guide to Tables in Mathematical Statistics* (1962). P. V. Sukhatme produced *Sampling Theory of Surveys, with Applications* (1954) from notes he developed for the Indian Council of Agricultural Research in the late 1940s and summer sessions conducted at the Indian Society of Agricultural Statistics in 1950 and 1951 and at Iowa State College in spring 1952. The second edition was co-authored with B. V. Sukhatme in 1970. Other books include *Topics in Intermediate Statistical Methods, Vol. I* by T. A. Bancroft in 1968 and *Probability, Statistics, and Data Analysis* by O. Kempthorne and J. L. Folks in 1971.

The Department of Statistics was instrumental in establishing Mu Sigma Rho as a national Statistics honorary fraternity stressing both theoretical and applied statistics. The Iowa Alpha Chapter was organized in 1968 at Iowa State University (Photo 6).

Photo 6 In a 1981 meeting, founding fathers of Mu Sigma Rho recall the beginnings in 1968. L to R: Oscar Kempthorne, T. A. Bancroft and David Huntsberger, seated. Standing: Herbert T. David, Wayne Fuller, and Don Hotchkiss (Holly Fuchs, not present)

The Department of Statistics and the Statistical Laboratory: The Next 50 Years

Under Bancroft's leadership, the range of department activities had expanded greatly from the early emphasis on agricultural experimentation and surveys. By the time T. A. Bancroft retired from administrative duties in 1972, he had developed a thriving statistics program with 32 permanent faculty, 2 collaborators in residence with the Mayo Clinic, 6 staff in the data processing service, and 33 research associates, statistical clerks, and support staff in the Statistical Laboratory. There were 86 graduate students with 25 supported by foreign governments and institutions or USAID agreements. Among the countries represented were Brazil, China, India, Iraq, Iran, Mexico, Nigeria, Peru, Thailand, Uganda, United Arab Republic, and Venezuela. Bancroft was extremely proud of the 5-fold increase in this type of support for international students during his tenure as Head. There were also 17 undergraduate majors in statistics and biometry programs.

In the subsequent 50 years the Department would be led by just three people. Hebert A. David came from the University of North Carolina in 1972 to assume the responsibilities of Department Head and Director of the Statistical Laboratory. Dean L. Isaacson served in the same capacity from 1984 through 2002. In 2002, the university abolished the office of Department Head, with its unlimited tenure, in favor of the office of Department Chair with fixed, renewable terms of no more than 5 years. Consequently, Kenneth J. Koehler assumed the roles of Department

Chair and Director of the Statistical Laboratory. As the number and types of statistics courses evolved and expanded over these 50 years, the Statistics Department continued as a central source of statistics education for the entire university. New undergraduate statistics courses were created to accommodate growing interests of engineering students and to accommodate the development of new programs in business. Other courses were added or modified to accommodate changing interests in the agricultural, biological, social, and behavioral sciences. Advanced undergraduate courses also served the needs of graduate students seeking to acquire quantitative research skills in a wide variety of disciplines. The graduate program continued to provide a good balance between theory and applications.

Developments in research and curriculum were aided by improvements in computational systems and statistical software. By 1972, responsibility for university computing resources had been shifted from the Statistical Laboratory to the University Computation Center and a separate Department of Computer Science had been created, but the Department of Statistics and the Statistical Laboratory retained university leadership in the acquisition and use of statistical software. Department faculty also led developments in statistical computing. William J. Kennedy and James E. Gentle co-authored the influential book *Statistical Computing* in 1980. Vincent Sposito fostered collaborations among Statistics, Engineering and Computer Science and published *Linear and Nonlinear Programming* in 1975 and *Linear Programming with Statistical Applications* in 1989. Increased access to digital computers opened up opportunities for their use in introductory statistics courses. Richard Groeneveld was among the early proponents, writing two books: *An Introduction to Probability and Statistics Using BASIC* in 1979 and *Introductory Statistical Methods: An Integrated Approach Using Minitab* in 1988. More recently, Dianne Cook and Heike Hofmann have developed an active research programs dynamic graphics and visualization of high-dimensional data, contributing software packages that have found significant applications in both research and teaching, and Ranjan Maitra has made significant contributions to clustering methods for large multi-dimensional data sets.

Continuing and expanding joint faculty appointments with other departments enabled the Department of Statistics to develop and maintain the expertise necessary to teach statistics to undergraduate and graduate students from a wide variety of disciplines and to provide statistical support for research in those disciplines. Statistics instruction and research collaborations in engineering and the physical sciences were strengthened by creating joint faculty appointments with departments in the College of Engineering and hiring other faculty with interest engineering applications of statistical methodology. Herbert T. David was given the first joint appointment between Statistics and Engineering in 1956 and held that position through his retirement in 1998. Joining the faculty in 1975, William Q. Meeker developed leading research programs in reliability, accelerated life testing, and nondestructive evaluation. His book *Statistical Methods for Reliability Data,* co-authored in 1998 with Luis Escobar, continues to have a substantial impact on the analysis of time to failure data. Stephen Vardeman, who began a joint appointment between Statistics and Industrial Engineering in 1981, has been a major contributor to engineering education, developing new courses on quality control and engineering statistics, authoring or co-authoring four textbooks, and contributing to research on decision theory and directional distributions. Max Morris, who began a joint appointment between Statistics and Industrial Engineering in 1998, collaborates with researchers in the forensic and material sciences and leads research in design of large scale factor screening experiments, spatial screening design and computational experiments. He published the book *Design of Experiments: An Introduction Based on Linear Models* in 2010. Other joint appointments were established in the early 1990s with the Department of Chemical and Biological Engineering and the Department of Aerospace Engineering.

The joint appointment of Krishna Athreya to the Statistics and Mathematics Departments in 1980 continued a long history of collaboration between those two departments that reaches back to Gerhard Tintner (1937–1962) and included Barry Arnold (1965–1980), Dean Isaacson (1968–2007), and James Cornette (1970–1979). Joint appointments with the Department of Economics also reach back to Tintner. Wayne Fuller held a joint appointment with Statistics and Economics from 1959 until his retirement in 2001. He made seminal contributions to three areas of statistics that are signified by his books: *Introduction to Statistical Time Series* in 1976, *Measurement Error Models* in 1987, and *Sampling Statistics* in 2009. Additional shared appointments were created with departments in the social and behavioral sciences, economics, genetics, and the College of Veterinary Medicine.

The Agriculture Experiment Station and the College of Agriculture and Life Sciences continued to provide partial salary support for about a dozen Statistics faculty who engaged in collaborative research in agriculture, genetics, meteorology, rural sociology, economics, and the biological, environmental, nutritional, and food sciences. Research programs in population genetics spearheaded by Oscar Kempthorne (1947–1989) and Edward Pollak (1964–2000) have evolved into a strong program in statistical genomics for both the plant and animal sciences led by Dan Nettleton, Volker Brendel, Karin Dorman, Peng Liu, and Chong Wang. Research developments in such diverse disciplines as precision agriculture, ecology, meteorology, geology, renewable energy, epidemiology, and medical imaging have been greatly aided by the development of statistical methodologies for spatial/temporal data. A program in spatial statistics was developed by Noel Cressie, who joined the faculty in 1983 and published his comprehensive book, *Statistics for Spatial Data* in 1991. Cressie moved to the Ohio State University in 1998, but younger faculty, Philip Dixon, Mark Kaiser, Petrutza Caragea, Dan Nordman, Zhengyuan Zhu, continue to develop statistical methodology for spatial temporal data and engage in collaborative research in the biological and environmental sciences, renewable energy, and precision agriculture. Faculty supported by the Experiment Station have also made significant contributions to creating statistical methodologies to improve research in human and animal nutrition and the food sciences. An important example is the survey methodology for national assessment of nutrient intake that has been deployed by the U.S. and many other countries. Corresponding methodologies for national assessment of physical activity are currently being developed to enable assessment of both nutrient intake and energy expenditure in monitoring trends national health status.

The Statistical Laboratory, currently administered by the Office of the Vice President for Research and Economic Development, continues to provide statistical support for university research. It has undergone significant reorganization in recent years and currently consists of the Center for Survey Statistics Methodology (CSSM), the Survey and Behavioral Research Services (SBRS), the Institute for Social and Behavioral Research (ISBR), a computational statistics and graphics group, and a group of faculty who collaborate with researchers in the engineering and physical sciences. The four faculty associated with CSSM maintain strong research programs in statistical foundations of survey research. In the rich tradition of the Master Sample, CSSM researchers continue to develop new survey methodologies for federal agencies and support the USDA in conducting surveys of natural resources in the U.S. and its territories. SBRS is a fee-for-service center that provides data collection services for both long- and short-term studies and expert advice on survey design and the analysis of survey data. ISBR promoted and supports sponsored research in the social and behavioral sciences.

Increases in undergraduate enrollments during the 1990s created an increased reliance on graduate students in teaching introductory courses in statistics that has persisted. Dean Isaacson engaged Richard Groeneveld and Bob Stephenson in the development of a mentoring system in which faculty coordinators of multi-section undergraduate service courses help graduate student instructors develop good teaching practises. Coordinators meet weekly with graduate student instructors to review lesson plans, discuss classroom performance, and develop instructional materials. This system also provides a means of developing faculty teaching skills. This initiative continues to evolve and led to many innovations in the undergraduate statistics education. There are currently four faculty members with primary interest in statistics education research and a joint PhD option in Statistics and Education that is available at Iowa State.

Online education in Statistics started in the early 1990s when the Department of Statistics entered into an agreement with the General Motors Corporation to provide statistical training for engineers. In the beginning, video tapes of course lectures were mailed to corporate continuing education centers, sometimes with a three- or four-day time lag. Students in the distance program were required to fulfill all requirements of the on-campus MS degree in Statistics, including a research paper that generally addressed a problem of interest to the company and was usually co-advised by a faculty member and a company engineer. NSF funding was obtained to facilitate visits of faculty to GM research centers that led to productive research collaborations and enabled faculty to incorporate important applications of statistical concepts into their courses. In 2002, the American Statistical Association presented the first award for Statistical Partnerships among academe, industry, and government to GM and the ISU Statistics Department for this collaboration. Eventually, this program expanded to other companies and research organizations and it evolved into the current online MS program in Statistics.

Photo 7 Dedication of the Renovated Snedecor Hall (June 3, 2009). L-R: Kenneth Koehler (Chair 2002–2012), Dean Isaacson (Head, 1984–2002), Herbert A. David (Head, 1972–1984), Gregory Geoffroy (ISU President)

In 2006, resources were found to completely renovate Snedecor Hall, the home of the department and the Statistical Laboratory. After relocating to an unused dormitory for 18 months, faculty, staff and students were able to move back into a completely renovated Snedecor Hall in May, 2009, 1 year late for the seventy-fifth Anniversary of the Statistical Laboratory. The renovation greatly improved the working environment, provided modern computer labs, and expanded office space for graduate student (Photo 7).

Summary

The Department of Statistics remains committed to providing students with a strong theoretical foundation in statistics and probability while introducing them to a wide range of applications and providing collaborative research experiences. Since degrees in statistics began to be offered in the 1930s, almost 500 PhD degrees and 1,200 MS degrees and over 500 BS degrees in statistics and biometry have also been awarded. Table 1 provides information on degrees awarded by decade. The size of the graduate program has been relatively stable over the last three decades with about 140 graduate students. Growth in the number of MS degrees awarded in Statistics over the last two decades is largely a result of the online MS program and graduate students in other disciplines at Iowa State adding a concurrent MS degree in Statistics. Interest in undergraduate degrees waned in the 1990s, but there has been a steady growth in the number of undergraduate statistics majors from 25 to over 100 in the past decade and the number of undergraduate majors is currently at an all time high.

Commenting on the early growth of the Statistical Laboratory at its twenty-fifth Anniversary in 1958, Charles Sarle wrote "The remarkable development of statistics at Iowa State, both in research and in teaching, came about in response to the basic statistical needs of U.S. agriculture research in both experimental and sampling design." The Department of Statistics has prospered because of its continuing commitment to research in statistical theory as well as developing new statistical methodologies needed to advance research in the agriculture and

Table 1 Statistics Graduates (BS, MS, and PhD)

Decade	Bachelor degrees	Masters degrees	PhD degrees
1931–1940	0	2	1
1941–1950	1	2	11
1951–1960	57	54	28
1961–1970	77	102	70
1971–1980	108	193	85
1981–1990	119	223	106
1991–2000	68	272	100
2001–2010	87	346	98
Total	517	1194	499

biological sciences, engineering, physical sciences, environmental sciences, behavioral and social sciences and economics. The commitment to collaborative research made by Snedecor that resulted in the founding of the Statistical Laboratory and establishment of a Statistics Section in the Agricultural Experiment Station was strongly reinforced and formalized in the first two decades of Department leadership provided by T. A. Bancroft and it was carefully nurtured by subsequent Department Heads. These efforts have had an important and sustained practical impact on collection and dissemination of information by government agencies, enhancing genetic research and expanding and improving food production, enabling environmental and climate research, development of procedures for assessing quality and reliability and improving industrial processes, and leading initiatives to improve and expand statistical education. These efforts have also firmly integrated the Statistics faculty into the research and instructional fabric of the entire university.

References

Bancroft TA (ed) (assisted by Susan Alice Brown) (1972) Statistical papers in Honor of George W. Snedecor. The Iowa State University Press, Ames, 328 pp

Bancroft TA (1974) George W. Snedecor, pioneer statistician, 1881–1974. Am Stat 28:108–109

Bancroft TA (1982) Roots of the Iowa State University Statistical Center: 1914–1950. Iowa State J Res 57:3–10

Cochran WG (1974) Obituary, Professor G. W. Snedecor. J R Stat Soc 137:456–457

Cox GM, Homeyer PG (1975) Professional and personal glimpses of George W. Snedecor. Biometrics 31:265–301

David HA (1998) Statistics in U. S. universities in 1933 and the establishment of the Statistical Laboratory at Iowa State. Stat Sci 13:66–74

David HA, David HT (eds) (1984) Statistics: an appraisal—proceedings of a conference marking the 50th anniversary of the Statistical Laboratory, Iowa State University. Iowa State University Press, Ames, 664 pp

Folks JL (1995) A conversation with Oscar Kempthorne. Stat Sci 10:321–336

Fuller WA (1984) The master sample of agriculture. In: David HA, David HT (eds) Statistics: an appraisal—proceedings of a conference marking the 50th anniversary of the Statistical Laboratory, Iowa State University. Iowa State University Press, Ames, pp 583–598

Hinkelmann K (ed) (1984) Experimental design, statistical models, and genetic statistics: essays in honor of Oscar Kempthorne. Marcel Dekker, New York

Iowa State University (1940) The Statistical Laboratory. Iowa State College Bulletin 50, May 15, 25 years 1915–1939

Kempthorne O (1974) George W. Snedecor, 1881–1974. Int Stat Rev 42:319–321

Nagaraja HN, Sen PK, Morrison DF (eds) (1996) Statistical theory and applications: papers in honor of Herbert A. David. Springer, New York, 334 pp

Johns Hopkins University Department of Biostatistics

Charles Rohde, Scott L. Zeger, Karen Kruse Thomas and Karen Bandeen-Roche

Biostatistics at Johns Hopkins: Founding Years

Overview and personalities. The Johns Hopkins Biostatistics department was established as the Department of Biometry and Vital Statistics in the Johns Hopkins School of Hygiene and Public Health in 1918 when the School was founded as the world's first of its kind. According to a comprehensive early history of the School (Fee 1987a) the Department started in the university's Old Physics Building. That building was destroyed by a fire in 1919, and subsequently space was borrowed from several places on the School's East Baltimore campus near the world-famous Johns Hopkins Hospital. The department was renamed the Department of Biostatistics in 1930 and moved to the third floor of the School of Hygiene building.

The department's first chair (1919–1925) was Raymond Pearl who had a strong interest in the mathematical modeling of natural populations. Pearl was a noted bon vivant and close friend of the legendary satirist H. L. Mencken.

In 1925, Lowell Reed succeeded Pearl as chair of Biostatistics, a position he held until 1948. Reed went on to serve as Dean of the School from 1937 to 1946. During that

Johns Hopkins University,
Department of Biostatistics,
615 North Wolfe Street,
Baltimore, MD 21205-2179, USA
C. Rohde · S. L. Zeger · K. K. Thomas · K. Bandeen-Roche (✉)
e-mail: kbandeen@jhsph.edu

S. L. Zeger
e-mail: szeger@jhsph.edu

K. K. Thomas
e-mail: kthoma48@jhmi.edu

C. Rohde
e-mail: crohde@jhsph.edu

A. Agresti and X.-L. Meng (eds.), *Strength in Numbers: The Rising of Academic Statistics Departments in the U.S.*, DOI: 10.1007/978-1-4614-3649-2_10,

period he and U.S. Surgeon General Thomas Parran guided American public health through a decade of depression and war. Reed served as president of the American Statistical Association, the Population Association of America, and the American Association of Schools of Public Health, which he founded in 1941. As chair of the U.S. Committee on Joint Causes of Death and the International Committee on Vital and Health Statistics of the World Health Organization, Reed oversaw the modernization of standard classifications for birth, death, and disease statistics, for which he received the Lasker Award. In the 1950s, Reed's prominence in public health was recognized with appointments to President Harry Truman's Commission on the Health Needs of the Nation and the National Health Advisory Council (Fee 1987b).

Notable faculty of the era included Charles Winsor, Alfred Lotka, and Margaret Merrell, who earned an Sc.D. under Reed in 1930. The term "winsorising" refers to methodology Winsor developed for reducing susceptibility of estimators to outlying values. Lotka wrote the first comprehensive text on mathematical biology. Notable students included Joseph Berkson, who founded the statistics group at the Mayo Clinic, Paul Densen, who chaired the Department of Community Medicine at Harvard, and Morton Kramer, who pioneered the use of statistics in Mental Health.

Research. Reed's early work with Pearl developed his lifelong interest in population growth. In 1920, Pearl and Reed published "On the Rate of Growth of the Population of the U.S. and its Mathematical Representation" in the Proceedings of the National Academy of Science (*PNAS*). The paper promulgated the logistic curve as a "law of nature," and so aroused considerable controversy. One author writes that "Pearl's invention survived partly through active promotion and partly through the fortuitous connection with people who were able to develop it properly..." (Kingsland 1982).

During the prewar years, Biostatistics began to incorporate ideas developed in statistics, notably statistical modeling and methods. Winsor spent time at Princeton during the war, and his work had great impact on John Tukey's exploratory data analysis (http://www.stat.berkeley.edu/~brill/Papers/jwtencyc.pdf). Departmental research initiatives benefited from funds the war pumped into the nation's health research efforts, including the National Research Council's cooperative clinical trials of penicillin in early syphilis (Marks 1997). Reed and Merrell worked closely with these efforts and also provided crucial advice on chemotherapy studies at Hopkins (Fee 1987c).

Merrell, as the principal statistician for the penicillin trials and statistical consultant to the Public Health Service (PHS) and the Surgeon General of the Army, emerged as a premier expert on statistical analysis for randomized controlled trials of therapeutic compounds. She and Reed helped to develop standard protocols that would ensure comparability among the results of cooperative studies of large numbers of patients. Despite her many collaborations with clinical medical researchers, Merrell remained true to her public health training. In an era when clinical trials were still often conducted on few patients without controls, she emphasized that "the final evaluation [of a drug] cannot be on patients but must be on the population... . A self-selected group of patients presenting themselves for observation, diagnosis, or treatment will not be representative of the general population. It may in fact give us almost no normal persons" (Merrell 1951).

Education: The Department of Biostatistics has been a leader in public health education since its inception. The high quality of teaching in Biostatistics, together with Epidemiology, was partly a result of their faculty's extensive consulting with agencies, such as the Baltimore City Health Department, PHS, and the Office of Naval Research. A Department of Epidemiology Chair of the era, Kenneth Maxcy, remarked that "The stimulation emanating from this [consulting] activity cannot be overrated. It keeps the [departments of epidemiology and biostatistics] in active touch with new studies which are going on in the field and it serves as a rich source of current material for teaching purposes" (Maxcy 1939). Students were required to analyze real-life raw data and case studies, often drawn from Baltimore. School of Hygiene alumni cited the courses as the best and most relevant in the curriculum (1936–1937 SHPH Catalog).

Lydia Edwards, a postdoctoral fellow in preventive medicine, remembered Reed as "one of the most brilliant teachers I ever had," who taught his students that "Just as X-rays had permitted us to see what was going on inside the human body, so statistics enabled me to see what was occurring within a population" (Edwards 2001). Edwards was one of the many students to learn the Reed-Frost epidemic model, which simulated how disease spread through a population. Reed and Frost represented this principle in the classroom by using a box containing three colors of marbles, corresponding to individuals with active cases of disease, those who were susceptible, and those immune. A fourth color represented barriers that created subgroups of marbles. By shaking the marbles and pouring them out, the marbles were randomly sorted. Between each set of barriers, susceptibles touching active cases were replaced with active cases, and active cases were replaced with immunes. The process was repeated to demonstrate how percentages of actively contagious, susceptible, and immune individuals changed through each stage of the epidemic. Reed even explained the epidemic model to television audiences in a 1951 episode of *The Johns Hopkins Science Review* (Epidemic Theory 2010).

In the earliest years Biostatistics students learned public health aspects of statistical analysis including sampling techniques and life tables (1936–1937 Johns Hopkins School of Hygiene and Public Health (SHPH) catalog). Biostatistics and epidemiology were also required for the M.D. at Johns Hopkins, and by 1938, the Department of Biostatistics taught as many students from the medical school, hospital, and science departments as the School of Hygiene (Reed, "Department of Biostatistics"; 1944–1945 SHPH catalog). Figure 1 displays a photo from this era.

Postwar Years

Overview and personalities. William G. Cochran became chair in 1948, a post he would hold for 10 years. In 1954, Cochran was the lead author of the American Statistical Association's book-length critique, *Statistical Problems of the Kinsey Report on Sexual Behavior in the Human Male*, which exposed numerous mathematical inaccuracies and sampling errors (Cochran et al. 1954).

Fig. 1 Johns Hopkins Department of Biostatistics, 1936–1937. Left to right, standing: Unknown, Morton Kramer, Paul Densen, unknown, Ed Crosby, Margaret Palmer. Left to right, seated: Velma Scruton, Lowell Reed, Margaret Merrell

In the late 1950s, the department was decimated after Cochran left to become a professor at the Harvard Department of Statistics. Cochran's successor, Jerome Cornfield, stayed only 2 years. The departure of two other senior faculty left the department with "a skeleton staff …". Margaret Merrell had planned to retire but stayed on as interim chair. She wrote, "It may be hard for anyone outside the field to realize the extreme shortage of persons with the qualifications we require and we are probably handicapped in this competitive area by the loss of prestige accompanying the simultaneous resignations of three senior members of the staff." Merrell proposed that Biostatistics should be reorganized as a joint department (Margaret Merrell to Ernest L Stebbins, Nov 6, 1957).

Amid the turmoil Merrell and Helen Abbey worked hard to maintain the high quality of Biostatistics teaching at Hopkins (1956–1957 Report of the Dean of the School of Hygiene and Public Health). Abbey received her D.Sc. degree and joined the Biostatistics faculty in 1950 as one of four assistants hired under a contract with the Office of Naval Research to develop statistical methods for biology and medicine (Cochran to Ernest Stebbins Mar 1, 1950).

In 1960, Allyn Winthrop Kimball accepted a joint appointment as Professor and Chair of Biostatistics in the schools of Hygiene and Medicine. Another full professor, David B. Duncan, also joined the department (JHU Trustees minutes, JHU Trustees Executive Committee minutes). Kimball was named chair of a third statistics department, established on the university's uptown "Homewood" campus in 1961. Geoff Watson joined the Homewood faculty in that same year, followed rapidly by Richard Jones, Joseph Gastwirth, and Leon Gleser. Kimball had been chief of the biological statistics section at Oak Ridge National Laboratory and published widely on statistics in genetics and in radiation biology. Other faculty of the era included Paul Meier, Charles Winsor, Irwin Bross, John Gart, Matt Tayback, and Rowland Rider. Students who went on to particularly notably further the field and department included Anita Kaplan Bahn, Glenn Bartsch, Ted Colton, Margaret Hoff, and Calvin Zippin.

Research. Cochran was an authority on sampling and the analysis of variance, just as multivariate analysis was emerging as a critical tool of risk-factor epidemiology (Rothstein 2003). Reed-Frost and other epidemiological models had been deterministic, but in the 1950s, Cochran was among the biostatisticians who developed stochastic process models, which introduced elements of uncertainty to explain more realistically the way disease spreads in a population (Cochran 1960).

Among publications of the era, two were particularly influential. In 1959, Cornfield and colleagues published one of the seminal treatises evidencing the link between smoking and lung cancer (Cornfield et al. 1959). A year earlier, Paul Meier copublished on the famous survival function estimator that bears his name together with Edward L. Kaplan (1958). In fact Meier and Kaplan submitted their work independently. An astute editor noticed that their findings overlapped and urged that they publish jointly, which they did.

Education. Biostatistics and epidemiology continued as the backbone of the Hygiene curriculum. During the 1950s the course content changed significantly along with public health practice, incorporating more advanced scientific, and technical concepts. The new standard for teaching was Statistical Methods in Epidemiology, an innovative joint course developed by Reed and Merrell. The course introduced statistical methods for investigations of epidemic and chronic diseases, particularly tuberculosis. Merrell assumed sole responsibility for the course after the war, and her students responded enthusiastically to her creative use of real-life public health data. One problem required students to measure survivorship and cumulative mortality, using data from a controlled longitudinal study of the effectiveness of BCG vaccine in American Indian children. Another problem involved time and age changes in male death rates from lung cancer, which had increased 5-fold from 1931 to 1951. Merrell corresponded with public health researchers and officials to solicit their case records for use in teaching. She also drew examples from articles in journals. Cochran requested a salary increase for Merrell, whose value to the department he found "difficult to overestimate... . She is the outstanding teacher in the department and carries the great bulk of the teaching load in addition to handling much of our most important consulting work and playing a prominent part in the direction of research by students" (William G Cochran to Ernest Stebbins Mar 1, 1950).

As the field of biostatistics matured, School of Hygiene courses added specialized methods for lab research and mathematical models for biological assays such as those used for measuring the strength of vaccines. Increasing numbers of students from the School of Engineering came to Hygiene to take biostatistics courses that complemented fields such as operations research (Charles Flagle, Sept 29, 2010 interview). Cochran introduced biostatistics courses on advanced sampling techniques and experimental design, and his textbooks on these subjects (Cochran and Cox 1950; Cochran 1953) went into multiple editions over the next three decades.

1960–1970s

Overview and personalities. In 1965, Allyn Kimball resigned as Chair of Biostatistics to become Dean of Arts and Sciences. Kimball's successor, Alan Ross, served as chief biostatistical consultant to the Federal Aviation Administration Office of Aviation Medicine as well as advisor to the WHO International Collaborative Study of Medical Care Utilization. NIH training grants for Biometry, initiated in the early 1960s, supported faculty and students throughout the decade.

Ross also was famed for introducing Kriegspiel, a blinded chess game in which players are separated by a partition and a "referee" adjudicates allowable moves. For nearly four decades lunchtimes in the department were punctuated by the outbursts of the players and the spectators in the fray. Annual holiday student skits were initiated in which dozens of faculty have been lampooned over the years, prominently including the authors of this piece. Early in Ross' chairmanship the department moved to a new wing of the Hygiene building.

The 1970s were difficult years for Biostatistics. NIH training grants for Biometry were discontinued in 1976, and the department relied on collaborative research for funds. The recruitment of new faculty was extremely limited. Nevertheless, the School's first IBM mainframe was purchased.

Biostatistics faculty of the 1960s and 1970s included David B. Duncan, Helen Abbey, Earl Diamond, John Gart, Alan Gittelsohn, Alan Ross, Michael Tallis, Miles Davis, Charles A. Rohde, Richard Royall, Susan Horn, John Sawyer, and James Tonascia. Students who went on to particularly notably further the field and Department included Philip Archer, Ray Chambers, Bill Dupont, Arthur Hartz, Jay Herson, Frank Hurley, Kathleen and Subrahmaniam Kocherlakota, Young Kim, Jeannette Yen Lee, Ellen MacKenzie, Maureen Maguire, Edmond Murphy, Eleanor Thomas, James Tonascia, Strother Walker, and Ann Zauber. As of this 2011 writing MacKenzie is the chair of the Johns Hopkins Department of Health Policy and Management. Hurley is chair and chief scientific officer of RRD International, a health care product development company that he cofounded in 2002 and whose aim is to bring safe and effective health care to people around the world. He and his wife, Catharine Dorrier, have particularly generously engaged in the life of the university and the department—among other contributions, endowing the professorship that supports the Department Chair.

The Homewood-based Department of Statistics grew during the 1960s. Faculty included Don McNeil, Alan Marcus, and Siegfried Schach. The department used joint appointments with the Biostatistics department to increase its course offerings. V. P. Godambe, C. R. Rao, Ted Hannan, Rupert Miller, Milton Vernon Johns, Jim Durbin, David Kendall, Jerome Cornfield, and E. J. G. Pitman spent sabbaticals at Homewood or in Biostatistics. Seminars were spirited and well attended. Leon Gleser recalls among his fondest memories statistical philosophy discussions between Godambe, Cornfield, Richard Royall, and himself. The department, though small, produced some outstanding students, including Rudy Beran, Ed Rothman, Ray Waller, Ayala Cohen, S. Sharma, Keith Eberhart, Bill Cumberland, Dan Bloch, Marcello Pagano, Bill Brelsford, Gary Chase, James C. Fu, Joseph Skwish, Woollcott Smith, and Dennis Dixon. Fu cofounded the journal *Statistica Sinica* and the International Chinese Statistical Association. Waller served as the Executive Director of the American Statistical Association, and Dixon served as the Chief of the Coordinating Centers and Biostatistics Research Branches in the Division of AIDS at NIAID. As of this 2011 writing Cumberland is the Chair of Biostatistics at the University of California, Los Angeles.

In the early 1970s, Geoff Watson's departure after only a few years as chair and a university budget crisis rocked the Homewood Statistics Department. The department was abolished in 1972, and Statistics, Operations Research and Computer Science were amalgamated into a new department of Mathematical Sciences. In 2004 that department was renamed as Applied Mathematics and Statistics. Prominent faculty have included Robert Serfling, Don Geman, Alan Karr, John Wierman, Jim Fill, Carey Priebe, and Daniel Naiman.

Research. In the 1970s the major area of impact for Hopkins Biostatistics was in sampling (Rohde 1976). Richard Royall and his students developed the robustness properties of the model-based prediction theory approach to finite population sampling. This put finite population sampling back into the mainstream of standard statistical inference and removed and explained most of the paradoxes of prior approaches.

In the Homewood department the major areas of impact were in the statistics of directions and in time series. Watson wrote highly cited papers in both areas, Ted Hannan finished an important book on Multiple Time Series while he was a visitor, and Dick Jones wrote several influential papers on random fields and time series. Joe Gastwirth produced influential work on robust nonparametric estimation.

Education. The department continued its outstanding teaching tradition, led by Helen Abbey. Abbey's illustrious teaching career in the department spanned over four decades. Her lectures were peppered with common wisdom, for example: "Cherish your outliers: they may constitute all you learn!" Her gentle manner, clear instruction, and generous open-door availability made her beloved by the many thousands of public health students whose lives she touched, in biostatistics and beyond. Late in Abbey's career it was said that if one parachuted Abbey anywhere in the world, there would be a former student to greet her within 50 miles of her landing. Successions of Abbey's dogs accompanied her to classes over the years and were equally known to the students as she.

1980–2008

Overview and personalities. These years were an age of growth and excellence for the department. Charles Rohde became chair in 1981. A period of intense recruitment was enabled when in 1980 he and Dean D. A. Henderson convinced the Mellon foundation to invest one million dollars over 10 years to strengthen biostatistics at Hopkins. Further gains were achieved when Rohde negotiated indirect cost recovery for the first time in the department's history, and a system was developed at the school to remunerate teaching approximately proportionally to students taught. In the first two years, Rohde and his faculty recruited five assistant professors: Hendricks Brown, Steven Self, Kung-Yee Liang, Ron Brookmeyer, and Scott Zeger. In the next several years Mei-Cheng Wang, Subhash Lele, Karen Bandeen-Roche, and Arieh Epstein joined them. In successive years (1990–1992) Liang, Zeger, and Brookmeyer won the APHA Mortimer Spiegelman Award for outstanding contributions to public health statistics made under the age of 40. Each ultimately assumed high academic leadership: Brookmeyer, as the Director of the Johns Hopkins Master of Public Health Program (2002–2008); Zeger, as the School's Senior Associate Dean for Academic Affairs (1991–1996) and Chair of Biostatistics (1996–2008) and the University Vice Provost for Research (2008-present); and Liang, as Vice President, National Health Research Institutes Taiwan (2003–2006) and President, National Yang-Ming University, Taiwan (2010-present).

Scott Zeger assumed the department chairmanship in 1996, and the department moved to new space on the third floor of the new building. In recruitments under Zeger's leadership, 11 faculty joined the department in as many years. They roughly grouped into the areas of: *statistical inference* (Tom Louis, Daniel Scharfstein, Constantine Frangakis, Brian Caffo, Ciprian Crainiceanu), *genomics* (Karl Broman, Rafael Irizarry, Ingo Ruczinski, Hongkai Ji), and *environmental science* (Francesca Dominici, Roger Peng). Each area became a strength of the department. Like their predecessors, these faculty became highly accomplished, with major recognitions, including the 1999 Snedecor Award for notable work in biostatistics (Scharfstein et al. 1997), the 2001 and 2006 American Statistical Association Gottfried H. Noether Young Scholar Awards recognizing excellence in nonparametric statistics (Irizarry, Crainiceanu), and the 2006 APHA Spiegelman Award (Dominici). Louis was elected as President of the International Biometric Society (2005–2008 executive term).

An innovation during the Zeger years was the creation of the Biostatistics Consulting Center. The Center provides biostatistical and information science expertise to clients within and outside the university. It arose in support of the suit filed by the State of Minnesota against several tobacco companies and related organizations in the late 1990s, in which the department partnered as consultant. It has served as a key resource of the department, school, and community ever since.

On the social scene, the late twentieth century saw the initiation, by Zeger, of the annual Chili Party welcoming in each academic year. Now running for over two decades, the event features a chili cook-off. It exemplifies the many opportunities the faculty, students, and staff of the department created to eat and drink together including frequent outings to Szechuan Best in nearby Reisterstown and Baltimore Orioles games and poker evenings welcoming all comers.

Among the era's many notable students who could be named: Paul Albert is Chief of the Biostatistics and Bioinformatics Branch at the NICHD. Marie Diener-

West has made renowned contributions to education in the Department, School, and beyond. Janice Dykacz established the departmental award given in recognition of an outstanding paper by a Biostatistics student in medical statistics. Patrick Heagerty was the 2001 COPSS Snedecor Award winner. Paul Rathouz was named Chair of the Department of Biostatistics and Medical Informatics, University of Wisconsin, in 2010.

Research. Building on the department's rich tradition in public health problems, foundations of inference, and model-based inference, key new research perspectives emerged from the department during Rohde's chairmanship. These contributed to the synthesis of biostatistical methods under the generalized linear model, partially observed data frameworks, model robustness, and partly parametric model specification and inference.

The first line of research was motivated by the preponderance of repeated measures and time-to-event data emanating from Johns Hopkins cohort studies. The objective of analysis could usually be formulated in generalized linear model terms, but modeling of the correlation among repeated values was necessary. Many faculty made important contributions. Particularly notably, Liang, and Zeger introduced "generalized estimating equations" or GEE as an approach that revolutionized the analysis of longitudinal data. Their foundational papers on the topic (Liang and Zeger 1986; Zeger and Liang 1986) were honored with the 1986 Snedecor Award. As of this writing their 1986 Biometrika article has been cited nearly 9,000 times.

Meanwhile, Brookmeyer and adjunct professor Mitch Gail, at NIH, led the statistical response to the HIV epidemic. In a decade of highly visible work, they introduced many key methods to accurately monitor HIV. Their development of the "back calculation method" to estimate the prevalence and incidence of AIDS and HIV, accounting for reporting delays, was essential to avoid mistakes like the one made by CDC and Johns Hopkins legend Alex Langmuir and colleagues in predicting that HIV would peak in 1988 (Bregman and Langmuir 1990). The method's impact was reflected in papers published in journals ranging from *JASA* (Brookmeyer and Gail 1988) to *Science* (Brookmeyer 1991). Brookmeyer also made noted contributions to the study of Alzheimer's disease.

Many other faculty made important contributions during this period. Royall published the seminal monograph on likelihood-based inference: *Statistical Evidence*, (Royall 1997) culminating a series of papers on the topic including one in which he connects the likelihood and frequentist strategies by introducing the "probability of weak evidence". Lele with Joan Ritschmeir developed a cogent theory of morphometrics. Wang developed the theory of recurrent event analysis. Bandeen-Roche, motivated by her growing engagement in science on human aging, developed a methodology for latent variable modeling and criticism. In 1997, she was named for the work as a Brookdale National Fellow for gerontological leadership. Liang, in addition to his methodological contributions, made noted contributions to the field of mental health. In 2010, he was recognized for these by The Rema Lapouse Award which honors excellence in psychiatric epidemiology and is given each year by the Mental Health, Epidemiology, and Statistics sections of the APHA.

During the Zeger years, public health and medical statistics were irrevocably changed by the explosion of information technologies, especially in genetics. Molecular biology became a quantitative science with the advent of affordable methods for genome sequencing, RNA expression measurement, and protein measurement. Of particular note among the department's contributions in this area was the work by Irizarry and his colleagues who initially focused on the analysis and preprocessing of microarray data, then moved on to second-generation sequencing, developing diagnostic tools, and discovering biomarkers. This research was published in diverse journals from *Biostatistics* (Irizarry et al. 2003) to *PNAS* (Kendziorski et al. 2005).

Many key papers representing the department's other areas of focus of the era could be cited; one from each is briefly summarized to give the flavor.

With his advisor Don Rubin, Constantine Frangakis in 2002 introduced principal stratification for drawing causal inference in the presence of postrandomization latent variables (Frangakis and Rubin 2002). This general approach is used, for example, to account for partial compliance, drop-outs or death in cohort studies, and clinical trials. As of this writing Google Scholar shows the methodology to have been cited more than 700 times.

The department's environmental biostatistics group was part of a larger collaboration involving epidemiologists, environmental engineers, toxicologists, and health policy scientists that addressed the health effects of urban air pollution. Many papers and in journals ranging from *JAMA* (Dominici et al. 2006) to *JASA* (Dominici et al. 2004), the group contributed substantially to the scientific foundation that EPA has used to regulate ozone and particulate air pollution.

Education. During the early 1990s, alumna Dr. Marie Diener-West was recruited back to the department to continue the department's tradition of outstanding educators. Led by Drs. Brookmeyer, Diener-West, and Royall, the entire statistics curriculum was modernized along what became four distinct tracks to meet the needs of varied students. The department introduced a Master of Health Science degree program which trains students earning doctorates in other public health departments in master-degree level Biostatistics. Training grants in Mental Health, Genomics, Aging, and Environmental Health were obtained. As of this writing the latter two continue to flourish, with the Aging training program enjoying its seventeenth year running. In 2004, the Helen Abbey and Margaret Merrell Professorship for Biostatistics Education was endowed by the generosity of Helen Abbey and numerous alumni. Diener-West was installed as its inaugural recipient.

2008 into the Future

Overview and personalities. In 2008, Karen Bandeen-Roche was named as the ninth chair of the Johns Hopkins Department of Biostatistics. Jeff Leek, Michael Rosenblum, and Han Liu have joined the faculty since then, adding expertise in genomics and risk prediction, causal inference, and the interface of statistics and

Fig. 2 Johns Hopkins Department of Biostatistics, 2010–2011

computing. The faculty continue to earn high recognitions including the 2009 APHA Spiegelman Award and the Presidents' Award given annually by COPSS for outstanding contributions to the profession before the age of 41 (Irizarry, in both cases), a 2010 Presidential Early Career Award for Scientists and Engineers (PECASE; Caffo), and election to Presidency of the International Biometric Society/Eastern North American Region (Bandeen-Roche 2012).

Figure 2 displays a current day photo of the department.

Research. The department continues its tradition of engagement at the interface of science and statistics. Irizarry, Ruczinski, Ji, and Leek continue to lead in quantitative discovery in genetics and molecular biology. Caffo and Crainiceanu are pushing the frontier of interpreting the massive data sets by which advanced research technologies like neuroimaging, real-time monitors, and electronic health records are informing health science. They are advancing discovery in functional data analysis, neurological health, and other areas. Zeger, Louis, Wang, Bandeen-Roche, and Peng continue to pursue biostatistics related to population health. Together with their colleagues they are advancing science related to chronic disease and aging, mental health, population surveillance for prevalent, incident and recurrent health insults, infectious disease, social determinants, and environmental health. Rohde, Louis, Scharfstein, Frangakis, and Rosenblum lead the department's engagement in inference and on study designs in which inferences on causal relations can be rationally grounded. Many benefits have resulted including for the treatment of trauma injuries and AIDS and for radiological therapies. Tonascia and Diener-West lead in the design and conduct of clinical trials to evaluate the effectiveness of treatments and interventions, and their translation into practice. Their contributions have been particularly notable in ophthalmology, asthma, gastric disease, and cardiopulmonary disease. The department targets moderate expansion in the coming years so as to grow its strength spanning the development of cutting-edge statistical methodology to its application to the advancement of health science.

Education. The department affirms its dedication to excellence in education. In 2008, Diener-West was named as the Chair of the School's Master of Public Health degree program. As of this writing nearly every student in the School takes biostatistics. The department has won 62 of the School's "Golden Apple" and "Advising, Mentoring and Teaching" awards for teaching excellence over the past 40 years—more than any other department despite being the third-smallest. In 2010, Diener-West and colleague John McGready were recognized by the American Public Health Association with ASPH-Pfizer Awards for excellence as educators.

Student enrollment stands strong with 37 PhD students, 10 primary master degree students, and a thriving MHS program for students primarily earning doctorates in other departments. The master degree program has been updated to more closely focus on modern statistical practice and now features an "internship program" in which students may intern in the Biostatistics Consulting Center in return for tuition and stipend.

Postscript. The present era is serving up spectacular problems for statistical science. It is a time of unprecedented opportunity for our field and for Johns Hopkins Biostatistics. We look forward to the advancements in statistical and health discovery we will be making in the years ahead, and the scientists whose training to those ends we will mentor.

Acknowledgments The authors are grateful to Leon Gleser and Dan Naiman for providing information on the Homewood Statistics Department and to Ashley Gilliam for her expert help with preparing the draft.

References

1936–1937 Johns Hopkins School of Hygiene and Public Health (SHPH) catalog, 32

1956–1957 Report of the Dean of the School of Hygiene and Public Health, JHU circular 77 (1958), p 135

1936–1937 SHPH Catalog, 32; report of sub-committee of applications and curriculum committee on alumni and their MPH curriculum suggestions, pp 12–13

Bregman DJ, Langmuir AD (1990) Farr's law applied to AIDS projections. JAMA 263:1522–1525

Brookmeyer R (1991) Reconstruction and future trends of the AIDS epidemic in the United States. Science 253:37–42

Brookmeyer R, Gail MH (1988) A method for obtaining short-term projections and lower bounds on the size of the AIDS epidemic. J Am Statist Assoc 83:301–308

Charles Flagle, Sept 29, 2010 interview with Karen Kruse Thomas

Cochran to Ernest Stebbins Mar 1, 1950, SHPH records, 3b: O.D. Financial, box 506104, "1949–1950 budget" binder

Cochran WG (1953) Sampling techniques. Wiley, New York

Cochran WG (1960) Review of elementary decision theory by Herman Chernoff and Lincoln E Moses, AJPH 50:417–418

Cochran WG, Cox GM (1950) Experimental designs. Wiley, New York

Cochran WG et al (1954) Statistical problems of the Kinsey report on sexual behavior in the human male. American Statistical Association, Washington

Cornfield J, Haenszel E, Hammond C et al (1959) Smoking and lung cancer; recent evidence and a discussion of some questions. J Natl Cancer Inst 22:173–203

Dominici F, McDermott A, Hastie T (2004) Improved semi-parametric time series models of air pollution and mortality. J Am Statist Assoc 468:938–948
Dominici F, Peng R, Bell M et al (2006) Fine particles air pollution and hospital admission for cardiovascular and respiratory diseases. JAMA 295:1127–1135
Edwards LB (2001) Life as a doctor: a Pesky Darling grows up (unpublished), Lydia B. Edwards papers, Alan Mason Chesney medical archives of the Johns Hopkins Medical Institutions, p 28
Epidemic Theory (2010) 1951 episode of the Johns Hopkins science review, Eisenhower library special collections, JHU. http://biostat.jhsph.edu/~nreich/reed.html. Accessed Jan 12, 2010
Fee E (1987a) Disease and discovery: a history of the Johns Hopkins School of Hygiene and Public Health, 1916–1939. Johns Hopkins University Press, Baltimore
Fee E (1987b) Disease and discovery: a history of the Johns Hopkins School of Hygiene and Public Health, 1916–1939. Johns Hopkins University Press, Baltimore, pp 138–146
Fee E (1987c) Disease and discovery: a history of the Johns Hopkins School of Hygiene and Public Health, 1916–1939. Johns Hopkins University Press, Baltimore, pp 64, 146
Frangakis CE, Rubin DB (2002) Principal stratification in causal inference. Biometrics 58:21–29
http://www.stat.berkeley.edu/~brill/Papers/jwtencyc.pdf
Irizarry RA, Hobbs B, Collin F et al (2003) Exploration, normalization, and summaries of high density oligonucleotide array probe level data. Biostatistics 4:249–264
JHU Trustees minutes June 1, 1959, 4632; JHU Trustees Executive Committee minutes May 2, 1960, 4714 and Dec 5, 1960, 4781
Kaplan EL, Meier P (1958) Nonparametric estimation from incomplete observations. J Am Stat Assoc 53:457–481
Kendziorski C, Irizarry RA, Chen KS et al (2005) On the utility of pooling biological samples in microarray experiments. Proc Natl Acad Sci USA 102:4252–4257
Kingsland S (1982) The logistic curve and the history of population ecology. Quart Rev Biol 57:29–52
Liang KY, Zeger SL (1986) Longitudinal data analysis using generalized linear models. Biometrika 73:13–22
Margaret Merrell to Ernest L Stebbins, Nov 6, 1957, Margaret Merrell papers, box 505378, "Office of the President" folder, AMC
Marks HM (1997) The progress of experiment. Cambridge University Press, Cambridge, pp 53–59, 98–103
Maxcy K (1939) "A review of the activities of the Department of Epidemiology," SHPH departmental reports, "Department Reports (Epidemiology)" folder
Merrell M (1951) Discussion. Am J Pub Health 41:76
Reed L, "Department of Biostatistics"; 1944–1945 SHPH catalog, 32–33; 1948–1949 Johns Hopkins School of Medicine Catalog, 115–116; 1954–1955 JHSOM Catalog, 124
Rohde CA (1976) Composite sampling. Biometrics 32:273–282
Rothstein WG (2003) Public health and the risk factor: a history of an uneven medical revolution. University of Rochester Press, Rochester
Royall RM (1997) Statistical evidence: a likelihood paradigm. Chapman & Hall, London
Scharfstein DO, Tsiatis AA, Robins AM (1997) Semiparametric efficiency and its implications on the design and analysis of group-sequential studies. J Am Stat Assoc 92:1342–1350
William G Cochran to Ernest Stebbins Mar 1, 1950, SHPH records, 3b: O.D. Financial, box 506104, "1949–1950 budget" binder
Zeger SL, Liang KY (1986) Longitudinal data analysis for discrete and continuous outcomes. Biometrics 42:121–130

Kansas State University Department of Statistics

John E. Boyer, Jr. and James W. Neill

The Department of Statistics at Kansas State University came into being under the authorization of the Kansas Board of Regents on July 1, 1959. Holly C. Fryer was named the initial head and there were five faculty members in the department. The department continues to be a vibrant and active entity today, with 14 faculty members, nearly 60 graduate students, and 3 staff positions.

The Formative Years

The history of statistics at Kansas State University begins with the hiring of Dr. Holly C. Fryer to a tenure track faculty position in the Department of Mathematics in 1940. Fryer was a native of Oregon and held degrees from both the University of Oregon and Oregon State University when he began doctoral studies at Iowa State University. He wrote his dissertation under the direction of Dr. George Snedecor and, although he was a student in the Department of Mathematics, his was the first doctorate awarded in the field of Statistics by Iowa State University. The title of his dissertation was *An Analysis of Group Differences Arising from a Poisson Distribution of Observations Obtained from an Irradiation Experiment*. Dr. Fryer had minors in genetics and mathematics, and, as he put it, "a little experience doing research with experiment station researchers".

Kansas State University,
Department of Statistics,
101 Dickens Hall,
Manhattan, KS 66506-0802, USA
J. E. Boyer, Jr. (✉) · J. W. Neill
e-mail: jboyer@ksu.edu

J. W. Neill
e-mail: jwneill@ksu.edu

A. Agresti and X.-L. Meng (eds.), *Strength in Numbers: The Rising of Academic Statistics Departments in the U.S.*, DOI: 10.1007/978-1-4614-3649-2_11,

Holly C. Fryer

In addition to a multitude of publications with university researchers over his career, Dr. Fryer authored two textbooks: *Concepts and Methods of Experimental Statistics* (1966, Allyn and Bacon) and *Elements of Statistics* (1964, Wiley). These books served as reliable entry points to statistics for many graduate students and scientists in the early years.

Dr. Fryer was hired at a salary of $2,000 for the 9-month academic year, and his assigned duties were: to be the statistical consultant for Kansas Agricultural Experiment Station (KAES), to start a computing lab under the auspices of KAES, to teach courses in mathematics and statistics, and to develop degree programs in statistics. He saw what seemed to be unlimited opportunity for the development of all facets of a statistical program at Kansas State College.

Along with a teaching load of 16 credits per year, Fryer quickly grew the consulting arm of the job. In 1941, he was appointed an agent of the USDA for the purpose of bringing in $100 per month so that a small consulting lab could be established. Its only technology in that first year was one partially manual Monroe desk calculator. By 1942–1943 Fryer had created seven different statistics courses which he offered on a 2-year rotating cycle; that same year he reported 253 statistical consultation sessions with faculty and students from 16 departments.

From May 1, 1944 to October 1, 1945, Dr. Fryer moved to New York City, where he joined the Applied Mathematics Group at Columbia University doing war research. The Applied Mathematics Group and the Statistical Research Group at Columbia were both housed in the same suite of offices in New York, bringing Dr. Fryer into close contact with such influential statisticians as Abraham Wald, Jacob Wolfowitz, Harold Hotelling, Allen Wallis, Churchill Eisenhart, and Milton Friedman.

On his return to campus in 1945, Dr. Fryer received assurances from the Dean of the Division of Arts and Sciences that he would be the next Head of the Department of Mathematics. A couple of months later, the dean reneged on the appointment and indicated his intention of appointing an old friend to the post. Fryer took the issue up with the university president (Milton S. Eisenhower, Dwight D. Eisenhower's youngest brother) who indicated that he would hold the dean accountable. The Dean of Agriculture, however, was concerned about Fryer's ability to run a department the size of mathematics and continue his heavy consulting load, and expressed that thought to the other administrators. A compromise was hatched which offered Fryer the leadership of a new entity called the Statistical Laboratory. Although it was to be housed within mathematics, its operation was totally out of the control of the department. Fryer accepted the compromise and took the position of Director of the newly established Statistical Laboratory on March 26, 1946. The initial budget for 1946–1947 contained $1500 and allocated 2/3 of one person's time (Fryer's) to the effort.

With a sympathetic ear in the university president's office and lots of work coming out of the Statistical Laboratory, it gradually grew. By 1947, there were three statisticians on the Mathematics Department faculty, and they were all listed as working with the Statistical Laboratory (although no mention is made of direct supervision by Dr. Fryer). Through the early and mid-1950s the number of faculty in statistics continued to grow gradually to its maximum in 1956–1957 when there were seven faculty members in that area. The number of consultations provided to experiment station scientists and students had grown to more than 1200 in the year. The first MS degree in Statistics was awarded in 1950 to Robert Cell. His thesis was entitled "*Sequential Analysis in Industrial Sampling*" and supervised by Dr. Fryer.

Creation of the Department

Although the program was growing, there were continuous questions of whose authority was to be consulted on various matters. Questions of the relevance of statistical publications in subject matter journals (especially within a department of mathematics), the question of publish or perish, and the value of consulting service to the university community continued to come up again and again. In 1959, the faculty in statistics petitioned the university president, Dr. James McCain, who had succeeded Dr. Eisenhower, for the creation of a separate Department of Statistics. This stemmed from their dissatisfaction with the environment in mathematics as well as a look forward to the kind of department they envisioned they could be, given their recent rapid growth. McCain rejected the proposal, not apparently from any animosity toward statistics, but because he felt that the climate on the Kansas Board of Regents was not right for creating any new entities at the time. At this point, some of the statisticians under the Department of Mathematics decided it was time to leave, Fryer among them. They started looking for other jobs, a fact which became widely known to the researchers on campus with whom they worked.

One evening early in March 1959, Fryer received a call from a personnel director at the Sandia National Laboratories in Albuquerque, saying that he wanted to invite Fryer to come for a job interview, but that he needed a name and telephone number of a person to call to ask the appropriate permission to make such an invitation. Fryer gave the director the name and telephone number of the university president, and told the director to indicate that Fryer was leaving because the new department was not being formed. The next day Fryer received a call from the president's office asking him to be in that office first thing on Monday morning. After some friendly opening conversation, the president told him that he would make the department happen, but that the time just was not right relative to the Board of Regents and urged Fryer, "You'll just have to trust me on this, Holly." Fryer indicated that he told the president, "We can wait two weeks until the next Board of Regents meeting." At the next meeting, on March 25, 1959, the Board of Regents authorized a separate Department of Statistics to be organized within the College of Arts and Sciences effective, July 1, 1959, with Dr. Fryer as its head. Even this was not without its difficulties; there was no budget for secretarial staff and the department would be housed side by side with the Department of Mathematics located in Cardwell Hall for a full year until space could be created for it in Calvin Hall, the College of Business building. Nevertheless, the first year as a separate department was a good one, with a growth in enrolments in both service courses and graduate students.

The Fryer Years

The new department began in 1959 with five full-time faculty members. Along with Dr. Fryer were Dr. Stanley Wearden (PhD from Cornell University), Dr. Arlin M. Feyerherm (PhD from Iowa State University), Robert Cochran, and Gary F. Krause. The latter two both held Masters Degrees at the time of the formation of the department, but both subsequently earned doctorates and had long academic careers.

By the summer of 1960, the department was housed in the building with the College of Business and was gradually expanding again. In 1963, the department received major funding in the form of a grant from the National Institute of General Medical Sciences to expand its graduate program. The grant provided $260,000 for stipends, tuition, fees, and dependents allowances for graduate students. This supported 6 students annually the first 3 years, and 12 in the succeeding years. This grant was renewed 5 years later for more than $400,000. This money was given under the expectation that the department would develop and create a PhD degree in Statistics. This was a major aid in expanding the graduate program toward the offering of a doctoral degree program. In 1964, the department hired its first two theoretical statisticians, K.S. Banerjee (PhD from the University of Calcutta), whose area of expertise was experimental design, and Jay Conover (PhD from the Catholic University of America), a nonparametrician, to aid in the expansion. A year later, Dr. Shelemyahu Zacks (PhD from Columbia University), with expertise in mathematical statistics was hired to further this expansion.

In addition, there was sufficient budget money to bring some prominent statisticians to the campus for extended visits. Jerzy Neyman, Robert Buehler, William Kruskal, and Regina Elandt-Johnson all made visits to the campus lasting multiple days in the mid-1960s.

In 1966, the department was authorized by the Kansas Board of Regents to offer a PhD degree program in Mathematical Statistics and Probability. The first degree was awarded in 1967 to Marvin Lentner, who subsequently took a faculty position at Virginia Tech University. His dissertation was entitled "*Application of Generalized Inverses for Optimal Design and Analysis of Randomized Fractional Replicates*" and supervised by Dr. Zacks. Over the 11-year period of the NIGMS grants, the department awarded 93 MS degrees and 41 PhDs. This was a time of great prosperity in the department, in terms of its graduate program and reputation.

In 1967, the Vice-President for academic affairs (provost) of the university indicated his intent to launch a Department of Computer Science. A good deal of debate was held on whether that unit ought to be housed within the College of Arts and Sciences or within the College of Engineering. Dr. Fryer was among those who argued for the Arts and Sciences location, saying that the wider scope of activities and research would benefit the department. His argument (just one of many, apparently) won the day, and the name of the Department of Statistics was changed to the Department of Statistics and Computer Science in the fall of 1967. By the spring of 1971, the joint department contained 22 faculty positions and Fryer had helped to grow the group in both areas. In the summer of 1971, a separate Department of Computer Science was created, and nine faculty members moved to the new department that fall. Although Fryer had no training in computing, it is clear that he had great respect for computer science, and at least a vague vision of its coming importance. It might be pertinent to point out that on the K-State campus, the Computer Science Department did eventually move to the College of Engineering in 1993.

Dr. Fryer was supposed to retire as Head of the Department and Director of the Statistical Laboratory on July 1,1974, due to university rules about ages of administrators that were in place at the time. A national search for a department head did not yield a successor, so he stayed on until January 1 of the next year. At that point Dr. Arthur Dayton, who was already on the faculty, was named interim Head of the Department and Director of the Statistical Laboratory, and the "interim" tag was removed 6 months later. Meanwhile, Dr. Fryer and his wife took sabbatical leaves at North Carolina State University (getting out of the way of the new head, we suspect!) in the spring of 1975.

Subsequent Growth and Development

The leadership of the department since the Fryer years has been provided by Arthur Dayton (1975–1985), George A. Milliken (1985–1990), James J. Higgins (1990–1995), Dallas E. Johnson (1995–2001), John E. Boyer, Jr. (2001–2008), and

Former department heads James Higgins, John Boyer, Jr., Dallas Johnson, George Milliken, with current head James Neill (seated) and wives of deceased former heads, Beth Fryer (Holly Fryer) and LaVonia Dayton (Arthur Dayton)

James W. Neill (2008–present). The following traces some of the developments and growth activities within the department during this time.

As computing capabilities have increased over the years, one common thread connecting the efforts of these department heads has been the work to accommodate increasingly computer-intensive research and computer-aided instruction. Upon notice from the university in the early 1980s that the department would move to different and somewhat larger quarters in Dickens Hall, Dayton reserved a large room in the new facility as a computing lab. With funding from NSF, a Harris mini-computer (a misnomer by today's standards as it was the size of a large refrigerator) was purchased and installed, along with 30 dumb terminals to be hard-wired to it (state of the art for those days). Dr. Kenneth Kemp, a faculty member who had a great talent for computer hardware and software issues, became the person in charge of the computing lab. The department's computing capabilities expanded dramatically and greatly enhanced the research and consulting efforts of the faculty and graduate students.

By 1990 the department had moved to personal computers and Sun workstations, facilitated by additional NSF funding. Prior to the World Wide Web, the department used a university-based network, called Bitnet, to send email among faculty. And by 1995, WWW, browsers, email, file-sharing, and the like were a part of our culture (for better or worse!). Since 2000, departmental computing has been based on ever-increasingly powerful personal computers, connected to a common department server. For especially computer-intensive statistical research projects, faculty and graduate students can, since the mid-2000s, utilize the university's high

Dickens Hall

performance computing cluster, called the Beocat. And just recently, in order to further facilitate communications, Dickens Hall became wireless. Throughout the years, many challenges were met for the modernization (including infrastructure and wiring) of Dickens Hall, making a building built in 1907 functional for today's computing and communications.

The use of computers has not only enabled computer-intensive statistical research and consulting, but computer-aided instruction has progressed as well. A major advance in the department was the development of the computer studio classroom environment. By the spring of 2000, Higgins and former faculty member Dr. Deborah Rumsey had created a curriculum for students of introductory statistics in the studio classroom with the main goal being to involve students in the whole process of scientific discovery. This contrasted with most traditional classes in introductory statistics where the primary emphasis had been simply instructing students in the basic methods of statistical analysis. As of spring term 2011, the department enrolled nearly 1,700 students in undergraduate general education classes, the majority with computer lab experiences. Online/distance learning opportunities have recently complemented the department's offering of undergraduate general education.

The department's graduate program also benefited immensely from the progress in computer technology. In order to effectively teach the methods of statistical science to graduate students, both within the department and those in subject matter areas of study, the importance of incorporating the use of appropriate statistical software in the classroom was recognized by faculty. To this point, several well-received textbooks were written by various faculty including *Analysis of Messy Data, volumes I: Designed Experiments* (2nd ed 2004, Chapman & Hall/CRC), *II: Nonreplicated Experiments* (1989, Chapman & Hall/CRC), and *III: Analysis of Covariance* (2001, Chapman & Hall/CRC), by Milliken and Johnson; *An Introduction to Modern Nonparametric Statistics* (2003, Duxbury Advanced Series) by Higgins; *Applied Multivariate Methods for Data Analysis* (1998, Duxbury Press) by Johnson; *Concepts in Probability and Stochastic Modeling* (1994, Duxbury Press) by Higgins and Keller-McNulty; *A SAS Companion for Nonparametric Statistics* (2005, Duxbury Advanced Series) by Richter and Higgins; *SAS for Mixed Models* (2nd ed 2006 SAS Institute, Inc) by Littell, Milliken, Stroup, Wolfinger and Schabenberger. The integration of foundational statistical theory and methodology in the department's graduate curriculum has, as of the end of 2010, resulted in 333 MS and 145 PhD graduates with the reputation of having the acumen for solving meaningful applied problems.

In addition to the on-campus graduate degree programs, distance/online learning has become an integral component of the department's graduate instructional program. In particular, the department initiated the Graduate Certificate in Applied Statistics in the online learning format during 2005. Also, the department has supported the MS in Operations Research offered through the Industrial and Manufacturing Systems Engineering department in the College of Engineering with online statistics courses since the early 1990s. More recently, beginning in 2004, the department has supported the Master of Public Health degree program in the College of Veterinary Medicine with online courses as well, and introduced an online biostatistics course during fall 2011 for accreditation of the MPH program. Looking forward, beginning in spring 2012, the department will participate in the online Master of Business Administration program in the College of Business Administration offered through the Kansas State University Olathe campus for professionals located in the Kansas City area.

Another common thread connecting efforts of the department heads has been the support of the annual departmental conference. Recollecting, Higgins and former faculty member Dr. James R. Schwenke, met for coffee in Aggieville (just off campus) in the late 1980s and hatched the idea of a conference on applied statistics with an emphasis on the problems faced by those who work in and with agricultural research. Milliken invited Dr. Ron Hocking from Texas A&M University to be the keynote speaker for the inaugural conference held in Manhattan in the spring of 1989. The conference was named the Kansas State University Conference on Applied Statistics in Agriculture and it was an immediate success. At the conclusion of the conference, plans began immediately for an encore event the next year. The conference has been held annually every year since 1989, and continues to attract academic, government, and industrial statisticians who work on agricultural problems of all sorts to Manhattan. Since 1992, the conference has included a technical workshop in an area of statistical application as well. Milliken, Johnson, and Boyer have all been instrumental in attracting prominent speakers to the conference throughout the existence of the event. A complete list of keynote speakers, along with address titles, can be found at the department's website.

Throughout the years, the department has had strong ties to the College of Agriculture and Research and Extension. At one point, there was support for four FTEs split between eight faculty members in the department for the purpose of collaboration with scientists across campus supported by Research and Extension. From the very beginning of a formal statistical presence on campus under Fryer, a signature component of the faculty's activities included consulting with graduate students and faculty from various departments in order to improve the quality of scientific research at the university. Consulting, along with the teaching of basic statistical methods courses, naturally involved statistics faculty as members of graduate committees for MS and PhD students across campus. Such involvement on graduate committees by statistics faculty often led to coauthorships of subject matter publications with other faculty and students, thereby also enhancing the level of scientific research on campus. Nowadays, statistical methods courses are offered by the department covering topics beyond the basic linear statistical model and design of experiments, including nonparametric statistics, categorical data analysis, multivariate statistical methods, bioassay, graphical methods, smoothing and regression analysis, computer-intensive methods of statistics, and the analysis of messy data. Such offerings are heavily populated by graduate students from other departments as well, and routinely rank the department at or near the top of graduate credit hours generated within the College of Arts and Sciences.

Although the number of KSRE FTEs has decreased gradually to roughly two as financial difficulties have grown in higher education, significant interdisciplinary collaborations have continued with support increasingly provided by faculty participating as coprinciple investigators on research grants with scientists from several departments. More broadly and consistent with the mission of the department, faculty members have provided statistical support for research faculty across all Colleges at Kansas State University. Based on this success and the university's initiatives in animal health and food safety/security, the department looks forward to new collaborations with scientists associated with the recently completed (2008) Biosecurity Research Institute, the anticipated (2016) National Bio and Agro-Defense Facility (both federal laboratory facilities located near campus), and the newly opened (April 2011) Kansas State University campus located in Olathe, KS with connections to the animal health industry corridor in the Kansas City region.

During the last 15 years, several long-time faculty members began to retire from the department, leading to several new faculty hires with unique research interests. As a result, the department's research focus began to evolve from traditional areas, such as linear models, design of experiments, probability, and stochastic processes to areas often involving computer-intensive techniques. Advances in computing have allowed removal of certain restrictive modeling assumptions and enabled faculty to work with researchers who generate massive datasets with complex structure, such as curve data, spectra, and DNA microarrays. Reflecting these advances in computing, from the 1990s onward, research interests of the faculty have grown to include hierarchical Bayesian modeling, generalized linear mixed models, robust nonparametric inference, high-dimensional data analysis, longitudinal and functional data analysis, and spatial and temporal modeling for the analysis of correlated data.

Past and present faculty at fiftieth anniversary. Bottom row (left to right): Juan Du, Haiyan Wang, Robert Cochran, Leigh Murray, Weixin Yao, Gary Gadbury. Middle row: Paul Nelson, Arlin Feyerherm, Suzanne Dubnicka, Ken Kemp, James Higgins. Top row: Dallas Johnson, George Milliken, John Boyer, Mike Rubison, Jim Neill, Ray Waller

Specialty areas of interest have included rank tests for experimental designs, mixture models, asymptotic methods, causal inference, reliability theory, and measurement error models. These research interests have resulted in theoretical and interdisciplinary research in the agricultural, biological, and engineering sciences with recent support from NASA, NIH, NSF, and USDA.

As a measure of departmental success over the years, several persons associated with the department have been named Fellow by the American Statistical Association reflecting significant scholarly achievements and service to the profession. This group comprised present and former faculty, including Holly Fryer (1966), Shelemyahu Zacks (1973), Jay Conover (1979), Dallas Johnson (1983), George Milliken (1989), John Boyer (1995), Ray Waller (1996), Sallie-Keller McNulty (1997), Lyman McDonald (1998), James Higgins (1999), Deborah Rumsey (2006), and Tom Loughin (2008), as well as graduates of the department, including Ron Iman (1982), Dennis Cook (1982), Richard Beckman (1989), and Ron Wasserstein (2003). At least five of the above have received the Founders Award from the American Statistical Association, the highest honor awarded by the Association, including Iman (1991), McNulty (2002), Johnson (2004), Waller (2004), and Boyer (2006). In addition, Iman and McNulty have served as President of the American Statistical Association in 1994 and 2006, respectively, while Waller and Wasserstein have served as Executive Director of the American Statistical Association.

On October 10, 2009, the department held an Open House in its Dickens Hall offices and a banquet in the K-State Alumni Center, celebrating its 50 years as a Department of Statistics. The department made special effort to invite the members of the original faculty to attend the celebration, and was honored with the presence of Drs. Arlin Feyerherm and Robert Cochran. Mrs. Beth Fryer and Mrs. LaVonia Dayton, wives of Holly Fryer and Arthur Dayton, were also present for the festivities and to renew old acquaintances. In addition, a good number of current faculty, former faculty, and graduates of the department participated in the event. The program opened with a welcome and perspectives on the discipline given by the current department head, James Neill. The previous department head, John Boyer, then presented a short history of the department and listed some of its accomplishments and those of people who taught or were educated within it. Several letters of congratulations from campus administrators were read, followed by a time of remembrances and reminiscing.

The faculty as of August 2011 is listed below, along with institutions at which they earned the PhD:

Nora Bello (Michigan State University, 2010)
John Boyer (Michigan State University, 1976)
Kun Chen (University of Iowa, 2011)
Juan Du (Michigan State University, 2009)
Suzanne Dubnicka (Penn State University, 1998)
Gary Gadbury (Colorado State University, 1998)
James Higgins (University of Missouri, 1970)
Abigail Jager (University of Chicago, 2006)
Leigh Murray (Virginia Polytechnic Institute and State University, 1981)
James Neill (Kansas State University, 1982)
Paul Nelson (Rutgers University, 1969)
Weixing Song (Michigan State University, 2006)
Haiyan Wang (Penn State University, 2004)
Weixin Yao (Penn State University, 2007)

Faculty members not mentioned previously, who spent several years as good faculty colleagues include: Shian-Koong (Peter) Perng, Raja F. Nassar, Shie-Shien (Winston) Yang, Doris L. Grosh, K. Michael Rubison, Richard A. Sundheim, Michael L. Deaton, David P. Hasza, James L. Hess, Mark S. McNulty, Hammou El Barmi, Jeffrey S. Pontius, and Minoru Siotani.

In closing, the faculties of the department have been, and are, pursuing meaningful solutions to interesting theoretical and applied problems, many of which are pushing the boundaries of classical statistical techniques. In addition, the department continues to strive to pass this knowledge on in its interdisciplinary consulting work and to its students in the classroom. In short, the Department of Statistics at Kansas State University is poised for another 50 years of excellence. Readers of this chapter will find much more history of K-State Statistics at the department's website, including more detailed research accomplishments over the years.

Michigan State University Department of Statistics and Probability

Dennis Gilliland and James Stapleton

The Agricultural College of the State of Michigan was founded in 1855. It became a land-grant institution under the Morrill Act of 1862 and reached university standing in 1955. The Department of Statistics was formed in 1955 around five faculty transferred from the Department of Mathematics. It was charged with responsibility for teaching statistics courses, providing statistical consulting, and for providing a center for routine desk calculations arising from statistical problems. On November 1, 1965, the department was renamed Statistics and Probability (S&P), a unique name that reflects the department's commitment to both probability and statistics.

There were five charter members of the Department of Statistics. W. Dowell Baten (PhD, University of Michigan, 1929) came to MSU as an Associate Professor in 1936 and had a partial appointment in the Agricultural Experiment

Gilliland and Stapleton are professor and professor emeritus, respectively. The authors thank their colleagues for many useful ideas and facts and Cathy Sparks for her research of department records to develop statistical information. Special thanks to the editors and to Bill Harkness and Ingram Olkin.

Michigan State University,
Department of Statistics and Probability,
A427 Wells Hall,
East Lansing, MI 48824, USA
D. Gilliland (✉)
e-mail: gilliland@stt.msu.edu

Department of Statistics and Probability,
Michigan State University,
A429 Wells Hall,
East Lansing, MI 48824, USA
J. Stapleton
e-mail: stapleton@stt.msu.edu

A. Agresti and X.-L. Meng (eds.), *Strength in Numbers: The Rising of Academic Statistics Departments in the U.S.*, DOI: 10.1007/978-1-4614-3649-2_12,

Station. Leo Katz (PhD, University of Michigan, 1946) joined in 1946; Ingram Olkin (PhD, University of North Carolina, 1951) joined in 1951; Kenneth Arnold (PhD, MIT, 1941) joined as an Associate Professor in 1952; and James Hannan (PhD, University of North Carolina, 1953) joined in 1953.

In the meantime, the department has granted 157 PhD's and over 760 master's degrees. In 2010, the tenure track faculty positions numbered 23.75 full-time equivalent positions (FTE's) and over 20,000 semester credits were earned in courses listed by the department.

This history is organized into seven sections. Section "Genesis and the Early Years" describes the genesis of the department in some detail. Section "Past Faculty and Visitors" lists some of its better known past faculty and visitors. Section "Chairpersons" is a chronology of its chairpersons. Section "Faculty Positions, Degrees Awarded, Enrollments, and Curricula" tracks growth in terms of numbers of tenure track positions, degrees awarded, and student enrollments leading to current faculty (Appendix A) and programs. Section "Some Graduates of the PhD Program" lists some of the graduates of the department's PhD program. Section "Statistical Consulting" gives the evolution of statistical consulting in the department. Section "Miscellany" includes what we believe are interesting anecdotes from the attempted hiring of Jerzy Neyman to the meeting of train and automobile carrying Professors Martin Fox and Herman Rubin. Appendix B lists some of the books authored or coauthored by faculty.

Genesis and the Early Years

The Department of Statistics and Probability started with a recommendation from an Administrative Group in 1955 that included J. Sutherland Frame, Head of the Department of Mathematics. On May 24, 1955 it came forth with this recommendation to the Board of Trustees:

> It is recommended that a separate department of statistics, to include a statistical consulting service, be formed in the Division of Mathematical and Physical Sciences of the School of Science and Arts at Michigan State College, effective July 1, 1955; and that an appropriate budget be provided for it.

The Group's report included a budget estimate for this new department and commented that five members of the Department of Mathematics would be transferred to the new unit, written "transferred to statistics teaching and consulting service."

The duties for the new unit were listed: "(1) To teach the statistics courses now taught in the Mathematics Department. (2) To provide a statistical consulting service on a campus-wide basis for staff and graduate students working on problems involving statistics. (3) To provide a center for a limited amount of routine desk calculating arising from statistical problems, and carried out by persons who might be trained by the Statistics Department and employed on the labor payroll with funds supplied by departments or individuals using the service. Eventually it may be better to transfer this aspect to another administrative unit".

The records of the Board of Trustees meeting of June 17, 1955 contain this proposal to the Board:

> Upon the recommendation of the Administrative Group, it is proposed to create a separate Department of Statistics to include a statistical consulting service in the Division of Mathematical and Physical Sciences, effective July 1, 1955. The total cost for the 1955–1956 fiscal year will be $32,800 for staff, supplies, and material, and this item is included in the tentative budget.

Perhaps the reader will find the rationale for the new department unremarkable. In May 1989, one of the coauthors sought more details and recorded notes of a discussion with Professor Emeritus J. Sutherland Frame, who served as Head of Mathematics in the 1940 and 1950s and as the Acting Head of the Department of Statistics for 1955–1956. Professor Frame mentioned these events and facts pertaining to the period leading up to the creation of the Department of Statistics: (1) There was much demand for statistics courses at an elementary level. (2) Mathematicians were not interested in teaching these courses or in doing consulting as needed about campus. (3) Statisticians were hired and their salary scale was higher than that of the mathematicians. (4) Those that consulted were given one fewer course to teach. Loads at that time were 12–13 hours per quarter, three or four courses.

Professor Frame indicated that these circumstances had led to friction and need for a separate department. He mentioned that all faculty in the Department of Statistics were to be given one fewer course to teach each term than their colleagues in the Department of Mathematics. This difference existed for several years after 1955 until the faculty in the Department of Mathematics had teaching loads reduced.

Leo Katz was one of the five faculty transferred from the Department of Mathematics and certainly was the driving force in creating the department at a time when there were relatively few such departments in the United States. Katz had a close relationship with faculty in Economics and Psychology, and those departments supported his efforts. Following the transition, he became the first chair of the department. He attended national and international meetings and was able to attract many distinguished visitors to the department in the 1950s and 1960s for talks or longer stays, including Fisher, Renyi, Revesz, and Mahalanobis.

Perhaps readers will find this James Hannan account of the visit from Sir Ronald Fisher in fall 1957 both interesting and informative (*A Conversation with James Hannan*, Statistical Science, **25**, No. 1, 126–144, D. Gilliland and R. V. Ramamoorthi). The interplay between Fisher and the fledgling department shows that faculty in the new department were most interested in the mathematical aspects of statistics. Here Gilliland is asking the questions. Q. We have heard the Sir Ronald Fisher visited Michigan State University in the late 1950s and gave a series of talks. What was that like? A. He gave a series of lectures. He had a new book (recent at that time) and wanted to talk about the book the way he thought about the book and, Kraft (Charles), in particular, was more interested in mathematical content. We had a conflict there. Kraft had definite theorems in mind, and he could never get anything like that out of Fisher. Q. So Kraft spoke up regularly during the lectures? A. Well, he did not badger him very much, but I knew what he had in mind. He worked with LeCam and worked at that level. Fisher was thinking that everything he did was more important than anything anyone else did. I think that Fisher had to miss a few dates and then he came back. He threw away his notes and proceeded to tell us what he thought of the kind of statistics we were doing. He said the *Annals* should be bundled up and deposited into "yon river," the Red Cedar.

The 1950s and 1960s were exciting years as the new department developed. In the late 1960s and early 1970s the department had up to 12 NSF grants per year for American doctoral students. There has been no similar support since. Early in the department's history, a rotating colloquium, based on the Berkeley-Stanford model, was held once a quarter with Wayne State University and the University of Michigan. Joseph Gani began the Journal of Applied Probability at MSU in 1964 before he moved to Manchester. The department record for largest lecture was set in about 1965 for the business statistics class STT 315 (575 students). David Blackwell was awarded an Honorary Doctor of Science degree at Michigan State University in 1969.

John Kinney was a highly respected professor in mathematics who spent time in the department 1965–1976. He would often speak in cryptic terms, and, with his back to the blackboard, would write on the blackboard.

Michael Waterman, now a professor at the University of Southern California, was Kinney's student in Statistics and Probability (PhD, 1969). His thesis is titled *Some Ergodic Properties of Multi-Dimensional F-Expansions*. From the USC website, "Professor Waterman works in the area of computational biology, concentrating on the creation and application of mathematics, statistics and computer science to molecular biology, particularly to DNA, RNA and protein sequence data. He is the co-developer of the Smith-Waterman algorithm for sequence comparison and of the Lander-Waterman formula for physical mapping." Waterman was inducted into the National Academy of Sciences in April, 2001. He is shown in Fig. 1 signing the Registry of Membership during the induction ceremony.

Past Faculty and Visitors

The department has had many prominent faculty who spent some parts but not the ends of their careers in the department. Here is a partial list with their years in the department given: Peter Brockwell (1971–1973), Stuart Either (1975–1986), Joseph Gardiner (1908–2008), Gopinath Kallianpur (1956–1964), John Kinney (1965–1976), Ingram Olkin (1955–1960), Herman Rubin (1959–1968), Robert Staudte (1968–1974), and Joel Zinn (1977–1983). Visitors who spent at least a year in total include Sergei Chobanyan, Joseph Gani, Chris Heyde, Ildar Ibragimov, Rafail Khasminskii, Andrzej Makagon, Desmond Nicholls, Ferdinand

Fig. 1 Waterman induction into the National Academy of Sciences

Osterreicher, P. K. Pathak, N. Portenko, N. U. Prabhu, Alfred Renyi, Pal Revesz, and Jayaram Sethuraman.

Chairpersons

Departmental Bylaws adopted in the 1960s call for the election of a chairperson every 3 years with the result going to the Dean of the College of Natural Science as a recommendation. Every once in a while an incumbent chair was willing to continue. The longest serving chairperson was Habib Salehi (1989–2005). He positioned the department well in the transition from the quarter system to the semester system in 1992 and oversaw a rapid growth in enrollments in statistics courses. The chairpersons were technically "heads" until fall 1963. The list of chairs follows with A denoting "Acting" (Table 1).

Faculty Positions, Degrees Awarded, Enrollments, and Curricula

Faculty. The department grew slowly in the period from 1970 to 2000, moving from about 13 to 17 FTE tenure track faculty positions. By 2010 the number had grown to 23.75 FTE. See Appendix A for a photograph and list of current tenure track faculty and professors emeriti. Also see Fig. 2. This river graphic is patterned after the one that Professor W. J. Beal created on the wall of his laboratory to represent the origins of Michigan Agricultural College and the growth of the student body from 1855 to 1913 (*History of the Michigan Agricultural College* by W. J. Beal, MAC Press, East Lansing, MI, 1915).

Table 1 Chairpersons of statistics and probability

Sutherland Frame 1955–1956 (A)	Leo Katz 1956–1961 and 1962–1963
Kenneth Arnold 1961–1962 (A) and 1963–1967	Martin Fox 1966 Fall (A)
James Stapleton 1967–1969 (A) and 1969–1975	V. Mandrekar 1975–1981 and 1982–1985
Hira Koul 1981–1982 (A)	Dennis Gilliland 1985–1989
Habib Salehi 1989–2005	Vincent Melfi 2005–2006 (A)
Mark Meerschaert 2006–2009	Hira Koul 2009–

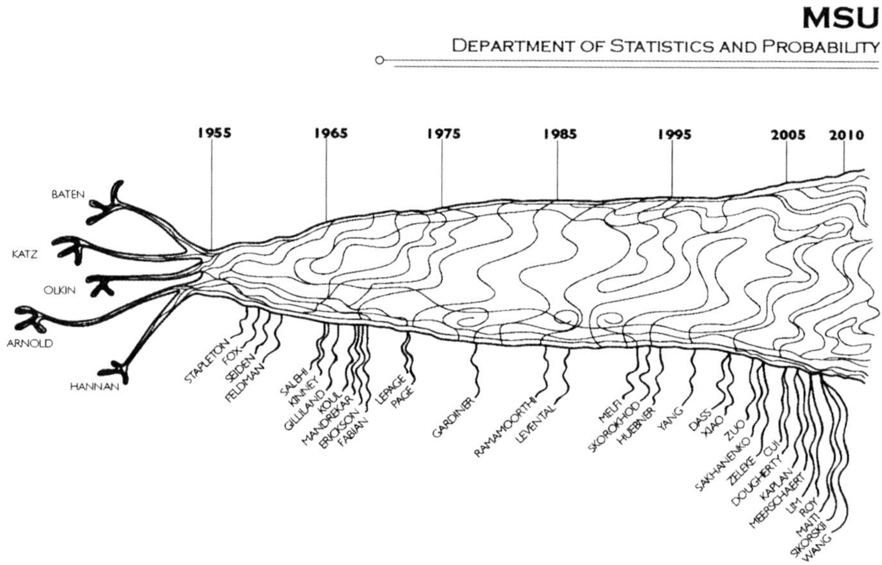

Fig. 2 River showing entry of the tenure track faculty now at or retired from MSU.[1] (Width of river is proportional to number of tenure track FTE positions)

The department has added six tenure track positions in the last 10 years. Entry level faculty positions have been added with the appointments of new chairs, following department reviews, and to meet the increasing demand for collaborative research and service courses. In the recent past, faculty positions have been targeted at specific research areas and collaborations. Mark Meerschaert came as full professor and chair in 2006 and Tapabrata Maiti joined as full professor in 2008.

Joseph Gardiner, after joining the department in 1978, gradually moved to the Department of Epidemiology in MSU's College of Human Medicine and is now full-time in that unit. In addition to heavy research and administrative responsibilities, he has served as a thesis advisor for nine doctoral students of the Department of Statistics and Probability, most of whom have taken positions in the drug or medical industry.

[1] Graphic by Jessica Crews, graphic designer, jlcrews25@yahoo.com

Research was strengthened with arrival of Academician Anatoli Skorokhod in 1993. With sadness we note that he passed away on January 3, 2011. According to Wikipedia, Skorokhod was the author of 450 scientific works, and was a member of the American Academy of Arts and Sciences.

Mandrekar and Koul have been honored by the university with distinguished faculty awards for their outstanding research. The three leading PhD thesis advisors to date, producing 66 of the total of 157 theses, are Koul 27, Mandrekar 20.5, and Hannan 18.5. Mandrekar has also directed or codirected students in other units, including mathematics. Salehi has had a 50% appointment in mathematics since joining MSU in 1965. Not only did he direct or codirect many PhD students in both departments, but he served on virtually all PhD committees in S&P.

The department has been enriched for most of the last ten summers from visits by Professor Ildar Ibragimov of the University of St. Petersburg. Ibragimov gives special topics courses for doctoral students.

Examples of professional service include the department's leadership in establishing the MidMichigan Chapter of the ASA in 1985. The inaugural address was given by Paul Meier of the University of Chicago. In 1990, Raoul LePage organized and chaired Interface 1990 together with a special meeting on the bootstrap method at Michigan State University sponsored by the Institute of Mathematical Statistics. *Statistics and Probability Letters*, published by North Holland, has offices in the department. Editors-in-Chief are Somnath Datta, 1988 department PhD graduate, now at the University of Louisville, and Hira Koul, currently department chairperson.

Undergraduate Program. The department has an undergraduate program that graduates fewer than 10 per year. A number of outstanding undergraduates took advantage of graduate courses and went on to earn PhD's at other institutions. These include D. van Dyk (PhD 1995 at the University of Chicago, now with Imperial College, London), T. Graves (PhD 1995 at Stanford, now at Los Alamos National Laboratory), S. Lalley (PhD 1981 at Stanford, now at the University of Chicago), Karl Rohe (PhD 2011 at UC-Berkeley, now at the University of Wisconsin), and G. Spaniolo (PhD 2000 at University of North Carolina, now at Genentech).

Doctoral Program. From 1955 to 2010, 157 PhD's were awarded with 4 in the 1950s, 21 in the 1960s, 28 in the 1970s, 34 in the 1980s, 34 in the 1990s, and 36 in the 2000s including 2010. See Fig. 3. In addition, faculty in the department have served as major professors for PhD graduates in mathematics or engineering.

Until the past several years, the course curriculum has been almost exclusively devoted to theory, either on statistical inference or on probability. The first year program includes two one-semester courses on measure theory-based probability and another two on "Lehmann type" statistical inference. The prelims, given at the beginning of the second year, have been, and still are, based on these courses. Doctoral students are now expected to take courses on applications, beginning with a linear models course. Recently, a course on computations has been added. Some thought is being given now to giving students a choice of two prelims from the three areas, probability, inference, and applications.

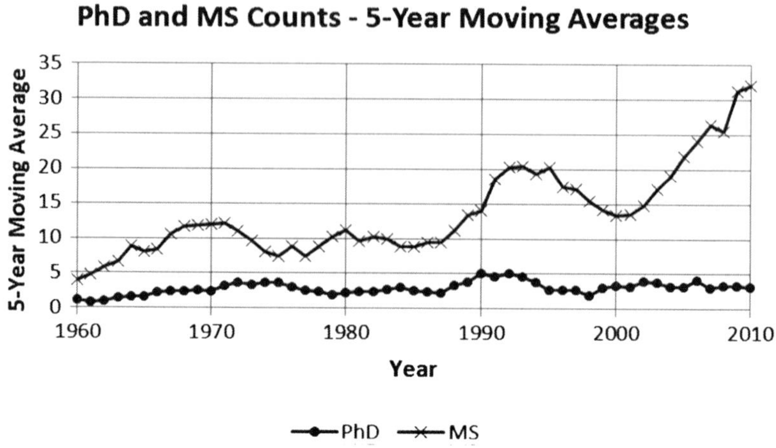

Fig. 3 Counts of PhD and MS Graduates. The latest dip in 5 year averages of numbers of MS degrees seems to have been caused by the Asian economic crisis of 1997–1998, bringing about significant decreases in the numbers newly enrolling in the Falls of 1997–2000 from Taiwan and South Korea

Perhaps the most significant changes in graduate enrollment over the last 29 years have involved Chinese students. Before 1982 a handful of Chinese students had come from Taiwan, just three for PhD's. The department's first mainland Chinese PhD graduate, Song Yang in 1988, has been followed by 37 others. Since 2001, 25 of the 32 PhD graduates have come from China. To date, 24 have remained in the US after graduation with 1 going to Canada. The big change among MS students began in 2006 when students bringing funds from China began study here, almost always for the MS degree. Currently the department has about 50 such students.

Masters' Programs. Over the years, the department has made its Masters' programs more accessible to a larger body of graduate students by removing certain mathematics requirements. These changes, the development of dual-master's programs, and the increased demand for the MS coming mainly from international students have led to an increase in interest and number of degrees granted (see Fig. 3).

Students admitted to the Master's program may have studied as little as 1 year of calculus plus a linear algebra course. That has changed little over the 55 years. There has seemed to be little relationship between a student's prior mathematical education and performance. Students have been expected to take two semesters on probability and inference (STT 861-2), plus courses on ε-δ analysis and linear algebra. Recently the math course requirement was dropped. Beyond these four courses, students were able to take any 800-level courses they wished, though most took courses on linear methods, sampling, time series, computing, and multivariate methods. All MS students have been required to pass an exam over STT 861-2 given during their second year of study.

The department is proud of the achievements of its MS graduates. It is feasible to list only a few of them. Examples are D. DuLaney, E. Nagy, N. Rudomino and M. Yao in actuarial science, G. Dilloway and R. Nero in business, S. Dorinski in government, and J. Samarias, who headed the technical side of the NASDAQ. Many of the MS graduates earned PhD's in other programs including statistics and biostatistics at other institutions.

Teaching. A major change occurred in 1992 that greatly affected the growth of the department. That year the university changed from the quarter system to the semester system, decided by a university faculty vote of 400–399. The authors, who voted "yes", felt important for a moment. Chairperson Salehi positioned the department well in the transition. For example, a new elementary statistics course with a computer component was introduced. Roy Erickson was its first instructor. Presently, students earn over 4,000 semester credits per year in the course. Immediately prior to the conversion, the quarter credits earned in the department's courses amounted to the equivalent of about 10,000 semester credits per year. In calendar year 2010, the total number of semester credits earned through the department was more than 20,000.

Teaching loads in the department are nominally three courses per year though some faculty have lighter teaching loads accompanying research and/or administrative responsibilities. There has been increasing demand for service courses in statistics. This is due to an emerging recognition of the importance of the subject to many disciplines and the fact that the university decided to allow some courses in precalculus probability and statistics to partially satisfy a university mathematics graduation requirement.

Administration. The Advisory Committee for the department includes all tenure track faculty plus two elected students. The committee determines curriculum, administers the prelim exams, conducts the election of chairpersons, and, in general, advises the chairperson. The subset of the AC consisting of associate and full professors advises the chairperson on personnel decisions.

Since 1975 the department has designated faculty as graduate directors or co-directors, with responsibilities for (i) admissions of graduate students and (ii) graduate assistant assignments. Professors Hannan (10 years), Salehi, Stapleton (20 years), Ramamoorthi, Yang, and (now) Maiti have held the position. For the last 5 years, some responsibility has been split off with Professors Dass and Levental administering (ii).

Some Graduates of the PhD Program

We list 30 of the 157 graduates of the doctoral program, organized by decade of completion of their degree. Choosing whom to include was a difficult task. Their last or current affiliation is indicated. There were four graduates in the 1950s including W. Harkness (Penn State). There were 21 graduates in the 1960s including S. D. Chatterji (École Polytechnique Fédérale de Lausanne, Switzerland),

V. Mandrekar (Michigan State), A. Oaten (UC-Santa Barbara), B.L.S.P. Rao (Indian Statistical Institute), J. Van Ryzin (deceased, chair at Columbia), and M. Waterman (Southern California, member of the National Academy of Sciences). There were 28 graduates in the 1970s including J. Boyer (Kansas State), J. Boyett (St. Jude Children's Research Hospital), G. Hamedani (Marquette), D. Ruppert (Cornell), V. Susarla (deceased, SUNY-Binghamton), and S. Vardeman (Iowa State). There were 34 graduates in the 1980s including S. Datta (Louisville), R. Karunamuni (University of Alberta), S. N. Lahiri (Texas A&M), M. Merkle (University of Belgrade), M. Pourahmadi (Texas A&M), A. Schick (SUNY-Binghamton), A. R. Soltani (Kuwait University), and T. N. Sriram (Georgia). There were 34 graduates in the 1990s including L. Chen (American Express), M. Geraldes (Bristol-Myers Squibb), K. Mukherjee (Lancaster, England), K. Podgorski (Lund University, Sweden), S. Rajagopalan (General Electric), and H. Zhang (Purdue). There were 36 graduates in the 2000s including J. Hannig (North Carolina), W. Song (Kansas State), and L. Wang (Georgia).

Statistical Consulting

The need for consulting was stated as a major reason for the creation of the Department of Statistics. At the outset, decisions to consult or not were left to the individual faculty and no formal consulting service was organized within the department. A Statistical Laboratory directed by Leo Katz did serve as an umbrella for outreach with a focus on grants, not campus-wide consulting. In the 1970s and 1980s a few individual faculty continued to push for and provide consulting while chairs asked for and sometimes received nonrecurring funds to support graduate students in consulting roles. In 1986, the Statistical Consulting Service was created in the department, and soon after that support for the director came in the form of a one course reduction in teaching per year. The university eventually put funds into the recurring budget for one graduate assistantship dedicated to consulting; the college, graduate school, and provost added nonrecurring funds for one or two more graduate assistants. For most of nearly 20 years, the Statistical Consulting Service was directed or codirected by Dennis Gilliland and Connie Page as demand for consulting and for support in grant proposal development increased greatly. In 2004, the MSU Foundation provided seed money to establish a centrally located and better funded consulting service. This led to the Center for Statistical Training and Consulting that is administratively housed in the Office of the Vice President for Research and Graduate Studies. The current director is not in Statistics and Probability, but many of the graduate assistants who provide consulting come from the unit and several faculty are on the Advisory Board.

The university did have and continues to have positions for statisticians in the College of Agriculture and Natural Resources with partial appointments for consulting in the Experiment Station. These individuals conduct basic research in statistics, provide vital support for research at the university, and teach courses in the Department of Statistics and Probability. Charles Cress (PhD, Iowa State University, 1966) was in the Department of Crop and Soil Sciences until 1993. Oliver Schabenberger (PhD, Virginia Tech, 1995, now at SAS, Cary, NC) filled the position until 1999 and Alexandra Kravchenko (PhD, University of Wyoming, 1997) currently serves. John Gill (PhD, Iowa State University, 1965) was in the Department of Animal Science until 1999. He was succeeded by Robert Tempelman (PhD, University of Wisconsin, 1999), who continues in that position.

Miscellany

(1) Esther Seiden joined the faculty in 1960, retired in 1978, and now at age 103 resides in Jerusalem.
(2) Pramod K. Pathak completed work on his PhD at the Indian Statistical Institute under the direction of Professors C. R. Rao and D. Basu in 1961 at the age of 20, probably an all-time record for youth. In 1962 he joined the MSU department as a research associate. After a career as professor at the Universities of Illinois and New Mexico, he returned to MSU in 1998 as a visiting professor with wife, Dorothy, a professor of epidemiology.
(3) While driving to Berkey Hall in about 1963 with fogged-up windows, the car driven by Martin Fox, accompanied by Herman Rubin, was hit by a train. Fortunately, the on-campus train, moving coal to the power station, was moving at 3 mph. The car, but not its occupants, was slightly damaged. The train survived.
(4) The department explored the idea that Jerzy Neyman might be induced to join following "mandatory retirement" at Berkeley in the early 1970s. This led to a meeting with the provost, but not the desired result. The department's faculty at that time included four graduates of Berkeley (Esther Seiden, Martin Fox, Dorian Feldman, and Hira Koul).
(5) Ken and Pauline Arnold, Dorian and Ann Feldman, and Hira and Shyama Koul have been particularly gracious hosts to many social gatherings over the years.
(6) The generosity of faculty and graduates of the department is shown through gifts and established endowments that benefit the department and its missions. After the death of Professor Emeritus Vaclav Fabian in 2007, his wife Alena made a gift to the department enabling it to completely refurbish the department's colloquia-coffee room, now the "Fabian Room." More recently, Bettie Hannan established an endowment to provide for James Francis Hannan Visiting Scholars in Theoretical Statistics and Probability. Other endowments are the Norbert Wiener, the John Kinney Memorial, the Van Ryzin-Susarla Scholarship, the Czech Students, the William Harkness Teaching Awards, and the James and Alicia Stapleton Travel Fund for Graduate Students.
(7) Recognition in the form of personal thanks from former students means a lot to the department. A call came to the department in December 2010 from a student wishing to thank Professor Katz for arranging credit that allowed her

to graduate some 40 years earlier. It serves as a reminder of the lasting and memorable impact that a professor can have. About 10 years ago a master's degree graduate of 1969, originally from Taiwan, out of the blue, sent the department $10,000 in gratitude for her success.

Appendix A: 2010 Tenure Track Faculty

Current faculty with full or partial appointments in the department are listed below. The PhD granting institution, the year of the PhD, and the entry year into a tenure track position at MSU are given.

Yuehua Cui (Florida, 2005, 2005)	Sarat Dass (Purdue, 1998, 2000)
Daniel Dougherty[a] (North Carolina State, 2002, 2005)	Dennis Gilliland (Michigan State, 1966, 1966)
Marianne Huebner (Southern California, 1993, 1994)	Jennifer Kaplan[a] (U. of Texas, 2006, 2006)
Hira Koul (Berkeley, 1967, 1968) Chairperson	Raoul LePage (Minnesota, 1967, 1972)
Shlomo Levental (Wisconsin, 1986, 1986)	Chae Young Lim (Chicago, 2007, 2007)
Tapabrata Maiti (University of Kalyani, 1996, 2008)	V. Mandrekar (Michigan State, 1964, 1968)
Mark Meerschaert (Michigan, 1984, 2006)	Vince Melfi (Michigan, 1991, 1992) Assoc Chair
Connie Page (Michigan, 1972, 1972)	R. V. Ramamoorthi (ISI, 1981, 1984)
Parthanil Roy (Cornell, 2007, 2007)	Lyudmila Sakhanenko (New Mexico, 2002, 2003)
Habib Salehi[b] (Indiana, 1965, 1965)	Alla Sikorskii (Michigan State, 2000, 2008)
Anatoli Skorokhod (Moscow State, 1956, 1993)	Lifeng Wang (Minnesota, 2006, 2008)
Yimin Xiao (Ohio State, 1996, 2000)	Lijian Yang (North Carolina, 1995, 1997)
Aklilu Zeleke[a] (Temple, 1997, 2003)	Yijun Zuo (U. of Texas–Dallas, 1998, 2002)
Roy Erickson[c] (Michigan, 1968, 1968)	Dorian Feldman[c] (Berkeley, 1961, 1961)
James Stapleton[c] (Purdue, 1957, 1958)	Esther Seiden[c] (Berkeley, 1949, 1960)

[a] Less than 100% in the Department.
[b] Joint appointment with the Department of Mathematics.
[c] Emeritus.

Seated from left to right are chairperson Koul and past chairpersons Melfi, Salehi, Gilliland, Mandrekar, Stapleton (E). Past chairperson Meerschaert is not pictured. Standing from left to right: Levental, Roy, Huebner, Lim, Wang, Sikorskii, Maiti, Kaplan, Feldman (E), Zuo, LePage, Sakhanenko, Dass, Page, Cui, Xiao, Dikong (V), Pathak (V), Sinha (V). Not pictured: Dougherty, Erickson (E), Ramamoorthi, Seiden (E), Skorokhod, Yang, Zeleke. E = emeritus, V = visitor

Appendix B: Example Books Authored or Edited by Faculty

Introduction to Probability and Mathematical Statistics, **Vaclav Fabian** and **James Hannan**, Wiley, NY, 1985.

Probability—the Mathematics of Uncertainty, **Dorian Feldman** and **Martin Fox**, Marcel Dekker, NY, 1991.

Weighted Empirical and Linear Models, **H. L. Koul**, Lecture Notes-Monograph Series, Vol. 21, Institute of Mathematical Statistics, Hayward, California. 1992.

Weighted Empirical Processes in Dynamic Nonlinear Models, 2nd Edition, **H. L. Koul**, Lecture Notes Series in Statistics, Vol. 166, Springer, New York, NY, 2002.

Frontiers in Statistics: A Collection of Refereed Papers by World Renowned Statisticians in Honor of Peter Bickel's 65th birthday, edited by Jianqing Fan and **H. L. Koul**, Imperial College Press, 2006.

Silver Jubilee Issue of Statistics and Probability Letters to Celebrate 25th year of the Journal and Felicitate seventieth birthday of the first Editor Richard Johnson, edited by M. Akritas, **H. L. Koul** and Anton Schick, July 2007.

Exploring the Limits of Bootstrap, edited by **Raoul LePage** and Lynne Billard, Wiley, NY, 1992.

Proceedings of the Norbert Wiener Centenary Congress 1994, Symposia in Mathematics, edited by **V. Mandrekar** and P. R. Masani, American Mathematics Society, 1997.

Stochastic Differential Equations in Infinite Dimensions with Applications to Stochastic Partial Differential Equations, L. Gawarecki and **V. Mandrekar**, Springer, 2011.

Limit Distributions for Sums of Independent Random Vectors: Heavy Tails in Theory and Practice, **M. M. Meerschaert** and H.P. Scheffler, Wiley, NY, 2001.

Mathematical Modeling, 3rd Edition, **M. M. Meerschaert**, Academic Press, 2007 (4th Edition coming in 2012).

Recent Developments in Nonparametric Inference and Probability, edited by Jiayang Sun, Anirban DasGupta, **Vince Melfi**, and **Connie Page**, IMS Monograph Series, Vol 50, 2006.

Bayesian Nonparametrics, J. K. Ghosh and **R. V. Ramamoorthi**, Springer-Verlag, NY, 2003.

Random Perturbation Methods with Applications in Science and Engineering, **Anatoli Skorokhod**, Frank Hoppensteadt and **Habib Salehi**, Springer-Verlag, NY, 2002.

Linear Statistical Models, 2nd Edition, **J. Stapleton**, Wiley, NY, 2009.

Models for Probability and Statistical Inference, **J. Stapleton**, Wiley, NY, 2008.

North Carolina State University Department of Statistics

Tom Gerig

Gertrude Mary Cox, Founding Head, 1941–1949

In the late 1930s, the US Department of Agriculture was moving to advanced statistical methods for the collection and analysis of data. By chance, early in 1940 a leading USDA statistician, W. F. Callander, was traveling on the same train as the President of the Consolidated University of North Carolina, Frank Porter Graham. During their conversation Callander described the need to establish a statistical center in the south. Graham, who had always been supportive of the College of Agriculture at NC State College, responded "We'll do it at North Carolina State College".

By summer 1940, the decision to establish a center at NC State had been made and a search for a director had begun. As chair of the search committee, G. W. Forster contacted several prominent statisticians soliciting their advice. George Snedecor of Iowa State responded by preparing a list of suitable candidates which he circulated among his colleagues for their reactions. The head of the Iowa State Computing Laboratory, Gertrude M. Cox responded "I don't see my name on there" to which he answered "That's an oversight, isn't it?" To his letter Snedecor added a postscript stating "If you are willing to consider a woman, I know of none better qualified than Gertrude M. Cox". Cox had a master's degree in mathematics from Iowa State and was enrolled in a PhD program. She had developed a course in design of experiments and taught it for several years.

Forster brought Cox to interview and during her visit she met with President Graham, establishing a valuable patron. She received an offer dated September 24,

North Carolina State University,
Department of Statistics,
2311 Stinson Drive, Campus Box 8203,
Raleigh, NC 27695-8203, USA
T. Gerig (✉)
e-mail: tom_gerig@ncsu.edu

A. Agresti and X.-L. Meng (eds.), *Strength in Numbers: The Rising of Academic Statistics Departments in the U.S.*, DOI: 10.1007/978-1-4614-3649-2_13,

Fig. 1 William Cochran, Gertrude Cox, and Harold Hotelling, Institute of Statistics, 1944

1940 which she accepted. Cox was successful in gaining agreement that the center would take the form of a department of statistics located in the College of Agriculture. She reported for work on November 1, 1940. On January 22, 1941, with the strong support of Graham, the Board of Trustees of the Consolidated University of North Carolina approved the establishment of the Department of Experimental Statistics and appointed Cox as a professor and head. This gave her the distinction of being both the first female professor (of any rank) and the first female head at the NC State.

Cox immediately began planning a consulting service and courses to be taught. During the department's first year she organized summer courses in applied and theoretical statistics that attracted students from around the country. Faculty included Harold Hotelling, George Snedecor, and Cox. More summer sessions would be organized in the years to come, adding to an already impressive list such notable faculty members as W. G. Cochran, R. A. Fisher, J. Wolfowitz, and C. I. Bliss. (Figs. 1, 2).

Her first hire was R. J. Monroe, who was Cox's assistant. She brought R. L. Anderson from Iowa State to a position in the mathematics department, soon after to move to statistics. J.A. Rigney, who was already in agronomy at NC State where he was teaching statistics, accepted a joint appointment in statistics. For A. L. Finkner, who worked for the USDA in Raleigh, Cox arranged a cooperative arrangement. Her fifth hire, R. E. Comstock, had a strong genetics background and could advise those working in crop and herd improvement.

From its outset, the department had a broad range of responsibilities including teaching, research, consulting, and providing computing services. Statistics was

Fig. 2 Statistics summer session, 1946. *First row* is G. Snedecor, J. Wolfowitz, G. Cox, R. A. Fisher, C. Bliss, W. Cochran, G. Casanove. Names of all pictured are at www.stat.ncsu.edu/information/history/photokeys

taught to undergraduates and graduate students, to extension workers, and to targeted groups from several disciplines. Research undertaken comprised a mix of methodological development and the application and adaptation of existing methods to new problems. Through its consulting the department supported researchers on campus and throughout the South and beyond.

For over 70 years the college of agriculture has supported a service consulting program in the department. Designated faculty and staff devote a portion of their effort to assisting clients with planning their experiments, analyzing their data, and interpreting their results.

The main focus of the department's research was on application in the agricultural and biological sciences, but soon broadened to include the social sciences and economics as well as industrial statistics. To support the researchers in these areas, the department organized numerous working conferences in such areas as animal science, plant science, economics, and quality control which were attended by invitation.

Institute of Statistics

Cox submitted a proposal to President Graham to organize an institute that would provide flexibility in the management of the state and external funds. She envisioned an institute that would administratively answer directly to the President's office. Graham received the proposal favorably and with the help of his ally, former Governor O. Max Gardner, would steer the proposal through the Executive Committee of the UNC Board of Trustees. On September 18, 1944 the committee approved the establishment of the Institute of Statistics and approved Cox as its director.

The goals of the institute of statistics were to provide a teaching, research, consulting, and service center for statistical work, to promote the use of efficient statistical techniques in diverse fields of application; and to advance statistics through the discovery of new techniques by theoretical investigations.

Following the establishment of the institute, Cox submitted a proposal for financial support to the General Education Board (a philanthropy created and funded by John D. Rockefeller). With Graham's support, on December 7, 1944, the board awarded $87,000 to the institute of statistics to provide for expansion of its programs.

It is useful to clarify the relationship between the institute and the department. The institute received state funds through the UNC system President's office while the department received funds through the usual university channels. The institute was a creature of the Consolidated University. At first funds in the institute were used solely to support activities in the department, including portions of faculty salaries. Cox was heard to say that the department was part of the institute. At this point she was the institute director and department head. In the early years of the department, faculty publications listed the institute of statistics as the home department of the author.

Effects of World War II

During World War II, several faculty members from the department served in uniform while others contributed their technical skills to training and research. The end of the war in 1945 brought a period of growth of academia. The institute of statistics (and the department) benefitted with a spate of new hires. Starting in 1946 these include W. G. Cochran (sampling and a Cox collaborator), Paul Peach (industrial statistics), H. F. Robinson (genetics), H. L. Lucas (animal science), F. E. McVay (social science and economics) in 1948, and H. F. Smith (crop science) in 1949.

Expanding the Institute of Statistics to Include UNC Statistics and Biostatistics

Cox recognized that to be effective the faculty should comprise a strong mix of theoretical and applied statisticians. In a recorded interview in 1980, D.D. Mason stated that

> ...(Cox's) dream was to have a complete unit here [at NC State] in which the core was a strong theory group and various applied statisticians relating to the various fields in teaching, consulting, and research around this theory group. Being part of the Consolidated University, the university officials said that the arrangement would have to be a little different and that the mathematical statistics group should be located at Chapel Hill and the applied group here.

Cox moved to expand the institute of statistics to include the UNC at Chapel Hill. Cox played a major role in attracting Harold Hotelling from Columbia as the founding head of a new theoretical statistics department established on July 1, 1946 at UNC. It is said that Hotelling's moving to Chapel Hill was W. G. Cochran's condition for his accepting a position with the institute in Raleigh.

In the institute's expanded format, Cochran was appointed associate director for the Raleigh branch, Hotelling that for the Chapel Hill branch and Cox remained the director. The graduate programs on the two campuses would be interdependent with NC State offering the applied courses and UNC the theoretical. Students from both programs would commute. Besides his new responsibilities with the institute, Cochran was given the task of building the graduate program in statistics.

In 1949, the institute continued its expansion when the Department of Biostatistics was organized at Chapel Hill. B. G. Greenberg, who had received his PhD from NC State that year, served as founding head.

Statistical Genetics

Through this period, two quantitative geneticists on the faculty, Comstock and Robinson, were gaining national recognition. In 1947, the Rockefeller Foundation provided a decade worth of support for the newly formed Quantitative Genetics Program. With this support, significant hires in statistical genetics would include C. C. Cockerham in 1953 and D. F. Matzinger and R. H. Moll in 1957. The Fulbright fellow, Ken-Ichi Kojima, was also a strong contributor.

In 1957, the college of agriculture organized a genetics department by bringing together existing faculty. The head of the new department would be Robinson from statistics. Also from statistics came Kojima, Matzinger, and Moll, while Cockerham and Comstock stayed behind.

In 1953, Cox was one of the principal organizers of what would become the Southern Regional Council on Statistics. Representatives from participating departments met periodically to discuss common problems, arrange short courses, and sponsor a summer research conference.

Cox Resigns the Headship

In 1949, Cox stepped down as department head, while retaining in her position as institute director for the next 10 years. In 1958, Cox undertook yet another challenge as she assisted with the establishment of the Research Triangle Institute (RTI) near Raleigh. In 1960, Cox retired from the institute directorship to become head of the newly established Statistics Research Division at RTI, which she held for 5 years.

Jackson A. Rigney, Head, 1949–1963

Dr. J. A. Rigney replaced Cox as head on July 1, 1949. Rigney was a graduate of Iowa State and was hired by NC State in the agronomy department in 1938, moving to statistics in 1941.

Among the faculty hired by Rigney, many would remain in the department well into the 1980 s. Hires with long tenure included R. J. Hader (1951, industrial), A. H. E. Grandage (1952, design), R. G. D. Steel (1969, design), J. C. Koop (1960, sampling), L. J. Herbst (1962, time series), T. D. Wallace (1959, economics), C. H. Proctor (1960, sociology), van der Vaart (1962, biomathematics), D. D. Mason (1953, agronomy), C. C. Cockerham (1953, genetics), J. O. Rawlings (1960, plant breeding), and D. W. Hayne (1962, zoology). Notable hires with shorter tenures included W. Mendenhall (1958, reliability), D. G. Horvitz (1953, sampling), D. F. Matzinger (1957, genetics), and R. H. Moll (1957, genetics).

During Rigney's tenure, two notable students received their doctorates from the department. Geoffrey Watson, an Australian who had a productive career at Princeton, and J. Stuart Hunter, a leading industrial statistician and member of the National Academy of Engineering.

Three significant programs were initiated during Rigney's term as head.

Southeastern Cooperative Fish and Game Statistics Project

In 1958, the Southeastern Cooperative Fish and Game Statistics Project (SCFGSP) was established in the institute through Finkner's effort. Its purpose was to provide statistical services and training to the State Fish and Wildlife Agencies in the southeast. The project was initially headed by Finkner, then by Hayne in 1962, by Pollock in 1980, and finally by Turner in 1987. At its peak year, 14 participating southeastern states entered into cooperative agreements with the institute to gain access to the services of the project. Much of the statistical work was carried out by graduate students, who benefitted from the experience. Through the years, state agencies developed their own in-house competencies, so the work of the project was terminated in 1995.

Biomathematics Program

Since its formation, the department had been heavily involved in applying statistical methods to biological problems. In the period leading up to 1960, H. L. Lucas and colleagues broadened this activity to include the mathematical modeling of biological phenomena. Lucas created a training and research program based on the combined fields of biology, mathematics, and statistics, now referred to as biomathematics.

Graduate degree programs in biomathematics were established in 1961 with degrees to be granted by the department. In 1966, an NSF program development grant supported graduate and postdoctoral stipends, and faculty salaries. With this stimulus, students and fellows were recruited and the department added biomathematics faculty including H. J. Gold (1965), D. C. Martin (1966), M. B. Williams (1967) and D. L. Ridgeway (1967).

Early Developments in Computing

Since its formation, an important element of the department's mission has been to provide computing for university researchers, especially to those in the Agricultural Experiment Station. At first these services were facilitated using (Monroe and Marchant) desk calculators. At the peak, the department supported a lab of 12 technicians carrying out clients' data analyses.

In 1956, the department acquired an IBM-650 computer, the first electronic digital computer in an educational institution in North Carolina. The acquisition led to pioneering efforts in writing computer programs for data analysis, an activity overseen by F. Verlinden.

Administrative Changes

In 1958, budgetary control of the institute was moved from the UNC system-wide director to the individual sections (Raleigh and Chapel Hill). From here on, the department head also served as local institute director.

In the post-Sputnik era, there was renewed emphasis on the sciences. This could account for the formation in 1960 of the School of Physical Sciences and Applied Mathematics, consisting initially of four departments: Chemistry, Applied Mathematics, Physics, and Experimental Statistics. In this reshuffle, the academic affairs portion of the department was moved from agriculture to the new school while a portion of its activities continued to be administered through agriculture, including statistical consulting and computing. The institute was unchanged.

David D. Mason, Head, 1963–1981

Mason earned his PhD in soil physics (with a minor in statistics) from NC State in 1948. He took a position at Ohio State where he developed and taught a course in design of experiments. Mason returned to NC State in statistics in 1953. When Rigney resigned his administrative positions, Mason was named head, a position he held until he retired in 1981.

During the early years of Mason's tenure, the university system relaxed the "allocation for function" rules which stipulated that statistics at NC State would hire applied faculty while that at UNC would hire theoretical ones. In a 1980 interview, Mason stated that because of this, he began hiring "...people who [are] oriented to the more theoretical aspects of statistics." Among those hired during this period were W.J. Hall (1960) only briefly, H.R. van der Vaart (1962, inference, biomathematics), B.B. Bhattacharyya (1963, general theory and OR), O. Wesler (1964, probability), C.P. Quesenberry (1966, inference), A. R. Gallant (1971, time series), and D. Boos (1977, nonparametrics). These were the first theoretically oriented faculty members hired since R. L. Anderson in 1942. They enabled the department to offer its own theoretical graduate courses including advanced inference, measure theory, and probability. In recognition of these changes, in 1970 the department renamed itself the Department of Statistics.

During this same period, the department built a program in time series and econometrics. Following the hiring of L. Herbst (1962), the department hired A. R. Gallant (1971) jointly with agricultural economics. Subsequent hires in this area included T. Johnson (1974), D. A. Dickey (1976), J. Monahan (1977), S. Pantula (1982), and P. Bloomfield (1983). Gallant, Dickey, and Pantula were all students of Wayne Fuller of Iowa State.

For several years, faculty from the departments of agricultural economics and statistics had been cooperating on econometric research. To support this effort, T. D. Wallace was hired in 1959 with a joint appointment. From this activity, the Joint Econometrics Program was established in 1964. Wallace resigned in 1974 and was replaced by Johnson. C. Little who joined the program briefly in 1968 and shortly after was replaced by A. R. Gallant.

Hires in design and linear models were: L. Nelson (1964, agronomy), J. A. Warren (1965, horticulture), A. R. Manson (1966), A. C. Linnerud (1967, animal science), T. M. Gerig (1969), J. H. Goodnight (1970), F. G. Giesbrecht (1971), B. Stines (1976), H. Kirk (1977) and W. Swallow (1980), and in nonparametrics: D. Boos (1977) and J. Deitz (1978). Other hires included: J. Wasik (1967, social science), T. Reiland (1977, OR), and K. Pollock (1980, quantitative ecology).

Biomathematics added H. Gold (1965), D. C. Martin (1966), D. L. Ridgeway (1967), M. B. Williams (1967), and F. A. Benford (1979).

The Quantitative Genetics Program headed by Cockerham added M. M. Goodman (1967), P. Burrows (1971), and B. Weir (1976). In 1963, this program received a 7-year NIH Program Project Grant supporting research in statistical genetics. The grant was directed for 27 years by Cockerham and for an additional 15 years by B. Weir.

Operations Research Program

In the 1964, an interdisciplinary program in Operations Research was established under the leadership of R. L. Anderson. In 1967, S. E. Elmaghraby was hired in industrial engineering to direct the OR program, which included interdisciplinary graduate degree programs. Participating departments included statistics, mathematics, and industrial engineering. Bhattacharyya was a founding member of the OR faculty, Reiland joined a decade later and C. E. Smith, S. Ghoshal and Z-B. Zeng subsequently became members.

Developments in Computing

The awarding in 1962 of a major NIH Computer Facilities Support grant with Mason as PI provided a significant boost to departmental computing. With successful renewals, this grant provided generous annual support for nearly 12 years.

More change in the computing environment occurred in 1965 with the establishment of the Triangle University Computing Center and its acquisition of an IBM System/360 computer. TUCC linked NC State, Duke, and UNC to a central computer located in Research Triangle Park. The 360 created the need to rewrite statistical software. By now, regional agricultural experiment stations were heavy users of statistical software and recognized the need for reprogramming but lacked the in-house expertise for doing so. In 1966, the University Statisticians of the Southern Experiment Stations (USSES) was formed, with participation of the 13 southern state universities, to coordinate their efforts. Since the department had the resources through its NIH computer grant and had experienced programming personnel, it played a leading role in this effort.

Early Development of SAS

Having written programs for the department to run on the IBM-650, Anthony James Barr was rehired in 1966 to reprogram his analysis of variance and regression software to run on the IBM-360. Later that year Barr made a presentation to a software committee of USSES, explaining his ideas for the development of statistical software. These ideas were a sketch of the architecture for what was to become SAS. In the months that followed, Barr wrote the fundamental structure and language of SAS. Subsequently James Goodnight, who had been hired as a student to write statistical software, began collaborating with Barr. Later, Goodnight programmed a general linear models procedure that would prove an important addition to SAS. Barr and Goodnight became co-leaders of the SAS Project housed in the department. The first formal release of the system was SAS 71. John Sall joined the project in 1973. A prolific programmer, Sall wrote procedures for time series, econometrics, categorical data, and for carrying out matrix algebraic manipulations.

After the NIH funding terminated, the SAS Project was supported by the USSES with the 13 coalition members making annual contributions. Demand for the system grew rapidly. In 1976, Barr, Goodnight, Sall, and Jane Helwig moved off campus and incorporated the SAS Institute, Inc.

Daniel L. Solomon, Head, 1981–1993

Daniel L. Solomon received his PhD from Florida State in 1968. That year, he began his career with the Biometrics Unit at Cornell and led the unit from 1977 to 1981. He came to NC State as head of the department on July 1, 1981.

At Solomon's arrival a significant proportion of the existing senior faculty had been department members for 30–40 years. Among these were six faculty members who worked with Gertrude Cox during the first decade. As the department turned 40 years old, R. J. Monroe, J. Rigney, D. D. Mason, D. Hayne, F. E. McVay, R. J. Hader, A. H. E. Grandage, and R. G. D. Steel retired. Along with this exodus of founding fathers went two key founding senior staff members, Helen Ruffin and Lillian Hunter, who had served the department well and kept the faculty in line for 35 years. Solomon would lead the department into its second 40 years.

During Solomon's 12 years as head some 20 tenure-track faculty members were hired.

Hires were made to add strength and depth to the department's holdings in time series and econometrics: P. Bloomfield (1983) and S. Pantula (1982) joining Gallant, Dickey, Monahan, and Johnson.

Two were hired to reinvigorate industrial statistics: F. Guess (1983) and J.C. Lu (1988) replacing the retiring Hader and Manson.

The statistical genetics program grew by S. Muse (1998), J. Thorne (1993), and S.-B. Zeng (1992) joining Cockerham and Weir.

The biomathematics program added S. Ellner (1986), T. Kepler (1993), and C. Smith (1983) joining R. van der Vaart, H. J. Gold, F. Benford, M. Wann, and D. L. Ridgeway.

Joining several faculty members in biometry, design, and quantitative ecology were K. Burnham (1983), J. Hughes-Oliver (1991), C. Brownie (1982) and M. Gumpertz (1989).

Four were hired in the area of general methodological and theoretical statistics: R. Berger (1983), M. Davidian (1987), D. Nychka (1983) and L. Stefanski (1986).

Cockerham's Retirement

In 1990 C. Clark Cockerham retired after leading the Statistical Genetics group since 1958. The first annual C. Clark Cockerham Lecture was held by the department in his honor in 1991. Weir replaced Cockerham as the head of the Statistical Genetics group.

Further Developments in Computing

In a first move toward the Unix world, in 1984, with NSF SCREMS support, a team of faculty purchased a DEC VAX-11/750 minicomputer. In 1991 the VAX was replaced by a network of several Sun workstations. As part of this network, the department developed a Statistics Instructional Computing Lab equipped with Sun workstations.

In 1991, the department opened a graduate student research computing laboratory largely made possible by the fundraising efforts of the graduate students.

Graduate Industrial Trainee Program

In 1983, the department initiated its highly successful Graduate Industrial Trainee (GIT) program. Nearly 30 years later these traineeships still provide graduate students with hands-on industrial experience, with a full financial package, and without interrupting their curriculum requirements. Some 150 students have benefitted from it. The department received the ASA/SPAIG award in recognition of the achievements of this program.

National Institute of Statistical Sciences

With Solomon heading the NC State contingent, in 1990, a Research Triangle consortium submitted the winning proposal that brought National Institute of Statistical Sciences (NISS) to the Research Triangle area. The department is collaborator with NISS and the faculty and students have benefitted greatly from its presence.

Solomon Becomes Dean

In 1993, Solomon resigned the headship to become an associate dean and in 2000, to became dean of the School of Physical and Mathematical Sciences.

Thomas M. Gerig, Head, 1994–2002

Thomas Gerig received his PhD from UNC and joined the department in 1969. Bloomfield and then Monahan served as interim department head beginning in August 16, 1993 until Gerig was appointed head September 1, 1994.

During Gerig's tenure, 12 faculty members were hired. Hires in the biological areas were four in biostatistics: M. Davidian (1997), A. Tsiatis (1997), W. Wang (1995), and D. Zhang (1998), two in biomathematics: T. Elston (1998) and S. Lubkin (1997) and one in biometry and design: J. Osborne (2001). In statistical theory and methodology there were hires in environmental statistics: M. Fuentes

(1998), in Bayesian statistics: S. Ghosh (1996) and S. Ghoshal (2001), in time series: M. Genton (2000), and in support vector machines: H. Zhang (2002). The department also made two teaching hires: T. Balan (1999) and P. Arroway (2001).

Biostatistics Program

In 1997, with the hiring of A. Tsiatis and the rehiring of M. Davidian, the department moved to establish a biostatistics program. Biostatistics courses were developed and a master's concentration in the area was offered. Affiliations with medical faculty at Duke have provided important access to problems and data. From the beginning, biostatistics proved to be popular with the students.

NSF VIGRE Grant 1999–2004

In 1999, the NSF advertized a training grant targeted at the Mathematical Sciences (and Statistics) promoting vertical integration of faculty and students in research and education. The department was awarded a VIGRE grant in the first round. A key element of the proposal called for faculty to form clusters focused on various research and training topics. Postdoctoral fellows, graduate students, and undergraduates participated in cluster activities thereby fostering vertical integration. Most of the clusters still function today.

Bioinformatics Program

Under the leadership of B. Weir, the Bioinformatics Research Center (BRC) was established in 2000. Its research addresses the theoretical and practical aspects of human, plant, and animal genetics. A graduate training component was developed as an interdisciplinary curriculum in bioinformatics to supplement the existing doctoral program in statistical genetics.

Statistical and Applied Mathematical Institute

Established in 2002, Statistical and Applied Mathematical Institute (SAMSI) is a DMS/NSF institute located in the Research Triangle. Each year workshops are organized from which working groups are formed around a research topic making use of both statistics and applied mathematics. NC State and the department are members of the consortium that created and manages the institute.

Sastry G. Pantula, Head, 2002–2010

Sastry Pantula earned his masters at the Indian Statistical Institute in Calcutta and his doctorate at Iowa State in 1982, and immediately joined the department. He assumed the headship following Gerig's resignation on September 1, 2002.

Some 20 faculty were hired during this period. Hires in the biological areas were in biostatistics: W. Lu, A. Maity, W. Sun, in bioinformatics: A. Motsinger-Reif (2007), E. A. Stone (2005), J.-Y. Tzeng (2003), and in biomathematics: K. Gross (2003), hires in statistical theory and methodology were in variable selection, high dimensional and complex data: H. Bondell (2005), L. Li (2005), Y. Wu (2008), A-M. Staicu (2009), H. Zhou (2010), in environmental statistics: B. J. Reich (2008), in stochastic processes: D. E. K. Martin (2007), and in quantile regression: H. J. Wang (2006), hires in Teaching and Consulting Faculty were teaching: J. Thompson (2006), K. Weems (2001), and R. Woodard (2003), and consulting: C. Arellano (2007) and a hire in statistics education: H. McGowan (2009).

Among those hired were seven females, two African–Americans and a Hispanic. The department was also successful in attracting minorities among staff members and students. In 2009, the department was recognized by the American Mathematical Society for its commitment to diversity.

During Pantula's tenure the department endowed three professorships (named for Gertrude Cox, Stu Hunter, and R. A. Fisher); thanks to the generosity of the department's donors.

Center for Quantitative Sciences in BioMedicine

Established in 2007, Center for Quantitative Sciences in BioMedicine (CQSB) is a multidisciplinary research center whose mission is to facilitate research in the biomedical sciences whose success requires collaborations among scientists in the quantitative and biological disciplines. CQSB is co-directed by M. Davidian (Statistics) and H. T. Banks (Mathematics).

Training Programs

Summer Institute for Training in Biostatistics (SIBS) at NC State: First offered summer of 2004, SIBS is a 6 week program offering qualified undergraduate students the opportunity to explore careers in biostatistics. Supported by NHLBI and headed at NCSU by Marie Davidian, SIBS enters its ninth year in 2012.

NCSU Computation for Undergraduates in Statistics Program (NCSU-CUSP): Supported since 2008 by the NSF/DMS, CUSP provides an applied computational statistics research experience to undergraduate statistics students. The goal is to encourage students to pursue graduate study and make informed career decisions.

VIGRE-2: Pantula served as a PI for the second 5-year period of the department's NSF/VIGRE training grant. The department also received an S-STEM grant in support of its mentoring at all levels.

Analytics Program

In 2007, in partnership with SAS, NC State established the Institute for Advanced Analytics and a Master Degree in Analytics (MSA). The department participated in the planning and its faculty members teach some of the courses.

Dedication of SAS Hall

On May 1, 2009, NC State dedicated a new building to serve as home for the mathematics and statistics departments. Construction was made possible by a bond referendum passed by North Carolinians in 2000, as well as by donors, including a substantial contribution from SAS.

Montserrat Fuentes, Head, 2011 to present

After Pantula resigned as the head in 2010 to become director, Division of Mathematical Sciences at NSF, Bloomfield served as an interim head for a year. During this period, Eric Laber was hired (2011, machine learning and dynamic treatment).

Montserrat Fuentes earned her PhD at Chicago in 1998 and joined the department. She was appointed head beginning July 1, 2011.

Post Scripts

Graduate enrollment in statistics has steadily increased over the decades from 57 in 1980, 109 in 1990, 116 in 2000, and 140 in 2012. Well over 1,000 graduate degrees have been granted.

The mathematical sciences at NC State have ranked in the top six nationally in federal funding since 1998.

As Mason took over as head, the department organized a fall weekend beach trip for faculty and their families. September 2012 will mark the fiftieth year of this event (save the occasional hurricane) which was broadened in the 1980s to include the graduate students (Figs. 3, 4).

Fig. 3 Departmental retreat, 2004. Names of all pictured are at www.stat.ncsu.edu/information/history/photokeys

Fig. 4 Retired faculty and staff with Sastry Pantula, 2007. Names of all pictured are at www.stat.ncsu.edu/information/history/photokeys

Highlighted Achievements Not Covered in Text

- Founding of Statistical Journals:

  Biometrics—G. M. Cox, Founding Editor
  Journal of Statistics Education—J. Dietz, Founding Editor

- Editorships:

  Journal of the American Statistical Association—Leonard Stefanski
  Biometrics—G. M. Cox, D. L. Solomon and M. Davidian
  Journal of Agricultural, Biological, and Environmental Statistics—Montserrat Fuentes
  Journal of Statistics Education—J. Dietz
  Biostatistics—A. Tsiatis

- Society Presidents:

  ASA—G. M. Cox (1956), S. Pantula (2010), M. Davidian (2013)
  Biometric Society—G. M. Cox (1968–1969)
  ENAR—M. Davidian (2004)

- Members of the National Academy of Sciences:

  G. M. Cox (1975)
  W. Cochran (1974)
  C. C. Cockerham (1974)
  M. M. Goodman (1986)

- Heavily Cited Publications (# Citations > 1,000)

David A. Dickey and Wayne A. Fuller (1979). Distribution of the Estimators for Autoregressive Time Series With a Unit Root. Journal of the American Statistical Association. Vol. 74.

Weir, B. S. and **Cockerham, C. C.** (1984) Estimating F-statistics for the Analysis of Population Structure, Evolution, Vol. 38.

David A. Dickey and Wayne A. Fuller (1981). Likelihood Ratio Statistics for Autoregressive Time Series with a Unit Root. Econometrica, Vol. 49.

Zhao-Bang Zeng (1994). Precision Mapping of Quantitative Trait Loci. Genetics, Vol. 136.

David A. Dickey and Said E. Said (1984). Testing for Unit Roots in Autoregressive-Moving Average Models of Unknown Order. Biometrika, Vol. 71.

Kenneth H. Pollock, James D. Nichols, **Cavell Brownie** and James E. Hines (1990). Statistical Inference for Capture-Recapture Experiments. Wildlife Monographs, Allen Press.

Burnham, K. P., D. R. Anderson, G. C. White, **C. Brownie** and **K. H. Pollock** (1987). Design and Analysis Methods for Fish Survival Experiments Based on Release-Recapture American Fisheries Society Monograph 5. American Fisheries Society, Bethesda, Maryland.

- Faculty Authored Books:

Anderson, Richard L. and T. A. Bancroft (1952). *Statistical Theory in Research*. McGraw-Hill Education.

Bloomfield, Peter (2000). *Fourier Analysis of Time Series: An Introduction, 2nd Ed.* John Wiley & Sons.

Bloomfield, Peter and William L. Steiger (1983). *Least Absolute Deviations: Theory, Applications, and Algorithms*. Birkhäuser, Boston.

Carroll, Raymond J., David Ruppert, **Leonard A. Stefanski** and Ciprian M. Crainiceanu (2006). *Measurement Error in Nonlinear Models: A Modern Perspective, 2nd Ed.* Chapman and Hall/CRC.

Casella, G. C. and **Roger Berger** (2001). *Statistical Inference, 2nd Ed.* Duxbury Press.

Cochran, William G. and **Gertrude M. Cox** (1957). *Experimental Designs* 2nd ed. Oxford, England: John Wiley & Sons (Original edition 1947).

Davidian, Marie and David M. Giltinan (1995). *Nonlinear Models for Repeated Measurement Data*. Chapman and Hall/CRC.

Gallant, A. Ronald (1987). *Nonlinear Statistical Models*. John Wiley & Sons, Inc., Hoboken, NJ, USA.

Gibson, G. and **S.V. Muse** (2009). *A Primer of Genome Science, 3rd ed.* Sinauer & Assoc, Inc. Sunderland, MA.

Giesbrecht, Francis G. and **Marcia L. Gumpertz** (2004). *Planning, Construction, and Statistical Analysis of Comparative Experiments*. Wiley-Interscience.

Monahan, John (2011) *Numerical Methods of Statistics, 2nd Ed.* Cambridge Series in Statistics and Probabilistic Mathematics.

Rawlings, John O., Sastry G. Pantula and **David A. Dickey** (1998). *Applied Regression Analysis: A Research Tool*. Springer-Verlag, New York.

Steel, Robert G D, Torrie, James H. and **David A. Dickey** (1997). Principles and Procedures of Statistics. McGraw-Hill (Original edition 1960).

Weir, Bruce S. (1996). *Genetic Data Analysis 2: Methods for Discrete Population Genetic Data*. Sinauer Associates Inc.

Weir, Bruce S. (1990). *Genetic Data Analysis*. Sinauer Associates Inc.

Evett, Ian W. and **Bruce S. Weir** (1998). *Interpreting DNA Evidence: Statistical Genetics for Forensic Scientists.* Sinauer Associates Inc.

Oregon State University Statistics Department

Daniel W. Schafer

1919–1957 and the Birth of the Department

In 1919, the Oregon Agricultural College offered its first two statistics courses—Mth 103, Elements of Statistical Methods, which was "an elementary mathematical treatment of graphic representation of data, frequency distributions and curves, averages, index numbers, and measurements of dispersion and correlation;" and the similarly titled ES 313, The Elements of Statistical Methods, in the Economics and Sociology Department. Other statistics courses were introduced in the next decade in the Departments of Mathematics, Business Administration, Education, Economics and Sociology, Farm Crops, and Agricultural Economics.

For its own majors, the Mathematics Department introduced the Mathematical Theory of Statistics sequence, Mth 441, 442, and 443 in 1932. The instructor, William Kirkham, had just published the book *Elements of Mathematical Theory of Statistics with Forestry and Economic Applications* in 1931. According to Jerome Li's notes from 1966, Henry Scheffé, who was then a young instructor of differential equations at Oregon State College (the new name, as of 1937), attended some of Kirkham's classes and "was so fired up by statistics" that he left Oregon State to retrain himself as a statistician with Samuel Wilks at Princeton. Other Mathematics instructors who turned their attention to statistics were Paul Hoel (who went on to UCLA after studying statistics in Oslo) and P.C. Hammer (who went to Wisconsin in 1947). Kirkham also started a university statistical consulting service in the 1930s, but the program petered out during the Depression and was discontinued.

Oregon State University,
Statistics Department,
44 Kidder Hall,
Corvallis, OR 97331-4606, USA
D. W. Schafer (✉)
e-mail: schafer@stat.oregonstate.edu

A. Agresti and X.-L. Meng (eds.), *Strength in Numbers: The Rising of Academic Statistics Departments in the U.S.*, DOI: 10.1007/978-1-4614-3649-2_14,

Fig. 1 Jerome Li

Hammer worked at Lockheed Aircraft and Kirkham served in the navy during World War II, but both returned to Oregon State after the war. In 1946, Kirkham again taught the Theory of Statistics sequence and Hammer taught Mth 341 and 342, "Applied Statistics," which was aimed at engineering students and required calculus. They were joined by new assistant professor, Jerome Li (Fig. 1), who taught Mth 109, a course in elementary statistics originally designed for Forestry students in 1932, but mainly attended by business and technology students after the war. All three of the statistics-teaching faculty members taught mathematics courses as well. According to Li, there was no doubt that statistics was only a minor sideline in the Mathematics Department at the time. Several events in 1946 and 1947 began to strengthen the statisticians' ties with the Agricultural College, though, paving the way for the birth of the Statistics Department.

First, Li's Mth 109 course faded away because the Business and Technology Department started offering its own statistics course. Li assumed two major responsibilities instead: he took over the Applied Statistics courses when Hammer left and he revived Kirkham's statistical consulting and computing service for the Agricultural Experiment Station. Apparently motivated by colleagues in Agriculture, Li dropped the calculus prerequisite from the Applied Statistics course to make it more accessible to agriculture graduate students. He did not receive any student interest at first and the course only survived because of the help from Robert W. Henderson, assistant to the Associate Director of the Agricultural Experiment Station who, "with great effort," recruited six graduate students—one from each of the six Agricultural College departments. The sequence did catch on and with GI's coming back from the war the enrollment increased steadily.

Jack Borsting's reflections on Jerome Li. Jack was an undergraduate student at Oregon State College from 1947 to 1950. He has served as Dean of Business Schools at the University of Southern California and the University of Miami, and as Assistant Secretary of Defense (Controller) under Presidents Carter and Reagan, and was president of the Operations Research Society of America (ORSA) and of the Military Operations Society of America (MORS).

"Jerome Li was instrumental in my decision to study statistics and get my advanced degrees in mathematics and statistics. The first statistics course I had as an Oregon State undergraduate majoring in mathematics was in 1949 from Professor Li. The course was a good mix of some theory and applications of statistics. Professor Li's way of analyzing data and applying statistics to real problems appealed to me. It was surprising that my interest in statistics was started with his first course because we had to calculate analysis of variance tables, etc. with a mechanical Monroe calculator and, of course, if one made mistakes they had to go back and do the calculations again."

"I talked to Jerry Li about going on to graduate school and studying statistics, and he suggested that one of the best places to go would be to the University of Oregon where I would study under Will Dixon and Frank Massey. One of the reasons Professor Li recommended going to Oregon was Dixon and Massey's blend of theory and application. I credit Professor Li for steering me on a very exciting and profitable career path. At that time, there was not a statistics department at OSU and Professor Li was the only statistician in the Mathematics Department so I obtained my degree in mathematics."

"Later in my career, when Lyle Calvin was the statistics department chairman, I was asked to be a visiting distinguished professor for one summer. I taught a course in discriminant analysis to PhD students and was able to interact with other faculty members and students in the department. In 2004, my wife and I endowed a fellowship in graduate statistics in the department. We have worked closely with two chairmen of the department, Dan Schafer and also Bob Smythe."

In 1953, Li was given his own budget and the Statistical Consulting and Computing Service became the Department of Statistical Service. Lyle Calvin (Fig. 2), the first faculty member at the University to have a PhD in Statistics (from North Carolina State University) was hired that year as an associate professor of Statistics in the Agricultural Experiment Station. (Calvin was hired at the associate level by mistake. OSU mistook his title of "assistant" at NCSU for "assistant professor," and offered him the promotion as part of the recruitment package.) Roger Petersen, also with a PhD in Statistics from North Carolina State, was added as an Experiment Station assistant professor in 1955. Now that Li had his own budget for statistical service and two additional assistant professors devoted entirely to statistics, the Department of Statistical Service became the center of statistics on campus. Given the inefficiency associated with duplicative statistical instruction—as symbolized by those first two statistics courses in 1919—it seems natural that the Department of Statistical Service would gain responsibility for coordination of statistical instruction.

In 1957, in fact, the Department of Statistical Service was organized as an instructional unit in the School of Science, independent of the Mathematics Department, and was given responsibility for coordinating statistical instruction on campus. The Department was physically located in Extension Hall at the home of the consulting and computing service. There were two subunits operating cooperatively—the instructional unit in the College of Science and the service unit within the Agricultural Experiment Station. Research activities were carried on within both units. Li, Calvin, Peterson, and Richard Link (one of John Tukey's students, who had been hired as a theoretical statistician by the Mathematics Department in 1955) were the original faculty members, with Li the first chairman. At that time, the Department offered a Master's degree in Statistics and minor degrees on Master's and PhD degrees offered in other departments.

1961–1981, Growth of the Department Under Lyle Calvin

In 1961, Oregon State College was renamed Oregon State University, the Statistics Department acquired its first computer—an IBM 1620, and Jerome Li served his last full year as chairman. In 1962, Li retired for health reasons, and moved to Taipei. In addition to starting and chairing the Department of Statistical Service, Li managed to write three books in a short time span: Introduction to *Statistical Inference* 1957; *Statistical Inference* 1964; and *Numerical Mathematics* 1966. Li died in 1969 at age 55.

Lyle Calvin was chair of the Statistics Department from 1962 to 1981. More than any other person, Calvin shaped the modern Oregon State University Statistics Department. In the early 1960s, Calvin envisioned an opportunity for future expansion and articulated a desire for additional faculty members who could participate in both consulting and research, with particular interest in areas of fisheries and wildlife ecology, system simulation, stochastic processes, biometry, and sampling techniques. While the department anticipated an increase in the training of mathematical statisticians, it is clear that Calvin placed a major emphasis on applied programs for instruction, research, and service; and on research into methodology

Fig. 2 Lyle Calvin

associated with consulting services projects. This vision was accompanied by remarkable growth. Sixteen tenured or tenure-track faculty members were hired in a 7 year span from 1962 to 1969: Donald Chapman, William Lowry, Donald Jensen, Edwin Hughes, Donald Guthrie, Kenneth Rowe, Scott Overton, David Faulkenberry, Donald Pierce, Thomas Lindstrom, Fred Ramsey, David Thomas, Charles Land, Norbert Hartman, Hugh Dan Brunk, and Justus Seely.

The first Master's Degrees in the Department were granted in 1962—to David R. Thomas and Arnold Schroeder. In 1965, a doctoral program was added and new minors were approved in Biometry and Operations Research. The first five PhD degrees in the Statistics Department were granted in 1969—to T.D. Burnett, J. Reinmuth, C.J. Phillips, A. Hirunaraks, and R. R. Fossum. In 1966, the Department moved from Extension Hall to the former library, Kidder Hall, where it resides today. Also in 1966, a campus-wide Computer Center was established and the computer facilities of the department were transferred to the center.

Recollections of the early years by Don Guthrie. Don was on the faculty from 1963 to 1973 before leaving to complete his illustrious career in the Biostatistics Department at UCLA. He was the major professor for six of the first eight PhD students to graduate from the Department and eight in all.

"The decade I spent at OSU was an exciting time, both for me and for the Statistics Department. I arrived while the department was still primarily involved with service teaching and consulting; what little graduate work was available ended with the Masters degree. Teaching loads were imposing and there was little time or encouragement for research. But we took advantage of the flourishing development of higher education during the 1960s, we hired new young faculty, we developed a new PhD program, and we expanded our extramural support. By the time I left, we had become recognized as a vibrant and growing program. The faculty focus broadened to include statistical research, we secured grant support directly to the department, we recruited bright and eager students, and our graduates moved into responsible careers. Disputes with university administrators notwithstanding, we had formed the nucleus of a vibrant program that reflected the goals of OSU and the statistics profession."

"Then, as now, the focus of OSU Statistics was on development and adaptation of statistical ideas in the context of applied research. Continuing associations with active research units provide motivation for statistical advances and for student experiences. Close association with Mathematics has continued to the present. The Computing Center developed one of the first truly interactive time-shared operating systems, and that facilitated our innovation in including computing in our service and our core graduate teaching. After leaving OSU, I discovered that many statisticians did not believe that we could integrate computing at this depth. Of course, this is now the norm across the profession; I'm proud that we were at the forefront."

"I supervised ten doctoral dissertations at OSU—eight in Statistics and two in Mathematics—and many Masters projects. The academic atmosphere was ideal for me, neither too theoretical nor too applied. I feel gratified that this spirit continues today."

Recollections about the first years of the graduate program, by Jim Reinmuth. Jim was one of the five students granted PhD degrees from the Statistics Department in 1969. He joined the University of Oregon's business faculty in 1967 and served as dean from 1976 to 1994.

"When first admitted to the MS program, I honestly didn't know the location of OSU. My first day in the office, I met Don Guthrie who was also new to the department. Don ultimately became my program advisor and subsequently, a good friend. As a Stanford graduate, Don was the outlier to the Iowa State and North Carolina State contingents who comprised the Statistics faculty at the time. As both an MS and PhD student, I was a classmate of Tom Burnett who spent most of his career with Portland General after a brief stint with the Naval Academy. Bill Maki and Hungfu Chiang were also MS classmates who did not continue on for a PhD. Bill spent his entire career with Weyerhaeuser, while Hungfu enjoyed a successful business career in Hong Kong."

"I completed my MS degree in 1965 at a time when written research papers were required. My paper was an application of linear programming to the securities market, and during my presentation, Mathematics Professor Harry Goheen fell asleep. At least I thought he was asleep until he startled me with a question that indicated he had heard every word I had uttered. To this day, I think Harry Goheen was a plant to see if I could be rattled during my presentation."

"I began my PhD studies in 1965 with Tom Burnett, Charles Phillips, Bob Fossum, and Samir Saad, an Egyptian student. We all enjoyed Samir as a classmate, since he had an uncanny ability to solve challenging assignments from Ed Hughes, Don Guthrie, and other faculty who were testing our patience. Better yet, Samir would always share his discoveries with us. For those who may remember, Samir was the guy who claimed someone had left a bomb on his doorstep disguised as a milk bottle. When the milk bottle turned out to contain Grade A pasteurized milk, Samir was sent home and we lost our in-house tutor."

"At the time, the Statistics Department was a very comfortable family, with Lyle Calvin as the leader. Lyle, Don Guthrie, Scott Overton, Ed Hughes (briefly), Dave Faulkenberry, Don Pierce, Dave Thomas, Ken Rowe, and Roger Peterson were supportive mentors and great friends. It was the Statistics Department lunchroom where I learned to play bridge, a pastime that has consumed a lot of my time over the past 47 years. I also recall fondly program assistants Peg Stuart and Dotty Jameson who were always available to help the students when we required assistance."

"My experience with the Statistics Department at OSU was life-changing for me. After earning an undergraduate degree in math from the University of Washington, I didn't know where to turn until a UW advisor recommended that I contact Lyle Calvin. My studies at OSU helped prepare me for an enjoyable, rewarding 44-year career with the University of Oregon. At OSU, I also met my wife (we've been married 45 years) and OSU is the alma mater for our two daughters."

Despite the rapid growth of the Department in the 1960s, its status was not altogether secure. The Dean of the College of Science did not see a need for statistics, which prompted intense discussions about the future of the Department. This came to a head in 1968, when Calvin commissioned an outside review by David Blackwell, William Kruskal, and Lincoln Moses. In their report, these eminent statisticians found much that was good with the Department, praised it for its collaboration with other units, and found the number one weakness to be a deficiency of mentors for PhD students. Their recommendation to hire a senior faculty member forced the dean's hand. The Department was allowed to continue and to search for a senior faculty member, which lead to the hiring of Hugh Daniel Brunk. Brunk's pioneering work in isotonic regression and mathematical statistics immediately enhanced the reputation of the Department.

Reflections of Don Pierce. As a faculty member from 1966 to 1996, Don was a top researcher, making contributions in many areas but most notably in methodology and application in radiation health epidemiology and in higher-order asymptotics.

"Although I cut my career at OSU somewhat short to work in Japan, I want to say that OSU Statistics was a great place for me in all respects. These mainly had to do with the

Fig. 3 Oregon State University Statistics Department, 1979. From left Ken Rowe, Dan Brunk, Paula Kanarek and child, Dave Thomas, Shirley Neyhart, Justus Seely, Lyle Calvin, Dave Faulkenberry, Dave Butler, Roger Petersen, Don Pierce, Dave Birkes, Fred Ramsey, Jeff Arthur

quality of the faculty and students, and the leadership of Lyle Calvin. Here I will mention only a few of many things, those coming first to my mind."

"Shortly after I joined the department, I found that OSU Stat was what turned out to be more than a decade ahead of its time in terms of statistical computing. Don Guthrie developed under an NSF grant a university-wide mainframe, time-sharing teaching system SIPS that used TTY machines distributed around the campus. This software was perhaps closest to the Minitab that came along about a decade later, and was simply revolutionary. In 1961-65 during my PhD, like most others I never actually did a multiple regression of any reality, but 3 years later we were having hundreds of students from all over campus doing this. The importance of SIPS was not only for teaching but for our more general statistical practice, and it was kept alive even beyond its time by Ken Rowe in experiment station work."

"The program of outstanding visiting statisticians mentioned later in this history had a profound effect on my professional development. Even today, I very often think of things I learned from those such as Bill Cochran and David Cox. Similarly, much of one's professional development derives from faculty colleagues, and I want to say that much of my "education" on statistical theory came from interactions with Dave Thomas. It seems that he learned more readily than I did what the theory of statistics was all about, e.g. the real meaning for applications and research of theory such as one finds in the advanced texts by Lehmann. His creativity is not adequately reflected by publications, but was widely recognized. For example early work in Cox regression often pointed to his unpublished technical reports as first discovering aspects of that theory, and Art Owen often credited a paper of Dave's for first using the idea of empirical likelihood."

"Partly since the author of this history will not be doing this, I want to comment on the importance of the text *The Statistical Sleuth* by Fred Ramsey and Dan Schafer. Those unfamiliar with this text should look at the review by G. W. Cobb in *The American Statistician*, 2000, pp. 78–81. This highly original text reflects the insights and diligence of the authors, and has had considerable influence in the teaching of the statistical craft, having been adopted for coursework by about 65 departments in the U.S."

The Oregon Fish and Wildlife Service was legally mandated to estimate the effects of fisherman and hunters on fish and wildlife populations, but lacked the expertise to do so themselves. With their funding, Calvin; Roy Young, the Dean of Research; and Bob Mason, created the Survey Research Center (SRC) in 1973. The SRC resided in the Statistics Department, with Calvin as director. This made sense because most survey centers that were forming in social science departments lacked the statistical expertise to produce credible results. At that time, Iowa State and Oregon State had the only Survey Centers that were embedded in Statistics Departments.

Reflections of Robert Mason, on the first years of the Survey Research Center. Bob was an editor for the Agricultural Experiment Station's news publication, but was enlisted to help start the Survey Research Center in 1973 because of his familiarity with questionnaire design. He continued to work for the Center until his retirement, and continued to help out in retirement until 2009.

"We received immediate attention due to our work on admitted tax evasion (supported by the Oregon Department of Revenue), in which one out of four in our Oregon sample admitted that they cheated on their taxes. The IRS immediately said that our study was wrong...that only about 5% were delinquent. The New York Times gave our study a full run, and I had to testify before Congress about our results. The head of the IRS had a very uncomfortable session, for he had no data to support how much cheating was going on. I was sweating our testimony but the House panel that heard the testimony agreed and told the IRS to get cracking on their own studies. I worked with Bill Lefbom, the IRS researcher in charge of their study. Their results showed that US taxpayers were cheating at about the 30% level and that most of it came from underreporting income, not overstating deductions or failure to file. In that respect, their results supported ours."

In the years 1965–1975 the Department had a Biometry Training Grant, which permitted visits of 1–3 weeks by a great many of the most prominent statisticians of that era. The chance to interact with these people had an important impact on the faculty. Among the visitors were David Cox, William Cochran, David Blackwell, Dennis Lindley, Nathan Mantel, R. L. Anderson, and Walter Smith. Also during this time, many of the Department's faculty members established overseas connections. Lyle Calvin spent time in Egypt and Scotland; Roger Peterson in Syria; Dave Faulkenberry in Yemen, Oman, and Egypt; Justus Seely and Dave Thomas in Egypt; and Don Pierce in Japan.

1982–2010 and the Recent Era

Dave Faulkenberry became the Department Chair in 1981 when Calvin stepped down to become Dean of the Graduate School, and Justus Seely took over from 1984 to 1997. When Seely retired, the Department was very short on senior faculty members and an external search led to the hiring of Bob Smythe, who had been chair at George Washington University. Smythe served as chair at OSU from 1998 until 2008. Dan Schafer followed him in 2008. The economic picture for higher education in Oregon was not great in this time span. The Statistics Department owes a great deal to Seely and Smythe for their perseverance in representing the Department in this period (Fig. 4).

The gender composition changed remarkably in these years. In 1982, all fourteen of the Department's tenure-track faculty members were male (and mostly named Dave, Dan, or Don). In 2010, there were four male and five female tenure-track faculty members. In 2009, Ginny Lesser became the first female promoted to full professor.

Fig. 4 Some members of the OSU Statistics community in December, 2010. Back row, from left Lisa Ganio, Associate Professor in Forest Ecosystems and Society (adjunct in the Statistics Department); Sarah Emerson, assistant professor; Ginny Lesser, professor; Paul Murtaugh, associate professor, John Molitor, Associate Professor in Public Health (adjunct in the Statistics Department); Yanming Di, assistant professor; Lan Xue, assistant professor; Jeannie Sifneos, research assistant. Front row, from left Cliff Pereira, research associate; Ken Rowe, professor emeritus; Bob Smythe, professor emeritus; Jeff Arthur, professor; Alix Gitelman, associate professor; Lisa Madsen, assistant professor; and Dan Schafer, professor

Environmental Statistics and Other Contributions

Early collaborations with the College of Forestry, the Department of Fisheries and Wildlife, and the Geography Department led to an emphasis on statistics associated with natural resources and environmental monitoring. Lyle Calvin worked on salmon returns to Oregon's rivers and streams for many years. W. Scott Overton (1965–1993) had a joint appointment with the Forestry College and worked closely with the Environmental Protection Administration (EPA), making major contributions to several aspects of EPA's National Surface Water Surveys conducted during the 1980s and to EPA's Environmental Monitoring and Assessment Program (EMAP). N. Scott Urquhart (1991–2001), a research professor, initially worked under Overton's EPA cooperative agreement and then with EMAP, and made many statistical contributions to problems originating from EMAP. Don L. Stevens (2001–2010), who joined the Statistics Department in 2001 as a research professor, obtained EPA funding for the Department through the Designs and Models for Aquatic Resource Surveys (DAMARS) program. He continued to refine and promote an environmental sampling methodology that he had developed with another former student, Tony Olsen.

Fred Ramsey (1966–2003) made major contributions to estimation of wildlife population sizes. In addition to writing many research papers on estimating wildlife abundance, he participated in major surveys of bird populations in the Hawaiian Islands and various South Pacific islands. One of Ramsey's PhD students, Steven Thompson has made major contributions to wildlife sampling in his own academic career at Penn State and Simon Fraser Universities. One of

Overton's students, Ken Burnham, has made similar contributions to wildlife sampling and abundance methodology at Colorado State University.

Current faculty members who have interests in environmental and ecological statistics are Paul Murtaugh (1992–) in statistical ecology, Virginia Lesser (1992–) in environmental sampling, Lisa Madsen (2003–) and Alix Gitelman (1999–) in spatial statistics, and adjunct professor Lisa Ganio (Forest Ecosystems and Society) in natural resources problems. Many doctoral graduates from the Department have had active careers in environmental areas, including Tony Olsen, EPA; Nancy Lo, National Marine Fisheries Service; Rich McArthur, Nevada Water Resource Center; Trent McDonald, West Statistical and Environmental Consultants; Kerry Ritter, Southern California Coastal Water Research Project; and Kathy Irvine, USGS.

Notable research contributions in other areas include Dan Brunk's pioneering work on isotonic regression and Don Pierce's major contributions to methodology for radiation health epidemiology. Some faculty members who have had important impacts on students were Dave Thomas, whose knowledge, sensibility, and generosity defined the character of the Department for many years; Justus Seely, whose commitment to the Department and students was exceptional (and often entertaining); and Dave Birkes, who has played an unheralded role in the completion of many PhD theses. Birkes has won the Statistics Students' teacher of the year award in 10 of the 39 years it has been given.

About one-fourth of the 122 PhD graduates of the department have gone into academic positions: Jim Reinmuth (University of Oregon), Robert Fossum (University of Texas), Charles Phillips (Cal State Sonoma), Jim Daly (Cal Poly, San Luis Obispo), Tom Burnett (Naval Post Graduate School), Len Deaton (Cal State University, Los Angeles), Brad Crain (Portland State University), Ken Burnham (Colorado State University), Steve Thompson (Penn State, Simon Frazer University), Dana Thomas (University of Alaska), Gary Sexton (Oregon Health Sciences University), Michael Jacroux (Washington State University), Nancy Carter (Cal State University, Chico), Don Stevens (Oregon State University), Yadolah Dodge (University of Neuchâtel, Switzerland), Robert Gray (Harvard School of Public Health), Loretta (Robb) Thielman (University of Wisconsin-Stout), Cliff Pereira (Oregon State University), Luis Cid (Universidad de Concepción, Chile), Rick Rossi (Montana Tech), Steve Stehman (State University of New York), Lisa Ganio (Oregon State University), Mohamed Limam (University of Tunis, Tunesia), Dawn Van Leeuwin (New Mexico State), DeWayne Derryberry (University of Puget Sound and Idaho State University), Shaun Wulff (University of Wyoming), Kathy Irvine (Montana State University), Pachitjanut Siripanich (The National Institute of Development Administration, Thailand), George Fegan (University of Portland and Santa Clara University), Michel Lejeune (University of Grenoble, France), David Bogdanoff (Cleveland State University), Len Haff (University of California, San Diego), Roza Sjamsoe'oed (Syiah Kuala University, Indonesia), Caryn Thompson (Acadia University, University of New Brunswick, and University of Louisville).

The Future

In many ways, the Oregon State University Statistics Department is at a crossroad. The last remaining faculty members who served with founding members Calvin and Petersen—Arthur and Schafer—will be retiring, and we are searching for three new assistant professors to complement the already strong nucleus of young faculty members Madsen, Xue, Di, and Emerson. We are currently fighting pressures from the University to drop the "Department" status of the Statistics Department. Although frustrating, this exercise usefully forces us to consider our relevance and our future. Not surprisingly, Lyle Calvin's vision of research into statistical methodology associated with collaborative applied work continues to make a great deal of sense. We envision statistical faculty as team members in large research projects, who assist with data analysis and conduct research motivated by their teams' data problems. We anticipate a more nimble curriculum structure that can respond rapidly to new training needs spurred by new data-producing technologies. Finally, we anticipate the further evolution of the lowest-level introductory course on campus (for *all* college students,) which transforms "intro stats" into the more practical "critical evaluation of conclusions from evidence."

Our niche in environmental statistics has suited us well both because it matches Oregon State University's strengths and objectives, and because we successfully attract graduate students who wish to apply their quantitative strength to environmental problems while living in the very nice environment that Corvallis offers. More recently, we have established close ties with biologists conducting genomics research. It seems natural that the Statistics Department will continue to respond to new research technologies for which existing statistical methodologies are inadequate.

Throughout its history, this department has been characterized by a warm camaraderie among faculty, staff, and students; a sensible approach to applied statistics; a commitment to excellent teaching at all levels; a serious attention to master's degree training of professional applied statisticians; PhD students who have assumed positions of substantial responsibility in government, industry, and academia; and strong collaborative ties throughout the University. We expect these trends to continue.

Penn State University Department of Statistics

William Harkness, Thomas Hettmansperger, Dennis Lin and James Rosenberger

If this chapter were being written by a writer for the New York Times, the headline might well have proclaimed "New Department of Statistics Created by Non-Statisticians at Penn State University" and it easily could have been defended as the truth. The writer would have elaborated as follows: "The Department of Statistics was created on July 1, 1968, with James B. Bartoo as its first head. Questions have been asked about how it came about and why. Who were the moving forces leading to its creation? What was their motivation? In this story answers to these questions are provided!" Our story may be typical of how other statistics departments were created. Regardless, these are the facts the writer could have cited to back up our story.

Five key dates and documents pinpoint and describe the early history of the department's creation:

1. February 14, 1957: Proposal for a Penn State Statistical Center.
2. December 28 1962: Proposal for the Coordination and Strengthening of Teaching and Research in Statistics at The Pennsylvania State University.
3. May 5, 1965: Proposal for Graduate Program in Statistics.
4. July 1, 1966: Section on Statistics within the Department of Mathematics Created and Graduate Program in Statistics Initiated.
5. July 1, 1968: Department of Statistics Implemented.

Penn State University, Department of Statistics,
326 Thomas Building, University Park, PA 16802-2111, USA
W. Harkness (✉) · T. Hettmansperger · D. Lin · J. Rosenberger
e-mail: wlh@stat.psu.edu

T. Hettmansperger
e-mail: tph@stat.psu.edu

D. Lin
e-mail: dkl5@psu.edu

J. Rosenberger
e-mail: JLR@psu.edu

A. Agresti and X.-L. Meng (eds.), *Strength in Numbers: The Rising of Academic Statistics Departments in the U.S.*, DOI: 10.1007/978-1-4614-3649-2_15,

1957: Proposal for a Penn State Statistical Center

An ad hoc committee appointed by the President of the Penn State Chapter (founded in 1954) of the American Statistical Association (ASA) wrote a report to the University President. The task set for the committee was "suggest the structure and function of a statistical center at the university." Results of a survey conducted by the PSU Chapter pointed out "the need for a statistical service on campus especially in the field of *Statistical Consultation.*" The ad hoc committee also discussed the structure of a center and indicated that it "should be flexible enough to include *Basic Statistical Research and Teaching.*" They made two (specific) recommendations:

1. "An investigation of the operation of existing statistical centers at other universities should be made."
2. "The consulting part of the statistical center be implemented as soon as possible." The primary impetus for this report was due to John C. Griffiths, a geo-mineralogist, pictured here.

1962: Proposal for the Coordination and Strengthening of Teaching and Research in Statistics at the Pennsylvania State University

The Vice-President for Resident Education appointed a "Committee on Statistics" in 1962 to "inquire into the teaching of statistics at Penn State and to make recommendations to him and to the Faculty Senate Courses of Study Committee"—"the committee extended its inquiry to include research aspects also." The six members of the committee consisted of one mathematician and five faculty members from other departments or colleges: Geo-mineralogy, Psychology, Industrial Engineering, Education, and Agriculture. The committee wrote: "This report proposes a beginning toward building an adequate statistics program at Penn State." It emphasized *the need for a statistics department* as follows: "(there is a need) for a core of competent statisticians who have sufficient autonomy and financial support to deal directly with Penn State's needs in statistics. A subordinate unit required formally or by the force of circumstances to approach statistics within the constraints imposed by the conventions of a classical mathematics department or of any field of application is not likely to be effective." The committee made a proposal for changes in the teaching of statistics, setting forth objectives and guidelines.

The Interim Period Following the 1957 and 1962 Reports

James Bartoo was the first PhD in Statistics to be hired at PSU, in 1952. He assumed the headship of Mathematics in 1961. By the end of 1962, course offerings in statistics had expanded to include two-term sequences in Mathematical Statistics and Advanced Probability Theory. William L. Harkness was hired as a new assistant professor (of Mathematics) in 1959. There was obvious interest in statistics among math graduate students, as evidenced by 13 students doing their Masters paper in Statistics from 1961 to 1966 and also three PhD theses in the period 1963–1966. In late December 1963 and early January 1964, G. P. Patil, a statistician, joined the mathematics department as a full professor. In the summer of 1963 he had organized an 'International Symposium on Discrete Distributions' hosted by McGill University. Additional statisticians were hired in the mathematics department prior to the formal birth of the department: Marllyn T. Boswell (1965), Robert Hultquist (1966), Thomas P. Hettmansperger (1967), and Peter Dress (1967—joint with Forestry).

1965: Proposal for Graduate Program in Statistics

As new faculty were appointed, a proposal for a Graduate Program was submitted on May 5, 1965. This proposal was designed to implement (to a large extent) the recommendations in the 1962 report summarized above. The proposal called for four degree programs: (i) Masters of Science in Applied Statistics, (ii) Masters of Arts in Mathematical Statistics (non-thesis option), (iii) Masters of Arts in Mathematical Statistics (thesis option), and (iv) PhD in Statistics.

Ingrid James, G.P. Patil, Robert Hultquist, C.I. Noll (Dean), James Bartoo, Tom Hettmansperger, Charles Antle, Thomas Ryan, Marllyn Boswell, William Harkness

1966: Section on Statistics Within the Department of Mathematics Created and Graduate Program in Statistics Initiated

In 1962, there was discussion about launching both a Department of Statistics and a Department of Computer Science. This followed exploration of a Department (or School) of 'Cybernetics' or similar type of organization (at the suggestion of John Griffiths). However, it was (apparently) decided to set up Computer Science first and delay Statistics until later. Subsequently, a graduate program in statistics was established within the mathematics department effective from July 1, 1966.

Department of Statistics Established

In 1968, James B. Bartoo left the headship of the Department of Mathematics to head the new department. The initial members of the faculty were: Professors James B. Bartoo (Head) and G. P. Patil; Associate Professors William Harkness and Robert A. Hultquist; and Assistant Professors Marllyn T. Boswell, Peter Dress (joint with Forestry), and Thomas P. Hettmansperger. Thus, the department started with seven faculty members including the Head. In addition, in 1968 there were two visiting professors: Charles E. Antle (from Missouri) and James Douglas (from Australia), with Ingrid (Sanders) James as the secretary. One year later Bartoo resigned to become Dean of the Graduate School and was succeeded by William Harkness who served as Head until December 31, 1987. An early visitor

1969: Bartoo (then Dean of the Graduate School), C.R. Rao, G.P. Patil, and W.L. Harkness

to the newly formed department was C. R. Rao, who about 20 years later joined our department and is shown in the picture to above.

Challenges and Obstacles

In the four decades after the graduate program was established (1966) and the department created (1968), what have been the biggest success stories and the most serious issues faced?

Success Stories

- Developing and launching the Minitab Statistical Computing Software Package and Statistical Consulting Program.
- Recruitment of C.R. Rao.
- Undergraduate Major and Minor.
- Master of Applied Statistics (MAS) Program.
- Rise of the program to prominence among departments of statistics.
- Anecdote: Teaching a student who had success in a high-paying career: Carl Gugasian.

Serious Issues

- Failure to receive adequate support for the department as its teaching load increased dramatically.
- Long-time failure of the administration to provide support for statistical consulting as an important function of a program in statistics.

Success Stories

The Minitab Project

In 1969, the Department of Statistics hired Thomas A. Ryan, Jr., a new PhD, with probability as his area of specialization. Soon after his arrival, however, Tom Ryan developed a strong interest in computing. In 1971, Brian Joiner was recruited from the National Bureau of Standards (now called National Institute of Standards and Technology). Brian brought with him Omnitab, a statistical package which was very user friendly but too large for practical student use. Brian had in mind to develop a smaller computer portable version. Penn State, at that time, was a leader in giving students quick turnaround so long as the punched card programs they submitted were thin. The quick turnaround that students got in those days was almost as close to interactive computing as anyone got elsewhere. Tom and Brian soon teamed up with Tom (the better programmer) doing most of the coding while Brian designed the user interface and wrote documentation. Barbara Ryan joined the team in less than a year to speed up work on what became the *Minitab Student Handbook*. Before long Minitab, being easy to use and having the ability to run on a wide variety of computers, was in use at other universities, and soon thereafter at businesses and government. The capabilities of Minitab, the number of customers, and the income all increased while Minitab was part of the department. Tom was the overall Director of the project (and later the first president of Minitab, Inc.). Tom, Barbara, and Brian's efforts resulted in Minitab being one of the most widely used statistical software packages for undergraduate instruction and six-sigma quality improvement, a major contribution to statistics.

The Statistical Consulting Program

The process of setting up a statistical consulting service was begun in the spring of 1970. Although without permanent funding, progress was made in beginning the service. Major strides were made through the hiring of Brian Joiner in the fall 1971. Joiner organized a Statistical Consulting Center that was widely used and effective. No direct university funding was provided for statistical consulting in the period from 1971 to 1974. The department underwent a program review in 1974. A 'Report on Operations of the Statistical Consulting Pilot Program Sept. 1, 1971–June 30, 1974', was prepared for the Program Review Committee, which performed an exhaustive, thorough assessment of the consulting program and gave an enthusiastic and thoughtful appraisal of it, ending with this conclusion: "The consensus is that the Statistical Consulting Service is essential to the graduate program of the Statistics Department. Even though the Statistical Consulting Service is quite young, it already seems to be gaining a reputation for excellence." Obviously, the department agreed, but received a negative response to a subsequent proposal in 1975 to the administration to implement the Graduate Program Review's recommendation. The department, however, still regarded the provision of consulting services as an important responsibility, but because of increased instructional demands, had to curtail the level of activity severely. This remained the situation for several years until a concatenation of events led the university to reassess its support for consulting.

In 1975, the department recruited James L. Rosenberger, who deserves credit for spearheading the creation of a strong, viable, lasting, and widely recognized Statistical Consulting Center (SCC). The source of funding for the center came primarily from the College of Agriculture, collaborative grant support, and several contracts with pharmaceutical companies—only later was permanent funding for the SCC achieved through a confluence of events.

Link Between Consulting and Minitab

In 1982, Minitab had outgrown the Department of Statistics and Tom and Barbara Ryan informed the university that they wished to set up a private company (Minitab, Inc.). Negotiations related to the Ryans' intentions were carried out under the aegis of R. G. Cunningham (Vice-President for Research and Graduate Studies). At about the same time, a new program, the Ben Franklin Partnership Challenge Program for Technological Innovation, was initiated by the State of Pennsylvania and the university was expected to submit a proposal to establish an Advanced Technology Center. Cunningham immediately recognized the potential value in linking the launching of Minitab, Inc., with the university's Ben Franklin proposal. However, the Ryans were undecided as to where the company, Minitab, Inc, should be located. An important consideration was that it be located at a site with substantial statistical expertise and activity. Two other locations, in addition to the State College area, were considered by the Ryans. Cunningham, with the intent, understandably, of exploiting the launching of Minitab as a private company as part of the university's Ben Franklin proposal, sought to determine what considerations were important in the Ryans' decision and measures the university could provide to entice them to remain in the area. The Ryans' response enumerated several concerns: first, the lack of a university supported SCC and second, the inadequate size of the Department of Statistics. The first of these concerns was addressed directly by Dr. Cunningham, with the second outside his control. He committed support for the SCC (presumably through Ben Franklin funds) and the Ryans decided to remain in the State College area. They were given incubator space and Minitab, Inc. came into existence on August 1, 1983. In December 1983, the department was given \$41,000 for consulting for the 1983–1984 fiscal year. These funds were made permanent and a longer term commitment was eventually reached in the spring of 1985.

C.R. Rao and the Eberly Chair in Statistics

In 1986, Penn State received one of its largest gifts: $10 million from the Eberly Family Charitable Trust of Uniontown, Pa. The gift established endowed professorial chairs in every academic department in Penn State's (renamed) Eberly College of Science, including Statistics. In 1988, C. R. Rao became the first person to hold the Eberly Chair in Statistics. Negotiations in recruiting Rao for the chair were led by then Department Head Thomas P. Hettmansperger. Here is the story on the negotiations to hire C.R.: "Jogesh Babu had been traveling to visit Rao at the University of Pittsburgh (Pitt) and indicated early in 1988 that Rao might be interested in leaving Pitt and would be receptive to an overture from Penn State. In February 1988, Rao made a visit to PSU and talked to our College Dean Zimmerman and Provost Richardson, both of whom would be instrumental in bringing Rao to PSU. There were several issues that needed to be resolved: transferring Office of Naval Research and Army contracts to PSU along with a large VAX computer, office space, relocating the Journal of Multivariate Analysis, the need for one or two secretaries for the Center, space for 12 or so summer visitors each year and six or so research assistants and PhD students. Rao requested a three year appointment (he would be 70 in September, 1990, and would have to retire). There were ongoing negotiations on salary, teaching loads and a few other matters. In late March, 1988, Professor and Mrs. Rao came for a visit. Negotiations continued. Rao was reluctant to make a final commitment. In early summer, 1988, after several phone calls to refine details, Tom Hettmansperger finally called him with instructions from Dean Zimmerman to make every effort to get a final answer from him. Rao answered the phone and Tom confirmed that PSU would meet all his requests and said that the Provost was eager to complete the appointment. Dr. Rao hesitated and then excused himself for a moment. Tom said "I could hear him in discussion with Mrs. Rao." He came back on the phone and accepted our offer! "Penn State, through Tom's efforts and others, made tremendous strides forward in its reputation, crediting the delicate politics of moving a living legend!"

He served as Director of the Center for Multivariate Analysis (which he had created at the University of Pittsburgh), from 1988 to 2001, and retired at age 80 as Professor Emeritus of the Eberly Family Chair in Statistics. He remained in the Penn State area until moving to Buffalo, New York in 2009. Among his many other honors, on June 12, 2002, Rao was awarded a National Medal of Science by President George W. Bush at the White House for his life-long achievements (See the photo below).

Undergraduate Major and Minor

Initially, the department felt that statistics should be exclusively a graduate program. When the head, James Rosenberger brought a proposal in 1992 to the faculty to consider an undergraduate major, the initial vote was 14-1 against it. Later, after realizing that a major could bring more external visibility, the department changed its view. In 1994, it was approved by the faculty with three options and Professor Arnold chaired the undergraduate committee. In 2000, a biostatistics option was added. Enrollment in the statistics major grew steadily and has been stable since 2000, with around 40 majors in total at any one time. With the addition in 2008 of an actuarial statistics option, the number of statistics majors in 2010 increased to the mid-50s, plus almost 100 minors in statistics.

Master of Applied Statistics Program

Although the statistics department had a thriving graduate program, it seemed that too many graduate students were joining the workforce after the Masters degree instead of pursuing the PhD degree. The strong demand for Master's level statisticians attracted many of our best domestic graduate students to take jobs in industry after completing their Masters degree. In 2000, the Eberly College of Science included the Statistics department as one of three departments to participate in a successful application to the Sloan Foundation to develop professional Masters degrees. The department obtained approval in 2001 to deliver the degree both in residence and online through the World

Campus of Penn State and launched the MAS program in the fall of 2002. At the same time, with support from the Eberly College the department began to develop courses for online delivery of the program. In 2006, the department launched a four-course online graduate certificate program which grew dramatically. The department announced the availability of the full online MAS in June 2009. By June 2010, more than 30 students were enrolled in the program, and enrollment in graduate courses soared. The target enrollment for the In-Residence MAS program was 15 students per year and the target has been met since 2002. In the 9-year period from 2002 to 2010, 87 students had graduated with the degree. Since the summer of 2009 when the opening of the MAS program for online delivery was announced, 30 new students are enrolling in this program per year. The degree requires about 3 years to complete, since most students are working full time and typically only take one course per semester.

Serious Issue: Lack of Support for the Department as its Teaching Load Increased

The period from 1980 to 1987 (the last years for Harkness as Head) was exciting but stressful and difficult. This period began with an excellent group of relatively new faculty with prospects for significant expansion of the department, based on their promise and rapidly growing instructional demands. But our plans were only partially successful. Big successes occurred in statistical consulting under James Rosenberger's leadership with university support and funding from various other sources. Faculty expansion was limited due to a lack of support from the administration, which provided year-to-year resources (supplemental funds) on an 'as needed basis' but not permanently. The bright future we saw for expansion was clouded by an administration unwilling to provide the resources for deliberate, careful planning. Our instructional load grew steadily (like a ***steep ladder***, close to linearly) from around 3,600 student credit hours (schrs) in 1968–1969, to about 19,400 in 1984–1985, an average increase per year of almost 1000. The regression equation of student credit hours per year (y) from 1968 to 1984 (x) is given by $\mathbf{y = 1832 + 970\,x}$. Although the administration believed our requests were completely merited, their response was piecemeal and responded only to urgent appeals for help. Supplements ranging from $116,000 in 1981–1982 to about $300,000 in 1986–1987 (compared to $598,086 in permanent departmental funds) were given. In these years, GA stipends averaged about $5,000–6,000 per academic year and Assistant Professors $23,000–25,000. Surely, the Head was stressed out by this and one can wonder how it was that he did not throw in the towel and turn the job over to someone else. In fact, he did resign in the fall of 1985 (probably made the Dean happy as well!), but stayed in the position until December 31, 1987 while a new head was sought.

Anecdote: Teaching a Student who had Success in a High-Paying Career

Carl Gugasian—'The Professional: America's Greatest Bank Robber' was briefly a doctoral student in statistics and probability at Penn State University. To read about this interesting character see http://en.wikipedia.org/wiki/Carl_Gugasian.

External Reviews and Assessments of the Department

In November 1995, an external review of our department was conducted. The overview of the department's self-study referred to a previous external review, conducted in 1984. An excerpt from the 1995 report states: "Their recommendations had a very large impact on the department's development over the following 5 years, primarily through their strong encouragement of an increase in the size of the department and its dependence on supplemental funding" (which was remedied in the period from 1988 to 1990). Regarding the NRC study of 1984 and 1995, the report from the 1995 external review committee went on to say: "The national ranking of the PSU Statistics Department improved from 31 (of 63) in 1982 to 19 (of 65) in 1993, the largest gain of any statistics program in the country, making one of a small group of PSU academic departments to be ranked in the top 20." They went on to say: "It is our firm belief that the Department is well positioned to continue its improvement, especially in the research area, and that if it continues to receive reasonable support from the Administration, it is likely that it will be ranked among the top 15 U.S. statistics departments a decade from now, and possibly that it will reach the top 10." The most recent rankings may indeed have verified this prediction, since in the latest NRC (2010) ranking of departments (based on 2005 data) Statistics at PSU could be ranked as predicted depending on criteria used (see www.nationalacademies.org).

Recent Years' Progress and Rise to Prominence

Many of our faculty and PhDs have achieved professional recognition, resulting in an enhanced reputation for the department. From our inception in 1968 with seven faculty members and eight graduate assistants we now have 23 full-time faculty, 13 joint appointments, 12 instructors, 78 traditional graduate students, and about 90 MAS degree students. The table below shows the growth in PhDs and Masters degrees over the past 40+ years in 5-year intervals except the first:

Number of Masters and PhD graduates

Degree	1968–1970	1971–1975	1976–1980	1981–1985	1986–1990	1991–1995	1996–2000	2001–2005	2006–2010	Total
PhD	6	8	17	10	15	21	30	30	46	183
Masters	13	54	52	46	50	46	43	84	77	465
Totals	19	62	69	56	65	67	73	114	123	548

Other Measures of Quality

Prizes and Awards

The *C.R. and Bhargavi Rao Prize* was established to honor and recognize outstanding and influential innovations in the theory and practice of mathematical statistics, international leadership in directing statistical research, and pioneering contributions by a recognized leader in the field of statistics. The Rao Prize is awarded by the Department of Statistics at Penn State University to a nominee selected by the members of the Rao Prize Committee. The first four Rao Prize recipients were Bradley Efron, Stanford University (2003), Jayaram Sethuraman, Florida State University (2005), Lawrence D. Brown, University of Pennsylvania (2007), and Peter Bickel, University of California at Berkeley (2009).

Lectureships

Rao created three named lectureships: one for the Rao Prize winner and two in honor of colleagues Khatri and Krishnaiah. A fourth lecture, The Clifford C. Clogg Memorial Lecture, was created to honor Clogg. He was an outstanding sociologist and statistician who passed away suddenly at the age of 45. The Clogg lectures were initiated in 1996. The first four lectures were given by Leo Goodman (1996), Donald Rubin (1997), Alan Agresti (1998), and Adrian Raftery (1999).

Note Awards and honors for C.R. Rao are too numerous to list here, as well as all of his presidencies of professional societies and other recognition. His most notable awards, in addition to the U.S. National Medal of Science conferred by President Bush in 2002, include membership in the U.S. National Academy of Sciences and the Guy Medal in Gold by the Royal Statistical Society.

Major Research Contributions to Field

Lindsay's statistical research includes likelihood-based statistical inferences, which are used widely in scientific data analyses. Lindsay also is recognized for methods he developed for working with mixture models, which are used when data are collected from a mixture of populations. His work in this area is recognized as a major contribution to the foundations of statistical theory. In addition, Lindsay develops statistical methods that are useful for research in other scientific disciplines; for example, he has constructed models and analyses that have been applied to biological data from genomic studies. Lindsay delivered the 2010 Fisher Lecture, "Likelihood: Efficiency and Deficiency—and the Special Role of Hidden Variables" in 2010.

Hettmansperger received the Noether Award for his research efforts in nonparametric statistics, significant contributor to the development and assessment of statistical methods for the linear model that are both efficient and robust, especially in the area of nonparametric rank tests and estimates. The extension of the median in one dimension to higher dimensional data and sign and rank tests was another major area of research, with interest on new methods that are invariant to linear transformations of the data and developing these methods for more complex experimental designs.

Clogg was nationally and internationally known for his work in quantitative methods and demography, particularly on the analysis of rates, standardization methods, and latent structure analysis. He wrote extensively on the statistical analysis of categorical data, covering loglinear models, cohort analysis, association models, and mobility tables.

Steve Arnold has made a substantial number of significant research contributions to the theory and applications of linear models and multivariate analysis. Steve's impact on statistical research, education, and practice has been exceptional, especially through his masterful textbooks *Mathematical Statistics* and *The Theory of Linear Models and Multivariate Analysis*, which have influenced two generations of statisticians, and his work on patterned covariance structures and repeated measures designs.

The collaboration of G. Jogesh Babu and Eric Feigelson (Astronomy and Astrophysics Department) is credited with starting the field of Astrostatistics and instrumental in its promulgation. The "Statistical Challenges in Modern Astronomy" conferences, held every 5 years since 1991, and the highly popular summer schools in statistics for astronomers (held at PSU and elsewhere) have become a main vehicle for technology transfer.

G.P. Patil and colleagues have had a major impact on environmental statistics and statistical ecology. He was founding Director, Center for Statistical Ecology and Environmental Statistics, the Statistical Ecology Section of Ecological Society of America; and a cofounder of the ASA Section on Statistics and the Environment. He served as Editor-In-Chief, Environmental and Ecological Statistics.

Michael Akritas has made significant contributions to bootstrapping for censored data, goodness-of-fit tests, rank statistics, nonparametric modeling, and thresholding and other high-dimensional testing techniques with emphasis on functional data and model checking. His 1990 JASA paper on the rank transform led to the development of nonparametric models for factorial, ANCOVA, and mixed and random effects designs.

Professional awards of faculty

Name	Year	Award	Organization
Clogg	1987	Paul F. Lazarsfeld award	Amer. Soc. Assoc.
Harkness	2007	Carver award	IMS
Hettmansperger	2004	Noether Senior Scholar award	ASA
Lin	2011	Donald Owen award	ASA
Lindsay	1990	Humboldt Senior Scientist award	Humboldt Foundation
Lindsay	1996	Guggenheim fellowship	Guggenheim Foundation
Lindsay	1997	Snedecor award	COPSS
Patil	1993	Distinguished achievement medal	ASA-SSE
Rosenberger	2011	Distinguished service award	NISS

Distinguished lectureships

Name	Year	Lecture title	Organization
Lindsay	2010	Fisher lecture	COPPS
Lin	2010	Youden address	ASQ
Lin	2011	The Isobel Loutit address	SSC
Richards	1991	Invited address	AMS-SSM

Dennis Lin has made major contributions on supersaturated designs in which there are more factors than the experimental runs. His 1993 Technometrics paper opened up a vital research area and has occupied an important place in the literature. Also of major significance are his consulting activities with government agencies and corporations.

Note All faculty listed as having made major contributions are senior full professors. We have a great group of young faculty whom we expect will be similarly recognized for their contributions in the future.

Other significant honors: professional societies

Babu	2007	Award for summer school for statisticians and astronomers	Mid-Atlantic Region University Continuing Education Assoc.
Harkness	2002	Sloan C award	Pew Foundation
R. Li	2004	NSF CAREER grant	NSF
Lin	2008	Mercator Professorship, Deutsche Forschungsgemeinschaft (DFG)	University of Giessen (Germany)
Lin	2008–2010	Chang-Jiang scholar	Chinese Government
Richards	2000–2001	School of mathematics institute for advanced Study	Princeton

In addition, other faculty including Hultquist, Heckard and Utts, and the Ryans, have written lower level texts. Many books have been edited as well.

Upper level books published by faculty

Name	Title
Arnold	The theory of linear models and multivariate analysis
Arnold	Mathematical statistics
Babu (with Feigelson)	Statistical challenges in modern astronomy
Babu (with Feigelson)	Astrostatistics
Clogg	Statistical models for ordinal variables (advanced quantitative techniques in the social sciences)
Clogg (Editor)	Categorical variables in developmental research: methods of analysis
Hettmansperger	Statistical inference based on ranks
Hettmansperger (with McKean)	Robust nonparametric statistical methods
Runze Li	Design and modeling for computer experiment
Lindsay	Mixture models: theory, geometry, and applications
Schafer	Analysis of incomplete multivariate data
Schafer	Developing statistical software in Fortran 95
Thompson	Sampling
Thompson	Adaptive sampling

Front Row: Bruce Lindsay[f], Laura Lindsay[f], Don Richards[f], Bill Harkness[r], Arkady Tempelman[f], Dennis Lin[f], G. P. Patil[r], Zhibiao Zhao[f], Jia Li[f] Second Row: Bob Hultquist[r], Jogesh Babu[f], Francesca Chiaromonte[f], Jennifer Shook[i], Pat Buchanan[i], Mosuk Chow[f], Naomi Altman[f], Yeonok Lee[p], Bonnie Cain[c], Alexandra Slavkovic[f], Michael Akritas[f], Cedric Neumann[f] Third Row: Runze Li[f], Steve Arnold[f], Durland Shumway[i], Andrew Wiesner[i], Murali Haran[f], Charles Antle[r], Rongling Wu[f], Bing Li[f], Yu Zhang[f] Fourth Row: David Welch[p], Trent Gaugler[i], Daniel Spencer[i], Bruce Lord[i], Barbara Freed[a], Laurie Roan[c], Jim Rosenberger[f], Linda Strauss[i]:"*f* faculty, *i* instructor, *p* postdocs, *a* administrative, *c* clerical". Photo taken in 2011

Left to right: Dennis Lin, Tom Hettmansperger, Jim Rosenberger, Bill Harkness

Department heads	
1968–1969	James B. Bartoo
1969–1987	William L. Harkness
1988–1990	Thomas P. Hettmansperger
1990–2006	James L. Rosenberger
2006–Present	Bruce L. Lindsay

Some Notable PhD Graduates

Jay Aubuchon, 1982. Design and Development at Minitab, Inc.
Ayan Basu, 1991. Full Professor at the Indian Statistical Institute, winner of 2006–2007 C.R. Rao National Award for Statistics in India
Mark Becker, 1985. President of Georgia State University
Marianthi Markatou, 1988. IBM, formerly Columbia University, past program Director of Probability and Statistics at NSF
Joe McKean, 1975 Western Michigan, directed many PhDs, co-author of graduate level text and many publications.
Kathryn Roeder, 1988, COPPS Presidents Award, outstanding young statistician
Jessica Utts, 1978. UC Riverside, Stat Education leader, Carver Award winner, and many other contributions to Statistics

Princeton University Department of Statistics (1965–1985)

David C. Hoaglin and Karen Kafadar

The Department of Statistics at Princeton lasted only 20 years, but its activities have had a lasting impact on the discipline. We recount the history of the department as best we can from our limited involvement (we were graduate students in 1966–1970 and 1975–1979, respectively) and from published articles, department annual reports, and a few interviews.

Background and Early Years

The story of statistics at Princeton begins more than three decades before the establishment of the Department of Statistics on July 1, 1965. (The Board of Trustees gave its approval in April 1965, and, in the words of the press release, "full development in terms of personnel and curriculum [was] not expected to be completed until July, 1966."). From *A Princeton Companion*, we have John W. Tukey's account of the period up through the department's early years (Tukey 1978):

> Statistics at Princeton was the creation of Luther P. Eisenhart [Dean of the Graduate School and Chairman of the Department of Mathematics and a former Dean of the Faculty] and Samuel S. Wilks. Its history falls naturally into four periods: Wilks alone, World War II, postwar under Wilks, the early years as a department.

73 Hickory Road, Sudbury, MA 01776, USA
D. C. Hoaglin (✉)
e-mail: dchoaglin@gmail.com

Indiana University Bloomington,
Department of Statistics, 309 North Park Avenue,
Bloomington, IN 47408, USA
K. Kafadar
e-mail: kkafadar@indiana.edu

A. Agresti and X.-L. Meng (eds.), *Strength in Numbers: The Rising of Academic Statistics Departments in the U.S.*, DOI: 10.1007/978-1-4614-3649-2_16,

Dean Eisenhart's vision brought Wilks to Princeton's Department of Mathematics in 1933. For the next eight or ten years, Wilks was the lone statistician in that department, building a sequence of courses, stimulating undergraduates, and starting to turn out Ph.D.'s. Much of his important research was carried out and published during this era. His influence on a variety of areas of application grew rapidly, and his editorship of the *Annals of Mathematical Statistics* carried that journal through its crucial and formative decades. His pattern of "write mathematics, and do applications" became well established and was thoroughly conveyed to his students.

During World War II Wilks became deeply involved in the activities of the National Defense Research Committee, both in the Applied Mathematics Panel and at Princeton, where SRG-P (Statistical Research Group-Princeton) and its branch at Columbia ("SRG-P Jr") involved graduate students T.W. Anderson, P.J. McCarthy, F. Mosteller and D.F. Votaw, and brought in R.L. Anderson, W.G. Cochran, A.M. Mood, L.J. Savage, J.W. Tukey, J.D. Williams, and C.P. Winsor for varying periods. (Many of these names were to become familiar to statisticians in the decades that followed.) [Wallis (1980) and Mosteller (2010) give further details.]

Merrill Flood became responsible for the Fire Control Research Group, involving in addition to mathematicians, physiologists, and engineers, C.P. Winsor, an engineer-turned physiologist-turned-statistician, J.W. Tukey, a chemist-turned-topologist-turned-statistician, and G.W. Brown, a mathematician-turned-statistician. [Fire Control Research's assignments included stereoscopic height and range finders, rocket powder, fire control from tanks, tactics for B-29 bombers, and leads for shooting at planes (Anscombe 2003).]

During these years the flavor was, of course: "do crucial applications, and any mathematics that can help."

This increase in the number of statisticians made statistics much more visible in Princeton (and vice versa) and must have played a part in the administration's approval of a Section of Mathematical Statistics in the Department of Mathematics [formed when Wilks returned to academic duties after World War II]. This section, which initially consisted of Wilks, Tukey (part-time Princeton, part-time Bell Telephone Labs), and usually a visitor, grew in 1956 by the addition of Francis J. Anscombe (a former visitor). The old flavor of "write mathematics and do applications" continued, although some were now much more willing not to look to mathematicians as the source of approval. Undergraduate courses and interest grew somewhat, and there were steady streams of Ph.D.'s and faculty research.

The Korean War stimulated the establishment of the Analytical Research Group, of which Forman S. Acton was director from 1952 to 1956, and about half of whose work was statistical. George E.P. Box was then brought to Princeton as director of the Statistical Techniques Research Group (1956–1959), whose participants included E.M.L. Beale, J. Stuart Hunter, Colin Mallows, Mervin E. Muller, and Henry Scheffé. Again there was a larger critical mass and mutual stimulation, with a further broadening of interests and a further recognition of mathematics as important but not all embracing.

After Wilks's death in 1964, with encouragement from some mathematicians and the administration, an independent Department of Statistics was formed in 1966. Tukey was its (part-time) chairman from 1966 to 1969 and after an administrative review and approval of expansion, he was succeeded by Geoffrey S. Watson in 1970. A new stream—undergraduate majors—joined the streams of Ph.D.'s, faculty research, and involvement in statistical applications; emphasis on undergraduate study became comparable with the emphasis on graduate study.

Frederick F. Stephan, who had come to Princeton in 1947 in Sociology, transferred to the new department, broadening its interests significantly. J. Stuart Hunter, who had returned to teach in the School of Engineering in 1968, also took an active role for a time. John Hartigan returned in 1964 to help hold the fort and, after Watson's arrival, the department was strengthened by Donald McNeil (1971–1976) and Peter Bloomfield (1971, acting chairman 1976–1977) among others.

Tukey's interest in data analysis (regarded as encompassing statistics [see Tukey (1962)]), now that the pressure of needing to teach "mathematics" was removed, led first to a Freshman course in Exploratory Data Analysis and then to a book (preliminary edition 1970–71, first edition 1977) emphasizing simple arithmetic and careful thinking. In December 1974, a PDP 11–40 computer was installed. Both these steps illustrate the continued broadening in the department's attitude, where very mathematical questions remain important but other Ph.D. theses focus on specific applications or even on the analysis of important sets of data.

Robustness (in the technical sense of that word) and time series continue to be strong interests in the department, as they have been for decades. More recently, collaboration with the Office of Population Research and the Center for Environmental Studies has been prominent. The department is proud that, both at Harvard and at Yale, the first two chairmen of the Department of Statistics came from Princeton and that, in the early seventies, those of its undergraduate majors who wished to do their graduate study in statistics were able to do so at Harvard, Stanford, or Yale. [from Alexander Leitch, A Princeton Companion. © 1978 Princeton University Press. Reprinted by permission of Princeton University Press.]

A number of conversations published in *Statistical Science* provide further glimpses into Princeton's impact on statistical activities in the U.S. and on the field more generally. S. S. Wilks's style of formulating a statistical problem in a rigorous mathematical fashion, solving the mathematical problem, and then using the result for applications (DeGroot 1986) and his unstinting work ethic (Mood 1990) made a profound and lasting impression on his students. As editor (1938–1949), he contributed strongly to establishing the *Annals of Mathematical Statistics* as the leading theoretical journal in the field; and, by having them serve as referees and proofreaders, brought his students into contact with current research (Flournoy 1993; Becker 2009). John Tukey's recruiting of George Box to head the Statistical Techniques Research Group brought Box permanently to the U.S., and his experience in STRG influenced his approach in building the Department of Statistics at the University of Wisconsin (DeGroot 1987). Many statisticians who received their PhD from the Department of Mathematics went on to play major roles in other universities and organizations. They include T. W. Anderson, Patrick Billingsley, David Brillinger, Joseph Daly, Arthur Dempster, W. J. Dixon, D. A. S. Fraser, David Freedman, Leo Goodman, Theodore Harris, John Hartigan, Alan James, Thomas Kurtz, Philip McCarthy, Paul Meier, Alexander Mood, Frederick Mosteller, Herbert Scarf, James Thompson, David Wallace, John Walsh, and Thomas Wonnacott. In the appendix we list statisticians and probabilists who received their degrees from the Department of Mathematics from 1940 through 1967.

One wonders what course S. S. Wilks would have charted for statistics at Princeton if he had not died suddenly, in his sleep, in March 1964. We found hints of a different direction in some of the memorial articles. T. W. Anderson mentioned that Wilks continued as Professor of Mathematics and Director of the Section of Mathematical Statistics until his death. He also wrote, "Although many other leading universities have subsequently set up separate departments of statistics, at Princeton mathematical statistics has been retained as part of the mathematics department" (Anderson 1965, p. 5). Two remarks by Stephan et al. (1965) are suggestive but ambiguous. First, "If he had seen a better chance for

Fig. 1 John Tukey in his office in the original Fine Hall. Photo from Princeton University Library

statistics to grow rapidly in Princeton as a well-regarded and mutual supportive part of a broadly conceived and totally excellent activity in mathematics, Princeton would indeed have been his Utopia" (pp. 939–940). And then, "Sam (we believe) could well feel, ever so privately, that most important things were being done, and that the tradition that statistics had much to do with both mathematics and the real world was being preserved and strengthened" (p. 943).

Composition of the Department

For the 1966–1967 academic year, the department's faculty consisted of Professor John W. Tukey (Fig. 1) and Assistant Professors Michael D. Godfrey (half-time in Economics) and John A. Hartigan. Its roster listed eight graduate students and ten undergraduate majors.

Each year several members of other departments or programs were associated with the department (e.g., William Feller in Mathematics, J. Stuart Hunter in Engineering, Richard Quandt in Economics, Frederick Stephan in Sociology). Those who provided the longest-term support were Forman S. Acton (Engineering), Ansley J. Coale (Office of Population Research and Department of Economics), and especially, J. Stuart Hunter (Engineering), whose contributions to statistics are well known. Others included Kenneth Deffeyes (Geology), Richard Holley (Mathematics), Jane Menken (Office of Population Research), James Poage (Director of the Computer Center), and Hale F. Trotter (Mathematics). Several of the department's courses (in the fall term 1966–1967, four of seven undergraduate courses and five of seven graduate courses) were cross-listings of courses taught in other departments. In the early years, a substantial portion of lower-level undergraduate instruction was handled by Visiting Lecturers (i.e., adjunct faculty) associated with other local institutions, such as Albert E. Beaton of Educational

Fig. 2 G. S. Watson at home in his study in December 1996, after the interview with R. J. Beran and N. I. Fisher. Photograph by N. I. Fisher

Testing Service and Samuel A. Schmitt of the Institute for Defense Analyses. These arrangements promoted connections with other fields and with the broader statistical community, and they allowed the department to maintain its service courses while it endeavored to grow.

The number of regular faculty remained at four from 1967–1968 through 1970–1971. After G. S. Watson (Fig. 2) arrived to serve as (full-time) chairman, the department grew modestly, to seven or eight in most years from 1973–1974 to 1984–1985. (The appendix lists members of the regular faculty, as well as the Visiting Lecturers.) Under his leadership the department maintained a strong undergraduate program and was at the forefront of statistical computing. Peter Bloomfield (Fig. 3) served as chair from 1979–1980 through 1981–1982.

The department attracted numerous visitors from other academic institutions, including Hirotugu Akaike, Julian Besag, Peter Bickel, Henry Daniels, James Durbin, Ruben Gabriel, Frank Hampel, Edward Hannan, Thomas Hettmansperger, Peter Huber, Alan James, Robert Kass, Kanti Mardia, Douglas Martin, George McCabe, Ingram Olkin, Daryl Pregibon, Eugene Seneta, Burton Singer, Bernard Silverman, S. D. Silvey, Terence Speed, William Steiger, William Strawderman, Paul Switzer, and Henry Wynn. The visitors contributed much to the research and educational programs of the department. The long-standing custom of tea in Fine Hall each weekday afternoon gave way in later years to perpetual coffee in the department's "Coffee Room" and helped to bring faculty, visitors, and students together for congenial conversation. The number of entering graduate students remained small, typically four or five per year. In the appendix we have included as complete a list, year by year, as we were able to assemble.

The department continued to attract new undergraduate majors; the number (juniors plus seniors) fluctuated somewhat, averaging around 14 in each year. In reviewing the department's annual reports, we saw a number of familiar names:

Fig. 3 Peter Bloomfield. Photo from Princeton University Library

Edward Binkowski, David Donoho, William F. Eddy, Mark Glickman, Shelby J. Haberman, David A. Lax, Sue Leurgans, David Pasta, William H. Rogers III, Joseph Romano, Nathaniel Schenker, Keith Soper, Michael A. Stoto, and Paul A. Tukey. We would also include Roy E. Welsch and Richard D. De Veaux, who graduated from the Mathematics Department in 1965 and 1973, respectively. Several of these statisticians have made important contributions to the field during their careers in academe, industry, and government.

From 1966 to 1985 the department had three chairs: John Tukey, Geoffrey Watson, and Peter Bloomfield. Tukey's research interests and contributions have received wide coverage, including the memorial articles by Brillinger (2002), Kafadar (2002, 2005), and McCullagh (2003); issues of *Technometrics* (August 2001), *Annals of Statistics* (December 2002), and *Statistical Science* (August 2003) were dedicated to his contributions and his lasting influence on the profession. Perhaps less well known is his style. He was readily approachable and willing to discuss statistical problems, from the point of view of those presenting them, and to offer suggestions, at any time and at any place (e.g., while birding with fellow statisticians Byron "Bill" Brown and Lincoln Moses), even in areas far from his research interests, such as Bayesian methods (DeGroot 1987). He was especially generous with students, though somewhat unorthodox in his choice of meeting locations: sometimes at his office in Fine Hall, but more often while chopping wood or doing yard work at the Tukeys' home, or at the filling station while his tires were being rotated, or even during trips to Spain, England, or Austria—his students learned to be efficient in their requests for his time and adaptable to meeting him wherever he was. Despite his frequent absences, his influence on the department was ever-present: statistical problems raised during his external engagements could serve as the basis for PhD theses, and substantial research funding involved Tukey and supported graduate students (e.g., the Twin

Rivers project from the Environmental Protection Agency and contracts from the Army Research Office). Some of these projects, described below, show him in operation. His varied interests outside statistics included birding expeditions all over the world (Bill Brown remarked, "For John, it's really just an excuse to discuss statistics."), whale watching, and frequenting mystery book stores, where he added to his impressive collection of detective and adventure fiction (bequeathed to Brown University).

G. S. Watson may be best known for the Durbin-Watson test for serial correlation, but that was not his favorite contribution (Beran and Fisher 1998). He made important early contributions in the field of paleomagnetism and pioneered research in directional statistics, the topic of his monograph *Statistics on Spheres* (Watson 1983). The Nadaraya-Watson (or "local constant") estimator in nonparametric regression was a forerunner of local polynomial methods for smoothing scatterplots.

Peter Bloomfield continued the department's interest in time series with a book (Bloomfield 1976) and a variety of papers on methods and applications. His research also dealt with topics in regression, including least-absolute-deviations fitting (Bloomfield and Steiger 1983).

Selected Projects

A number of projects based in the department made lasting contributions. We briefly describe four, recognizing the likelihood of important omissions because of the limitations of our sources.

Citation Index. Funded by the National Science Foundation and led by John Tukey, this effort aimed to remedy the uneven coverage of statistics and probability in the available citation indexes. The database that it assembled led to a citation index and several other reference tools, including a permuted title index (Dolby and Tukey 1973; Tukey 1973; Ross and Tukey 1973, 1975a, b). The project served as the inspiration for the development of *Current Index to Statistics*.

Princeton Robustness Year. For the 1970–1971 academic year, G.S. Watson invited a number of statisticians who had been working on robustness to visit Princeton. The visitors were Peter Huber, Peter Bickel, and Frank Hampel. The year-long seminar attracted Princeton staff and students, as well as participants from other universities and Bell Labs. (Hampel 1997 reproduces the handwritten original list of participants.) The fall term produced vigorous discussion, but little unanimity on interpretations of robustness or on the important problems. Building on the evident enthusiasm, David Andrews suggested a comparative survey, which focused on point estimation of the center of a symmetric distribution. Over 60 estimators were studied, some well known, many developed during the study. They were compared on properties in large and small samples, including variances, percentage points, breakdown bounds, influence curves, and sensitivity curves. Data for finite samples were drawn from over 30 sampling situations, making substantial use of Monte Carlo techniques. The work produced several new asymptotic results, as well as new methods for summarizing the large body of data. Most of the results were reported in the book by Andrews et al. (1972). Open problems served as the basis for a number of PhD and undergraduate theses in subsequent years, often involving substantial computation.

Exploratory Data Analysis. The developments that led to John Tukey's 1977 book (Tukey 1977), preceded by the limited preliminary edition (Tukey 1970a, b, 1971), were well along by the mid-1960s. Tukey used an early draft for an undergraduate course as early as 1969–1970. Other members of the department subsequently taught that course, and some produced books that broadened its influence (e.g., McNeil 1977; Siegel 1988). Reflecting the department's emphasis on computing, many EDA techniques had their first software implementation in SNAP-IEDA, a local package whose development was begun by Michael D. Godfrey. Directly and indirectly, attitudes and techniques of EDA have had a profound impact on data analysis and statistics. Ironically, EDA was viewed with great skepticism by mathematical statisticians in the 1960s and 1970s, some of whom called it "data snooping". Today, the "data mining" philosophy has strong overtones of EDA and has become natural in analyzing data.

Cancer Mapping. In 1977 the second-year graduate students collectively felt a need to work with faculty on a "meaningful" project, which they explained in a letter to the department. Fortunately, Tukey had taken an interest in a recent acquisition, *The Atlas of Cancer Mortality* (Mason et al. 1975). It contained "patch maps" that showed the mortality in U.S. counties from various forms of cancer, indicating those counties whose rates (averaged over the years 1950–1969) were "statistically significantly higher" than the national average. Tukey recognized the benefits of smoothing such rates, particularly for counties with relatively fewer person-years, in identifying *regions* of the country with high rates. Within a few days he met with the graduate students and distributed five sets of mimeographed notes, discussing issues with these data (how to smooth rates defined with latitude and longitude coordinates, how to calculate a valid uncertainty of smoothed rates, which sites to combine, which regions of the country to consider, computational issues). The class divided into five groups and worked on these subprojects, focusing ultimately on three regions with high rates for several cancers: the northeast corridor (counties within 200 miles of New York City, which included Boston, Philadelphia, and D.C.); the Great Lakes (counties within 300 miles of Toledo, which included Chicago); and the Gulf of Mexico (counties within 200 miles of Baton Rouge).

For smoothing the rates, Tukey proposed a two-dimensional version of "3R" that he called "head banging" (later published with Paul Tukey [Tukey and Tukey 1981], and applied by Katherine Hansen in her PhD thesis 1989 and by Mungiole et al. 1999 for cancer mortality rates). Tukey recognized that, if a geographical component in cancer could be found, one would first have to adjust the rates for obvious effects (Tukey 1979), such as the contribution of smoking to lung cancer (or, as a proxy for smoking, urbanization; see Kafadar and Tukey 1993). Many researchers followed with proposals for "nonparametric regression" and adjustment of rates for obvious covariates. None of these results would have been possible without the extensive use of the computational resources available at Princeton at that time.

Final Years

We do not have access to detailed information surrounding the closure of the department. It seems likely that it died of "natural causes." The small number of graduate students made it difficult for them to have a community. The total enrollment in the department's undergraduate courses was variable but not growing. John Tukey was scheduled to retire in 1985, and G. S. Watson was likely to retire a few years later. After spending 1982–1983 at the University of North Carolina on sabbatical, Peter Bloomfield moved to North Carolina State University. Watson had made it clear in his annual reports "that there was no way Princeton could now have a good Statistics Department unless they had four well-known professors" (Beran and Fisher 1998). Faced with a choice between committing additional resources to statistics and investing those resources in emerging areas, the administration chose the latter.

The academic years after 1984–1985 were a period of transition. G. S. Watson moved to the department of mathematics. J. Michael Steele, Elvezio Ronchetti, and Colin Goodall moved to the School of Engineering and became a statistics and operations research group; they were subsequently joined by Erhan Çinlar and Richard De Veaux. The department apparently continued to have a presence in the general catalog, but it was managed by acting chairs from other departments (for 1985–1987, George F. Pinder of Civil Engineering and Elias M. Stein of Mathematics; for 1987–1990, Arthur S. Wightman of Mathematics; for 1990–1991, Simon B. Kochen of Mathematics). The catalog for 1991–1992 listed the Committee for Statistical Studies, an interdepartmental committee chaired by Richard De Veaux with representatives from ten departments. Thus, statistics continued to have a presence at Princeton, but not as a department.

Acknowledgments Without the generous help of others, we could not have written even this limited account. We thank J. Stuart Hunter for researching the composition of the Department, Associate Dean David N. Redman of the Graduate School for supplying a copy of the material in *A Princeton Companion* and for valuable suggestions, University Archivist Daniel J. Linke and the staff of the Seeley G. Mudd Manuscript Library for access to the Department's annual reports for 1969–1970 to 1976–1977 and 1978–1979 to 1984–1985, and Richard D. De Veaux, James J. Filliben, John A. Hartigan, J. Michael Steele, and James R. Thompson for helpful conversations.

Appendix

From the Department's annual reports for 1969–1970 through 1984–1985, we assembled the following lists of regular and adjunct faculty with their years of service.

Regular Faculty

Dan Anbar, 1971-72; David F. Andrews, 1968-71; Peter Bloomfield, 1971-82; Henry I. Braun, 1973-79; Joseph G. Deken, 1976-79; Luisa T. Fernholz, 1980-84; Noreen Goldman, 1984-85; Colin Goodall, 1984-85; John A. Hartigan, 1965-68; Nicholas P. Jewell, 1979-81; Lawrence S. Mayer, 1974-79; James McBride, 1976-80; Donald R. McNeil, 1971-76; Gary W. Oehlert, 1981-84; Antonio M. Possolo, 1983-84; Elvezio M. Ronchetti, 1983-85; Glenn R. Shafer, 1973-76; Andrew F. Siegel, 1979-83; Gary A. Simon, 1971-75; J. Michael Steele, 1983-85; Leonard M. Steinberg, 1967-70; Frederick F. Stephan, 1965-71; Mary Lou Thompson, 1979-81; John W. Tukey, 1965-85; James Veitch, 1983-84; Geoffrey S. Watson, 1970-85; Allan Wilks, 1981-84; Thomas H. Woteki, 1976-78

Statisticians with Local Non-academic Affiliations Who Served as Visiting Lecturers

Albert E. Beaton (Educational Testing Service), 1966-1977; Richard A. Leibler (Institute for Defense Analyses), 1966-1969, 1971-1975; Richard F. Link (consultant), 1966-1970; Donald B. Rubin (Educational Testing Service), 1971-1974; Samuel A. Schmitt (Institute for Defense Analyses), 1966-1969, 1971-1975

Statisticians and Probabilists Who Received Their PhD from the Department of Mathematics, 1940 through 1967

Theodore W. Anderson, Jr., 1945; Harvey J. Arnold, 1958; Patrick P. Billingsley, 1955; Edward S. Boylan, 1962; David R. Brillinger, 1961; George W. Brown, 1940; Morton B. Brown, 1965; Bradley D. Bucher, 1957; Donald S. Burdick, 1961; Kai-Lai Chung, 1946; Gordon B. Crawford, 1962; Joseph F. Daly, 1940; Arthur P. Dempster, 1956; Wilfrid J. Dixon, 1944; Richard M. Dudley, 1962; A. Ross Eckler, 1954; Donald A. S. Fraser, 1948; David A. Freedman, 1960; David Gale, 1949; Donald P. Gaver, 1956; W. Morven Gentleman, 1966; Jay R. Goldman, 1965; Leo A. Goodman, 1950; N. Roy Goodman, 1957; Theodore E. Harris,

1947; John A. Hartigan, 1962; Gilbert A. Hunt, Jr., 1948; Alan T. James, 1953; Edward L. Kaplan, 1951; Samuel Karlin, 1947; Frank B. Knight, 1959; Joseph B. Kruskal, 1954; Robert A. Kurtz, 1967; Thomas E. Kurtz, 1956; James A. Lechner, 1959; Richard F. Link, 1954; Philip J. McCarthy, 1946; Henry P. McKean, Jr., 1955; Paul Meier, 1951; Alexander M. Mood, 1940; Frederick Mosteller, 1946; R. Bradford Murphy, 1951; Peter B. Nemenyi, 1963; Melvin P. Peisakoff, 1950; Loren D. Pitt, 1967; Francis M. Sand, 1963; Herbert E. Scarf, 1954; Lawrence A. Shepp, 1961; Bernard Sherman, 1950; Martin L. Silverstein, 1965; James G. C. Templeton, 1957; James R. Thompson, 1965; Hale F. Trotter, 1956; David F. Votaw, Jr., 1947; David L. Wallace, 1953; John E. Walsh, 1947; David M. G. Wishart, 1960; John W. Woll, Jr., 1956; Thomas H. Wonnacott, 1963; Ralph Wormleighton, 1955

Graduate Students in the Statistics Department, by Year of Entry

Year of entry	Name (year of PhD)
1965–1966	James J. Filliben (1969)
1966–1967	George J. Hariton, David C. Hoaglin (1971), Howard J. Hoffman, Jon Helge Knudsen (1971), Lynn Roy LaMotte, Charles Lewis (1970), Lincoln Polissar (1974), James J. Schlesselman (1971)
1967–1968	Neil Bromberg, Alan M. Gross (1973), Stanislaus M. D'Souza (1971), Richard C. Madden (1971), Stephen J. Finch (1974)
1968–1969	Roderick Montgomery
1969–1970	Thomas L. Corwin (1973), Gunnar T. Gruvaeus, Anita G. Nowlin (1973), Gregory R. Traversy
1970–1971	Dennis D. Ash, Edward S. Binkowski (1974), Robert J. Epp (1973), Theo A. Gasser, Richard B. Landau, John W. Sawyer, Jr.
1971–1972	Ruth W. Arad (1975), David W. Hunter (1975), Tony K.S. Quon (1976), Glenn Shafer (1973), Paul F. Velleman (1975)
1972–1973	Noel A.C. Cressie (1975), James E. Ertel (1975), Fred M. Hoppe (1975), Karen M. Kirby (1975), John C. Turner (1976)
1973–1974	Farouk Kazim (1976), Lois A. Kellerman
1974–1975	Pierre C. Delfiner (1978), Alice K. Gerb, Petrus Jakobs, Clifton J. Lancaster, Kuang-Hsien Lin, Michael D. Schwartzschild (1979), Donna F. Stroup (1980), David E. Tyler (1979), Hania Zlotnik (1978)
1975–1976	Susan P. Arthur (1979), John Brian Carney, Jr., Anne M. Houtman (1981), Karen Kafadar (1979), Louis K. Kates (1981), Donna L. Mohr (1981), Arthur R. Silverberg (1980), Terry L. Watt
1976–1977	Yoav Benjamini (1981), Francine Blanchet, Mark Gordon, Roberta Guarino (1981), Allan R. Wilks (1980)

(continued)

(continued)

Year of entry	Name (year of PhD)
1977–1978	Flavio C. Bartmann (1982), Kaizo I. Beltrao (1982), David E. Coleman, Miriam L. Goldberg (1982), Jenepher M. Gosling, Paul S. Horn (1981), Katherine B. Krystinik (1981), Robert A. Stine (1982), Patrick L. Zulke
1978–1979	Jeffrey A. Raynes, Scott L. Zeger (1982)
1979–1980	Javier F. Cabrera (1983), Joanne Goldberg
1980–1981	Dhammika J. Amaratunga (1984), Clifford M. Hurvich (1985), William H. Lilliefors, Carlos E. Mendoza (1984), Stephan Morgenthaler (1983), Fanny L. O'Brien (1984), E. Seidel, Christine E. Soteros
1981–1982	George S. Easton (1985), David I. Rubin
1982–1983	Eugene G. Johnson (1988), Ha Hung Nguyen (1986)
1984–1985	Katherine M. Hansen (1989), Edward A. Ipser, M. Semenovsky, Peter J. Slasor

References

Anderson TW (1965) Samuel Stanley Wilks, 1906–1964. Ann Math Stat 36:1–23

Andrews DF, Bickel PJ, Hampel FR, Huber PJ, Rogers WH, Tukey JW (1972) Robust estimates of location: survey and advances. Princeton University Press, Princeton

Anscombe FR (2003) Quiet contributor: the civic career and times of John W. Tukey. Stat Sci 18:287–310

Becker MP (2009) A conversation with Leo Goodman. Stat Sci 24:361–385

Beran RJ, Fisher NI (1998) A conversation with Geoff Watson. Stat Sci 13:75–93

Bloomfield P (1976) Fourier analysis of time series: an introduction. Wiley, New York

Bloomfield P, Steiger WL (1983) Least absolute deviations: theory, applications, and algorithms. Birkhäuser, Boston

Brillinger DR (2002) John W. Tukey: his life and professional contributions. Ann Stat 30:1535–1575

DeGroot MH (1986) A conversation with T. W. Anderson. Stat Sci 1:97–105

DeGroot MH (1987) A conversation with George Box. Stat Sci 2:239–258

Dolby JL, Tukey JW (1973) The statistics cumIndex. The R&D Press, Los Altos

Flournoy N (1993) A conversation with Wilfrid J. Dixon. Stat Sci 8:458–477

Hampel F (1997) Some additional notes on the "Princeton Robustness Year." In: Brillinger DR, Fernholz LT, Morgenthaler S (eds) The practice of data analysis: essays in honor of John W. Tukey. Princeton University Press, Princeton

Hansen KM (1989) Some statistical problems in geophysics and structural geology. PhD thesis, Department of Statistics, Princeton University

Kafadar K (2002) The influence of John Tukey's work in robust methods for chemometrics and environmetrics. Chemometr Intell Lab Syst 60:127–134

Kafadar K (2005) John Wilder Tukey. In: Armitage P, Colton T (eds) Encyclopedia of biostatistics. Wiley. doi:10.1002/0470011815.b2a17150

Kafadar K, Tukey JW (1993) U.S. cancer death rates: a simple adjustment for urbanization. Int Stat Rev 61:257–281

Mason TJ, McKay FW, Hoover R, Blot WJ, Fraumeni JF Jr (1975) Atlas of cancer mortality for U. S. counties: 1950-1969. DHEW publication no. (NIH)75-780. U. S. Government Printing Office, Washington

McCullagh P (2003) John Wilder Tukey 16 June 1915—26 July 2000. Biogr Mem Fellows R Soc 49:537–555

McNeil DR (1977) Interactive data analysis: a practical primer. Wiley, New York
Mood AM (1990) Miscellaneous reminiscences. Stat Sci 5:35-43
Mosteller F (1964) Samuel S. Wilks: statesman of statistics. Am Stat 18(2):11–17
Mosteller F (2010) Fienberg SE, Hoaglin DC, Tanur JM (eds) The pleasures of statistics: the autobiography of Frederick Mosteller. Springer Science+Business Media, New York
Mungiole M, Pickle LW, Simonson KH (1999) Application of a weighted head-banging algorithm to mortality data maps. Stat Med 18:3201–3209
Ross IC, Tukey JW (1973) Index to statistics and probability: locations and authors. Information access series, vol 5. The R&D Press, Los Altos
Ross IC, Tukey JW (1975a) Index to statistics and probability: permuted titles, A-microbiology. Information access series, vol 3. The R&D Press, Los Altos
Ross IC, Tukey JW (1975b) Index to statistics and probability: permuted titles, microclimatic-Z. Information access series, vol 4. The R&D Press, Los Altos
Siegel AF (1988) Statistics and data analysis: an introduction. Wiley, New York
Stephan FF, Tukey JW, Mosteller F, Mood AM, Hansen MH, Simon LE, Dixon WJ (1965) Samuel S. Wilks. J Am Stat Assoc 60:939–966
Tukey JW (1962) The future of data analysis. Ann Math Stat 33:1–67, 812. Reprinted in The collected works of John W. Tukey, volume III: Philosophy and principles of data analysis, 1949–1964 (Jones LV (ed)). Wadsworth & Brooks/Cole Advanced Books & Software, Monterey (1986), pp 391–484
Tukey JW (1970a) Exploratory data analysis, limited preliminary edition. vol I. Addison-Wesley, Reading
Tukey JW (1970b) Exploratory data analysis, limited preliminary edition, vol II. Addison-Wesley, Reading
Tukey JW (1971) Exploratory data analysis, limited preliminary edition, vol III. Addison-Wesley, Reading
Tukey JW (1973) Index to statistics and probability: citation index. Information access series, vol 2. The R&D Press, Los Altos
Tukey JW (1977) Exploratory data analysis. Addison-Wesley, Reading
Tukey JW (1978) Statistics at Princeton. In: Leitch A (ed) A Princeton companion. Princeton University Press, Princeton, pp 449–451
Tukey JW (1979) Statistical mapping: what should not be plotted. In: Proceedings of the 1976 workshop on automated cartography, pp 18–26. DHEW publication no (PHS) 79-1254. Reprinted in The collected works of John W. Tukey, volume V: Graphics, 1965–1985 (Cleveland WS (ed)). Wadsworth & Brooks/Cole Advanced Books & Software, Pacific Grove (1988), pp 109–121
Tukey PA, Tukey JW (1981) Graphic display of data sets in 3 or more dimensions. In: Barnett V (ed) Interpreting multivariate data. Wiley, Chichester, pp 189–275, Chaps 10–12. Reprinted in The collected works of John W. Tukey, volume V: Graphics, 1965–1985 (Cleveland WS (ed)). Wadsworth & Brooks/Cole Advanced Books & Software, Pacific Grove (1988), pp 189–288
Wallis WA (1980) The Statistical Research Group, 1942–1945. J Am Stat Assoc 75:320–330
Watson GS (1983) Statistics on spheres. Wiley-Interscience, New York

Purdue Statistics: A Journey Through Time

Sanvesh Srivastava and Rebecca W. Doerge

Since its modest inception as the Statistical Laboratory in 1947, the Department of Statistics, Purdue University has grown to one of the largest and most diverse in the country supported by a distinguished list of alumni, outstanding contributions in research, and major advances in statistical education. Its current (2011) size of 62 faculty, 125 graduate students, and almost 400 undergraduate students reflects its commitment to developing statistical sciences research for the present and the future, and to providing high quality education to students, both in statistics and in other disciplines. Historically, the Department of Statistics at Purdue University has been an important center for diverse areas of statistical research. Its strong presence in probability, theory, and education set the stage for its expansion in the mid-1990s. As the field of statistics expanded to include many interdisciplinary areas that require specialization (statistical genetics and bioinformatics, computational finance, machine learning, etc.), Purdue Statistics engaged in an aggressive program of hiring well-prepared faculty with diverse backgrounds who are playing leading roles in the development of the field as it expands its scope. Today, Purdue Statistics stands strong as the highest ranked department in the College of Science at Purdue University, and is enjoying its place among the top ranked departments in the United States.

Purdue University,
Department of Statistics,
250 N. University Street,
West Lafayette, IN 47907-2066, USA
S. Srivastava · Rebecca W. Doerge (✉)
e-mail: doerge@purdue.edu

S. Srivastava
e-mail: sanvesh@purdue.edu

A. Agresti and X.-L. Meng (eds.), *Strength in Numbers: The Rising of Academic Statistics Departments in the U.S.*, DOI: 10.1007/978-1-4614-3649-2_17,

Fig. 1 Carl Kossack (standing) (1947–1959) started the Statistical Laboratory

The Beginning and Formative Years (1962–1970)

The Department of Statistics at Purdue University traces its origin back to 1947 when Carl Kossack (Fig. 1) (1947–1959)[1] started the Statistical Laboratory (Stat Lab). Later, the name of the laboratory was changed to Statistical and Computing Laboratory. The Stat Lab was located on the second floor of the Engineering Administration Building on Purdue University's West Lafayette campus. Most of the faculty members that were associated with Stat Lab were professors of Mathematics; a small number of whom also taught Statistics. Kossack served as the Stat Lab director until 1956; Virgil L. Anderson (1951–1987) took over the position until the Stat Lab ceased to exist in 1966.

In 1962, Felix Haas, Head of the Department of Mathematical Sciences at Purdue University, interviewed Shanti S. Gupta (1962–2002) for a faculty position. Like many events that shaped the history of Purdue Statistics, it was our good fortune that Haas had great foresight. It was his vision of mathematical sciences that provided the first opportunity for developing a program in probability and statistics at Purdue. Although Shanti had another offer from the University of Illinois at Urbana-Champaign, he decided to join Purdue Mathematical Sciences as he too had a vision. Shanti was hired primarily in anticipation that his theoretical and practical experience would promote the growth of statistics at Purdue. He received his PhD in theoretical statistics from the University of North Carolina at Chapel Hill in 1956 and worked at Bell Telephone Laboratories until the time he joined Purdue's faculty (McDonald 1998).

From 1962 to 1968, the core of what would become the faculty of Purdue Statistics joined the Department of Mathematical Sciences as professors of both

[1] The entries in parentheses denote the duration of association with the Department of Statistics, Purdue University.

Fig. 2 Late 1960s going away party for Cletis Smith, a secretary in the Department of Statistics. Pictured left to right: Mildred Kroger, Virgil Anderson, Harlley McKean, Paul Randolph, Shanti Gupta, Myrna Studden, Irving Burr, Cletis Smith, K.C.S. Pillai, Marcel Neuts, and Steve Samuels

Mathematics and Statistics. Marcel Neuts (1962–1976) and K.C. Shreedharan Pillai (1962–1985) were the first to join. At that time Pillai was a Senior Statistical Advisor for the United Nations, and was also instrumental in founding the Statistical Center at the University of Philippines. Today, the Pillai Lecture is named in his honor. The following year, Y.S. Chow (1963–1969), Paul Holmes (1963–1967), and Stephen M. Samuels (1963–2003) joined the department to expand the research focus in probability. Prem S. Puri (1966–1989) and William J. Studden (1964–2005) were also hired, working mainly in statistics and applied probability. Today, the Puri Lecture and the William J. Studden Publication Award are named in their honor. Later, Glen Baxter (1965–1983), Norm Johnson (1967–1970), Eugene Klimko (1967–1974), David S. Moore (1967–2004), Herman Rubin (1967–present), and James W. Yackel (1966–1987) were hired to cover the remaining areas of statistical research. Virgil Anderson, who was the Director of the Stat Lab at the time, also joined the nascent department, as did Irving Burr (1941–1974) and Lou Cote (1960–1991), who were both members of the Department of Mathematics. Glen Baxter, Lou Cote along with Harlley McKean and Henry Teicher pursued research in probability and stochastic processes. During this period, the senior most faculty member, in the soon to become Department of Statistics, was Irving Burr; he was a leading expert in quality control and industrial statistics (Fig. 2).

Interestingly, in 1962 the Department of Mathematics briefly moved from the School of Science, Education, and Humanities to the School of Engineering. In 1963, the Division of Mathematical Sciences was created to include Mathematics, Statistics, and Computer Science as its units. Each unit later went on to form its own department; the Department of Computer Science being the first in the United States. Organizationally, the Division of Statistics shared its budget with the Department of Mathematics, before becoming an independent entity in 1968. The newly formed Department of Statistics was housed on the fifth floor of the Mathematical Sciences Building, which continues to be one of its homes today. During these formative years Shanti originated many of the departmental traditions that remain in place today and that continue to keep students and faculty socially engaged. Most notable are the fall picnic, the Holiday party (now called the

'After-Holiday party', held in Februrary) and the welcoming environment that continues to host many eminent statisticians.

By 1968, Felix Haas, the initial visionary for Statistics at Purdue University, was the Dean of the School of Science, Gupta was the Head of the newly created Department of Statistics, and history was in the making. The initial group of faculty members gained further strength when David Root (1968–1974), James N. Arvesen (1968–1974), and Myra Samuels (1968–1992) joined the department. Samuels was the first female hired in Purdue Statistics. She earned her PhD in Statistics from the University of California, Berkeley, under the direction of Jerzy Neyman. She originated biostatistics research in the department by focusing on both conceptual issues and statistical applications, and later became Associate Director of the Statistical Consulting Service in the Department of Statistics. Her text, "Statistics for the Life Sciences" remains in use today, and the annual Myra Samuels Lecture is named in her honor.

During 1969–1970 the Vietnam War was raging, and an application for occupational deferment from the draft board was second only to gaining employment. George McCabe, having just completed his PhD in Mathematical Statistics with Herbert Robbins, heard good things about Purdue Statistics from both Robbins, who had visited Purdue in 1966, and an instructor who was a graduate student with Steve Samuels. With the draft board looming, George McCabe applied, interviewed, and accepted a 12-month appointment to Purdue Statistics. Although Purdue University applied for an occupational deferment from the draft board on George's behalf, this was also the time that President Lyndon Johnson announced there would be no more occupational deferments of this type. With Purdue's deferment application request (envelop unopened!) on the desk of the draft board, a series of anxious exchanges between Purdue and the draft board took place concerning whether an unopened letter qualified as an application in process. George did receive a deferment, and arrived to Purdue Statistics in 1970; a new life was breathed into the Stat Lab, which had ceased to exist in 1966.

George McCabe organized the Statistical Consulting Service (SCS), serving as the Director, to address the statistical consulting needs of researchers at Purdue by focusing on the design of experiments and methods for data analysis. At the time the SCS was supported by the Purdue University Computing Center (PUCC) and employed two Statistics graduate students, a Psychology graduate student, and a PUCC employee, Mary Ann Ross. PUCC funded 40% of McCabe's time and the two half-time Statistics graduate students. In addition to offering assistance with experimental design and data analysis, software help was also available for SPSS and BMD. Today, the SCS continues to serve the Purdue research community and staff, and presently consists of two faculty members (Director Bruce Craig and an Assistant Director), a manager (Regina Becker), a secretary, and 10–12 graduate student consultants. George McCabe is currently the Associate Dean for Academic Affairs of College of Science, after serving as the Director of Statistical Consulting Service for 34 years.

Shanti Gupta's Era: Building the Foundation (1971–1995)

When people reflect on how Shanti Gupta went about 'creating from scratch a major statistics department in a cornfield' (source of this quote is unknown) they often smile and then start telling stories about his hiring tactics, hidden pots of money, his ways of getting things done, and how he showcased Purdue Statistics as 'the place to be'. He knew that attracting young faculty to a landlocked state was a great challenge, so he often sought out graduate students, well before they really thought about where they wanted to work; he planted a seed about Purdue Statistics. He was well connected, very well liked, and was known to make phone calls to friends to discuss PhD students who were about to finish. In fact, one call to Jack Kiefer (Cornell Math) resulted in David Moore being hired (1967). Shanti's approach and yield were both truly amazing. In the spring of 1973 Jim Berger (personal communication) met Shanti at a conference in Ithaca, New York. It was then that Shanti planted the seed that Jim should keep Purdue Statistics in mind when he started applying for jobs. With his research allowing him to unexpectedly graduate in the fall of 1973, and the fact that he was supporting his family on a $300/month stipend, Jim decided to find a faculty position that would allow him to start in January 1974 (he was 23 years old) instead of waiting until August. Every place that he contacted was encouraging, but asked that he apply on cycle during the regular process (i.e., interviews in the spring with a start date in the fall). That is, everyone except Shanti. He called Jim and offered him a job on the phone. While he did offer to host Jim for a visit, Jim did not have time (he was busy writing his dissertation) so he accepted the job offer, and started at Purdue Statistics in January 1974 where he remained until 1997 when he relocated to Duke University.

The main areas of research in the department's early years were decision theory, design of experiments, multivariate statistical analysis, optimal designs, ranking and selection, reliability, and stochastic processes, especially optimal stopping and the secretary problem. Prem K. Goel (1971–1983) strengthened the research focus of the department in both the theoretical and practical aspects of Bayesian hierarchical modeling in various domains. Leon Gleser (1973–1990) was hired as a full professor specializing in multivariate statistics, and in part to liaison with experimental scientists in both biology and psychology. Jim Berger (1974–1997) and Mary E. Bock (1974–present) further advanced the department's focus on statistical decision theory; both doing foundational work in admissibility and minimaxity. Later, Jim's research turned to the foundations of statistical decision theory, specifically Bayesian methodology.

Research in theoretical and applied probability, specifically stochastic processes, was enhanced by the work of faculty members, such as Marcel Neuts (1962–1976), Burgess Davis (1974–present), Andrew Rukhin (1977–1987), Philip Protter (1978–2002), Stanley Sawyer (1978–1984), Thomas Sellke (1984–present), and Steven Lalley (1986–2000). Tom Sellke has contributed significantly to the research in hypotheses testing problems and the use of p-values in applied statistics. Lalley contributed significantly in the field of probability and random processes. In 2000, he relocated to the University of Chicago. Ker-Chau Li (1981–1984), William Notz (1978–1983), Jun Shao (1987–1989), and Arif Zaman (1981–1982) were faculty members for a short time. Anirban DasGupta (1983–present), Tom Kuczek (1987–present), and Chong Gu (1990–present) joined the department in the 1980s, and in doing so expanded the research in applied probability, asymptotic statistics, computational methods for statistical inference, experimental design, quality control, modeling, and model selection. Protter's research focused on stochastic calculus, simulation and approximation of solutions for stochastic differential equations, and related areas in probability theory. He later served as Associate Director of the Actuarial Science program until 2002, before moving to Cornell University.

In 1988, the undergraduate Actuarial Science Program was founded jointly by the Department of Mathematics and the Department of Statistics. This was preceded by the establishment of Actuary Club in 1983 with Keith Schwingendorf as the club's sponsor, and students Kent Somers, Kirk Baumgartner, and Craig Hanford as the founding members. Today the Actuarial Sciences undergraduate program remains shared equally between the Departments of Mathematics and Statistics, and is one of the largest programs in the country with upwards of 300 students.

Through the 1980s and 1990s Purdue Statistics was a major center for research in foundations of statistics and Bayesian methodology. Many researchers specializing in these or related areas visited the department. Among them, Anthony O'Hagan, a prominent subjective Bayesian, Susie Bayarri, a leading researcher in foundations of statistics, and Arif Zaman, who specialized in Markov chains and computer simulations. Despite the strong focus on theoretical statistics, the Department of Statistics kept up with the computational advances of the 1980s. If not for the demand of mathematical purity of its faculty, Purdue Statistics may have been more successful at diversifying into applications of statistics long before the mid-1990s. This aside, computing was considered of value, and in 1983 the department purchased its first in-house computer, a VAX 11/780. Although statistical software packages such as SPSS and BMD had been used in courses for many years (as early as 1975), the establishment of in-house computing for the department would turn out to be another visionary decision. By the early 1990s, the department had two file-servers with X-Windows terminals in every faculty, graduate student, and staff office.

In addition to being the architect of Purdue Statistics, Gupta organized the first *Purdue Symposium in Statistical Decision Theory and Related Topics* in 1970. He continued to be one of the main inspirations and organizers behind the 1976, 1981, 1986, 1992, and 1998 symposia. Shanti was fondly remembered for his contributions in the 2003 symposium after his death in 2002. The Purdue symposia remain highly regarded for their quality and participants, and continues today with the *8th International Purdue Symposium on Statistics: Diversity in Statistical Sciences in the 21st Century* held in June 2012. The establishment and growth of the Department of Statistics flourished under Gupta's unique leadership. The foundations that were built more than 40 years ago continue to provide a strong base for the ever broadening field of Statistics. In 1995, after 27 years as the Head, Shanti stepped down and continued his thriving research career. He remained an active member of the faculty until his death in 2002.

Mary Ellen Bock's Era: Adapting to the Changing Times (1995–2010)

Mary Ellen Bock is the first female full Professor in the Department of Statistics, and became the second head of the department in 1995; she was the first female Head of department in the College of Science, Purdue University. She continued the tradition of excellence that was established by Shanti Gupta, and was visionary in her view of what was to become a change point in the field of statistics. While many departments in the nation remained rooted in their belief that theory, probability, and methodology were the only worthwhile activities for statisticians, Mary Ellen realized an opportunity to advance Purdue Statistics through interdisciplinary areas that depend heavily on both statistics and computing for theoretical and methodological research. The common themes of massive data and computational complexity played a role in all of the interdisciplinary areas in which the department invested, namely statistical bioinformatics, computational finance, statistical consulting, machine learning, computational statistics, and environmental statistics. Mary Ellen's vision was timely in that Purdue President Martin Jischke's (mid-2000s) strategic plan included hiring 300 new faculty members on the West Lafayette campus. Purdue Statistics responded to the initiative by hiring and establishing joint faculty appointments for junior faculty across numerous departments and colleges. In less than 10 years, the number of students and faculty in the Department of Statistics doubled.

With the increased use of statistics in many fields of research, there followed an increased need for advanced service courses for non-majors. In the spring of 1995, the department taught 2,257 students. Today the department teaches (Fig. 3) the most graduate level service classes in the College of Science, and approximately 4,000 students each semester in both undergraduate and graduate level courses. As the department continues to gain strength in numbers, online teaching (distance learning) under the guidance of Continuing Lecturer Ellen Gundlach is providing education in statistics to many students beyond the walls of Purdue's West Lafayette campus.

At the end of the twentieth century, the availability of powerful and relatively cheap computing accompanied the advancement of technologies in experimental sciences. As the size of data continued to increase it became apparent that statistical methodology and computational strategies needed to adapt. There was a huge surge in the demand and appreciation of data analysis skills in industry, government, and medical sciences. Toward this end, Purdue Statistics was well positioned to meet these needs with its in-house computing, Information Technology at Purdue (ITaP), and interdisciplinary reputation in areas of statistical applications, such as statistical bioinformatics, genomics, massive data analysis, machine learning, visualization, and computational finance. To date, the increasing importance of statistics in interdisciplinary fields has led to departmental faculty hires with joint appointments in Agronomy, Biology, Child Development and Family Sciences, Civil Engineering, Computer Science, Earth and Atmospheric Sciences, Electrical and Computer Engineering, Forestry and Natural Resources, and Mathematics.

Fig. 3 Purdue Statistics Faculty (2009). *Row one*: Ellen Gundlach, Jun Xie, Bo Li, Xuanyao He, Rebecca Doerge, Min Zhang, *Row two*: Michael Zhu, Hao Zhang, Guang Cheng, Lingsong Zhang, Alex Gluhovsky, *Row three*: Bruce Craig, Xiao Wang, Kris Jennings, Tom Howell. *Row four*: Jian Zhang, Tonglin Zhang, Bill Cleveland, David Moore. *Row five*: Tom Sellke, Sergey Kirshner, Mark Ward, Michael Levine, Tom Kuczek. *Row six*: George McCabe, Lou Cote, Chuanhai Liu

With the changing times and the increasing application of statistics in other scientific disciplines, the department hired Rebecca Doerge (1995–present) to increase its presence in statistical genetics (now referred to as statistical bioinformatics) and Bruce Craig (1996–present) to expand its focus in applied statistics and consulting. As with many of the faculty appointments that followed, Rebecca's appointment, which remains today, is joint between the Department of Statistics (75%) and the Department of Agronomy (25%) in the College of Agriculture. Jayanta K. Ghosh, who had been a regular visitor in the department since 1989, joined the department in 1996. His presence continues to enhance the research in the foundations of statistics and Bayesian methodology, especially Bayesian nonparametrics. Alexander Gluhovsky (1996–present) and, later Hao Zhang (2007–present) were hired with joint appointments in the Department of Earth and Atmospheric and Planetary Sciences and the Department Forestry and Natural Resources, respectively.

In response to the increasing demand for quantitative skills in finance the Computational Finance Program was established in 1998. Frederi Viens (2000–present) joined the Purdue Statistics faculty in 2000. He specializes in quantitative finance, has a joint appointment with the Department of Mathematics, and is currently the Director of the Computational Finance Program. Three faculty members in the Department of Statistics are among the ten Purdue faculty affiliated with the Computational Finance Program. José Figueroa-López (2007–present), Michael Levine

Fig. 4 Built in 1909, Haas Hall was once home to the Department of Computer Science, and originally the Purdue Memorial Gymnasium. In 2006 more than half of the Department of Statistics moved to Haas Hall named after longtime supporter of the Department of Statistics, Professor Felix Hass

(2003–present), and Viens have research programs and teaching responsibilities that support the program.

In 2001 the graduate students of the department established the organization known today as "Statistics in the Community" (StatCom) to provide pro bono statistical consulting services to local governmental and nonprofit groups. StatCom, the brainchild of Purdue Statistics PhD student Nels Grevsted, is the first such graduate student run volunteer community outreach organization in the United States. Despite its modest start at Purdue, StatCom has now become an international movement with StatCom branches located around the world.

As society continued to become more data-driven, the need for statistical analysis of massive data demanded innovative computational approaches. William S. Cleveland and Chuanhai Liu were hired from Bell Laboratories in 2004 and 2005, respectively. Their skill-sets advanced Purdue Statistics in data visualization and computational statistics, and resulted in Purdue Statistics hiring several faculty members with specialties in data mining and machine learning. Many of these faculty members hold joint appointments with the Department of Computer Science (CS). Specifically, Jennifer Neville (2006–present) holds a joint appointment between CS (75%) and Statistics (25%), as do S.V.N. Vishwanathan (aka Vishy) (2007–present; 25% CS; 75% Statistics), and Yuan (Alan) Qi (2007–present; 75% CS; 25% Statistics). Bowei Xi and Sergey Kirshner joined the computational statistics group in 2004 and 2008, respectively.

By 2006, the Department of Statistics was bursting at its seams. Its expansion was challenging the limits of the allotted space in the Mathematical Sciences Building. In August 2006, the Department of Statistics main office and more than half of the faculty and some of the graduate students relocated to Felix Haas Hall. To date, the information technology support for the department, the Statistical Consulting Service, along with many faculty and graduate students remain in the (ground, second, and fifth floors) Mathematical Sciences Building. Given his early vision and support for establishing statistics at Purdue, it seems fitting that the Department of Statistics should expand into Felix Haas Hall (Fig. 4). Built in

1909, Haas Hall was once home to the Department of Computer Science, and originally the Purdue Memorial Gymnasium.

Under Mary Ellen's leadership the Department of Statistics continued to flourish. In cooperation with the College of Science and the university, the department continues to excel in the dissemination of statistical education and has considerably expanded its reputation for interdisciplinary collaborations and focus. The balanced focus that the department has maintained with respect to interdisciplinary research, probability, and theory promoted the department to the 10th ranked Statistics Department in the U.S. in 2009 and the highest ranked department in the College of Science at Purdue University.

Selected Research Accomplishments

One of the main foci for research since the inception of the department has been statistical decision theory, in particular admissibility (Jim Berger, Mary Ellen Bock), Bayesian inference, robustness, and decision theory, foundations of statistics (Herman Rubin, Jim Berger, Jayanta Ghosh), and ranking and selection (Shanti Gupta). Another main focus was on applied probability, specifically sequential selection, optimal stopping (Steve Samuels), optimal designs, and Tchebysheff systems (Bill Studden).

Herman Rubin's presence and influence in Purdue Statistics' history are legendary. He is passionate about educating the next generation of statisticians, and for many years had a strong hand in the formation and maintenance of the graduate program in statistics at Purdue. Herman arrived at Purdue in 1967 and is still an active member in the department having recently (October 27, 2011) celebrated his eighty-fifth birthday at a party in his honor hosted by Rebecca Doerge in her home. Among the many, many, many toasts and speeches Mathematics Professor Leonard Lipshitz said it best, "Herman, you are the kindest person I have ever met." Herman was among the original cast of characters who originated the department, and although he has made significant contributions to the fields of asymptotics, decision theory, and probability (to name only a few), his reputation for asking questions at the end of seminars and his ability to see the answer years before others is second to none. His innate ability to generalize all problems to abstract mathematics continues to make him Purdue Statistics' most important resource. One day he walked into an office (the door was open) to find two researchers staring at a complicated double infinite series on the blackboard. They had been struggling for days trying to decide if it converged or not. Herman looked at the expression and said, 'two', and then walked out of the room having said nothing more. Yes, he was correct.

As the department evolved its research focus has undergone major changes. Today the department's major research areas include applied probability (Anirban DasGupta, Herman Rubin, Mark Ward), asymptotics (Anirban DasGupta, Jayanta Ghosh, Guang Cheng), Bayesian statistics (Bruce Craig, Jayanta Ghosh, Bill Cleveland, Sergey Kirshner), applied and theoretical statistics (Sharon Christ, Xiao Wang, and Lingsong Zhang), statistical bioinformatics (Bruce Craig, Hyonho

Chun, Rebecca Doerge, Olga Vitek, Jun Xie, Dabao Zhang, Min Zhang, Michael Zhu), computational methods, data mining, and machine learning (Bill Cleveland, Chong Gu, Sergey Kirshner, Jen Neville, Alan Qi, S.V.N. Vishwanathan, Mark Ward), statistical consulting (Bruce Craig, George McCabe, Tom Kuczek), experimental design (Bruce Craig, Bill Studden, Micheal Zhu), quantitative finance and time-series (Frederi Viens, Jose Figueroa-Lopez, Micheal Levine), spatial analysis (Bo Li, Tonglin Zhang, Alex Gluhovsky, Hao Zhang), and statistics education (Bruce Craig, George McCabe, David Moore).

Although greater detail is provided at www.stat.purdue.edu, Purdue Statistics faculty (both past and present) continue to author influential books. Two examples are Jim Berger's seminal book in decision theory and Bayesian analysis, *Statistical Decision Theory and Bayesian Analysis,* and Jayanta Ghosh's (co-authors Delampady and Samanta) *An Introduction to Bayesian Analysis*. In fact, some of the fundamental research work in the asymptotic aspects of Bayesian nonparametrics was conducted by Jayanta and his collaborators at Purdue University and the Indian Statistical Institute, and culminated in the first book by the same name focusing on the theoretical aspects of Bayesian nonparametrics. Chuanhai Liu recently co-authored *Advanced Markov Chain Monte Carlo Methods*, and Anirban DasGupta recently published a comprehensive coverage of asymptotics in Statistics in *Asymptotic Theory of Statistics and Probability* and applied probability in *Probability for Statistics and Machine Learning*.

Apart from important contributions in mainstream Statistics, the department continues its important contribution in statistics education. Over the years many influential texts in introductory statistics for students from varied backgrounds have been penned. The initial efforts of David Moore made it feasible for students from nonmathematical backgrounds to study statistics. *Statistics: Concepts and Controversies* is currently in its seventh edition and now co-authored with William I. Notz. *Introduction to the Practice of Statistics*, by Bruce Craig, David Moore, and George McCabe, emphasizes the analysis of real data to strengthen and reinforce the learning of statistical methods. With Layth Alwan, Moore, McCabe, and Craig have published *The Practice of Statistics for Business and Economics*. Rebecca Doerge and former PhD student, Martina Muelbach Bremer, recently published *Statistics at the Bench: A Step-by-Step Handbook for Biologists,* a manual that assists biologists and researchers in related fields in analyzing their experimental data.

Faculty, Alumni, and Student Recognition

Several Purdue Statistics current faculty and staff have received significant awards and honors based on academic achievement and professional service. David Moore and Mary Ellen Bock each served as the President of American Statistical Association (1998 and 2007, respectively). David Moore also served as the President of the International Association for Statistical Education. In 2010, Mary Ellen received the National Institute of Statistical Sciences (NISS) Distinguished Service Award. Virgil Anderson, Mary Ellen Bock, Bill Cleveland, John Deely, Rebecca Doerge, George McCabe, David Moore, and Steve Ruberg are Fellows of the American Statistical Association. Mary Ellen Bock, Bill Cleveland, Anirban DasGupta, Burgess Davis, Jayanta Ghosh, David Moore, Tom Sellke, and Bill Studden are Fellows of the Institute of Mathematical Statistics.

The Department of Statistics at Purdue, in addition to the Purdue Symposia, is the birthplace of the popular bi-annual international objective Bayes workshop, O'Bayes. Starting in 1996, the workshop emphasized recent research in objective Bayesian methods, facilitated collaborations, and opened new directions for future research. In addition, there are a number of invited named lectures and colloquiums in honor of former faculty. Specifically, the Pillai Lecture honors K.C.S. Pillai, the Puri Lecture is in memory of Prem S. Puri, and the Myra Samuels Lecture in Biostatistics is in honor of Myra Samuels. Further, there is an endowed distinguished professorship in honor of Shanti Gupta (previous held by David Moore, and currently held by William Cleveland).

The Department of Statistics at Purdue University has alumni with distinguished careers in industry and government. In 2007, in honor of both Statistics and Actuarial Science Outstanding Alumni an award with this name was established. Among the first receiving this award was the department's first female PhD student, Mrdulla Gnanadesikan (1964), advised by Shanti Gupta. The College of Science also honors alumni with the Distinguished Alumnus Award in recognition of contributions and leadership within their professions. A list of alumni receiving these awards can be found at www.stat.purdue.edu.

The Department of Statistics maintains a number of awards that support both undergraduate and graduate education and recruitment. The Virgil L. Anderson and David S. Moore Scholarships are given once a year to an undergraduate demonstrating potential in statistics and its applications. The Bob and Marjorie McLean Scholarship, I. W. Burr Award (PhD only), L. J. Cote Excellence in Statistics Award (MS only), and the William J. Studden Publication Award (PhD only) acknowledge graduate students at various levels for their success and potential to contribute to the discipline of statistics.

The Frontier of Statistical Sciences

On July 1, 2010, Rebecca Doerge (Fig. 5) assumed responsibilities as Head of the Department of Statistics, Purdue University. Under her leadership the department will maintain and grow the established traditions of advancing the frontiers of statistical science education and research in both theory and applications. As the landscape of twenty-first century educational institutions adjust to accommodate budgetary limitations and technological advances, Purdue Statistics remains

Fig. 5 Left to right the three Department of Statistics, Heads: Professor Shanti Gupta (1968–1995), Professor Mary Ellen Block (1995–2010), and Professor Rebecca W. Doerge (2010–present)

committed to being a leader in the statistical sciences and in addressing the needs and challenges of the research, teaching, and societal engagement.

Acknowledgments We are grateful to Purdue Statistics PhD students Faye Zheng and Doug Baumann for helpful comments on an earlier version of this manuscript, and to Professors Mary Ellen Bock and George McCabe for their editing expertise and opinions. The invaluable experiences of Professor James O. Berger, Professor Mary Ellen Bock, Professor Jayanta Ghosh, Professor George McCabe, and Professor Herman Rubin helped to incorporate some undocumented events in the history of Purdue Statistics. We are also very grateful to three anonymous reviewers who guided us through details that may have otherwise been lost. Complete information about Purdue Statistics, and its history is available at: www.stat.purdue.edu.

References

McDonald GC (1998) A conversation with Shanti Gupta. Stat Sci 13(3):291–305

Some Frequently-Cited Articles Written by Purdue University, Statistics Faculty (Current Purdue Statistics Faculty Underlined; Citation Count in Parentheses)

Anderson TW, Rubin H (1956) Statistical inference in factor analysis. In: Proceedings of the third Berkeley symposium on mathematical statistics and probability, vol V (549)

Berger JO, Sellke T (1987) Testing a point null hypothesis: the irreconcilability of P values and evidence. J Am Stat Assoc 82(397):112–122 (442)

Bhattacharya RN, Ghosh JK (1978) On the validity of the formal Edgeworth expansion. Ann Stat 6(2):434–451 (431)

Brown LD, Cai TT, DasGupta A (2001) Interval estimation for a binomial proportion. Stat Sci 16(Part 2):101–116 (433)

Churchill GA, Doerge RW (1994) Empirical threshold values for quantitative trait mapping. Genetics 138(3):963 (2872)
Cleveland WS (1979) Robust locally weighted regression and smoothing scatterplots. J Am Stat Assoc 74(368):829–836 (4405)
Davis B (1990) Reinforced random walk. Probab Theory Related Fields 84(2):203–229 (Springer) (160)
Gupta SS (1963) Probability integrals of multivariate normal and Multivariate t_1. Ann Math Stat 34(3):792–828 (326)
Karlin S, Studden WJ (1966) Optimal experimental designs. Ann Math Stat 37(4):783–815 (185)
Liu C, Rubin DB (1994) The ECME algorithm: a simple extension of EM and ECM with faster monotone convergence. Biometrika 81(4):633 (225)
Moore DS, Spruill MC (1975) Unified large-sample theory of general Chi-squared statistics for tests of fit. Ann Stat 3(3):599–616 (113)
Pillai KCS (1955) Some new test criteria in multivariate analysis. Ann Math Stat 26(1):117–121 (173)
Puri ML, Ralescu DA (1986) Fuzzy random variables. J Math Anal Appl 114(2):409–422 (802)
Teegarden D, Proulx WR, Martin BR, Zhao J, McCabe GP, Lyle RM, Peacock M, Slemenda C, Johnston CC, Weaver CM (1995) Peak bone mass in young women. J Bone Mineral Res 10(5):711 (213)
Viswanathan SVN, Smola AJ (2004) Fast kernels for string and tree matching. Adv Neural Inf Process Syst (15):585–592 (201)
Wahba G, Wang Y, Gu C, Klein R, Klein B (1995) Smoothing spline ANOVA for exponential families, with application to the Wisconsin Epidemiological Study of Diabetic Retinopathy. Ann Stat 23(6):1865–1895 (149)

Some Textbooks Written by Purdue University, Statistics Faculty

Alwan L, Moore DS, McCabe GP, Craig BA (2011) The practice of statistics for business and economics, 3rd edn. W.H. Freeman, New York
Anderson V, McLean RA (1984) Design of experiments: a realistic approach. Marcel Dekker Inc., New York
Cleveland WS (1993) Visualizing data, 1st edn. Hobart Press, Summit
Cleveland WS (1994) The elements of graphing data, 2nd edn. Hobart Press, Summit
DasGupta A (2010) Asymptotic theory of statistics and probability, 1st edn. Springer, New York
DasGupta A (2010) Fundamentals of probability: a first course, 1st edn. Springer, New York
Doerge RW, Bremer MM (2010) Statistics at the bench: a step-by-step handbook for biologists, 1st edn. Cold Spring Harbor Laboratory Press, Spring Harbor
Ghosh JK, Delampady M, Samanta TK (2010) An introduction to Bayesian analysis: theory and methods, 1st edn. Springer, New York
Gu C (2010) Smoothing spline ANOVA models, 1st edn. Springer, New York
Liang F, Liu C, Carroll RJ (2010) Advanced Markov chain Monte Carlo methods: learning from past samples, 1st edn. Wiley, Chichester
Moore DS (2009a) The best practice of statistics, 5th edn. W.H. Freeman, New York
Moore DS (2009b) Essential statistics, 1st edn. W.H. Freeman, New York
Moore DS, Notz WI (2008) Statistics: concepts and controversies, 7th edn. W.H. Freeman, New York
Moore DS, McCabe GP, Craig BA (2007) Introduction to the practice of statistics, 6th edn. W.H. Freeman, New York

Rutgers University Department of Statistics and Biostatistics

Joseph Naus

Introduction

The Rutgers Statistics Department began in 1952 with four full-time faculty. As of 2011, the department had 21 faculty, 117 undergraduate majors, 163 graduate students, and a 4,600 annual statistics course enrollment.

The department evolved through cycles of demand, entrepreneurship, excellence, collaboration, and diversification. It was a period of challenge and excitement brightened by a culinary adventure in brews, bagels, picnics, pizza, and parties. This history traces the evolution and the adventure.

Demand and Entrepreneurship: 1947–1967

Rutgers started as a colonial college in 1766. The hot curriculum was divinity, learned languages, and liberal arts including mathematics. Classes met at a pub called Sign of the Red Lion. This was 10 years before the American Revolution, and the British saw pubs as centers of radicalism. Classes were forced out of the pub, students and faculty boarded with farmers, and still spirited learning continued.

In 1809, Rutgers hired its first mathematics professor, Robert Adrain who found that measurement errors had a bell curve distribution, and gave the first proof of

Rutgers University,
Department of Statistics and Biostatistics,
Hill 506, Busch Campus , 110 Frelinghuysen Rd,
Piscataway, NJ 08854-8019, USA
J. Naus (✉)
e-mail: naus@rci.rutgers.edu

A. Agresti and X.-L. Meng (eds.), *Strength in Numbers: The Rising of Academic Statistics Departments in the U.S.*, DOI: 10.1007/978-1-4614-3649-2_18,

the Gaussian Distribution.[1,2] New Jersey authorized a lottery to pay for Adrain's salary and building cost overruns (see footnote 2). This foreshadows the present day New Jersey lottery's financing for higher education.

In 1864, Rutgers became a Land Grant College with an Agricultural Experimental Station and a College of Engineering, and in 1945 it became the State University of New Jersey. The main campus was in New Brunswick, and it had two undergraduate liberal arts colleges: Rutgers College for men and Douglass College for women. University College (UC) was a college for part-time students, and served the main campus in New Brunswick, and the Newark and Camden satellite campuses. In 1947, there were separate mathematics departments at Rutgers, Douglass, and UC, but no statistics department.

Within an hour's radius of Rutgers there were major centers for statistical research and employment that shaped the development of the department, recruitment of faculty, and placement of graduate students. New Jersey was the center of the pharmaceutical industry, and Johnson & Johnson was a block from the campus. Bell Laboratories of AT&T, a magnet for world class scientists and mathematicians, had centers nearby in Murray Hill and Holmdel.

After World War II there was a growing demand from these and other New Jersey companies for part-time statistics education for their employees. They turned to Rutgers University, their longtime neighbor, to meet that demand. Rutgers had just become the State University of New Jersey and was prepared to devote resources to further economic development in the State.

The UC Math Department was a natural place to provide part-time statistics courses to meet the industry's demand. They recruited an academic with industry credentials to be chairman of the UC Math Department. Ellis Ott obtained his PhD in Mathematics from the University of Illinois, and was a professor of mathematics at the University of Buffalo[3] from 1934 to 1946. He then worked for the National Union Radio Corp in Newark, NJ where he was in charge of applying statistical quality control methods.

Ellis Ott was a superb salesman and entrepreneur for statistics. He had excellent contacts in industry and was a founding member of the American Society for Quality Control, wrote a book on process control, consulted to General Electric and Bristol Myers, headed a United Nations team to teach quality control in India, and made similar teaching trips to Mexico and Japan.

Bell Labs had a strong group in industrial statistics with Walter Shewhart, Harold Dodge, Harry Romig, and Paul Olmstead who together laid the foundations of statistical quality control and acceptance sampling. Ellis Ott was able to lure

---

[1] Adrain, R., "Research concerning the probabilities of the errors which happen in making observations," *Analyst* 1 (1808), 93–109. Reprinted in S. M. Stigler, *American Contributions in Mathematical Statistics in the Nineteenth Century*. 2 Vols. New York, Arno Press, 1980.

[2] From a history of Mathematics at Rutgers. See expanded discussion there. http://www.math.rutgers.edu/~weibel/history.html.

[3] At that time a private College that in 1963 became part of the SUNY system.

Harold Dodge[4] and Martin Wilk[5] from Bell Labs. By 1952 there was an approved graduate program in Applied and Mathematical Statistics, and a Statistics Center set up with Ott as the Director. The Center faculty members were Ott, Romig, Wilk, Roger Pinkham,[6] and Mason Westcott. Wilk was Director of Research from 1959 to 1963. At the time, most other leading statistics PhD programs were strongly theoretical. Rutgers was strongly applied with PhD students receiving training in theory and applications.

Richard Trout (a student, and later a Professor, of the Statistics Department) shares some memories:

> Ellis had a huge impact on my, and many other students', professional development. He brought in many well known people but kept the seminars at a level to inspire the students. During the early days Harold Dodge was one of the most interesting [faculty], primarily because of his influence in the field of acceptance sampling and all the wonderful stories that he told, which made the field come to life in his classroom.

David Hogben was the first PhD and went on to a career at the National Bureau of Standards. Sam Shapiro was the second PhD, and the Shapiro–Wilk statistic to test for normality became a citation classic and won the Youden prize. The following are from some of Sam's recollections.

> The department was located on the Agricultural campus which allowed us to buy milk and ice cream at reduced prices. The statistics department at that time had a night time program for students from local industry. Ellis was responsible for starting the program and due to his reputation in quality control was able to get outside funding from local corporations.
>
> With the presence of Wilk and Pinkham the department began to institute a more theoretical program with a stress on applied statistics. In 1961 the department was moved to the main campus of Rutgers. Unfortunately the offices of the department were placed in a house scheduled for demolition. Martin Wilk refused to move into these quarters and was given an office in the main campus building.

The Statistics Center offices in New Brunswick were a sharp contrast to the excellent facilities Martin Wilk had when working at Bell Labs. The New Brunswick statistics offices were in converted army barracks on Morrell Street. On the positive side, the offices were centrally located off College Avenue, a block from the library, gym, and cafeteria. Best of all it was a few steps from Patti's Bar and Grill with its ready access to beer and pizza. A few blocks away was Buccleuch Park where faculty and students played ball and had picnics. Ellis Ott and his gracious wife Virginia had barbecues at their home for students and faculty.

---

[4] http://qualitygurus.com/gurus/list-of-gurus/harold-f-dodge/par

[5] See http://www.ssc.ca/en/about/history-ssc/martin-bradbury-wilk. A detailed interview with Martin Wilk appears in *Statistical Science 25* (May 2010), pp. 258–273.

[6] A referee noted the interesting fact that Pinkham's major paper dealt with Benford's law of leading digits.

Fig. 1 Ellis Ott and Horace Andrews, January 20, 1965. Special Collections and University Archives, Rutgers University Libraries

Wilk left the department in 1963 and Pinkham followed. The program needed faculty who did and could direct PhD research in statistics and probability theory.

From 1963 to 1965 Ott hired four new assistant professors and one associate professor. The associate professor, Horace Andrews, was hired from Swift and Company where he was head of the Statistics Division and specialized in experimental design. Figure 1 shows Ellis with Horace demonstrating a response surface analysis.

The assistant professors were Arthur Cohen, Joseph Naus, Richard Gundy, and Saul Blumenthal; all brought a combination of theory and application. Art Cohen, a Columbia PhD, had worked for the Public Health Service. Joe Naus had been an operations researcher while finishing his PhD thesis at Harvard on a problem in applied probability. Dick Gundy had a PhD in Experimental Psychology from Indiana and a PhD in Statistics from Chicago. Saul Blumenthal's Cornell PhD in Industrial Statistics was in the area of reliability and failure distributions. Saul stayed for 3 years, but left for an Associate Professorship at NYU. Art and Dick went on to be pillars of the department in theory, Art in mathematical statistics, and Dick in theoretical probability.

Excellence Leads to Funded Lines: 1968–1977

In the early years, student enrollments were a major justification for additional faculty. Many departments and graduate programs taught their own statistics courses, and gained resources from the additional "student equivalent hours." Statistics faculty served on doctoral committees, advised on statistical designs of studies, and developed and taught courses in collaboration with faculty in specific disciplines.

To increase "student equivalent hours" taught, the Statistics Center provided service teaching of basic statistics courses to undergraduate students in the biological sciences, engineering, and graduate courses in the Schools of Arts & Sciences, Social Work, Education, and Library Science.

The university had the goal of being a top research university, and the quality of a department's research began to play an increasing role in getting additional faculty lines. At the same time, worldwide inflation pushed up university costs much higher than the state's support. Faculty lines were allocated to the highest quality research departments, and the link between enrollments and faculty lines was broken. Departments with introductory undergraduate statistics courses had to either increase teaching loads for faculty or increase already large class sizes. The standard teaching load at the time was three courses per semester. Economics and Business were quick to let the Statistics Center teach their statistics courses.

Art Cohen became Director of the Statistics Center from 1968 through 1973 and served during a critical juncture in the department's rise to excellence. With the reorganization of the university, the center became a Statistics Department in 1974. The center director's job had combined the duties of chairperson and graduate director. Planning had begun on a new building to house Mathematics, Statistics, and Computer Sciences and services. The groundbreaking of Hill Center took place in September, 1969 (Fig. 2).

From 1974 to 1977, Cohen assumed the Department Chairman's position and Naus served as Graduate Director. The department's faculty put strong emphasis on quality of research (theoretical or applied) in all its new hires and promotions. This also helped the department in its growth and resources, and support for reduced teaching load, by gaining the confidence of the deans and administration.

By 1968, Art Cohen, Joe Naus, and Dick Gundy had all obtained tenure. In 1968, Harold Sackrowitz, Burgess Davis, Henry Teicher, and Richard Trout were hired, with Teicher as a full professor. Teicher, Davis, and Gundy gave a strong presence in probability. Teicher, Cohen, and Sackrowitz became the core of an outstanding group in theory of statistics. Richard Trout received his PhD from Rutgers, and served as director of the Statistics Consulting Group for the Agricultural Experiment Station. The department had built up excellence in mathematical statistics and probability, while maintaining strength in applied areas.

In 1970, the department continued to build up strength in statistical theory with the hiring of Robert Berk and Bill Strawderman. Berk came to Rutgers from the University of Michigan, and specialized in sequential methods,

Fig. 2 Art Cohen at Hill Center building site. September 25, 1969.* (left to right) President Mason Gross, Doug Eastwood (head Computer Center), and department chairs: Art Cohen (Statistics), Ken Wolfson (Math), Saul Amarel (Computer science). Special Collections and University Archives, Rutgers University Libraries

nonparametric statistics, and large sample theory. Strawderman was an operations research and applied statistics analyst at Bell Labs before coming to Rutgers, and did his PhD thesis in the area of decision theory working under Cohen. Larry Brown joined the department from 1972 through 1979 and together with Cohen, Sackrowitz, and Strawderman formed a world class research group in decision theory. During that period the influence of the Department's research was rated as third among the U.S. statistics departments.[7]

At the start of 1972, the department moved to its new offices in Hill Center on the Busch Campus in Piscataway, across the river from the New Brunswick Campus. The Busch Campus was the center for the mathematical and physical

[7] American Council of Learned Sciences, ACE, NRC, and SSRC. *An Assessment of Research-Doctorate Programs in the United States: Mathematical and Physical Sciences*. 1982. Based on Item (16) 'Estimated "overall influence" of published articles attributed to the program, 1978–1979," Rutgers tied with Stanford at third place.

sciences including engineering, physics, chemistry, biology, and psychology. The Waksman Institute funded by the discovery of streptomycin was nearby, and later so were the medical school and the library of science and medicine. Hill Center housed the Statistics, Mathematics, and Computer Science Departments, and had a joint library on the first floor. The mainframe computing facilities were in the basement, together with a statistics laboratory with Marchant calculators. The Statistics Department was, and still is, on the fourth and fifth floors.

Diversification and Meeting Demand: 1977–1996

Bill Strawderman was chair from 1977 to 1983 and from 1989 to 1992. Joe Naus was acting chair during 1981–1982 when Bill was on sabbatical, became chair during 1984–1986, and was acting chair again during 1988/1989 and 1992/1993. Bob Berk was chair from 1986 to 1988 and David Tyler was chair from 1993 to 1996. Harold Sackrowitz was Graduate Director from 1977 to 1995 to be followed in this position by Kesar Singh during 1995–2000.

From 1977 through 1984 the department made sharp hires to keep on the cutting edge. Kesar Singh wrote the first published paper that derived the asymptotic theory of the bootstrap. David Tyler was an expert in robust methods, multivariate analysis, and time series. Computationally intensive methods were beginning their spectacular rise. Regina Liu did pioneering work in data depth, resampling, and data mining. Javier Cabrera, an expert in computational statistics, did research on graphical methods.

The department's Office of Statistical Consulting (OSC) was established in 1983 with faculty and students working with researchers from a broad range of disciplines. Cabrera, Berk, John Kolassa, and then Minge Xie ran the OSC consulting, and Richard Trout and Dave Tyler ran the forerunner statistical consulting for the Ag Experiment Station at Cook. Today, faculty collaborate on research grants with the Department of Genetics, the Institute of Health Care Policy and Aging, the Discrete Mathematics Center (DIMACS), the Center for Alcohol Studies, University of Medicine and Dentistry-Robert Wood Johnson Medical School, the Cancer Institute of New Jersey, and surrounding industries. Students working on consulting projects at OSC gain knowledge and hands-on experience with real-life statistical problems.

In 1987, the statistics department started a Masters program in Quality Assurance Science, jointly offered by Industrial Engineering and Statistics. A core of courses together with additional specialized courses led either to a Masters in Industrial Engineering with an option in quality and reliability engineering or a Masters in Statistics with an option in quality and productivity management.

There were many new initiatives, programs and hiring, and departmental meetings proliferated. Food was an important component in having department meetings proceed smoothly. Joe Naus brought bagels to every meeting. To meet the New Jersey pharmaceutical industry's demand for biostatisticians, the Department developed a Biostatistics track for the Masters degree. The department also got approval and start-up funding to establish a Biostatistics Institute, with an industrial advisory committee of pharmaceutical professionals. The department set out to hire top research biostatisticians when lines became available. At the same time, the department wanted to maintain its strength in probability and statistical theory.

In 1985, the department hired Johannes (Joop) Kemperman[8] and Herbert Robbins.[9] Joop was an outstanding researcher in mathematics and probability. Sackrowitz and Cohen would always say "If you don't know the answer to a question, Ask Joop!" (see footnote 8.) Robbins was a member of the National Academy of Sciences, a super lecturer, and a stimulating colleague for 12 years until his retirement. He was Director of the Institute of Biostatistics, and he met with executives from pharmaceutical industries to generate funds for faculty research funding, post-doctoral, and graduate student support. When Robbins retired, ZhiliangYing, David Madigan, and Javier Cabrera subsequently became Director of the Institute.

In 1987, Cun-Hui Zhang and Yehuda Vardi joined the department. Zhang, a brilliant researcher in mathematical statistics and probability, was attracted because Robbins had come to the department. Zhang also served as Graduate Director for many years. Vardi came from Bell Labs and immersed himself in scientific problems, whether image restoration, survival analysis, sample selection bias, or other areas. He interacted and stimulated research with faculty and students.

The department had built up strength in mathematical statistics and probability theory, and set as a high hiring priority applied areas including computational statistics and biostatistics. To fill these needs, the department hired Ramanathan (Ram) Gnanadesikan[10], and Zhiliang Ying. Ram was a Director of Statistics at Bell Labs, and was an expert in multivariate analysis who had developed widely used graphical techniques. Zhiliang Ying, a rising star in biostatistics, came to Rutgers for 7 years from 1994 through 2001.

Expanding and Meeting Demand, 1996–2005

Yehuda Vardi was the chairman during this period. He was a keen judge of quality of research and researchers. He would sometimes emphasize his views by graphically putting a picture of a giraffe on a blackboard. "If it looks like a giraffe then it is a giraffe." Nobody could figure out why a giraffe and not the standard duck.[11] But then again, Yehuda was always original.

---

[8] A detailed interview with Kemperman appears in *Statistical Science* Volume 15, Number 4 (2000), 396–408.

[9] See detailed interview with Herbert Robbins in *Statistical Science* 1 (1986), 276–284.

[10] See detailed interview with Ram Gnanadesikan in *Statistical Science* 16 (2001) 295–300.

[11] http://en.wikipedia.org/wiki/Duck_test.

Yehuda helped recruit Larry Shepp in 1997. Larry deeply felt that methodology applied to a scientific problem should be built to fit the problem, and not the other way around. Larry is a constant source of challenging problems which he generously shares with faculty and students.

Each department at Rutgers is periodically reviewed by a visiting committee of distinguished scientists in the field. The 2000 review committee recommended biostatistics and computational statistics as important areas for the department to develop. New Jersey is a major center for the pharmaceutical industry, and biostatistics is a natural area to build up courses and programs. At the same time the department wanted to maintain strength in traditional areas while expanding in computational statistics and data mining. This required the university administration to provide additional support. It helped that the review rated us in the top five statistics departments among state universities. Under Yehuda the build-up took a giant step forward. Between 1998 and 2000 the department hired Steve Buyske, Don Hoover, John Kolassa, and David Madigan. Hoover came from Merck, where he served as Associate Director of Epidemiology. He added strength to the department's biostatistics track and programs, through his teaching and research in clinical trials, epidemiology, and group randomization. He interacted as a core member of the Institute for Health, Health Care Policy, and Aging Research. In 2002 the department hired Rebecka Jornsten, who brought strength to bioinformatics.

John Kolassa came to Rutgers from the Department of Biostatistics at the University of Rochester. He is an expert in saddlepoint methods, asymptotics, and biostatistics. John has served the department as Graduate Director since 2005 during a time of unparalleled growth.

Rebecka Jornsten has a PhD in Statistics from the University of California at Berkeley. Her research interests span clustering, computational biology, image processing, and information theory.

Steve Buyske had a PhD in Mathematics from Brown University when he came to Rutgers to complete his PhD in Statistics. He consulted at the Rutgers Statistical Consulting Office and NJ Agricultural Experimental Station, and collaborates heavily with the Rutgers Genetics Department. With Hoover, Kolassa, Buyske, Ying, and Jornsten, the department had developed a strong core in biostatistics

The department also had a hiring priority in computational statistics. David Madigan came to Rutgers in 2001 with a research agenda in Bayesian models, machine learning, graphical methods, and data mining. David had excellent administrative skills, and was made associate dean of mathematical and physical sciences from 2005 to 2007. He left Rutgers in 2007 to become professor of statistics at Columbia.

Minge Xie has served as Director of the Office of Statistical Consulting and is an expert on statistical models. He has a natural talent for developing new statistical methodologies and theories for problems stemming from interdisciplinary research. One of his many projects involves research on Port Security under a grant from the Department of Homeland Security.

Yehuda Vardi negotiated with the administration for funds for the department's new computer lab on the fourth floor of Hill Center and for additional office space and seminar rooms for the department. Yehuda was a great negotiator. His death in 2005 at age 58 was a great loss to us all.

Harold Sackrowitz was Undergraduate Director from 1995 to 2010. In addition to his theoretical research in decision theory, he also worked in Sports Statistics. Football coaches eager for a win agonized whether to risk going for the two-point conversion by passing or running from the two-yard line, or taking the safer 1-point kick. Harold designed a card that coaches can use to answer this question based on the score and time left in the game.[12] The Rutgers coaches, vigilant for every edge, met with Harold and listened carefully to his message. Football at Rutgers has a storied tradition. In 1869, the first intercollegiate football game was played at Rutgers across from where the department's Morrell Street offices later stood. Today, Rutgers football fans have tailgating parties in the Hill Center parking lot.

Surviving in a Tough Economy: 2005–2011

Regina Liu became department chair in 2005. Regina had done a great job of convincing the administration that they should find money for hires to keep the department up to strength. This included both faculty and administrative–secretarial staff.

The department needed to maintain strength in biostatistics/bioinformatics and computational statistics/data mining. Jornsten, with her expertise in bioinformatics, had come as a new assistant professor, received tenure in the department, but later returned to Norway. Ying had left earlier to go to Columbia, to be followed later by Madigan. Under Regina Liu, the department hired outstanding new faculty to build up strength: Tong Zhang (from Yahoo Inc.) in computational statistics, machine learning, and data mining; Rong Chen (from University of Illinois at Chicago) in time series, bioinformatics, and financial statistics; Zhiqiang Tan (from Johns Hopkins University) in Monte Carlo, causal inference, and missing-data problems; Ying Hung (from Georgia Institute of Technology) in computer experiments and experimental design; and Lee Dicker (from Harvard University) in inference in high-dimensional data, and analysis of proteomic and genomic data.

Many new faculty members have impressive research grant records. In fact, over three-quarters of faculty are Principal Investigators on major research grants, many with multiple grants. Several statistics faculty are members of interdisciplinary graduate programs in Computational Molecular Biology, Mathematics Education, Vision and Perception, and Public Health. Figures 3 and 4 show many of the current faculty.

The department currently offers an undergraduate major in Statistics, and a joint Mathematics/Statistics major, a PhD in Statistics, and Masters degree specializations in Industrial Quality Management, Biostatistics, Data Mining, and a new program in Financial Statistics and Risk Management. Rong Chen took

[12] NY Times, January 16, 2000, "In football, 6 + 2 often equals 6" by David Leonhardt.

Fig. 3 2010 picture in the department coffee room. Left to right: Ying Hung, John Kolassa, Mingie Xie, and Larry Shepp

the lead, and together with Zhang, Kolassa, and Liu developed this new program, which has just been approved by the Graduate School and is part of our department's and discipline's tradition of entrepreneurship in finding new ways to serve society.

The department continues to uphold the tradition of strong research and education. The scope of research areas and the extent of the involvement of interdisciplinary research among the statistics faculty have never been greater. This strength of the faculty and the new areas of growing importance in statistical science are well reflected in the new curriculum.

We are in a world exploding with data, an interdisciplinary world that demands statistical and computational tools to explore it all. There are exciting opportunities to contribute through research to develop the tools and through teaching to train future tool developers and users. The department's goal is to seek out and meet the challenge of these opportunities. Our approach to this goal is to have a faculty that combine excellence and entrepreneurship in research, teaching, and interdisciplinary consulting and interaction.

Graduates of the Department

Many of our PhD and MS graduates have gone on to prominent positions in education, industry, research, and government. In education, our graduates have taught at over 50 universities, colleges, and medical schools. Our PhD graduates have won their university's outstanding teacher and researcher awards, received professional recognition such as Fellows of ASA and IMS, and national

Fig. 4 Current statistics faculty at Rutgers (composite of two pictures taken on successive Wednesdays) September 2011. Left to right, Back row: Richard Gundy, Han Xiao, Donald Hoover, Lee Dicker, Kesar Singh. Middle Row: Zhiqiang Tan, Cun-Hui Zhang, William Strawderman, John Kolassa, Tong Zhang, Rong Chen. Front row: Harold Sacrowitz, Steve Buyske, Regina Liu, Joe Naus, Arthur Cohen. Missing: Javier Cabrera, Ying Hung, David Tyler, Minge Xie

recognition such as the 2008 AAUP Excellence in Research Award. A partial list of our graduates' college and university employment in the United States and abroad will appear on our departmental website.

Our Masters graduates have had outstanding and exciting careers, as highlighted by Mathew Goldstein and Jun Liu. Matt was appointed President of Baruch College, and later became Chancellor of the City University of New York. Jun is a professor at Harvard, and won the 2002 COPSS Award given by the Committee of Presidents of Statistical Societies annually to a statistician under age 40.

Our graduates in industry worked for, managed, or directed statistics groups at major industrial corporations, research & development, and financial institutions. They have had successful careers in state and federal government agencies. Other graduates started their own statistical consulting companies, and PhD Ray Huntington founded together with his wife a national chain of learning centers. Our graduates have served as directors, managers, and chief statisticians at major companies such as Pfizer, Dupont, Kraft Foods, Exxon, and IBM. Many have won awards for work in industrial statistics such as the American Society for Quality Control (ASQC) Shewhart, Deming, Hunter, and Brumbaugh Awards. The department's programs in biostatistics bore fruit, with many of our graduates establishing careers at the major pharmaceutical companies and in biostatistics groups in government agencies. Other graduates worked at research & development or in quality and process control at many non-pharma companies. For details see the departmental website.

Faculty

Steven Buyske, Research Associate Professor, PhD Brown, PhD Rutgers
Areas: statistical genetics, biostatistics, psychometrics, and experimental design.
Javier Cabrera, Professor, PhD Princeton
Areas: Biostatistics, Data mining, Genomics, statistical computing and graphics.
Book: Analyzing Micro-Array Data. Fulbright Scholar.
Rong Chen Professor, PhD Carnegie Mellon
Areas: nonparametric time series, Monte Carlo, bioinformatics, financial statistics IMS treasurer. Fellow: ASA and IMS.
Arthur Cohen, Professor, PhD Columbia.
Areas: statistical inference, decision theory, multivariate analysis
Former Editor *Annals of Statistic*s and AE for *JASA*. Fellow: ASA, IMS, ISI.
Lee Dicker Assistant Professor, PhD Harvard
Areas: high-dimensional data, design genomic and proteomic studies.
Richard F. Gundy Professor, PhD Indiana (Psychology); PhD Chicago (Statistics).
Areas: probability theory, harmonic analysis. Fellow: IMS.
Donald R. Hoover Professor, PhD Stanford, MPH Epidemiology UCLA
Areas: clinical trials, epidemiology, longitudinal methods, multiple comparisons.
Core member, Institute for Health, Health Care Policy and Aging Research.
Ying Hung Assistant Professor, PhD Georgia Institute of Technology
Areas: Experimental design, computer experiments.
John Kolassa Professor and Graduate Director, PhD University of Chicago
Areas: asymptotics. Biostatistics, saddlepoint approximations
Former AE *JASA*. Fellow: ASA, IMS.
Regina Y. Liu Professor and Chair, PhD Columbia
Areas: data depth, text mining, nonparametric multivariate, resampling, extremes, statistical process control, aviation safety and risk management.
Former Ed. *J. Multiv. Anal.* AE *Annals. Stat., JASA, TEST.* Fellow: ASA, IMS, ISI.
Joseph I. Naus Professor, PhD Harvard
Areas: scan statistics, data editing, bioinformatics.
Books: Scan Statistics, Data Quality Control & Editing. Fellow: ASA.
Harold B. Sackrowitz Professor and Undergraduate Director, PhD Columbia
Areas: statistical inference, decision theory, sports statistics. Fellow: ASA, IMS.
Lawrence Shepp Professor Emeritus, PhD Princeton
Areas: math finance, probability, closed-loop diabetes management tomography
Fellow: National Academy of Science, IOM, AAAS.
Kesar Singh Professor, PhD Indian Statistical Institute
Areas: asymptotics, resampling, data depth, angular data, mathematical finance
Pioneering paper on asymptotic theory of Bootstrap. Fellow: IMS, ISI

William Edward Strawderman, Professor, PhD Rutgers.
Areas: decision theory, Bayesian analysis, multivariate analysis.
AE *JASA*, *IMS Lect. Ser.*, former AE *Ann. Stat.*, IMS Council Member. Fellow: IMS, ASA

Zhiqiang Tan, Associate Professor, PhD University of Chicago
Areas: semi-nonparametrics, causal inference, missing data, sampling, MC integrate.
AE. *Biometrics, Annals of Statistics*. NSF Career Award.

David E. Tyler Professor, PhD Princeton
Areas: multivariate, robust stat., directional data, computer vision and time series.
AE *J. Royal Stat. Soc. B, J. Stat. Plan. Inf.* Fellow: IMS.

Minge Xie Professor & Director, Office Stat. Consulting, PhD U. Illinois Urbana
Areas: inference, asymptotic statistical models, and interdisciplinary research. Fellow: ASA, ISI.

Cun-Hui Zhang Professor, PhD Columbia
Areas: high-dimension-data, empirical Bayes, MRI, networks, semi-&nonparametrics
AE *Statistica Sinica*, and *Annals of Statistics*. Fellow: IMS.

Tong Zhang Professor, PhD Stanford (Computer Science)
Areas: machine learning, data mining.

Southern Methodist University Department of Statistical Science

Richard F. Gunst, William R. Schucany and Wayne A. Woodward

Visionary Founding

Paul D. Minton (Fig. 1) graduated with Bachelor of Science and Master of Science degrees in Mathematics from Southern Methodist University (SMU). Following the completion of his doctoral degree in Statistics from the University of North Carolina (Chapel Hill) in 1951, he returned to SMU along with a vision of a unique Statistics Department within a university that had as its primary mission a broad-based liberal arts undergraduate education. Minton's vision focused on his belief that statisticians were collaborators and that there was no limit on opportunities for collaboration in Statistics, regardless of the nature and mission of an academic institution. The key to achieving this vision was the collaborative merging of interests and opportunities within the university and throughout the local region.

Recognizing the growing importance of computing to the implementation of statistical theory and methods in industrial, agricultural, governmental, and social applications, Minton founded the university's Computing Laboratory in 1957 and remained its head until 1962. The Computing Laboratory maintained or provided remote access to commercial state-of-the-art mainframe computers and provided the earliest access for the university community to scientific computing and

Southern Methodist University,
Department of Statistical Science,
PO Box 750332,
Dallas, TX, USA
R. F. Gunst (✉) · W. R. Schucany · W. A. Woodward
e-mail: rgunst@mail.smu.edu

W. R. Schucany
e-mail: schucany@mail.smu.edu

W. A. Woodward
e-mail: waynew@mail.smu.edu

A. Agresti and X.-L. Meng (eds.), *Strength in Numbers: The Rising of Academic Statistics Departments in the U.S.*, DOI: 10.1007/978-1-4614-3649-2_19,

Fig. 1 Early brochure picture. Left to right: Richard Bland, Paul Minton, and Jack Webster

computing/statistical support. Simultaneously, he built university support for the creation of a free-standing department of experimental statistics. SMU approved the *Department of Mathematical and Experimental Statistics* and a master's degree program in Statistics in 1961. The program grew out of courses and services in Statistics offered by the Mathematics Department and by the Computing Laboratory. Minton chaired the new department which initially consisted of two faculty members, himself and Vanamamalai Seshadri, and 15 graduate students. The first graduate of the Statistics Department was Vann B. Parr who was awarded a Master of Science degree in 1963.

Early Growth

From 1964 to 1966 the Department of Mathematical and Experimental Statistics began to establish its teaching and research credentials. Most of its faculty members were quarter-time research collaborators at the *Southwest Center for Advanced Studies* (the precursor of the *University of Texas at Dallas*) working in the Division of Mathematical and Stochastic Systems headed by Donald B. Owen. These joint appointments not only reinforced Minton's emphasis on developing a department with a strong emphasis on collaborative research, it also enabled the department to attract like-minded colleagues who would not otherwise have been attracted to the region.

In 1966, the newly renamed *Department of Statistics* became one of the first at SMU to offer a doctoral program. Major assistance in establishing the program came from a *National Institutes of Health* training grant in Biometry and was the culmination of early collaboration between the department and the *University of*

Texas Southwestern Medical School at Dallas. Additional graduate student and research funding was secured by Professor John E. Walsh from the *Mobil Research and Development Corporation.* In addition to Minton (1961–1972) and Walsh (1967–1972), Professors Richard P. Bland (1963–1988), Chandrakant H. Kapadia (1963–2001), Anant Kshirsagar (1968–1971), Donald B. Owen (1966–1991), William T. Tucker (1966–1972), and John T. Webster (1962–1987) joined the faculty during this early period of growth (Fig. 1). The first PhD graduate of the department was Lowell Gregory in 1968.

In 1968, a THEMIS contract was obtained from the *Department of Defense*, and continued through the *Office of Naval Research* until 1975 with total funding of about $1M. There was no cost sharing on this contract and over $700,000 went into the development of the Statistics Department. The intent was to make SMU a center of excellence in the field of statistics. The THEMIS support by the Department of Defense implied a commitment on the part of SMU to provide continued support and development of the Statistics Department. The department typically graduated about five PhDs per year in the period from the late 1960s to the mid 1970s.

A major change in facilities accompanied this early period of departmental development. Minton's emphasis that statistics was a discipline important to a wide range of departments and programs on campus, especially research-oriented doctoral programs, enabled him to convince administrators that the Department of Statistics should be included in the planning for a new building. In 1969, the department joined two other PhD programs, Anthropology and Geology, in moving to the new Heroy Science Hall. The design of the first floor of this building included classrooms, a departmental office, faculty offices along the outer wall, and graduate student offices along the inner wall for the Department of Statistics. The placement of the faculty offices relative to the graduate student offices continues to serve its initial intended purpose, to facilitate routine and frequent interactions between faculty and graduate students.

The departmental faculty reached its peak size in the mid 1970s. In addition to Bland, Kapadia, Owen, and Webster, the faculty included J. Wanzer Drane (1968–1985) and Campbell B. Read (1969–1999) who had joint appointments with *University of Texas Southwestern Medical Center at Dallas* (UTSW) until 1976. Also added were U. Narayan Bhat (1969–2004), joint with the *Operations Research Department*, William R. Schucany (1970–2011), Richard F. Gunst (1971–present), Jagdish Patel (1972–1975), Henry L. (Buddy) Gray (1973–2004, *C. F. Frensley Professor of Mathematical Sciences*), joint with the *Department of Mathematics*, Harold J. Hietala (1973–2002, joint with the *Department of Anthropology*), and Wayne A. Woodward (1974–present). These hires solidified the founding vision of collaborative activities across the university and region.

Fig. 2 Faculty group in 1979. From left to right: Gray, Webster, Gunst, Woodward, Schucany, Boyer, Drane, Kapadia, Bland, and Owen

Evolution and Impact on the University

Owen succeeded Minton as Chair in 1972, and with a core group of research scholars in the department, he led the faculty to fulfill a new university mandate of increased scholarly productivity and visibility. The department's national visibility received a major boost when the department hosted the *1974 Symposium on the American Mathematical Heritage to Celebrate the Bicentennial of the United States of America*. This symposium brought distinguished architects of the foundations of statistical theory and methods in the United States to campus, including William G. Cochran, Herman Chernoff, H.O. Hartley, Oscar Kempthorne, Jersey Neyman, and Elizabeth L. Scott. This was a very prestigious symposium at the time and the presence of so many renowned scholars provided the relatively young faculty opportunities for personal interactions and exposure at a critical juncture in their development into recognized scholars within the field of statistics.

In 1979, a *Division of Mathematical Sciences* was created in order to streamline the administration of the mathematical sciences on campus and to foster a synergy between the Departments of Mathematics and Statistics in the School of Humanities and Sciences and the Department of Operations Research in the School of Engineering. U. Narayan Bhat was appointed to be the Director of the Division. It rapidly became clear that this organizational change was not serving its intended purposes. Simultaneously with Bhat being selected to be Vice Provost and Dean of Graduate studies in 1980, the Department of Operations Research administratively returned to the School of Engineering. Schucany was named Director of the reduced Division, and Gunst was named Chair of the Department of Statistics. This lasted one more year, at which time the two departments mutually agreed to return to the School of Humanities and Sciences as separate programs. Gunst remained Chair of the Department of Statistics (Fig. 2).

The department's faculty voted in 1986 to change the department's name to the *Department of Statistical Science* in order to emphasize the scientific nature of the discipline. Doing so focused the image of the department on its role in creating novel statistical theory and methods that are based firmly on scientific principles.

Fig. 3 Faculty group in 2000. Left to right: Read, Gray, Schucany, Kapadia, Woodward, Bhat, Natarajan, Gunst, Christensen, Guerra, and Sain

Administrative support for the university has long been an accepted responsibility of department faculty. In addition to Minton and Owen, department faculty who administratively led the program, each with more than one term as chair, are Bhat (1979–1980, 1987–1989), Gunst (1980–1984, 2000–2006), Schucany (1984–1987, 1989–1990, 1997–2000), and Woodward (1990–1997, 2006–present). Gray, in addition to his responsibilities as the Frensley Chair holder, served in several administrative positions in Dedman College, the renamed School of Humanities and Sciences: Associate Dean (1980–1988), Dean ad-interim (1988–1989) and Dean and Vice Provost (1989–1991). While serving in these positions he was instrumental in promoting scholarly research and strong graduate programs.

Faculty members not mentioned previously who spent several years as faculty colleagues include John Boyer, Randall Eubank, Paul Whitney, Georgia Thompson, Robert Brunell, Rudy Guerra, Yuli Koshevnik, Sabyasashi Basu, Roberto Gutierrez, Steve Sain, William Christensen, and Raji Natarajan (Fig. 3).

Research and Graduate Program

The *Statistical Research Laboratory* (Stat Lab) was founded in 1968 to provide expert statistical consulting services to the academic and business communities while at the same time providing consulting experiences for the department's graduate students. Schucany served as the first manager of the Stat Lab while he completed his doctoral research. The structure of the Stat Lab has gone through a number of changes over the succeeding decades. From 1970 to 1979 the manager of the Stat Lab was a permanent full-time faculty member who taught a statistical consulting course and infrequently taught other courses within the curriculum. In addition to Schucany, Stat Lab managers during this time were William H. Frawley, Gunst, and Woodward.

In 1980, the Stat Lab manager became a salaried staff position. The manager was expected to bring in sufficient income from clients, short courses, grants, and contracts to meet the budgeted salary. Because of fluctuations in income generation, a third style of organization was implemented in which the Stat Lab manager received little if any salary from the university. The department provided an office and computing support and allowed the manager to conduct business as a statistical consultant. In return the manager taught a graduate-level statistical consulting course and facilitated actual consulting experiences for the students. Currently, there is no functioning Stat Lab with a manager. Graduate students continue to receive consulting training and experiences through the statistical consulting course, participation with faculty who are all heavily involved in collaborative research projects on and off campus, and a wide array of summer internships.

Owen's experiences at Sandia National Laboratories instilled in him a keen interest in numerical tables of probability distributions. In addition, military requirements of World War II and the Cold War led to continual funding opportunities for his work. His 1962 *Handbook of Statistical Tables* was used throughout the nation as the source of high quality tables of distributions and other quantities needed in contemporary statistical applications. One reason for the popularity of the tables was his insistence that he would not publish tables unless he had at least two very different algorithms for the calculations, a self-imposed check to ensure accuracy. This was a difficult and laborious task both in finding dual algorithms and in performing the computations themselves in the period just before the widespread use of computers. Even when mainframe computers became available he would insist on two or more different computing algorithms, and not simply coding the same algorithm in two or more different ways. Much of Owen's later work concentrated on computational algorithms for the developing field of what was then termed *quality control*. This was a natural transition to industrial product and process improvement that had its early emphasis in wartime military needs.

Perhaps one of Owen's greatest contributions to research came from his desire to alleviate the inordinately long time delay from the completion of statistical research efforts to ultimate publication, a problem that persists today. His frustration with this lengthy process led to the creation of a new, corporately financed journal which he titled *Communications in Statistics*. The novel feature of this journal was the transfer of responsibility and authority for acceptance of submitted articles to a carefully selected group of reviewers, the journal's International Editorial Board. Board members could accept an article either after personally reviewing it or after having it reviewed by a referee more familiar with the topic of the article. The obvious concern about the quality of the refereeing process was mitigated by publishing the names of reviewers along with the article itself. This novel approach to scientific publication achieved its goal of rapid publication in the field of statistics.

Walsh's international stature was solidified by the publication of the three volumes of his *Handbook of Nonparametric Statistics* in 1962, 1965, and 1968, which was the state-of-the-art reference for nonparametric statistics at the time. The success of both Walsh and Owen in their research programs attracted large numbers of graduate students to the department.

Linear model theory and applications (including regression) and multivariate analysis were also strengths of the department at this time. Webster and Kapadia, from their roots at North Carolina State and Oklahoma State, respectively, formulated core courses in linear models and experimental design. Kshirsagar also contributed to the linear models research and teaching but is mostly associated with his work in multivariate analysis and discriminant analysis, topics that led to his 1972 *Multivariate Analysis* text. This text was the second in the new Marcel Dekker series of statistics publications, a series edited by Owen. The "renegade" in the department at his time was Bland, who taught and promoted the widely shunned topic of Bayesian methods. All of these research areas led to the creation of courses and to doctoral dissertations.

When Bhat joined the Department of Statistics full-time in 1979, he brought a well-developed research program in stochastic processes and queuing theory. His 1971 *Elements of Applied Stochastic Processes* formed the basis for graduate-level courses in his fields of expertise, courses which served graduate students in the department and in the Engineering School. His research program, combined with those of Gray and Woodward, formed a strong emphasis on theoretical and applied research in stochastic processes and time series.

Gray was conducting basic research in improved convergence of numerical sequences when he first joined the department in 1973. Owen encouraged Gray and Schucany to adapt this work to the bias reduction feature of the jackknife, enabling them to generalize this distribution-free approach to bias reduction. While Schucany continued work in bias reduction, Gray, soon joined by Woodward in a long and highly productive collaboration, redirected his efforts to research in time series modeling and analysis. They created a research team involving graduate students over the ensuing three decades in the areas of spectral estimation, ARMA model identification, long-memory time series, and the analysis of time series with time-varying frequencies.

Over this same time span, Schucany continued work on bias reduction and other nonparametric research areas, including rank statistics, kernel smoothing, and resampling methodologies. His research program resulted in major publications on the jackknife, bootstrap intervals, concordance of rankings, goodness of fit, minimum-distance estimation, adaptive bandwidths for nonparametric regression, simulation methodology, permutation tests, and wavelet resampling.

Gunst, building on his early mentoring by Webster, diversified the department's research programs in linear models and regression. Initial emphasis was on biased regression estimators, including ridge regression, principal components regression, and latent root regression. He and his students also contributed to the growing interest in influence diagnostics for regression models. Recognizing the tenuous nature of assuming non-stochastic covariates and independent errors in regression

Fig. 4 Faculty in 2010/11. Front row left to right: Harris, Stokes, Wang, and Cao. Middle left to right McGee, Gui(visiting), Butler, and Ng. Top row left to right: Schucany, Woodward, and Gunst

modeling, he pursued work in measurement error modeling and spatial modeling, respectively. All of these topics continue to be taught in advanced graduate courses in regression.

Read was an important contributor to the department's research and graduate programs during this period, and he became first an Associate Editor and later a co-Editor, with founders Norman L. Johnson, and Samuel Kotz, of Wiley's nine-volume *Encyclopedia of Statistical Sciences* in 1982. Read not only edited many of the articles contributed to the encyclopedia, now 16 volumes, but he also wrote a large number of them.

Evolving Research Programs

In addition to Gunst and Woodward, current full-time tenured or tenure-track faculty are Butler (2007), Jing Cao (2005), Ian R. Harris (2001), Monnie McGee (2002), Hon Keung (Tony) Ng (2002), S. Lynne Stokes (2001), and Xinlei (Sherry) Wang (2003) (Fig. 4). Butler became the *C.F. Frensley Professor of Mathematical Sciences* following Gray's retirement. He is internationally recognized for his ongoing research program on saddle-point methods. Other areas of major activity are bootstrap methods, survival analysis, systems theory, and stochastic networks.

A major research emphasis over the last decade by Stokes, Harris, and Cao has been in collaboration with colleagues in the *Simmons School of Education and Human Development*. Areas of active research include studies of the relationship between low motivation and test-taking behavior, the predictive ability of teacher fidelity intervention measures on student performance, and statistical modeling and assessment of the outcomes of large-scale reading intervention programs. Over her career, Stokes has worked on problems of modeling and measuring non-sampling errors in surveys. Her early work focused on measurement of interviewer variance and effects of other interviewer errors, including undercount, in Census data. She also studies methods for improving data collection efficiency, for example by using ranked set sampling and other methods making use of auxiliary information, and methods for ensuring privacy and confidentiality in data collection.

Ng is building a prolific research program that emphasizes censoring, precedence testing, and survival modeling and analysis. He is applying this work not only to industrial applications but also in epidemiology with colleagues at *Baylor Medical Center at Dallas*. Tony has assumed the role as our computing guru. In this position, he teaches the graduate course on Statistical Computing to our first-year graduate students and oversees departmental hardware and software.

Currently, the most extensive areas of collaborative research activities involve research groups of department faculty and graduate students with colleagues at UTSW. McGee collaborates with colleagues in the UTSW Department of Pathology on developing methods for determining whether genes are significantly overrepresented in biological pathways of interest. A goal of this activity is to discover new genes involved in important molecular processes and to categorize the functions of genes. She has also developed an active program to develop methods for preprocessing microarray data. Wang is collaborating with the Quantitative Biomedical Research Initiative Group in the UTSW Department of Clinical Sciences on developing Bayesian hierarchical spatial models, integrative Bayesian analysis for investigating molecular mechanisms of cocaine addiction, and the development of preprocessing methods for high-throughput data. Cao is interacting with colleagues in the UTSW Departments of Clinical Sciences and Cell Biology on simultaneous significance testing in high-throughput data analysis. She is also investigating gene set enrichment analysis and statistical issues relating to procedure volume and surgical outcomes investigations. Gunst, Schucany, and Woodward have worked for several years developing statistical methods for the analysis of brain imaging data. This work is being conducted with colleagues in the UTSW Departments of Internal Medicine and Clinical Sciences in support of a major study of Gulf War syndrome involving veterans from the 1991 Persian Gulf War.

A by-product of this extensive collaboration between SMU and UTSW is the creation of an internship program through UTSW's Department of Clinical Sciences. This program has provided funded support for Department of Statistical Science graduate students to provide statistical research support for UTSW scientists. These students have contributed to research projects in the Center for Human Genetics-McDermott Center for Human Growth and Development, Division of Translational Pathology, Simmons Cancer Center, Comprehensive Center for Molecular, Computational, and Systems Biology, Department of Clinical Sciences, and the Department of Internal Medicine.

Undergraduate Instruction

The contributions of the Department of Statistical Science to SMU's undergraduate teaching mission have primarily focused on introductory service courses for students in all of the undergraduate colleges of the university. This is consistent with Minton's vision for the department and with the interdisciplinary focus of the field of statistical science. The department provides courses for approximately 800 undergraduates per semester, most of whom are first- and second-year students. The department offers courses to a small number of majors and minors; however, the greatest enrollments are in our first-year *Introduction to Statistics* course and our second-year *Statistics for Modern Business Decisions* course. The first-year course is one of several options that satisfy the university's quantitative literacy requirement. The business statistics course is one of seven courses all pre-business students must take prior to becoming a major in the *Cox School of Business*.

In order to take best advantage of its small number of faculty, the teaching paradigm that the department has used throughout most of its history in undergraduate service courses has been to teach large lecture sections with full-time faculty twice weekly. The third weekly sessions are small laboratory or discussion/review sessions taught by statistics graduate students. The intent (which is consistent with SMU's philosophy) is to ensure that virtually all undergraduate students attending the department's Statistics classes are taught by experienced tenure and tenure-track faculty. To accommodate both our graduate and undergraduate courses, a two-course per semester load for faculty members typically includes a course within our graduate curriculum and an undergraduate service course.

The department's dedication to providing outstanding undergraduate service teaching is attested to by two of its faculty members, Harris and Woodward, recently receiving the *Altshuler Distinguished Teaching Professor Award*. Only four university professors are selected each year.

Distinguished Graduates

Through calendar year 2010, 159 Master of Science and 157 PhD degrees have been awarded to graduates of the Department of Statistical Science. Due to page limitations, only very few of the department's distinguished graduates' careers can be highlighted here.

Stephen L. George (1969) began his career in Biostatistics at MD Anderson Hospital and subsequently moved to St Jude's Children's Research Hospital and then to Duke University, where he is a Professor of Biostatistics and Bioinformatics. Don Wheeler (1970) owns a statistical consulting and publishing firm, SPC Inc., and has extensive experience assisting clients in statistical process control.

Two of the department's earliest doctoral students, Schucany (1970) and Gunst (1972), joined the SMU faculty immediately upon graduating and have had long and distinguished careers. Robert L. Mason (1971) holds the title of *Institute Analyst* at Southwest Research Institute and is nationally renowned for his decades-long activities in automotive research. He served as President of the American Statistical Association in 2003. Dwight B. Brock (1971) began his career with the Department of Health Statistics and rose to be the Head of Biometry in the National Center on Aging. He currently works for Westat, Inc.

After graduating with his doctorate, William L. Lester (1974) returned to his faculty position at Tuskegee Institute. He later rose to the position of Provost, a position he held for over 20 years. Joan Reisch (1974), Chief, Division of Biostatistics at UTSW, heads a large contingent of biostatisticians several of whom are SMU alums. Edward Mansfield (1974) and Michael Conerly (1982) are long-time faculty members and have each served as Head of the Department of Information Systems, Statistics, and Management at the University of Alabama.

James L. Hess (1977) has had a distinguished career with DuPont Chemical in Delaware and for the past 15 years with Leggett & Platt, where he is currently Vice-President for Operations Services. William C. Parr (1978) entered academia in Statistics at Texas A&M University (TAMU) upon graduating, then continued his career at the University of Florida, Harris Semiconductor, and the University of Tennessee. He is currently a Professor in the International School of Management in Shanghai, China.

Jeffrey D. Hart (1981), honored as Fellow of IMS and ASA is Professor of Statistics at TAMU. Mani Y. Lakshminarayanan (1984), also a fellow of ASA, has spent his entire career in the pharmaceutical industry, first at Hoechst-Roussel, then at Pfizer and Centocor, and is now with Merck & Co. Lori Thombs (1985) began her academic career at the University of South Carolina. She is now the Director of the Center for Social Statistics at University of Missouri, Columbia.

Pat Gerard (1993) has been on the faculty at Mississippi State and at Clemson, where he is Professor in the Applied Economics and Statistics Department. Alan Polansky (1995) is on the faculty at Northern Illinois University and is the author of the recent book *Observed Confidence Levels: Theory and Applications.* Krista Cohlmia (2003) is professor and chair of the Department of Mathematics at Odessa College.

Continuing Vision

Minton's vision of a Statistics Department that would provide scholarly collaboration in teaching, research, and applications within SMU and throughout the surrounding region is being realized. The Department of Statistical Science is producing nationally recognized masters and doctoral graduates who are sought by academic, governmental, and industrial institutions, and is also providing statistical education for employees of regional industrial companies and medical research labs. The department continues to emphasize graduate and undergraduate courses that are available to diverse academic disciplines at SMU. Although he could not foresee the involvement of the current faculty's research and methodological developments in fields as diverse as census confidentiality, brain imaging, nuclear test monitoring, genomics, accelerated life testing, and vehicle emissions modeling, his vision of the interaction of department faculty with scientists and engineers within the university, regionally, and nationally has been achieved. This history is dedicated to that vision, a vision that continues to evolve with emerging changes in our collaborative activities.

References

Balakrishnan N, Ng HKT (2006) Precedence-type tests and applications. Wiley, New York

Bhat UN, Miller GK (2002) Elements of applied stochastic processes, 3rd edn. Wiley, New York

Butler RW (2007) Saddlepoint approximations with applications. Cambridge University Press, Cambridge

Cao J, Moosman A, Johnson VE (2010) A Bayesian goodness-of-fit test for censored data. Biometrics 66:426–434

Chen Z, McGee M, Liu Q, Scheuermann RF (2007) A distribution free summarization method for Affymetrix GeneChip® arrays. Bioinformatics 23:321–327

Gray HL, Schucany WR (1972) The generalized jackknife statistic. Marcel Dekker, Inc, New York

Gray HL, Zhang N-F, Woodward WA (1989) On generalized fractional processes. J Time Ser Anal 10:233–257

Gunst RF, Mason RL (1980) Regression analysis and its application: a data-oriented approach. Marcel Dekker, Inc, New York

Harris IR (1989) Predictive fit for the exponential family. Biometrika 76:675–684

Kotz S, Balakrishnan N, Read CB, Vidakovic B (eds) (2006) Encyclopedia of statistical sciences, 16 vols, 2nd edn. Wiley, New York

Mason RL, Gunst RF, Hess JL (2003) Statistical design and analysis of experiments, 2nd edn. Wiley, New York

Owen DB (1962) Handbook of statistical tables. Addison Wesley Publishing Co., New York

Patel JK, Read CB (1996) Handbook of the normal distribution, 2nd edn. Marcel Dekker Inc, New York

Schucany WR, Gray HL, Owen DB (1971) On bias reduction in estimation. J Am Stat Assoc 66:524–533

Stokes SL, Sager TW (1988) Characterization of a ranked-set sample with application to estimating distribution functions. J Am Stat Assoc 83:374–381

Thombs LA, Schucany WR (1990) Bootstrap prediction intervals for autoregression. J Am Stat Assoc 85:486–492

Wang X, Stokes SL, Lim J, Chen M (2006) Concomitants of multivariate order statistics with application to judgment post-stratification. J Am Stat Assoc 101:1693–1704.

Webster JT, Gunst RF, Mason RL (1974) Latent root regression analysis. Technometrics 16:513–522

Woodward WA, Gray HL (1981) On the relationship between the s-array and the Box-Jenkins approach to ARMA model fitting. J Am Stat Assoc 76:579–587

Woodward WA, Gray HL, Elliot AE (2011) Applied time series analysis. Chapman and Hall, New York

Stanford University Department of Statistics

Ingram Olkin and Guenther Walther

The teaching of statistics has a long history at Stanford prior to the formation of a statistics department. The university was founded in 1891, and starting in the early 1920s, courses in statistics appeared in a number of departments, each emphasizing a different aspect of the field.

The Early Years: 1920–1930

In 1920, the School of Education listed a course called *EDUC 3*: *Educational Statistics* with the following description: "This course will deal with the theoretical and practical aspects of statistical methods as applied to education, including scale, units, and standards." The instructor was Truman Lee Kelley, who had just joined the faculty and who taught courses in statistics and psychometry (a precursor to psychometrics) until 1931 when he migrated to Harvard. Kelley was a joint author of the Stanford Achievement Test Battery and the author of the book *Statistical Methods* in 1924. This was also the early development of mental testing that was pioneered by Lewis Madison Terman, the inventor of the Stanford-Binet IQ test. Terman was also a leading developer of the field of educational psychology, now the area responsible for statistics in schools of education. In 1921, Terman started a longitudinal study of gifted children titled *Genetic Studies of Genius*. This study has continued to this day.

Stanford University,
Department of Statistics,
390 Serra Mall,
Stanford, CA 94305-4065, USA
I. Olkin (✉) · G. Walther
e-mail: iolkin@stat.stanford.edu

G. Walther
e-mail: walther@stat.stanford.edu

A. Agresti and X.-L. Meng (eds.), *Strength in Numbers: The Rising of Academic Statistics Departments in the U.S.*, DOI: 10.1007/978-1-4614-3649-2_20,

From 1916 to 1925, the Department of Psychology introduced a number of statistics courses. One such (*PSYCH 8*) stated that "sound knowledge of algebra is a prerequisite." One of the books to be used was that of G. Udny Yule. This course was "a lecture and seminar course covering the theory of chance, and the application of statistical methods developed from the mathematical theory of probability to experimental psychical research." In 1924, there is the first appearance of an advanced course, *Advanced Statistical Methods*, taught by Truman Kelley; this included a discussion of the analysis of correlated data, multiple correlation, and categorical data.

The Department of Economics introduced a course in 1919/1920, *ECON 6*: *Introduction to Statistics* with the description: "A study of elementary statistical methods appropriate for dealing with problems in the social sciences and business." The Department of Political Science introduced its statistics course in 1925/1926. *POL.SCI119*: *Quantitative Studies in Politics and Administration* was aimed at students "interested in the problem of methods in the measurement and analysis of political phenomena."

Eugene L. Grant joined the Engineering faculty in 1929 and introduced the course *Civil Eng 132*: *Statistics in Engineering* which focused on statistical methods in the inspection of manufactured products. This was later expanded into a department of industrial engineering.

The year 1924 is singular in that Harold Hotelling joined the faculty in the Food Research Institute, and subsequently in the Department of Mathematics. Hotelling was one of the statistical luminaries of the twentieth century, known for the development of principal components, canonical correlations, and the multivariate generalization of Student's *t*-statistics. Hotelling's course was *Math 175*: *Theory of Probability and of Statistical Inference,* and may be one of the earliest courses in Mathematical Statistics. The description includes "considerable material relating to statistical methods and theories which is found in memoirs but in no textbook". One of the topics was "Law of Great Numbers," which is Hotelling's translation of Poisson's paper "La loi des grands nombres" and is now generally called Law of Large Numbers. It may be of interest to note that Holbrook Working joined the Food Research Institute and he and Hotelling developed simultaneous confidence regions for a regression line, a procedure named after them; see *JASA 24*: 73–85 (1929). Hotelling left for Columbia in 1931, where he was instrumental in attracting an eminent group of statisticians.

The mathematical foundation of probability and statistics was strengthened when James Victor Uspensky joined the mathematics faculty in 1929. His 1937 book "*Introduction to Mathematical Probability*" is one of the early textbooks on the mathematics of probability. He regularly taught the course *Theory of Probability*.

The Pre-department Years: 1930–1948

By 1930, statistics courses appeared in the departments of education, psychology, economics, political science, civil engineering, and mathematics. Because statistics courses were not centralized, the 1930 catalog has a listing labeled "Statistics" with this comment: "As a result of the increasing interest in the field of statistics, the following courses presented in several departments are listed for the information of students in this field. No formal departmental or divisional organization is involved."

In 1934, Stanford offered its first degree program in statistics in the form of a minor subject to a Doctor of Philosophy. The task of administering curricula for that degree was delegated to a Committee on Instruction of Statistics. This committee consisted of John B. Canning (Economics) as chairman, Theodore J. Kreps (Graduate School of Business), and James V. Uspensky (Mathematics). The composition of this committee varied slightly from year to year among the faculty with a strong interest in Statistics: Harold M. Bacon (Mathematics), Quinn McNemar (Psychology and Education), Frank W. Weymouth (Biological Sciences), Holbrook Working (Economics), Eugene L. Grant (Economics and Civil Engineering), and, starting with his arrival at the Stanford Mathematics Department in 1942, George Pólya was continuously involved with this committee.

Pólya taught a course on mathematical statistics almost every year and is considered to be one of the greatest teachers in the field of mathematics. His books on problem solving and plausible reasoning are classics in mathematical education. When he retired in 1953, he was called back to service and taught until his ninety-first year. The mathematics department had a regular offering of courses related to statistics, growing from two courses in 1933 (*Theory of Probability* by Uspensky and *Statistical Inference* by Bacon) to four courses in 1948 (in addition to the above, there was the course *Mathematical Statistics* and a *Colloquium in Statistics*). The departmental listings of 1946 already include Albert Bowker and Herbert Solomon, although Bowker taught statistics courses in a variety of departments, such as a course in economics in 1946 and *Advanced Statistics in Engineering* in 1947.

The department with the largest offering of statistics courses at that time was Economics, where in 1936 statistics was listed as one of nine major subfields. The *Stanford Bulletin* for that year lists the courses *Elementary Statistical Operations*, *Introduction to Statistical Inference I* and *II*, *Advanced Statistics I* and *II*, which dealt with topics such as goodness of fit and the fitting of frequency curves, and a *Seminar in Advanced Statistics*. Faculty included John Canning and Holbrook Working. This center for statistics was further strengthened in 1938 with the arrival of W. Allen Wallis, but the Economics Department resisted the creation of a separate Statistics Department. As reported by Stigler (1999), as late as 1947, Holbrook Working wrote, "It seems to me that Stanford has little excuse for entering into competition with Berkeley in offering intensive training in mathematical statistics," and a report from the Chairman of the Economics Department concluded:

> It is our feeling that it is very important that instruction in the applications of statistical technique should continue to be under the departments in whose subjects the applications of statistical technique are made.... Both statistics and economics gain if statistical work is done by economists.

The controversy of statistics distributed across fields of application versus statistics collected as a central core continues to this day. The 1941 paper by Hotelling on "*The teaching of statistics*" was very influential in the move toward a central core. Jerzy Neyman, the founder of the Department of Statistics at Berkeley, cites this paper as providing the basis for the centralization of statistics in U.S. universities.

The effort to create a separate department suffered an additional setback when Wallis left in 1942 to run the Statistical Research Group at Columbia. However, before he left, Wallis hired Al Bowker, and then, as Stigler (1999) reports, "recommended adding Abe Girshick to Al to form the nucleus of a new group, admitting that Girshick had 'no administrative ability. Since Bowker does have, however, they should make an excellent team.'" Bowker had studied under Harold Hotelling at Columbia and North Carolina.

Another stronghold of statistics at that time can be found in Education and in Psychology, with the offerings *Elementary Statistical Methods*, *Introduction to Statistical Methods*, and *Advanced Statistical Methods I* and *II*. Those courses were partly taught by Quinn McNemar, whom the *Bulletin* of 1934 lists as an Instructor in Education and Psychology.

The Department of Biological Sciences lists courses named *Biometry* and *Biometrical Analysis* taught by Frank Weymouth, and in the Department of Civil Engineering, E. Grant was teaching the courses *Statistics in Engineering, Advanced Statistics in Engineering*, and, starting in 1945, *Quality Control by Statistical Methods*. Felix Bloch, who in 1952 would share the Nobel Prize in Physics, annually taught a course in statistical mechanics, beginning upon his arrival at Stanford in 1934. Although not a statistics course, it had a component of stochastic processes and combinatorics and was of interest to physicists. It was later that analysts recognized that it involved large deviations with dependent variables.

Formation of the Department of Statistics

The history of the formation in 1948 of the Department of Statistics has been documented in several sources. Royden (1989) provides a history of mathematics and statistics. The department celebrated its fiftieth anniversary in 1998, at which time Stephen Stigler gave a talk on the origins of the department (Stigler 1999). In brief, Allen Wallis joined the Stanford Department of Economics in 1938, with responsibility for the teaching of statistics. With the start of the war he moved to Columbia to run the Statistical Research Group (SRG). Bowker was a student at Columbia and worked on sampling inspection problems with the SRG. Before leaving Stanford, Wallis urged the Mathematics Department to hire Bowker, and he joined the faculty in 1947. Wallis also recommended hiring Abraham Girshick (then at the Rand Corporation) with the suggestion that Bowker be chair of a department of statistics. The department was founded in 1948 and by 1950 had a faculty of five: Bowker and Girshick, Quinn McNemar (joint with Psychology), Kenneth Arrow (joint with Economics), and Herman Rubin. David Blackwell was a visitor, and it was during this time that he and Girshick wrote their book on decision theory (Blackwell and Girshick 1954). By 1956, Bowker brought Herman Chernoff, Charles Stein, Lincoln Moses, Gerald Lieberman, and Samuel Karlin into the department. Thus, in less than a decade the department reached adulthood.

The next 5 years (1956–1961) saw astounding growth in the university. The Medical School was moved to the campus and the Stanford Linear Accelerator was created. The department was part of this growth when Emmanuel Parzen, Vernon Johns, Herbert Scarf, Herbert Solomon, William Madow, Rupert Miller, Harvey Wagner, Kai Lai Chung, Patrick Suppes, Hirofumi Ozawa, and Ingram Olkin all joined the faculty. Bowker's administrative genius was to recognize that statistics alone would not be able to sustain a large department. However, by generating a liaison with other departments in the form of joint appointments, the department could have an impact in the university and also carry out a research agenda in various substantive fields.

Over the years, there have been joint appointments with Economics (Anderson, Arrow, Romano), Mathematics (Candes, Dembo, Diaconis, Karlin), Earth Sciences (Rajaratnam, Switzer), Education (Olkin), School of Medicine (Efron, Hastie, Johnstone, Lai, Miller, Moses, Tibshirani, Wong), Operations Research (Lieberman), Symbolic Systems (Holmes), SLAC (Friedman), Electrical Engineering (Cover, Montanari), and Psychology (McNemar).

The following is an excerpt from a letter written by Bowker to Wallace Sterling on May 1951, when Sterling was President of the university:

> Our Statistics Department has been integrated quite successfully into the general university program. Professor McKinsey in Philosophy and Professor Grant Ireson in Industrial Engineering have been brought to Stanford by funds provided by our projects; Professor Hans Lewy, a very distinguished applied mathematician works on one of our programs. Faculty from the Departments of Mechanical Engineering, Civil Engineering, Economics, Philosophy, and Mathematics are all associated with research programs we have developed, and members of our staff have worked either as collaborators or as statistical consultants with faculty from the Medical School, the Graduate School of Business, the Hoover Library, the Food Research Institute, the School of Mineral Sciences, as well as the Departments of Physics, Mathematics, Philosophy, Economics, Psychology, Sociology and Anthropology, and Biology.

A second factor that impacted the growth of the department was the support of the Office of Naval Research (ONR). As noted by Mina Rees in 1980 when she was Director of ONR, support of Mathematics was provided to NYU, MIT, Stanford, Berkeley, and Tulane. Bowker recommended the establishment of a laboratory, the Applied Mathematics and Statistics Laboratory (AMSL) that was a channel for obtaining funding from the Federal government. The AMSL supported a broadly defined field of mathematical sciences that included core and applied mathematics, statistical decision theory, game theory, mathematical economics, inventory theory, mathematical psychology, as well as general mathematical statistics and applied statistics.

The combination of an exciting department, proximity to the activities at Berkeley, and the California weather created a domain of attraction at Stanford.

There were many postdocs who went on to become well known. A large number of visitors on sabbatical leave chose to spend a year at Stanford. A perusal of publications written jointly with the Stanford faculty in the early years will vividly show the influence of the department.

Students

The first doctorates were awarded to Herbert Solomon (1950) and Lincoln Moses (1951). Since then, over 400 doctorates and over 1600 Master's degrees have been awarded.

Decade	Master's degrees	PhD degrees
1948–1960	75[a]	29
1961–1970	403	84
1971–1980	275	74
1981–1990	281	68
1991–2000	255	69
2001–2010	344	84
Totals	1,633	408

[a] Data available beginning with academic year 1953–1954

Our students have had a great impact on the profession. Many have been chairs, and it is hard to find a campus that does not have a faculty member influenced by our department either as a student, or as a faculty who visited our campus.

The department offered a Bachelor's degree for a number of years, but the number of students was small. Instead, a joint degree that included mathematics, statistics, applied mathematics, and computer science was created. This degree, called "Mathematical and Computational Science", has been highly successful, with a 2010 graduating class of 22.

With the growth of computing and data analysis, the master's degree has become very popular. Also, students in doctoral programs in other fields find a master's degree in Statistics to be very useful and advantageous.

The university established a series of fellowships in 1993 in honor of Gerald J. Lieberman. The fellowships are awarded to outstanding advanced doctoral students who intend to pursue a career in university teaching and research. In 2011, the department established the Charles Stein Fellowship in Statistics that is designed to be a post-doctoral career-building step for new scholars.

Faculty

The list of faculty is long and is provided on the website for this book. The present faculty, as of December 2010, listed in order of longevity within the department (dating from first appointment) are: Charles Stein (1953, emeritus); Ingram Olkin

(1961, emeritus); Patrick Suppes (1961, courtesy/emeritus); Bradley Efron (1965); Paul Switzer (1966, emeritus); Theodore Anderson (1967, emeritus); Richard Olshen (1967; 1990, courtesy); David Siegmund (1967); Thomas Cover (1970); Persi Diaconis (1974); Jerome Friedman (1981, emeritus); Iain Johnstone (1981); Art Owen (1985); Joseph Romano (1986); Tze Leung Lai (1987); David Rogosa (1987, courtesy); Helena Kraemer (1988, courtesy/emerita); Amir Dembo (1990); David Donoho (1991); Trevor Hastie (1994); Guenther Walther (1994); Balasubramanian Narasimhan (1996); Charles Chui (1997, consulting); Susan Holmes (1998); Robert Tibshirani (1998); Jonathan Taylor (2001); Wing Hung Wong (2004); Andrea Montanari (2006); Nancy Zhang (2006); Simon Jackman (2007, courtesy); John Chambers (2008, consulting); Philip Lavori (2008, courtesy); Emmanuel Candès (2009); Jacqueline Meulman (2009, visiting); Balakanapathy Rajaratnam (2009); Chiara Sabatti (2009, courtesy), Hua Tang (2010, courtesy).

When Bowker became Dean of Graduate Studies in 1958, Herbert Solomon became Chair of the department. Solomon served vigorously as Chairman from 1959 to 1964, and thereafter the department moved to a 3-year, non-self-succeeding chairmanship. To date, 14 faculty have served as Chair: Bowker, Chernoff, Efron, Friedman, Hastie, Johnstone, Lai, Miller, Moses, Olkin, Siegmund, Solomon, Switzer, and Wong. This has had an advantage in that each Chair has been able to negotiate some feature to improve the department.

After Rupert Miller's untimely death in 1986, the department created a lectureship in his honor. The first speaker was Sir David Cox in 1987, followed by John Tukey in 1989, and Fred Mosteller in 1980. Other speakers were Norm Breslow in 1994 and Anastasios Tsiatis in 2002.

Biostatistics

Lincoln Moses joined the faculty in 1953 as a joint appointment between Community Medicine and Statistics. At the time, the School of Medicine was housed in San Francisco. In 1959, the School of Medicine moved to the Stanford campus, and biostatistics became a division in what was the Department of Community Medicine. Rupert Miller joined the Division, again with a joint appointment. Byron Brown and Bradley Efron were later added to the group. They were successful in being awarded an NIH Training Grant (with Miller as principal investigator) that supported a number of students interested in biostatistics. This was a very vibrant group. Although these students "majored" in biostatistics, their training was identical to that of the other students. There was no distinction based on the field of interest, whether biostatistics, psychology, education, and so on, in the training that students received. In 1988, the Division of Biostatistics became one of three arms in the Department of Health Research and Policy, where it remains today. Its faculty has grown considerably, with much interaction between Biostatistics and Statistics. The current key personnel who are affiliated with the Department of Statistics are Bradley Efron, Trevor Hastie, Iain Johnstone, Philip

Lavori, Balasubramanian Narasimhan, Richard Olshen, Chiara Sabatti, Robert Tibshirani, and Wing Wong.

Doctoral Courses

The following is a list of doctoral-level courses 25 years apart. This provides a trajectory of the development of Statistics at Stanford and reflects the changes in the profession.

Courses in 1960	
220a,b	Estimation and testing hypothesis
224a,b	Multivariate analysis
230a,b,c	Advanced probability
234	Time series analysis
235a,b	Non-parametric statistical inference
236a,b,c	Decision theory and statistical inference
242a,b,c	Stochastic processes
244a,b	Large-sample theory

Courses in 1985	
230a,b	Advanced probability
233a,b,c	Applied statistics
236a,b,c	Theoretical statistics
314	Inequalities: Theory and applications
317, 318	Stochastic processes
324	Multivariate analysis
326	Sequential analysis
332	Asymptotic methods in statistics
333	Robust estimation
336a,b	Decision theory and statistical inference
343a,b	Time series analysis
350	Topics in probability theory
351	Geometric probability and applications
358	Queueing theory
359	Applied probability
360	Applied stochastic processes: Control and information
361	Statistical pattern recognition and robustness
362	Information and statistics
364	Topics in gambling and investing
376	Information theory

Courses in 2010	
300a,b,c	Theory of statistics
305	Introduction to statistical modeling
306a,b	Methods of applied statistics
310a,b,c	Theory of probability
314	Advanced statistical methods
315a,b	Modern applied statistics: Learning, data mining
316	Stochastic processes on graphs
317	Stochastic processes
320	Heterogeneous data with kernels
322	Function estimation in white noise
324	Multivariate analysis
329	Large-scale simultaneous inference
330	Introduction to compressed sensing
345	Computational algorithms for statistical genetics
351a	Introduction to random matrix theory
352	Spatial statistics
362	Computational biology
367	Statistical models in genetics
370	A course in Bayesian statistics
375	Inference in graphical models

Sequoia Hall

The Department of Statistics was housed in a block-style building that was part of the early construction on campus. As Stanford was conceived to be a coeducational institution from the start, Sequoia Hall dated from 1891 and was designed to be the first dormitory for women. As such, it had large parlors, high ceilings, and wide hallways. The reinforced-concrete structure included three stories, but in later decades the upper two floors were closed for occupancy due to seismic safety concerns and then removed altogether.

From 1948 until 1964, Sequoia Hall was home to some mathematicians (e.g., Paul Garabedian, George Forsythe, Stephen Bergmann) and all the statisticians. The Mathematics Department moved to a central part of the campus in 1964, thereby leaving Sequoia Hall to Statistics. Early on there was sufficient room for students and faculty, but the building was increasingly in disrepair. The old Sequoia Hall was demolished in 1996 and a new Sequoia Hall was dedicated on nearly the same site in 1998, the first building to be completed as part of the new Science and Engineering Quad. New Sequoia Hall aligns with the historic Main Quad, with an updated stone veneer and clay tile roof that reflects that early design. There is now a high-ceilinged lobby with a curving staircase leading up to the second floor beneath a skylight, as well as a library, lounge, and classroom.

The wide hallways were maintained (after considerable discussion with the architects) and they still provide easy collaboration between colleagues.

The effects of an architectural structure may not seem to be important at first glance. However, in the present case, this building is one of the features at Stanford that is almost always remembered for generating a friendly atmosphere within its walls.

Joint Colloquia with Berkeley

There has always been a sense of camaraderie (both socially and scientifically) between the two Statistics departments. Erich Lehmann in writing his reminiscences describes many of the connections between the two departments (Lehmann 2008). The Stanford/Berkeley (and Berkeley/Stanford) Joint Colloquia have become a tradition that is well remembered by faculty and students as well as the many visitors to the two Bay Area institutions. In the early days, there were three joint colloquia every quarter; now there are three events per year, held in alternating locations.

Honors and Awards

Members of the department, past and present have been recognized in many ways, and only a few honors and awards are cited here. Many of the faculty have given named lectures such as the Wald, Fisher, and Medallion lectures.

National medal of science	Arrow, Efron, Karlin
Nobel Prize in Economics	Arrow
Norbert Wiener Prize in Applied Mathematics	Donoho
McArthur Fellows	Diaconis, Donoho, Efron
National Academy of Sciences	Anderson, Arrow, Diaconis, Donoho, Efron, Friedman, Johnstone, Siegmund, Stein, Wong
National Academy of Education	Olkin
Norwegian Academy of Science and Letters	Anderson
French Academy of Sciences	Donoho
Guggenheim Fellow	Anderson, Arrow, Olkin, Olshen, Siegmund
George Pólya Prize	Candès
Guy Medal in Silver	Johnstone
COPSS President's Award	Donoho, Johnstone, Lai, Tibshirani, Wong
National Science Foundation Waterman Award	Candès

Role in the University

Although the department is a relatively small one, it has been very influential in the university. Indeed, many members of the department have served the university in a number of administrative roles:

Provost	Lieberman
Vice Provost and Dean of Graduate Studies and Research	Lieberman
Dean of Graduate Studies	Bowker
Associate Dean of Humanities and Sciences	Efron, Johnstone, Lieberman, Moses, Siegmund
Vice-Dean for Academic Planning, H&S	Johnstone
Director, Public Policy Program, H&S	Moses

Research Areas of Faculty

Biological applications (Narasimhan, Zhang)
Bootstrap methods (Efron, Romano, Tibshirani)
Classification (Olshen)
Combinatorics (Diaconis)
Communication theory (Cover)
Computational biology (Wong)
Computational methods (Narasimhan)
Computational statistics (Donoho, Holmes, Walther)
Data mining (Hastie, Holmes, Friedman)
Decision theory (Stein)
Econometrics (Anderson, Romano)
Empirical likelihood (Owen)
Environmental statistics (Switzer)
Financial mathematics (Lai)
Genetics (Siegmund, Zhang)
Gaussian processes (Taylor)
Graphical models (Montanari, Rajaratnam)
Inequalities (Olkin)
Information theory (Cover, Dembo, Montanari)
Imaging sciences (Candès)
Longitudinal data (Olshen)
Meta-analysis (Olkin)
Monte Carlo methods (Diaconis, Owen)
Machine learning (Friedman, Hastie, Tibshirani)
Multivariate analysis (Anderson, Johnstone, Olkin, Rajaratnam, Stein)

Multiple comparisons (Efron, Taylor)
Nonparametric analysis (Owen, Romano, Switzer, Walther)
Sequential analysis (Lai, Siegmund)
Spatial statistics (Switzer)
Stochastic processes (Dembo, Lai, Siegmund)
Signal processing (Candès, Donoho, Johnstone)
Statistical inference (Wong)
Time series (Anderson)

Books Published by Faculty 2000–2010

Anderson, T.W. (2003). *An Introduction to Multivariate Statistical Analysis*, 3rd ed. Wiley Series in Probability and Statistics. Wiley-Interscience, New York.
Efron, Bradley (2010). Large-Scale Inference: *Empirical Bayes Methods for Estimation, Testing, and Prediction*. Institute of Mathematical Statistics Monographs I, Cambridge University Press, Cambridge.
Hastie, T., R. Tibshirani and J. Friedman (2001, 2009). *The Elements of Statistical Learning: Prediction, Inference, and Data Mining*, 1st and 2nd ed. Springer Series in Statistics. Springer-Verlag, New York.
Lai, T.L. with V. de la Pena and Q.M. Shao (2009). *Self-Normalized Processes: Limit Theory and Statistical Applications*. Probability and its Applications. Springer, New York.
Olkin, I. with A.W. Marshall (2007). *Life Distributions: Structure of Nonparametric, Semi-Parametric, and Parametric Families*. Springer Series in Statistics. Springer, New York.
Olkin, I. with A.W. Marshall and B. Arnold (2010). *Inequalities: Theory of Majorization and Applications*, 2nd ed. Springer Series in Statistics. Springer, New York.
Owen, A.B. (2001). *Empirical Likelihood*. Monographs on Statistics and Applied Probability 92. Chapman and Hall/CRC Press, Boca Raton, FL.
Owen, A.B. and L'Ecuyer, P. (eds) (2008). *Monte Carlo and Quasi-Monte Carlo Methods*. Springer-Verlag, New York.
Romano, J.P. with E.L. Lehmann (2005). *Testing Statistical Hypotheses*, 3rd ed. Springer Texts in Statistics. Springer, New York.

Further History

History often changes with the perspective of the author. The publication of conversations with individual faculty will provide further details of the growth of the department.

Three faculty, Kenneth J. Arrow, David Blackwell, and Abraham Girshick (from left to right), at the time the Department of Statistics was formed.

Key figures in the development of the Statistics Department were (from left to right) Albert Bowker, Gerald Lieberman, Lincoln Moses, and Herbert Solomon.

Published Interviews with Stanford Faculty

DeGroot, M.H. (1986). A conversation with T. W. Anderson. *Statistical Science* **1**, 97–105.

DeGroot, M.H. (1986). A conversation with Persi Diaconis. *Statistical Science* **1**, 319–334.

DeGroot, M.H. (1986). A conversation with Charles Stein. *Statistical Science* **1**, 454–462.

Olkin, I. (1987). A conversation with Albert H. Bowker. *Statistical Science* **2**, 472–483.

Phillips, P.C.B. (1986). The ET interview: Professor T.W. Anderson. *Econometric Theory* **2**, 249–288.

Switzer, P. (1992). A conversation with Herbert Solomon. *Statistical Science* **7**, 388–401.

Brown, B.W., Jr. and Hollander, M. (1999). An interview with Lincoln Moses. *Statistical Science* **14**, 338–354.

Holmes, S., Morris, C., and Tibshirani, R. (2003). Bradley Efron: A conversation with good friends. *Statistical Science*, **18**, 268–281.

Sampson, A.R. (2007). A conversation with Ingram Olkin. *Statistical Science* **22**, 450–475.

Articles that Relate to the Statistics Department

Olkin, I. (1991). A conversation with W. Allen Wallis. *Statistical Science* **6**, 121—140.

Rees, M. (1980). The mathematical sciences and World War II. *American Mathematical Monthly* **87**, 607–621.

Hollander, M. and Marshall, A.W. (1995). A conversation with Frank Proschan. *Statistical Science* **10**, 118–133.

Bather, J. (1996). A conversation with Herman Chernoff. *Statistical Science* **11**, 335—350.

Newton, H. J. (2002). A conversation with Emanuel Parzen. *Statistical Science* **17**, 357–378.

Acknowledgments The authors are grateful to the Institute of Mathematical Statistics for permission to reprint the photograph of our founding faculty.

References

Blackwell D, Girshick MA (1954) Theory of games and statistical decisions. Wiley, New York

Lehmann EL (2008) Reminiscences of a statistician: the company I kept. Springer, New York

Royden H (1989) A history of mathematics at Stanford. In: Duren PL, Askey R, Merzbach UC (eds) A century of mathematics in America, Part II. American Mathematical Society, Providence, pp 237–277

Stigler SM (1999) The foundations of statistics at Stanford. Am Stat 53:263–266

SUNY at Buffalo Department of Statistics

Alan Hutson

Statistics at the University at Buffalo (UB), The State University of New York, has a long and interesting history dating to 1925 when the University was a private institution and known as the University of Buffalo. The first two formal statistics organizational structures to form at UB were the Bureau of Business and Social Research founded in 1926 and the Department of Statistics and Insurance founded in 1930. Over the course of several decades the Department moved across multiple schools and has had several interesting turning points in its history, including a well-known and acrimonious split in 1973. We chronicle the key points in the Department's history to date. The current Department of Biostatistics, housed in the School of Public Health and Health Professions, which is a direct descendent of the original Department of Statistics and Insurance, is stronger than ever and currently offers a Minor in Statistics, an MA and PhD in Biostatistics, an MPH with an Emphasis in Biostatistics, and an MS in Bioinformatics and Biometrics. The Department currently has 55 graduate students and continues to teach over 1000 students yearly in terms of statistics service teaching.

The Early Days

In April 1925 a committee was appointed by Chancellor Capen to "consider a general university program of research in business and the underlying social sciences." A year later, in April 1926, the Bureau of Business and Social Research

University at Buffalo, School of Public Health and Health Professions, Department of Biostatistics, 706 Kimball Tower, 3435 Main St, Buffalo, NY 14214-3000, USA
A. Hutson (✉)
e-mail: ahutson@buffalo.edu

A. Agresti and X.-L. Meng (eds.), *Strength in Numbers: The Rising of Academic Statistics Departments in the U.S.*, DOI: 10.1007/978-1-4614-3649-2_21,

was established under the direction of Dr. Oliver C. Lockhart "for the dissemination of the results of such studies" (Lockhart 1928). The Bureau was designed to "serve as the center of statistical research in the University and its community and through the gradual accumulation of dependable information contribute to the understanding and solution of a variety of problems of general interest" (Lockhart 1928). The Bureau presented statistical studies that spanned the fields of banking, business and industry, employment, real estate, and vital statistics and presented the results of these studies in their publication the *Statistical Survey.*

An example of their early research was a study comparing the "life expectancy" of staying in business for a drug store relative to a grocery store during the period of 1918–1927, where the research was highlighted in the following headline: "Drug Store Has 75 Chances In 100 Of Staying In Business A Year; Grocery Has 40 Chances In 100" (1929). Another research example of the Bureau was a graphic of a geospatial analysis of the ratio of aliens-to-citizens in the City of Buffalo in 1927.

Dr. Robert Riegel, first Chair of the Department of Statistics and Insurance (1930–1936)

In 1928, Dr. Robert Riegel, an actuarial scientist and business statistician, came from the University of Pennsylvania's Wharton School of Business to establish the Department of Statistics and Insurance and to head the Bureau of Business and Social Research in the then private University of Buffalo. Dr. Riegel's specialty was rate making for insurance. In 1924, while at the University of Pennsylvania he wrote the first ever book written in business statistics. Later, Dr. Riegel wrote a book on insurance (co-authored with Miller), which became the dominant textbook on insurance in America. Dr. Riegel directed the Bureau until 1942 when it closed due to financial problems caused by World War II.

The actual Department of Statistics and Insurance was formed in 1930 and was housed in the School of Business Administration. Professor Riegel introduced courses in insurance, statistics, and business mathematics. These were service courses, because the department had no degree program in statistics. In those early days, mathematical probability was taught in the Department of Mathematics, while courses in business and financial mathematics were taught by Dr. Riegel and his staff.

Concurrent with his chair duties Dr. Riegel also served as the varsity tennis coach in 1931 even though he smoked up to a dozen cigars a day. He was also known to be so frugal that most who knew him believed he was quite poor, yet, when he passed away he left an estate of almost 2 million dollars. Dr. Riegel was a strong believer in the use of statistics with respect to making business decisions, as noted by his quote in an article in the 1935 *Buffalo Times* "The 'hard-headed' businessmen of olden days, relying on the homely virtues of common sense and native shrewdness and scorning statistics, indices and economic theory, has been rendered as extinct as the dodo by this current depression."

One of the early students who benefited from the courses in the Department of Statistics and Insurance was Robert E. Rich, Sr. His illustrious career and his company, *Rich Products*, and *Coffee Rich*, are well known throughout the world.

By 1936, the demands of the Bureau on Dr. Riegel were so great that Dr. Martin Brumbaugh, also from the University of Pennsylvania's Wharton School of Business, joined the Department as Chairman. His book, *Business Statistics*, written at Buffalo, was the largest-selling book in statistics for several years. While at Buffalo, Dr. Brumbaugh founded the *Journal Industrial Quality Control* of the *American Society for Quality Control,* which was succeeded by *Quality Progress* and the *Journal of Quality Technology*. The society's prestigious "Brumbaugh Award" is named to honor his vital contributions to the war effort in WWII. In an editorial published in 1944, Brumbaugh predicted "Quality control will be a powerful force in the postwar period" (ASQC Website 2010).

In 1946, graduate courses in statistics were introduced, but these served only as part of a program in the School of Business Administration. In 1949, Dr. Zenon Szatrowski joined the Department as Chair and established the masters and doctoral programs in Statistics, within the Graduate School. The establishment of these two graduate degree programs marked a turning point in the history of the Department of Statistics. A few students were attracted to these programs, but at that time the emphasis was on teaching loads. The Statistics faculty found it difficult to have graduate students because of the conflict between the pressure to teach large service courses and their desire to offer graduate courses, which would be for only a few students. Thus, the graduate program did not flourish and in more than a decade, only one student received a Masters degree in Statistics through the Graduate School—Alfred Blumstein, now a professor at Carnegie-Mellon University and Executive Director of the famous "National Consortium on Violence Research," a multi-disciplinary, multi-institution effort. The graduate program in Statistics languished until 1959 when Dr. Norman C. Severo joined the Department and became Chairman.

Dr. Richard Schmidt joined the faculty full-time in 1947. Dr. Schmidt gave the first lecture on computers at UB on December 26, 1956. He said, "I lectured for an hour or more and afterward a couple of my friends said, 'You're crazy.'" At the time Dr. Schmidt did his data processing on a UNIVAC I. In 1963, Drs. Schmidt and Meyers wrote the widely acclaimed book *Electronic Business Data Processing*. To this day Dr. Schmidt serves the Department at the age of 94 as Professor Emeritus, and is the only faculty member to serve in every configuration of the department.

The Glorious 1960s

In 1961, Dr. Norman Severo recruited Dr. Willard H. Clatworthy and revived the graduate program. By 1962, despite the Department's harmonious relationship with the School of Business Administration, it became evident that a graduate program in statistics could not flourish because their need was for service courses involving the application of statistics to management. With the blessing and encouragement of the School of Business Administration, Drs. Willard Clatworthy, Gopal Mohanty, Richard Schmidt, and Norman Severo joined the Department of Mathematics in 1963 to form an autonomous Division of Statistics and to expand the statistical offerings within the graduate program.

A key event in the history of the department occurred relative to the formation of the State University of New York System. In September 1962, the "University of Buffalo" was selected by New York State to become the "State University of New York at Buffalo." The stated aim was to mold UB into a major graduate center. At the time the saying was that UB was to become the "Berkeley of the East." During this period "Big Budgets" became available and the Division had continued to flourish. By 1965 the members of the Division felt they must form a separate department outside of the Department of Mathematics if statistics was to be viable as a discipline. After considerable negotiations the Department of Mathematical Statistics was formed in the School of Arts and Sciences. At that time Dr. Mohanty left and Drs. Desu and Geisser joined the Department. Dr. Seymour Geisser became Chair, coming from the National Institutes of Health. One of the first things he did was change the name of the Department from "Mathematical Statistics" to "Statistics". In quick succession, Drs. Charles Mode, Sidney Addelman, Marvin Zelen, Peter Enis, James Dickey, Millu Rosenblatt-Roth, Marcello Pagano, Emanuel Parzen, Ross Prentice, Dave Sprott, William Schotz, and John Kalbfleish joined the Department.

By 1967, the faculty had grown to the size where it could offer an undergraduate program. Preparations were begun and a full BA program was implemented in the fall of 1968. Also in 1968 a PhD degree with a special emphasis on biometry was established. Essentially, this was UB's first biostatistics PhD degree. The first PhD in Statistics at UB was awarded to Mr. John Slivka in 1968.

However, even though there was tremendous growth in Statistics during this time Dr. Geisser was unhappy in Buffalo (Christenson and Johnson 2007) and in 1971 left to build a statistics group at the University of Minnesota.

The Tumultuous and Productive 1970s

From 1965 through 1972 the Department had its major growth, increasing from 4 faculty members to 13 faculty members. By 1971, an excellent undergraduate program was also under way. By the early 1970s the Department had obtained a high national ranking, generally reputed as being among the top 10. This department, among the highest ranked in the university, brought high honor to UB. In 1972 the Department awarded 8 PhD degrees, 27 MA degrees and 6 BA degrees. Some notable graduates from the 1970s included John Blessing (GOG Executive Director), Richard Kryscio (Chair of Biostatistics, University of Kentucky), and Joseph Newton (Dean of Science, Texas A&M).

In 1971, Dr. Emanuel Parzen became Chair of Statistics, arriving from Stanford University. At this time some important developments had occurred, particularly in the biostatistical arena. In 1972, Dr. Zelen established the first clinical cancer coordinating center in the country. He was also responsible for advising the Eastern Cooperative Oncology Group, the Radiotherapy Oncology Group, and the Veterans Administration Lung Cancer Group on all aspects of statistics as it affects clinical trials research. Despite the strong national reputation of the Department, it was clear however that all was not rosy. Below, we provide a taste of some of the events that led to a major historical split in the Department that occurred in 1973, which resonated throughout the statistical community.

In 1972, Drs. Jerome Cornfield, J. H. B. Kemperman, and Kenneth Magill filed a review of the graduate program of the Statistics Department (Cornfield et al. 1972) in which they noted that the "Department of Statistics is among the top dozen in the country. The quality of the present faculty is good to excellent." However, early signs of trouble were apparent as noted in the same report that "When we visited SUNY Buffalo on April 12 and 13, 1972 we expected to devote our time entirely to evaluating the training and research activities of the Department of Statistics.......However, we were very quickly made aware during our visit of a serious factionalism within the Faculty of the Department which in a major way interferes with the day to day business of the Department. As we have explained, the Department has in principle a great future. Moreover, the substantive disagreements which we have been able to identify seem relatively unimportant to an outsider." That said, the report then goes on to state that "Nevertheless, we judge the situation to be relatively serious. The Department is at present an unhappy one and the schism, even aside from raising doubts about the ability of the Department to achieve its future goals, is adversely affecting day to day performance." "We were unable to determine the fundamental basis of the disharmony. Some five faculty members appear to be in constant disagreement with the Chairman, Dr. Parzen." A review of several departmental memos appear to point to issues related to time on funded grants relative to teaching loads, particularly as it pertained to Dr. Rosenblatt-Roth. The reviewers continue on with respect to a suggestion of some sort of arbitration "Failing this step, or some other constructive step which has not occurred to us but may occur to others, the situation will probably deteriorate. The Department would tear itself apart.........In view of the great promise of the Department such an outcome would be little short of tragic." Parzen's response to Dean Hull (Parzen 1972) to the outside reviewers provides some insight as to his own frustrations, where he writes that "The Evaluators' Report is incorrect in implying that there was once a state of faculty harmony to which we can aspire to return. The Department has grown despite feelings of personal frustration expressed by a few faculty members occasionally since its founding in 1965 and frequently since 1969........I arrived in September 1970 to find without warning this unhappy situation."

By 1972, things had deteriorated so badly that the President of the University had to intervene. We see some very interesting points made in this memo from President Ketter to Provost Nancollas (Ketter 1972) as it pertains to the acrimony within the Department, where the President writes that "First, there is no question but that bad blood exists between Dr. Addelman and Dr. Zelen." This memo is followed up with a 1972 response memo from Provost Nancollas to President Ketter (Nancollas 1972) "I am very much aware of the problems concerning the Department of Statistics....As you probably know, many of these troubles arose long before Dr. Parzen became Chairman of the Department. They are accentuated at the present time since attempts are being made within the Department to alter the thrusts of the program in order to correspond with more modern developments in Statistics........I have been concerned with finding a solution to these problems for some time now and have met, both collectively and individually, with the statisticians. Many of us......are concerned with finding a solution to these problems." Some of the reflections about this situation from Dr. Parzen may be found in his interview by Dr. Newton (2002).

In 1972, Drs. Parzen and Zelen put forth a proposal (Parzen and Zelen 1972) that was very forward thinking relative to the future of statistics in that they expected (relative to the future direction of statistics). "(A) A large increase in almost all disciplines, especially those in social, biological, and physical sciences, and in medicine, education, law, management, and engineering. (B) An increased demand for statistical, computational, and analytical services and training at the Bachelors and Masters levels. (C) An increase in the amount of participation by statisticians in collaborative research." Within their proposal Parzen and Zelen developed one of the early formal definitions of *Statistical Science.* However, many of the faculty were in opposition to this shift in philosophy from a traditional statistics degree to more of an applied statistics degree, particularly as it pertained to the core set of courses. Another clear sticking point within the department was the relative teaching loads of the various faculty members who were funded on grants. In January 1973 the differences were not resolved and the proposal that Drs. Parzen and Zelen generated created a roadmap for a new division of statistics within the Department of Computer Science. There was roughly a 50/50 split of department resources (Nancollas 1973) "in order to provide an immediate solution to a very serious division which already exists in the Statistics Department."

In 1973, there were 13 tenured or tenure-track members in the Department of Statistics. Seven members, namely, Drs. Dickey, Kalbfleisch, Pagano, Parzen, Schmidt, Schotz, and Zelen, joined the Department of Computer Science in the Division of Statistical Science. Dr. Severo began a 6-year term as Chair of Statistics, where the remaining faculty of Drs. Addelman, Rosenblatt-Roth, Severo, Clatworthy, Desu, and Enis remained.

The beginning of the end of the Division of Statistical Science occurred around 1977, when Dr. Marvin Zelen, who had developed and led an outstanding biostatistics center in the Division, had a final conflict with the top administration over funding and cost-sharing. Although he liked living in Buffalo and strongly desired to remain here, he finally decided to take his entire biostatistics group to Harvard. This group of researchers was oftentimes referred to at Harvard as "Marvin's baseball team." This was a great loss to UB but soon a boon to Harvard, where Dr. Zelen continued to expand his group such that it became one of the world's great statistical centers. Some of the names from UB's biostatistics group that trained under Dr. Zelen and have gone on to have outstanding careers at other institutions included Drs. Steven Lagakos, Rich Gelber, David Schoenfeld, Ken Stanley, Jack Kalbfleisch, Ross Prentice, Stuart Pocock, and Colin Begg.

Dr. Parzen resigned from UB effective from May 22, 1978. By 1978, the faculty of the Division of Statistical Science within the Department of Computer Science had all left the university with the exception of Dr. Schmidt who rejoined the Department of Statistics. Thus began a long struggle for the Department in terms of their battles over funding. Even more tragic, in July 1979, Dr. Sidney Addelman, who had just assumed Chairmanship of the Department suddenly died. Dr. Peter Enis agreed to serve as Chair.

Under Attack in the 1980s

The Department never fully recovered from the historic 1972–1973 split, and subsequently came under attack twice in the 1980s as an easy target of the administration during tight fiscal times. As of 1982, seven tenure-track faculty made up the Department: Drs. Kuang-Fu Chen, Mahamunulu Desu, Peter Enis, Ralph Russo, Richard Schmidt, Norman Severo, and Ching-Ming Yeh. Dr. Schmidt recalled that while serving as Acting Chair on the Ides of March in 1982 that he was called to a meeting in which the Associate Vice-President for Academic Affairs told him, "We decided to close Statistics!" This was incomprehensible to the Faculty. The Department had a heavy workload, being the top department in student/faculty ratios. Statistics was the highest or second highest in most student evaluations. Thus, on the record of work effort and merit, the Department ranked very high. Apparently, the cause of the attack was the fact that the Department did not have *big grants*.

In 1983 a task group, which from the Department's perspective was rigged by the upper administration, developed a report (1983) in which the committee had the following charge: "What would be the impact with respect to the Department of Statistics: (1) Suspend admission to the graduate program in Statistics for the foreseeable future; (2) Maintain undergraduate degree programs and service instruction in statistics; and (3) Merge the Statistics faculty into the Department of Mathematics, or some other appropriate department?" Based on this charge the task force made the recommendation that they believe "….. that there is little justification for an expansion of the existing statistics department during a period of university-wide contraction……….the most reasonable action would be to merge the existing faculty either with another department such as mathematics or a larger unit such as a department/division of mathematical sciences. This option might be initially unsettling to faculty but, given the reality of diminished resources, would create opportunities for programmatic advancement not possible if continued independence were attempted." The attempt to close the Department was a great controversy for a couple of years, then things reverted to normal for a while.

In 1987 the Department was again under attack. Dr. Schmidt recalled that in February 1987, Dean Thomas George called a faculty meeting of the department and opened his remarks with three statements:

- I'm going to close Statistics,
- I'm going to do so in a week, and
- The probability is a small epsilon above zero that I will fail.

State University of New York at Buffalo

Reporter

April 23, 1987 Volume 18, No. 27

A solution for Statistics?

The department's master's program shouldn't be deactivated, but the Ph.D. program should be, Vice Provost Rennie will recommend to the provost.

"My main aim was to deactivate the Ph.D. program," Dean Thomas George said. He indicated it's fine with him if the department wants to accept master's students.

It's up to Statistics Chair Peter Enis to decide whether he will accept students into the program. The master's would then be the terminal degree in that department.

April 23, 1997 UB Reporter

In April 1987, Dean George modified his stance and made the following statement in the *UB Reporter* that (Stofko 1987) "My main aim was to deactivate the PhD program" and that "He explained that he had come around to this view the more the issue was discussed." There was also a proposal to merge the Department with the Department of Mathematics and the Department of Computer Science. Ultimately, the Department, and in particular the Chair, Peter Enis, fought a public battle in the school paper and rallied the support of the faculty senate. Dr. Enis should be given a great deal of the credit for keeping the Department afloat during these times. In 1987, President Sample created a commission to review the matter, which is now referred to as the "Triggle Report." Two telling recommendations in the 1988 Triggle report (Giese et al. 1988), that would dictate the next chapter in the Department's history, were: (1) "That two lines be filled in Statistics. One appointment will be the Chair jointly appointed with Health Sciences. This will be crucial to the effective implementation of all further recommendations" and (2) "That a biostatistics unit be established within the faculty of Health Sciences as a component of a Department of Statistics and that this unit be headed by the jointly appointed Department Chair." In addition, the report went on to state that "We recommend that adjunct appointments be made between the Department of Statistics and Departments in other Faculties and Institutions (including Roswell Park)."

It was clear from the *Triggle Report* that there was the overall theme toward transforming what was a traditional mathematical statistics department into a biostatistics department. As it turns out, the plan was put into motion to move the Department to the School of Medicine and Biomedical Sciences and to maintain the graduate program. This move was not without its critics. The arguments, both pros and cons were presented through a series of articles in the *UB Reporter*. Faculty Senator Ralston noted (Snyderman 1988) that "there was one merit to the proposed move. 'The only advantage of the move for the department would be to remove it from the control of a dean (Thomas George) who has treated Statistics like dirt and has shown his contempt for them because they did not have grants'.......He added that 'the resources available to the department in the Faculty of Natural Sciences and Mathematics had been systematically withheld to sap the department's strength'." Ultimately, the upper administration usurped the Faculty Senate and the Department moved to the School of Medicine and Biomedical Sciences in January 1990.

The 1990s: Mirage of Resources in the Medical School

After all of the dust settled from the battles of the 1980s, a fresh carpet was laid and fresh paint applied to a new office suite for the Department of Statistics in the School of Medicine and Biomedical Sciences. Although in some sense the Department was forced to move to medicine, there was a glimmer of hope that Statistics would finally be provided the appropriate resources to maintain its graduate and undergraduate programs, as well as serve as a biostatistical consulting hub for the health sciences. The ultimate goal of the administration was to transform a traditional mathematical statistical department to a biostatistics department.

In 1993, the Department hired Dr. Irwin Guttman as Chair. Dr. Guttman had a stellar reputation as a mathematical statistics researcher and was promised several new lines as part of the Department's original move to medicine. However, those promises were quickly reneged upon. The Department was forced to operate with roughly five lines through the first half of the 1990s, while maintaining an undergraduate and graduate program, carrying out substantial service teaching, and providing statistical consultation to the university as a whole.

In 1996, problems began to surface as two biostatisticians who focused primarily on grant collaborations within the Department of Social and Preventive Medicine left. At this point it appeared publicly that a higher priority was placed on collaborative grant activity by the administration in terms of statistical support rather than that of the education and research activities of the Department. The process of what to do with Statistics played out behind closed doors, and at the urging of the Chair of Social and Preventive Medicine, Maurizio Trevisan, Dean Naughton of Medicine and Provost Headrick, it was decided that the only viable solution would be to absorb the faculty with biostatistics expertise into the Department of Social and Preventive Medicine. The remaining faculty members would be moved to the Department of Mathematics. Ultimately, after all of the dust settled in 1999, only one faculty member and one lecturer remained in Social and Preventive Medicine from the original Department of Statistics.

What was not public knowledge at the time was that there was a move to reinvigorate the statistics group as what now might be referred to as a traditional biostatistics department. Behind the scenes there was also a nascent move to develop a school of public health at UB and it was the administration's view that the statistics department did not grow into a biostatistics department as originally anticipated with the move to Medicine. In 1998, a working group on the plight of statistics (Triggle et al. 1998) noted that, "There is no effective leadership in the department for change, nor any that can obviously be recruited to lead and implement new structures or directions. The period of uncertainty in the past months has not helped the climate for regrouping or recruitment" and that "Finally, the proposal to incorporate statistics into Social and Preventive Medicine (SPM) was badly handled originally to the extent that it was not seen or announced as a process by which a biostatistics unit of integrity was to be maintained within SPM; rather it was seen, rightly or wrongly, as a process by which lines were simply to be transferred to SPM for use as it saw fit." The committee had three basic recommendations, with the basic plan to rebuild the department, and "to not reappoint the current chair". Ultimately, the administration squeezed out the current faculty members who were more traditional mathematical statisticians through attrition, transfers, and retirements.

Even though in some ways history bears out that the administration's plan was successful, their approach to this process left much to be desired with respect to the treatment of faculty and students. As Irwin Guttman noted in 1997 (Triggle et al. 1998) that "Dean Naughton (medicine) once remarked to me that Statistics was not neglected, it was abused." At a December 8, 1998 Faculty Senate meeting, for which the merger was discussed, Professor George noted that "the issue of Statistics has been a twenty-five year history of colossal mismanagement….." The Department, and particularly the Chair, did not go down without a public fight with several well-publicized editorials. Ultimately, as in the past two decades the administration did as it pleased, and in 1998 the department was decommissioned and its graduate and undergraduate admissions were suspended. Dr. Irwin Guttman joined the Department of Mathematics until retirement, while the remaining faculty members left for other institutions or moved to the Department of Social and Preventive Medicine. One graduate PhD student, Yachuan Huang, remained to complete his studies. He ultimately graduated in 2004 with the last degree in Statistics awarded at UB.

However, there was still hope on the horizon. In December 1998, Drs. Kiang Liu, Thomas Louis, and Robert Woolson visited UB at the request of the Chair of Social and Preventive Medicine, Maurizio Trevisan, and generated an important report (Liu et al. 1999) regarding the establishment of a formal Division of Biostatistics with the Department of Social and Preventive Medicine. A key point that was part of this report was that "SUNY Buffalo faculty and administrators recognize the need for re-creating a (bio)statistical presence on campus." What was unclear from the report at the time was whether or not there would ever be a PhD program in statistics or biostatistics at UB in the immediate future. In 1999, a telling letter (Holm 1999) from the Senior Associate Dean of Medicine and Biomedical Sciences, Dr. Bruce Holm, to Provost Triggle indicated progress was being made and that there was glimmer of hope for a rebirth of the department. In this letter Dr. Holm noted that "a new initiative is progressing to recruit a Division Chief of Biostatistics and six new faculty members in addition to four new faculty members at Roswell Park Cancer Institute." It took 3 years to fill this position. Ultimately the decisions of 1998 debilitated the department for about 6 years. At the end of the 1990s one could argue that Statistics at UB was at its lowest point in its long-standing history.

Current: Rebirth of a Department

In 2002, Dr. Alan Hutson joined UB from the University of Florida as Chief of the Division of Biostatistics within the Department of Social and Preventive Medicine. At the time that he arrived there was one faculty member and one lecturer left in the Division from the former Department of Statistics and no active graduate program. He was promised the ability to hire several new faculty members with expertise in biostatistics and encouraged to investigate developing a graduate program in biostatistics. In that same year, he began to develop close ties with the Roswell Park Cancer Institute Department of Biostatistics, chaired by Dr. Jim

Kepner, as well as the Gynecological Oncology Group (GOG) Statistical and Data Center, and its Executive Director, John Blessing.

Nancy Barczykowski, Dick Schmidt, and Alan Hutson

A key development in the history of the Department was that a new School of Public Health and Health Professions was formed in 2003 at UB through the merger of the Department of Social and Preventive Medicine and the School of Health Professions. The Division of Biostatistics was moved into the new school as part of this merger. With the support of the upper administration, the Department of Biostatistics was officially created at UB on 25 February 2003. A critical agenda item for Dr. Hutson was to re-establish a graduate presence at UB for the newly formed Department. A key to moving this process forward was the recruitment of Dr. Randolph Carter from the University of Florida. Drs. Carter and Hutson were the primary architects of new MA and PhD programs in Biostatistics. The actual degree programs were a merger of the old MA and PhD programs in Statistics with a newer set of biostatistical courses integrated into the program. On 25 January 2004, an outside review committee of Drs. Ralph D'Agostino, Vernon Chinchilli, and John Klein reviewed the Department's proposal to reinstate the graduate program and in a written report gave the program their blessing. On 12 March 2004 the graduate MA and PhD program was officially reactivated. Also during this time an M.P.H. degree with an Emphasis in Biostatistics was established.

Department of Statistics 1969. *Top Row* Sidney Addelman, Irwin Bross, Willard Clatworthy, Seymor Geisser and Manavala Desu *Bottom Row* Charles Mode, Robert Roine, Richard Schmidt, Norman Severo, and Marvin Zelen

In 2005, Dr. Hutson agreed to chair both the UB and Roswell Park Cancer Institute Department of Biostatistics, such that the Department of Biostatistics essentially now spans two institutions. Also, in 2005 Dr. Randy Carter took over as Director of the Population Health Observatory, a key research center in the School, which has a mission very similar to the original Bureau of Business and Social Research.

By 2006, things were moving forward nicely for the Department. The first four MA degrees in Biostatistics were awarded, and the faculty numbered 10 tenure or tenure-track faculty members, its highest level of faculty support since 1972. In July 2006, Dr. Richard Schmidt, who was originally Professor, Department of Statistics and Insurance, and served in every iteration of the Department to date, celebrated his ninetieth birthday. The Department now celebrates Dr. Schmidt's long-standing service with the annual *Richard Schmidt Award for Outstanding PhD Student*. In addition, the Department established the *Sidney Addelman Award for Outstanding M.A. Student*. In 2007, the first graduate PhD degree in Biostatistics was awarded to Antara Majumbdar. One highlight of that day is that Dr. C. R. Rao served as an honorary dissertation committee member. The Department has been in continual motion since 2002. In 2009–2010 there were 55 graduate students in our program and we taught well over 1000 undergraduates through our service teaching. Recently, we have developed a Minor in Statistics with the first new students enrolling in Fall 2010. We have also established a new MS degree program in Bioinformatics and Biometry that started in Fall 2010.

Departmental Chairs

From 1930 to 1963, the Department was in the School of Business Administration. In 1963, it became an autonomous division in the Department of Mathematics. From 1965 to 1990 it was in the Faculty of Natural Sciences and Mathematics. From 1990 to 1998 the Department was an integral part of the School of Medicine and Biomedical Sciences. From 1998 to 2003 the Department was reconfigured as the Division of Biostatistics. In 2003 the Division became the Department of Biostatistics, which currently resides in the School of Public Health and Health Professions. The chairs and their dates of service are listed below.

1930–1936	Dr. Robert R. Riegel
1936–1947	Dr. Martin A. Brumbaugh
1947–1948	Dr. John Smith
1948–1949	Dr. Robert R. Berner
1949–1955	Dr. Zenon Szatrowski
1955–1959	Dr. Richard N. Schmidt
1959–1963	Dr. Norman C. Severo
1963–1965	Dr. Willard Clatworthy
1965–1971	Dr. Seymour Geisser
1971–1973	Dr. Emanuel Parzen (Head of Statistical Science Division 1973–1978)
1973–1979	Dr. Norman C. Severo
1979–1979	Dr. Sidney Addelman (Died shortly after assuming Chair duties)
1979–1987	Dr. Peter Enis
1987–1991	Administrative supervision
1991–1993	Dr. Peter Enis
1993–1998	Dr. Irwin Guttman
1998–2002	Administrative supervision
2002–present	Dr. Alan Hutson

Top Five Most Cited Articles by UB Statistical Scientists Since 1971 (first author, rank ordered)

Pocock, S. J., Simon, R. (1975) Sequential Treatment Assignment with Balancing for Prognostic Factors in Controlled Clinical Trial. *Biometrics*, **31** 103–115.
Zelen, M. (1973) A Keynote Address on Biostatistics and Data Retrieval. Cancer Chemotherapy Reports Part 3, **4** 31–42.
Kalbfleisch, J. D., Prentice, R. L. (1973) Marginal Likelihoods Based on Cox's Regression and Life Model. *Biometrika*, **60** 267–278.
Parzen, E. (1974). Some Recent Advances in Time Series Modeling. *IEEE Transactions on Automatic Control*, **AC19** 723–730.
Prentice, R. L. (1973). Exponential Survivals with Censoring and Explanatory Variables. *Biometrika,* **60** 279–288.

Acknowledgments I would especially like to thank Professor Richard Schmidt for his careful proofreading of this document. I would also like to thank University at Buffalo Archives Director John Eden and Staff Member William Offhuas for their expert guidance, Terry Sikorski for providing secretarial assistance, and for the time and advice provided by the two external reviewers.

References

ASQC Website (2010) http://www.asq.org/about-asq/who-we-are/bio_brumbaug.html

Christenson R, Johnson W (2007) A conversation with Seymour Geisser. Stat Sci 22:621–636

Cornfield J, Kemperman JHB, Magill K (1972) Report of the evaluation and review committee for the graduate degree program in statistics, State University of New York, June, 1972, University at Buffalo, University Archives, Buffalo

"Drug Store Has 75 Chances In 100 of Staying In Business A Year; Grocery Has 40 Chances In 100," Retail Ledger (1929)

Giese RF, Graham S, Griffith DA, Nichols RC, Priore R, Tedesco L, Ratchford BT, Triggle DJ (chair) (1988) Report to provost Greiner regarding the future role of statistics teaching and research programs on this campus, April 1, University at Buffalo, University Archives, Buffalo

Holm BA (1999) Letter to Provost Triggle regarding faculty recruits in biostatistics at UB, May 17, Department of Biostatistics records

Ketter RL (1972) Memo from the president to Provost Nancollas regarding bad blood in statistics, November 27, University at Buffalo, University Archives, Buffalo

Liu K, Louis TA, Woolson R (1999) Findings and recommendations: regarding establishing the Division of Biostatistics, Department of Social and Preventive Medicine, School of Medicine and Biomedical Sciences, State University of New York at Buffalo, Department of Social and Preventive Medicine records

Lockhart OC (1928) Report of the Director of the Bureau of Business and Social Research to Dean Marsh, 14 November 1928, University at Buffalo, University Archives, Buffalo

Nancollas G (1972) Memo to President Ketter on problems concerning the Department of Statistics, December 5, University at Buffalo, University Archives, Buffalo

Nancollas GH (1973) Memo to Vice President of Academic Affairs, Dr. Bernard Gelbaum pertaining to the program in statistical science, January 29, University at Buffalo, University Archives, Buffalo

Newton HJ (2002) A conversation with Emmanuel Parzen. Stat Sci 17:357–378

Parzen E (1972) Memo to Dean Hull of the Graduate School in response to the report of the Statistics Graduate Program Evaluation Committee, July 10, 1972, University at Buffalo, University Archives, Buffalo

Parzen E, Zelen M (1972) A proposal for a section of statistical science in the Department of Statistics, July 6, University at Buffalo, University Archives, Buffalo

Report of the task group on statistics and mathematical sciences, January, 15, 1983, University at Buffalo, University Archives, Buffalo

Stofko C (1987) A solution for statistics. UB Reporter 18(27):2

Snyderman DM (1988) Faculty vote to help decide fate of Statistics Department. UB Reporter 21:4

Triggle DJ, Holm B, Levy K, Goodman N, Nickerson P, Thompson M (1998) Memo to Provost Headrick: statistics—a recommendation, May 11, Department of Social and Preventive Medicine records

Texas A&M Department of Statistics

Simon J. Sheather and Jennifer South

1962: Graduate Institute of Statistics Formed

H. O. Hartley

The Department of Statistics was formed in 1962 as the Graduate Institute of Statistics with the mandate of providing statistical research, consulting, and instruction for Texas A&M University. The Institute was authorized from its inception to grant MS and Ph.D. degrees. Prior to 1962, a number of different departments provided the few statistics courses taught at the undergraduate and graduate level. Herman Otto Hartley arrived in the fall of 1963, and was the Institute's first Director. Mrs. Grace Hartley helped establish a friendly yet proper

Texas A&M University,
Department of Statistics, 3143 TAMU,
College Station, TX 77843-3143, USA
S. J. Sheather (✉) · J. South
e-mail: sheather@stat.tamu.edu

J. South
e-mail: jennifer@stat.tamu.edu

A. Agresti and X.-L. Meng (eds.), *Strength in Numbers: The Rising of Academic Statistics Departments in the U.S.*, DOI: 10.1007/978-1-4614-3649-2_22,

(British inspired) social atmosphere among the graduate students with diverse backgrounds. Her backyard was like a rose garden, where many breakfasts were sponsored for faculty and students. A Stat Wives Social group was organized. The Hartleys also bought their 100 acre "Ranch" in Snook to become true Texans.

Dr. Hartley was short in stature, and in giving lectures he would often use an audience icebreaker by asking "Can you hear me? Can you see me?" Many distinguished statisticians came to visit Texas A&M during Hartley's tenure. One was Egon Pearson, a slight man of "considerable height", who had numerous honors including the title of Commander of the British Empire. He had worked with Hartley on many projects including some pioneering work on using the range statistic. H. O. Hartley, standing side by side with this colleague who towered above him, introduced him by noting "Never were there two more appropriate statisticians to work on the concept of range statistics".

Ralph Kodell would often entertain at stat socials by mimicking Dr. Hartley, much to Hartley's delight. Ralph could copy Hartley's mannerisms and British accent near perfectly, though he could not mimic his height, as Ralph is rather tall. Under Hartley's leadership, the department engaged in considerable interdisciplinary work, with funding from agriculture and numerous joint projects with other disciplines.

Also during this time, the department (then the Institute of Statistics) had an NSF 'start-up' grant that provided funds to regularly have distinguished guest speakers. William Cochran (Harvard) gave one of those week-long series of lectures. He was the only statistician of the U.S. Surgeon General's panel of 10 experts that affirmed the link between cigarette smoking and lung cancer. During the Q&A after one of those lectures he stopped and lit a cigarette (yes, you could smoke in A&M buildings then). To the puzzled audience that had just heard him discuss the connection of smoking with cancer, he commented that "habits trump science."

By the fall of 1964 the department consisted of a faculty of five; R. J. Freund (1962–1991), H. O. Hartley (1963–1976), R. R. Hocking (1963–1970), E. Dayhoff (1963–1964), and G. Darroch (1964–1975) and hosted twelve graduate students. In addition, this same year, Lee H. Smith was awarded the first Ph.D. degree in the Institute of Statistics. From the mid 1960s to mid 1970s, there were many prominent faculty that joined the department, including Omer Jenkins (1965–1997), Larry Ringer (1965–2003), Charles Gates (1966–1992), William B. Smith (1966–2003), James Matis (1973–2007), and Thomas Wehrly (1976–).

Freund Hocking Darroch Jenkins Ringer Gates Smith Wehrly

The department was first located in the Animal Industries Building. The building had a slaughtering room, from which the squeals of swine were often heard. The department soon moved to the Olin E. Teague Building in 1966. The construction of this building was greatly assisted by an NSF Center for Excellence grant obtained by Professor Hartley and others. The university computing center also resided in the Teague Building. The rapid expansion of the university during the 1970s and subsequent demands for space by the computing center resulted in the department moving to its current location on the fourth floor of the John R. Blocker Building in 1981. In the fall of 2004, the department acquired additional office space on the fifth floor of the Blocker Building. The Graduate Institute of Statistics was included in 1966 as a member of the newly formed College of Science and in 1984 acquired its current name, the Department of Statistics.

1977: William B. Smith Succeeds H. O. Hartley as Department Head

In 1977, Professor Hartley retired and was succeeded by William B. Smith. During Dr. Smiths's tenure, Emanuel Parzen (1978–2009) joined the faculty as Distinguished Professor, and Michael Longnecker (1977–), H. Joseph Newton (1978–), P. Fred Dahm (1979–), Jeffrey Hart (1982–), and Daren Cline (1984–) were all appointed Assistant Professors. All of them, with the exception of Emanuel Parzen who retired in 2009, are still currently working in the department. In addition, R. R. Hocking returned to the department as Professor. Although other faculty appointments were made, those individuals have moved to positions of leadership at other institutions, including government and industry. Under Bill Smith's 9 years of leadership, great changes were made. These changes include the Graduate Institute of Statistics becoming the Department of Statistics, the change in location from the Teague Building to vastly expanded offices (classrooms, computing, and laboratory space) in the John R. Blocker Building, an increase in undergraduate enrollment in statistics classes from a few hundred to about 5,000 students per year, a rise in graduate majors from 30 to about 65, an expansion of research funding (including TAES and TEES funding) to include most faculty members, and lastly, creation of the department computer network with industry funding. By the early 1980s the department had grown to 18 faculty members with 40–50 graduate students. Larry Ringer served as interim department head in 1986/1987, after William Smith stepped down from the position.

Parzen

Longnecker

Newton

Dahm

Hart

Cline

1987: Raymond J. Carroll Appointed as Department Head

After a national search, Raymond J. Carroll was appointed Department Head (1987–1990). During Raymond Carroll's tenure, a period of intense faculty recruiting occurred, resulting in the tenured and tenure-track faculty increasing to 22. Cliff Spiegelman (1987–) was one of these notable hires. In addition, during this period, a change was made from 12- to 9-month state funded faculty salaries and Dr. Carroll instituted teaching loads that are still used today. Randall Eubank (1988–2005) and Suojin Wang (1990–) were hired.

Spiegelman

Eubank

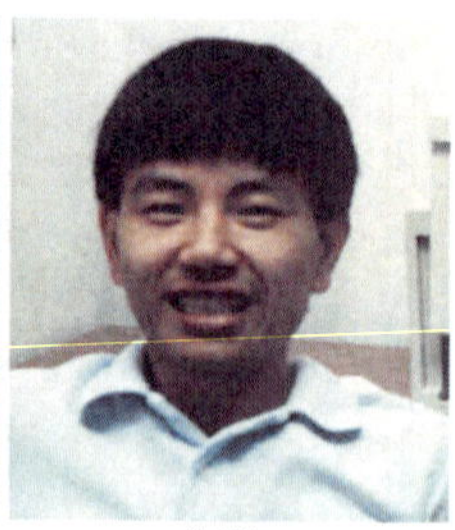
S. Wang

1990: H. Joseph Newton Succeeds R. J. Carroll as Department Head

R. J. Carroll (1987–) was succeeded by H. Joseph Newton in 1990. The faculty by this time had grown to 25 members with 60–65 graduate students. Along with the size increase of the faculty and graduate student population, was a concurrent increase in research productivity and funded research. Another indicator of the university's high regard for the department was the designation of Manny Parzen and Raymond J. Carroll as Distinguished Professors. In addition, Dr. Newton had many more contributions to the department, including the publication of the first ever *StatLinks* "News from the Department of Statistics at Texas A&M University to link its Friends and Former Students", the formation of the Herman and Grace Hartley Endowed Fellowship (1994), as well as the Emanuel and Carol Parzen Prize for Statistical Innovation (1994). In the fall of 1992, the department celebrated a thirtieth anniversary with a 2-day technical conference and evening festivities in which more than 100 current and former students, staff, and faculty participated. Also this same year, James Calvin (1991–) and Naisyin Wang (1992–2009) were hired. Long time faculty members and friends retired: Rudi Freund (1991), Chuck Gates (1992), and Ronald R. Hocking (1993). In 1994, to fill the job positions from the previously retired, Dr. Newton appointed F. Michael Speed as Associate Professor with tenure (1994–) and Michael Sherman as Assistant Professor (1994–). This same year in 1995, the TAMU Department of Statistics was ranked as thirteenth best U.S. University in Statistics by the National Research Council. Prior to this ranking, the department was ranked twenty-fourth in 1982. After serving two terms as Department Head (1990–1997), H. Joseph Newton went on to become Dean of the College of Science, which is the position he still currently holds.

  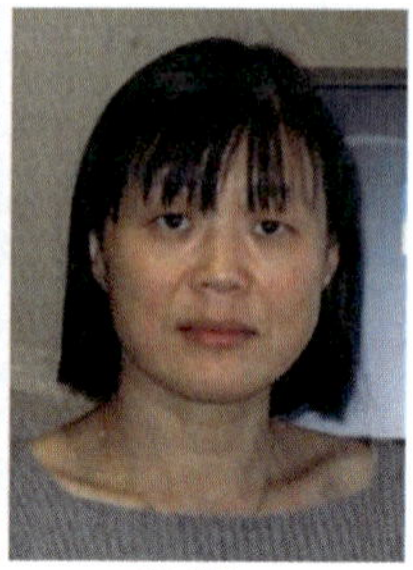

Speed Sherman N. Wang

1998: James Calvin Appointed as Department Head

In 1998, James Calvin was appointed to head the department, which at that time consisted of 26 faculty members with 60–65 graduate students. During this same year, Bani Mallick and Marina Vannucci were hired and Omer Jenkins retired. In 1999, the Statistics Department Former Student Fellowship fund was formed. This fund is still providing fellowships to outstanding students wishing to pursue a Ph.D. in statistics. In 2000, Larry Ringer stepped down as Assistant Department Head and Michael Longnecker filled this position. 2001 proved to be a successful recruiting year with the hires of Assistant Professors Willa Chen and Ruzong Fan. In October 2002, the department celebrated its fortieth anniversary with a 2-day conference entitled "Frontiers in Statistical Research". The conference reflected the wide diversity in research interests of its faculty. The topics of the sessions were mixed linear models, generalized linear models, bioinformatics, time series, and smoothing. Also in 2002, Faming Liang was hired as an Assistant Professor from the National University of Singapore. In 2003 Bill Smith retired after 35 years of service to Texas A&M University and shortly thereafter, Larry Ringer also retired (2004). Bill Smith moved on to serve as Executive Director of the American Statistical Association. In 2004–2005, an additional six faculty joined our team, including David Dahl (Assistant, 2004–), Marc Genton (Associate, 2004–), Erning Li (Assistant, 2004–2011), Yanyuan Ma (Assistant, 2004–), Samiran Sinha (Assistant, 2004–), and Jianhua Huang (Associate, 2005–). Dr. James Calvin stepped down in 2004 as

Department Head of Statistics to accept the position of Executive Associate Vice President for Research. Michael Longnecker became Interim Head until the appointment of Simon J. Sheather in March, 2005.

Mallick Vannucci Chen Fan Liang Dahl

Genton Li Ma Sinha Huang

2005: Simon J. Sheather Appointed as Department Head

With the appointment of Simon Sheather as Department Head in March 2005, the department has had numerous innovative changes. Under Dr. Sheather's leadership, an annual faculty retreat, staff retreat, and new graduate student conference were instituted. An Alumni Advisory Board was formed in 2008, the Statistics Graduate Student Association (SGSA) was formed in 2009, in 2010 an Industrial Affiliates Program was formed. An external company, Texas A&M Statistical Services LP, was established in February 2012 to provide training and expertise in statistics, business analytics, and business process improvement, to companies across the US. These services are to be delivered by faculty contracted from the Department of Statistics at Texas A&M University, and other sub-contractors. In the area of development, in 2008 the department received a planned gift from a former student valued at 1.5 million. In 2009, the Raymond J. Carroll Young Investigator Award was established and in 2010, the Anant M. Kshirsagar

Endowed Fellowship Award, as well as the Margaret Sheather Memorial Award were established. Simon Sheather has made many changes in the way of marketing the department. This includes updating the departmental logo, completely revamping the departmental newsletter into a full color magazine and launching a new department website in 2009.

The Texas A&M University System Board of Regents approved the Master of Science degree in Statistics for distance delivery on September 22, 2006. Given below is an excerpt from President Robert M. Gates' presentation to the Texas A&M University System Board of Regents on that day:

> Texas A&M University is requesting authority to offer our existing Master of Science degree in Statistics via Distance Education. The Texas A&M University Department of Statistics has ... been putting together a master's degree in statistics that is extraordinary in its quality and for the way it uses technology to enhance student learning.
>
> What makes these statistics courses far superior to the old-fashioned talking heads video lectures that many of us think about when we hear the term distance education is that local students and distant students are simultaneously enrolled in the exact same classes. The lectures are live presentations of information for the local students in the class, but all lecture information is projected from a computer onto a screen rather than using a blackboard or other external media. Lecture notes, impromptu drawings and remarks, video demonstrations—all of these are done within an integrated system. The result? What the student at a distance gets (and the local student too) is not a video of a professor at the front of a classroom with an unreadable blackboard in the background. Instead, the student gets replayable video and audio files that are focused on what is most important for the student to be paying attention to at a given instant in the class. For example, if a student in the class asks a question during the lecture, and the teacher answers the question by highlighting something on a graph or showing a demonstration on screen, a student who is watching a replay of the class at a distance will be able to listen to the dialogue and simultaneously see the highlighting on the graph or the on-screen demonstration just as clearly as the local student does. Through the way they record their classes, our statistics faculty are doing some remarkable things to capture the essence of the classroom experience. The second aspect of these statistics classes that is impressive is the way statistics supplements classes through its integrated on-line system. Specifically, they provide all students in these classes with an on-line bulletin board where a student can post questions related to the lecture and homework and get a posted on-line response from a faculty member or teaching assistant within 24 hours. This is just one of a number of technology enhanced learning activities that are done in these courses to make sure that students get the support they need to learn the material of the course.

The department's first cohort of students in the Distance Learning MS in Applied Statistics started in fall 2008. In 2008, the Transcripted Certificate of Applied Statistics was introduced, while a joint Texas A&M University/SAS certificate was introduced in 2009. As of December 2011, there are over 400 students enrolled in the distance program and 27 professionals have completed their MS degree in Statistics. The distance program is expected to produce just over one million dollars in revenue in the academic year 2011/2012.

Many faculty were hired under Dr. Sheather's leadership including, Mikyoung Jun (Assistant Professor, 2005–), Alan Dabney (Assistant Professor, 2006–), Ursula Müller-Harknett (Assistant Professor, 2006–), Suhasini Subba Rao (Assistant Professor, 2006–) and most recently Huiyan Sang (Assistant Professor,

2008–), Lan Zhou (Assistant Professor, 2008–), and Jamis Perrett (Assistant Professor, 2008–2013). Notable senior hires included Webster West (2006–), Soumendra Lahiri (2007–) and Mohsen Pourahmadi (2008–). The department celebrated the retirements of both friends and colleagues, Drs. James Matis (2007) and Emanuel Parzen (2009). In 2010, U.S. News & World Report ranked Texas A&M University tied for twelfth among all U.S. Departments of Statistics and Biostatistics, in terms of the quality of its graduate program.

Jun Dabney Müller-Harknett SubbaRao Sang Zhao Perrett

West Lahiri Pourahmadi

From its inception 43 years ago, the department has grown to a faculty with 34 tenure/tenure-track positions, 5 lecturers, and nearly 100 graduate students. The department has awarded close to 500 MS and nearly 300 Ph.D. degrees.

Sponsored Research

The department from its inception has been substantially involved in collaborative research. Professor Hartley and the faculty had research grants and contracts from ONR, NASA, Army Research Office, and National Center for Toxicological Research. For over 20 years funding for statistical consulting was provided by the Texas Agricultural Experiment Station. The department was actively involved in the development of the Center for Environmental and Rural Health, an NIEHS funded research center. Also, several faculty members greatly assisted in obtaining the competitive renewal of Texas A&M University's EPA funded Superfund Basic Research Program. In 2001, Professor Carroll received a grant from NCI to establish a 2-year training program in Bioinformatics and Nutrition. This grant has been renewed through 2011. In 2006, Bani Mallick received approval and funding for a Bayesian Bioinformatics Lab. This laboratory is still currently operating and focuses on research and development of Bayesian modeling and computational tools for the rapidly emerging fields of Bioinformatics, a discipline which deals with the analysis of data generated by high-throughput techniques in genomics, molecular, and cellular biology. In particular, the laboratory has extensive publications in analysis of gene expression data, proteomic data, and protein structure determination. It has been funded by National Cancer Institute (NCI). In fall 2009, Texas A&M University was selected as one of four universities to receive a $25 million grant from the King Abdullah University of Science and Technology (KAUST). Faculty members James A. Calvin and Raymond J. Carroll have been selected to lead KAUST's Global Research Partnership in an effort to aid in support of a research center for applied mathematics and computational science, which was one of four priority areas identified by KAUST. As such, since 2009, the IAMCS or Institute for Applied Mathematics and Computational Science was established within the Department of Statistics. Also in 2009, Marc Genton started a Program in Spatial Statistics (PSS). This program focuses on teaching, research, and interdisciplinary collaboration in spatial statistics. It was formed by members of the Department of Statistics at Texas A&M University, as well as affiliates from other departments.

Distinguished Lecture Series

The department has established three major lecture series. The H. O. Hartley Memorial Lecture series was established in 1988 to honor the memory of Herman Otto Hartley. The speakers (1988–2010) were Peter J. Diggle, Bradley Efron, E. J. Hannan, Sir David R. Cox, Wayne Fuller, Adrian Raftery, Peter Hall, Terry Speed, James Berger, Edward George, and Regina Liu. The Parzen Prize for Statistical Innovation was established in 1994. The awardees (1994–2010) were Grace Wahba, Donald P. Rubin, Bradley Efron, C. R. Rao, David R. Brillinger, Jerome H. Friedman, Alan E. Gelfand, Nancy Reid, Marvin Zelen, and Roger Koenker. The third lectureship, the Ronald R. Hocking Lecture Series, was established in 2002 to recognize exceptional contributions to the field of linear models and their generalizations. The speakers have been Ronald Hocking in 2002, David Harville in 2003, Ronald Hocking, Dallas Johnson, Ramon Littell and former student, Michael Kutner in 2007, Brian Marx, Tim Hesterberg, and Oliver Schabenberger in 2009. All three of these lecture series have provided our graduate students with the opportunity to interact with pioneers in the statistics profession.

Teaching

Although the department does not offer an undergraduate degree in statistics, undergraduates can obtain a concentration in statistics. The department developed, in conjunction with the mathematics department, a statistics option within the BS degree in Applied Mathematical Sciences. An option for non-math majors is to obtain a minor in statistics. The department's undergraduate course offering has grown to over ten courses per semester with over 4,500 undergraduates enrolled per academic year. Besides the courses offered to its MS and Ph.D. students, the graduate service courses have an enrollment of nearly 1,000 per academic year.

From 1996 to 2004, James Matis organized a Statistics Advanced Placement Summer Institute on campus for high school mathematics teachers. This institute provided an opportunity for the department to interact with AP statistics teachers. It is our hope that the message about the career potential for students in statistics is conveyed by these exceptional teachers to their students. The AP Summer Institute continues to this day with Jamis Perrett as the current director.

Notable Alumni

Lee Smith '1964

Dr. Lee Smith received the very first Ph.D. degree awarded in the Institute of Statistics in 1964 under the direction of H. O. Hartley. At age 39, Dr. Smith became one of the youngest presidents of a major university in the U.S. when he was appointed president at Southwest Texas State University (now Texas State University) in 1974. He remains a loyal former student of Texas A&M, having given the keynote address at the Aggie Muster in 1979 and later, with his wife Eva, having established an Endowed Scholarship for graduate students in the Statistics Department. Dr. Smith is also very proud of being inducted into the Academy of Distinguished Former Students in the College of Science as well as receiving the H. O. Hartley Award in the Statistics Department. The H. O. Hartley Award is given annually to a former student of the Department of Statistics at Texas A&M University for distinguished service to the discipline of statistics.

Michael Kutner '1971

Dr. Michael Kutner received his Ph.D. in 1971 under the direction of Professor Ronald R. Hocking. He is the former Director of Biostatistics, Epidemiology, and

Research Design for the Atlanta Clinical and Translational Science Institute and is currently the Biostatistics Core Director for the Center for Aids Research. Dr. Kutner is a Fellow of the American Statistical Association (ASA) and is the author of approximately 135 research articles in refereed journals and coauthor of a widely used statistics textbook. In 1996, he was one of four statisticians to receive the ASA's Founder's Award, an award that recognizes members who have rendered distinguished service to the Association and to the discipline. In 1997 he was inducted into the College of Science Academy of Distinguished Graduates which is an honor bestowed upon a former student "who has brought honor to their profession in mathematics, the sciences, and medicine." Most recently Dr. Kutner was the recipient of the 2011 Charles R. Hatcher, Jr. MD Award from Emory's Woodruff Health Sciences Center which is awarded to those who, through their lifetime of work, exemplifies excellence in public health.

Ersen Arseven '1974

Ersen Arseven received his Ph.D. from Texas A&M University in 1974 under the direction of Professor Anant Kshirsagar. Dr. Arseven then began his career as a statistician with American Cyanamid Corporation. From 1984 to 1992, he worked in the statistics group at Boehringer Ingelheim Pharmaceuticals. By 1990, he had risen to Group Director for U.S. Operations in the Scientific Affairs Division and supervised a group of 83 professionals. In 1993, he left Boehringer to find a private consulting company offering statistical services and consulting to the pharmaceutical industry. For his many outstanding contributions to the field of statistics, in 2003 Dr. Arseven was awarded the H. O. Hartley Award given by the Texas A&M Department of Statistics. More recently, Dr. Arseven, along with former student Luisa Sia, provided the funds to establish the Anant Kshirsagar Endowed fellowship.

Ralph Kodell '1974

Dr. Ralph Kodell, a distinguished graduate of the Department of Statistics, received his Ph.D. in 1974. Based on his outstanding academic record, in 1973 Dr. Kodell received the Department of Statistics W. S. Connor Outstanding Graduate Student Award. In 1983 he received our top alumnus award, the H. O Hartley Award, for his distinguished service to the discipline of statistics and in 1992 the American Statistical Association made him Fellow. In 2003, Dr. Kodell was recognized once again by becoming a member of the 2003 Class of the Academy of Distinguished Graduates. Dr. Kodell is known by many of his peers as one of the world's leading researchers in statistical methods for toxicology. Dr. Kodell was formerly the Director of the Division of Biometry and Risk Assessment at the National Center for Toxicology Research in Jefferson. He is currently a Professor of Biostatistics at the University of Arkansas for Medical Science located in Little Rock, Arkansas.

Joshua Baker '1985

Dr. Joshua Baker received his doctorate and master's degrees in statistics from Texas A&M University in 1985 under the direction of Professor Tom Wehrly. Dr. Baker now serves as the Managing Partner of Statistical Associates LLC. He is an industry executive with more than 20 years of pharmaceutical management and drug development experience. Prior to this appointment, Dr. Baker held many senior positions, including President and CEO of PPGx, Inc., Executive Vice President and Senior Vice President of Global Operations for PPD Development, Inc. A quote from Carolina Newswire reads "during his tenure, PPD Development became one of the top four companies in the global clinical research industry in terms of revenue and size".

Roland Acra '1986

Roland Acra has been the President and Chief Executive Officer of Arch Rock Corporation, now part of Cisco, since December 2005. He holds a Master of Science degree in Statistics from Texas A&M University, a master's degree in Engineering from the Ecole Nationale Superierue des Telecommunications in Paris and a degree in Mathematics and Physics from Ecole Polytechnique in Paris. His prior work experience included working for 13 years at Cisco Systems, where he held several senior management positions in the U.S. and Europe.

Katherine Ensor '1986

Dr. Katherine Ensor received a doctorate degree from Texas A&M University in 1986 under the direction of Professor H. Joseph Newton. She is currently serving as Professor and Chair for the Department of Statistics at Rice University in Houston. In 1985, Dr. Ensor received the William S. Connor Award for her outstanding role as a graduate student and thanks to her exceptional contributions to the discipline of Statistics, Dr. Ensor also received the H. O. Hartley award in 1998. In 2000, the American Statistical Association bestowed upon her the honor of Fellow of the ASA.

Ching-Yun Wang '1993

Dr. Ching-Yun Wang received his Ph.D. in 1993 under the direction of Professor Raymond J. Carroll. He was recognized in 2002 by being awarded our top alumnus award for distinguished service to the discipline, the H. O. Hartley Award. Dr. Wang currently holds several appointments, one of the most prestigious being a Full Member at the Fred Hutchinson Cancer Research Center where he works in the Biostatistics division. Also in 2005, he was appointed as Affiliate

Professor in Biostatistics at Feng Chia University, as well as the University of Washington.

Jeffrey Morris '2000

Dr. Jeffrey Morris received a master's in statistics from Texas A&M University in 1997 and Ph.D. in 2000 under the direction of Professors Raymond J. Carroll and Naisyin Wang. After graduating from Texas A&M University, he was hired as an Assistant Professor at MD Anderson Cancer Center Department of Biostatistics and Applied Mathematics. Dr. Morris's outstanding research accomplishments landed him several awards including the Mitchell Prize for outstanding paper in 2003, the University of Texas MD Anderson Cancer Center E.N. Cobb Faculty Scholar Award, and among the most prestigious awards, he was honored as the 2005 American Statistical Association Noether Young Scholar Award. Most recently, in 2010, Jeffrey Morris was promoted to full Professor at the MD Anderson Cancer Center, Department of Biostatistics and Applied Mathematics.

Faculty Book Publications (Books with 200 or more citations according to *Google Scholar*)

Parzen, E., (1960) *Modern Probability Theory and Its Applications*, John Wiley and Sons, New York.

Parzen, E., (1962) *Stochastic Processes*, Holden Day.

Carroll, R.J. and Ruppert, D., (1988) *Transformation and Weighting in Regression*. Chapman and Hall, London.

Newton, H.J., (1988) *TIMESLAB: A Time Series Analysis Laboratory*, Wadsworth & Brooks/Cole, Pacific Grove, California.

Staudte, R.G. and Sheather, S.J. (1990) *Robust Estimation and Testing*, Wiley, New York.

Hart, J.D., (1997) *Nonparametric Smoothing and Lack-of-Fit Tests*, Springer-Verlag, New York.

Dennison, D., Holmes, C., Mallick, B. and Smith A.F.M. (2002) *Bayesian Methods for Nonlinear Classification and Regression,* Wiley, New York.

Lahiri, S.N., (2003) *Resampling Methods for Dependent Data*, Springer, New York.

Ruppert, D., Wand, M.P. and Carroll, R.J., (2003) *Semiparametric Regression*. Cambridge University Press.

Carroll, R.J., Ruppert, D., Stefanski, L.A. and Crainiceanu, C.M., (2006) *Measurement Error in Nonlinear Models: A Modern Perspective, Second Edition.* Chapman and Hall CRC Press.

Ott, L., Longnecker, M., (2010) *An Introduction to Statistical Methods and Data Analysis*, 6th Edition.

Faculty Journal Publications (Articles with 200 or more citations according to *Web of Knowledge*)

Hartley, H.O., (1950) The Maximum F-Ratio as a Short-Cut Test for Heterogeneity of Variance. *Biometrika,* 37, 308–312.

Hartley, H.O., (1961) Modified Gauss–Newton Method for Fitting of Non-Linear Regression Functions by Least Squares. *Technometrics,* 3(2), 269–280.

Parzen, E., (1962) Estimation of a Probability Density Function and Mode. *Annals of Mathematical Statistics*, 33(3), 1065–1076.

Hartley, H.O., and Rao, J.N.K., (1967) Maximum-Likelihood Estimation for Mixed Analysis of Variance Model. *Biometrika, 54,* 93–108.

Hocking, R. R., (1976) Analysis and Selection of Variables in Linear Regression, *Biometrics*, 32(1), 1–49.

Parzen, E., (1979) Nonparametric Statistical Data Modeling. *Journal of the American Statistical Association*, 74, 105–121.

Ruppert D. and Carroll, R.J., (1980) Trimmed Least Squares Estimation in the Linear Model. *Journal of the American Statistical Association*, 77, 828–838.

Davidian, M. and Carroll, R.J., (1987) Variance Function Estimation. *Journal of the American Statistical Association*, 82, 1079–1092.

Wu, M.C. and Carroll, R.J., (1988) Estimation and Comparison of Changes in the Presence of Informative Right Censoring by Modeling the Censoring Process. *Biometrics*, 44, 175–188.

Sheather, S.J., & Jones, M.C., (1991) A Reliable Data-Based Bandwidth Selection Method for Kernel Density Estimation, *Biometrika*, 53(3), 683–690.

Ruppert, D., Sheather, S.J., & Wand, M.P., (1995) An Effective Bandwidth Selector for Local Least Squares Regression, *Journal of the American Statistical Association*, 90, 1257–1270.

Kipnis, V., Subar, A.F., Midthune, D., Freedman, L.S., Ballard-Barbash, R., Troiano, R. Bingham, S., Schoeller, D.A., Schatzkin, A. and Carroll, R.J. (2003) The Structure of Dietary Measurement Error: Results of the OPEN Biomarker Study. *American Journal of Epidemiology*, 158, 14–21.

A Brief History of the Statistics Department of the University of California at Berkeley

Terry Speed, Jim Pitman and John Rice

The early history of our department was dominated by Jerzy Neyman (1894–1981), while the next phase was largely in the hands of Neyman's students, with Erich Lehmann (1917–2009) being a central, long-lived, and much loved member of this group. We are very fortunate in having the biography *Neyman—From Life* written by Constance Reid (1918–2010), and Erich's *Reminiscences of a Statistician: The Company I Kept* and other historical material[1] documenting the founding and growth of the department, and the people in it. In what follows we will draw heavily on these sources, describing what seems to us a remarkable success story: one person starting[2] "a cell of statistical research and teaching not being hampered by any existing traditions and routines," and seeing that cell grow rapidly into a major force in academic statistics worldwide. That it has remained so since its founding is a testament to the strength of Neyman's model for a department of statistics.

[1] There are published conversations with and volumes of papers honoring many former and current Berkeley Statistics faculty, and obituaries of our deceased colleagues. We refer to *Statistical Science* for the first, and http://www.stat.berkeley.edu/people/memorials for the last mentioned. The remainder are readily accessible, e.g., through PROJECT euclid or JSTOR. We give birth dates only for deceased individuals at the first appearance of the name, and in those cases include the date of death. Moore (2007) is a further valuable reference.

[2] Letter from Neyman to Deming (written c1937), quoted in Reid (1982), p. 151.

University of California, Berkeley,
Department of Statistics,
367 Evans Hall 3860,
Berkeley, CA 94720, USA
T. Speed · J. Pitman (✉) · J. Rice
e-mail: pitman@stat.berkeley.edu

T. Speed
e-mail: terry@stat.berkeley.edu

J. Rice
e-mail: jariceiii@gmail.com

A. Agresti and X.-L. Meng (eds.), *Strength in Numbers: The Rising of Academic Statistics Departments in the U.S.*, DOI: 10.1007/978-1-4614-3649-2_23,

The Years Before the Department (1938–1955) and the Department

In 1938, at the age of 44, Jerzy Neyman left the Department of Applied Statistics at University College, London and arrived in Berkeley to take up a position as Professor of Mathematics. His status as a leading scholar of his generation was firmly established by his paradigm-setting work in areas including hypothesis testing, confidence intervals, and sampling design. Shortly after Neyman's arrival, he founded the Statistical Laboratory (Stat Lab), and this was to be the center of Statistics at Berkeley until the creation of a separate Department of Statistics in 1955, and the center of his activities until his death in 1981.

As well as throwing himself into theoretical and applied statistical research, Neyman devoted a great deal of energy to teaching—for example, during the second semester of 1939–1940 he increased his voluntary teaching load to 25 hours per week. In 1939, he recruited Elizabeth (Betty) Scott (1917–1988, BA 1939), an astronomer who had been auditing his course, as a research assistant, while in 1941 Evelyn Fix (1904–1965), a high-school mathematics teacher who had attended a summer session at Berkeley, joined the Lab as a technical assistant. Statistical work contributing to the U.S. effort in World War II soon became the principal focus of the Lab, in particular extensive research on bombing patterns.

When the war ended in 1945, Neyman organized a symposium on mathematical statistics and probability[3] "to contribute to the revival of scientific work in mathematical statistics and allied fields". This landmark gathering featured contributions from Hotelling, Pólya, Wolfowitz, Doob, Hsu, Feller, and Neyman's student, Erich Lehmann, as well as papers on the application of statistics to physics, astronomy, psychology, economics, forestry, animal breeding, rainfall, and insect populations. Hotelling's keynote address[4] "The Place of Statistics in the University," and the ensuing discussion are as relevant today as they were then.

The *Berkeley Symposium on Mathematical Statistics and Probability* was a great success, and was to be the first of six, held roughly every 5 years, whose proceedings grew from a single volume of 500 pages published in 1949 to six volumes totaling well over 3,000 pages published in 1972. These symposia were an[5] "enormous achievement for Neyman, a testament to his imagination, energy and organizing ability. During the 25 years of their existence, they functioned as the most important international statistics meeting, and put Berkeley on the map as a world center."

Neyman returned with enthusiasm to teaching in fall 1945, assembling a varied group of individuals to cover ten statistics courses. He led the team, and was

[3] Letter from Neyman to Provost Deutsch (written c1944), quoted in Reid (1982), p. 197.

[4] Neyman (1949), p. 21.

[5] Lehmann (2008), p. 27.

assisted by Pao-Lu Hsu (1910–1970), his former student and later colleague from University College, London. At that time, they were only two instructors with PhDs, and they were joined by Erich Lehmann, by then a graduate student in mathematics, who received his PhD under Neyman in 1946, as did Evelyn Fix (PhD 1948), John Gurland (1917–1997, PhD 1948), and Mark Eudey (1912–2007, PhD 1949). Two others came from outside statistics to complete the team: Elizabeth Scott, who received her Astronomy PhD in 1949, and Edward Barankin (1920–1995), who completed a Mathematics PhD on linear algebra in 1946.

Four of the fall 1945 instructors went on to become faculty in Mathematics and later Statistics, Lehmann in 1946, Barankin in 1947, and Fix and Scott in 1950, and another Neyman student from this era, Joseph L. Hodges, Jr (1922–2000, PhD 1949) in 1951. Lucien Le Cam (1924–2000, PhD 1952) was the last of Neyman's students to be appointed (in 1952) to the faculty. These six Berkeley graduates were joined by four outstanding outside appointments. Charles Stein (Columbia PhD 1947), a student of Abraham Wald, was appointed in 1947, Michel Loève (1907–1979), a student of Paul Lévy came in 1948 from a position at the University of London and a visiting appointment at Columbia, Henry Scheffé (1907–1977) whose PhD was in Mathematics, but had moved into Statistics, joined in 1953, and David Blackwell (1919–2010), a student of Joseph Doob, visited Berkeley in 1954–1955, and joined the department in 1955. We think of these ten as the department's "founders", as all but one was present when the department was founded in 1955, and remained there until their retirement or death. One notable event prior to formation of the department was the 1949 California oath controversy, where faculty were required to sign an anticommunist loyalty oath, which had a profound impact on the Stat Lab, and resulted in the permanent loss of Charles Stein to the University of Chicago and later Stanford.

The story of Neyman's relentless push for the creation of a Department of Statistics separate from Mathematics has been told before,[6] and we will not recount it again here. Suffice it to say that Griffith C. Evans (1887–1973), the long-running chair of Mathematics at Berkeley who had brought Neyman to Berkeley in the first place,[7] had an opposing vision, one of a broad-based Department of Mathematics encompassing all of the mathematical sciences. On Evans' retirement in 1949, the way became clear for the formation of a separate department, and it eventually came into being in the fiscal year 1955–1956.

---

[6] Reid (1982, pp. 213 *et seq*), Lehmann (1996, pp. 142–143), Moore (2007, Chap. 11), Lehmann (2008, p. 91).

[7] Evans had been keen to recruit a leading figure in Statistics to his department. He brought R. A. Fisher to Berkeley in September 1936 as Hitchcock Lecturer, but the visit was not a success, see Moore (2007, p. 73). Later, Raymond T. Birge (1887–1980), chair of the Berkeley Physics Department recommended Neyman on the basis of the high opinion his collaborator W. Edwards Deming (1900–1993) had of him. Deming had been instrumental in organizing Neyman's 1938 *Lectures and Conferences*. He wrote a major paper with Birge, see Deming and Birge (1934).

The Department

Less than a year into the life of the new department, its Chair, Jerzy Neyman offered his resignation, to be effective from July 1, 1956. Again, this is a story that has been told well elsewhere,[8], and again we pass over the details, though the following gives a hint as to the reasons[9]

> his [Neyman's] tenure [as chair] was short and bumpy. There were conflicts with the new dean...over a variety of issues including teaching loads and the role of statistical consulting. In addition a breach had opened up between Neyman and some of his oldest colleagues over some internal issues.

When the inevitability of Neyman's resignation became apparent, the unanimous view was that David Blackwell should become Chair, and so he did, to the complete satisfaction of everyone involved. Neyman continued as Director of the Stat Lab until his death.

Eric Lehmann has described the department in the early days as being "constantly on the move". The graduate students typically had office space in a collection of so-called temporary buildings, left over from the World War II years, which lasted until the 1990s. In 1971 Evans Hall opened, with the department on the third and fourth floors. The graduate students are now all housed together with the faculty and staff. A lounge on the tenth floor, christened the Neyman Room in the early 1980s, serves as both a statistics seminar room and as a venue for social functions. The lounge on the third floor serves as a lunch room, as a coffee room, and as a gathering place for informal interactions.

The Statistics Students Graduate Association (SGSA) has long played an important role in the department. It is a voice for the interests of graduate students, provides information to new students, organizes seminars and social events, and sponsors an annual t-shirt design competition.

Biostatistics

The Interdepartmental Group Biostatistics was created in 1955 by Neyman and Jacob Yerushalmy (1904–1976), linking the biostatisticians in Berkeley's School of Public Health (SPH) with statisticians in the Department of Statistics. For a period the Group had co-chairs, one from the SPH and one from Statistics. The initial ones were Chin Long Chiang who completed his Statistics PhD under Neyman in 1954, and Betty Scott. Today the Group continues to intertwine the SPH and the Department, both students and faculty.

---

8 Reid (1982, p. 247), Lehmann (1996, p. 143), Lehmann (2008, p. 97).

9 Moore (2007, p. 162), based on Reid (1982, pp. 243–250).

In the next three sections, we will summarize the department's efforts at teaching, research, and consulting, calculating and computing, from this early period up until the present day. Then we will describe where we are now, and where we are going.

Statistics Teaching at Berkeley

It was clear from the outset that for Neyman, the teaching of statistics at Berkeley was not to be a rehearsal of cookbook-style recipes for the stylized analysis of data. In the Introduction to his 1950 book *First Course in Probability and Statistics* "for a one-semester basic course for beginners" he wrote "As conceived in this book, the theory of statistics is a section of the theory of probability." It is of some interest to note that Neyman's *First Course* began with an introductory chapter concerning the scope of the theory of probability and statistics which referred to inductive behavior, statistical decision functions, admissible hypotheses, and performance characteristics, and which was followed by a chapter introducing probability in a more or less traditional manner. However, this second chapter ended with a quite serious 25-page section entitled *Evaluation of Competing Risks*. Chapter 3 was entirely devoted to *Probabilistic Problems of Genetics,* and began with an explanation of the laws of heredity, including recombination.

Neyman's book was written for students taking Stat 1, which for many years was the "theoretical" introductory course, in contrast to the "non-theoretical" introductory course Stat 2. Lehmann recounts Scheffé's memorable explanation of the distinction[10]: "the more theoretical course (Stat 1) was for students who needed to understand statistics but were not planning to use it, while the methods oriented cookbook-style course (Stat 2) was for students who planned to use statistics but did not need to understand it."

Undergraduate statistics texts at Berkeley, of which Neyman's was the first, show how introductory statistics teaching there has evolved. Joe Hodges was a superb and inspiring teacher who was particularly concerned with undergraduate teaching. In this connection he co-authored the elementary texts *Basic Concepts of Probability and Statistics* (1964, 2nd ed. 1970) jointly with Lehmann, which was translated into Danish, Hebrew, Italian, and (fairly recently) Farsi, and *Stat Lab—An Empirical Approach* (1975) jointly with the psychologists Krech and Crutchfield.

Hodges and Lehmann's *Basic Concepts* were a successor to Neyman's book, arguably one more accessible to the Stat 1 audience.[11] Blackwell's *Basic Statistics* (1970) was for Stat 2, while *Stat Lab* was another introduction, to a different audience again. By that time Stat 1 had been split into separate courses in

[10] Lehmann (2008), p. 44.

probability and statistics, but Stat 2 has always remained. The most recent and by far the most successful of all these introductions is *Statistics* by David Freedman (1938–2008), Robert Pisani and Roger Purves, first edition 1978, and now in its fourth (2007) edition. This unique text is neither "methods-oriented" nor "cookbook-style," but while making minimal mathematical demands on its readers, it[12] "teaches a different approach: thinking." For more than 30 years now, the book has been used by over 2,000 students at Berkeley each semester, in Stat 2 and other introductory courses.

In addition to those works mentioned above, faculty at Berkeley have continued the tradition begun by Neyman of writing textbooks, at levels ranging from introductory to graduate, strongly reflecting their personal views of subject matter and pedagogy. A full list of books by the founders and selected books by later generation faculty is appended below.

The teaching of large lower division courses such as Stat 2 has rarely been a popular assignment among regular faculty, notable exceptions including Blackwell, Freedman, and Jack Kiefer (1924–1981). As a result, the department has appointed lecturers, that is, faculty whose principal duty is teaching, to fill this role. Roger Purves and Ani Adhikari are our current senior lecturers and Hank Ibser our continued lecturer, providing our service teaching at a very high standard. Many of the ladder faculty are superb teachers as well, notable examples being Blackwell, Freedman, and Deborah Nolan. Adhikari and Nolan have both received the extraordinarily competitive Distinguished Teaching Award, given each year to between one and five members of Berkeley's ~1,500 full-time faculty. In 2007, Philip Stark presented the first online course at Berkeley. He continues to develop his online statistics course, called SticiGui, and adapt it to different introductory audiences.

The Major

The department has offered an undergraduate major since 1955–1956, and for most of that time, majors in a given year numbered in the teens. Over the last decade, enrolments have exploded—there are currently about 250 majors. Nolan has been instrumental in building our statistics major. She has published several books on teaching statistics and in the mid-2000s pioneered a very popular upper division course called Concepts in Computing with Data. In addition to this course, and courses on standard topics, there are small interactive seminar courses on special topics proposed by individual faculty, which emphasize student projects and presentations. With the impetus of an NSF VIGRE program, many undergraduates participate in faculty research, often working alongside graduate students and postdocs.

---

11 Lehmann (2008), p. 114.

12 Freedman et al. (1978), p. xv.

The Quals

Over the years, the form of the doctoral qualifying exam (quals) has evolved. In the early 1960s the first year curriculum for PhD students was uniform: students enrolled in two year-long courses (Probability Theory and Theoretical Statistics), and took oral exams in *three* areas: Probability, Theoretical Statistics, and Applied Statistics. Preparation for the last-mentioned was the reading of Scheffé's *The Analysis of Variance*. In the early 1980s a year-long PhD-level course in Applied Statistics was introduced. Incoming PhD students now enrolled in two of the three year-long courses, and took written quals in their two courses at the end of the summer following their first year. Students who failed the quals exams twice left the program.

Since about the year 2000, it became increasingly apparent that this one-size-fits-all curriculum was a poor match to the increasing diversity of faculty and student interests and of preparation of students entering the program. Also, we found that only a very small number of students who had left the program did so because of failing the quals twice (students did leave, but for other reasons). In 2005 the uniform curriculum and written quals of the first phase of the program were replaced by a more flexible system with evaluations in each semester being based on course work and coordinated by the Head Graduate Advisor (see http://www.stat.berkeley.edu/programs/phd for details).

Research at Berkeley

Probability

A student of the distinguished Russian probabilist Serge Bernstein, Neyman had studied Lebesgue's theory of measure and integration before Kolmogorov built the measure-theoretic foundations of modern probability theory in 1937 on an abstract form of Lebesgue's theory. Neyman's interests shifted quickly from measure theory to the foundations of statistics, but he continued while at Berkeley to work with Scott on probability models for the clustering of point processes, motivated by applications in astronomy. Neyman maintained an appreciation of the mathematical theory of probability, and attracted to Berkeley representatives of the two main schools of probability theory in the early twentieth century: Blackwell from the USA and Loève, from France. Both held joint appointments in Statistics and Mathematics, and probability thrived at Berkeley at the interface of these two departments. Further appointments built up the probability group at Berkeley to include Lester Dubins (1920–2010), David Freedman, and P. Warwick Millar. Blackwell, Dubins, and Freedman collaborated during the 1960s to produce a series of influential results around inequalities for martingales and related stochastic processes, while Millar developed these ideas to provide foundations for the theory of stochastic integration for processes with independent increments.

By the 1970s, Berkeley was established as one of the best known departments internationally for research and exposition of probability. This reputation attracted a next generation of probabilists to Berkeley in the late 1970s and early 1980s, when Michael Klass, Jim Pitman, and David Aldous joined the department. During the 1980s, Klass obtained definitive results around the law of the iterated logarithm. Aldous's work on mixing times of finite Markov chains provided inspiration for a renaissance of the theory of finite Markov chains and its applications to computer science and computational statistics through MCMC. Pitman collaborated extensively with Marc Yor at Paris VI on the theory of Brownian motion and related stochastic processes. In the late 1980s and early 1990s Aldous developed his general theory of continuum random trees as limits of combinatorial models of random trees. While on sabbatical in Paris in 1989, Pitman worked with Jacques Neveu and Jean-Francois Le Gall to develop a theory of random trees embedded in Brownian excursions. It was some years before the confluence of these ideas was fully appreciated through collaboration of Aldous and Pitman on asymptotic theory of random trees and associated tree-valued and partition-valued random processes. Motivated by applications to Brownian motion and Bessel processes, Pitman and Yor developed enrichments of the theory of random discrete distributions (Dirichlet distributions), with roots going back to the work of Blackwell and Ferguson in the early 1970s. Steven Evans, who joined the faculty in 1989, has worked in diverse areas, including the theory of Markovian superprocesses, probability on algebraic structures, population biology, and biodemography. Aldous, Evans, and Pitman worked on various aspects of processes of coagulation and fragmentation processes and their relations to random trees. Yuval Peres joined the department in late 1992. His research across a wide range of modern probability theory, including particle systems and random walks, Brownian motion, percolation, and critical phenomena, attracted a large number of students and injected much energy into the Berkeley probability group in the early 2000s. More recent appointments in probability at Berkeley include Elchanan Mossel (2003) and Allan Sly (2010), with interests at the interface of probability and computer science, and Sourav Chatterjee (2006, Stein's method, theory of spin glasses, random matrices).

In 1992 the *Line and Michel Loève International Prize in Probability* was created in honor of Michel Loève, by his widow Line. Awarded every 2 years, the Loève prize is intended to recognize outstanding contributions by researchers in mathematical probability who are under 45 years of age. This biennial award is a notable recurrent event involving the Berkeley probability group.

Statistical Theory at Berkeley, the Founders

Theoretical statistics at Berkeley in the decade from the mid-1940s was dominated by the contributions of Erich Lehmann, Joe Hodges, and Lucien LeCam, together with Henry Scheffé and Charles Stein. Underlying themes included the continuation of Neyman's agenda to find tests that were optimal in multiparameter settings, the development of testing and later estimation within Wald's emerging framework of statistical decision theory, and an increasing focus on large sample properties of tests and estimators, arising in part from a searching critique of the method of maximum likelihood.

Neyman's main contribution was to develop the theory of Best Asymptotically Normal Estimates, for both estimation and testing, with a major paper being presented at the 1945 Berkeley Symposium. Later, he devised the class of $C(\alpha)$ tests for testing in the presence of nuisance parameters, a topic he was to continue working on until the end of his life. Interestingly, this class of tests has recently (2011) reappeared providing an approach to uncovering rare genetic risk factors from DNA sequence data.

Lehmann's and Scheffé's collaboration produced their now classic work on completeness, similar regions, and unbiased estimation. Lehmann also discovered the notion of complete class of decision rules in the course of proving results concerning admissibility of tests, and later discussed general unbiasedness. In 1948, Neyman and Scott provided some compelling examples that reinforced an observation made earlier by Wald on the inconsistency of the maximum likelihood estimator in a context where the number of nuisance parameters grows along with the number of observations. This was followed not long afterwards by Hodges' 1951 discovery of a superefficient estimator of the mean of a normal distribution with known variance. This counterexample, together with the seminal work of Wald, paved the way for Le Cam's PhD thesis. There, he proved that for 1D parameters, Bayes estimates exhibit local asymptotic minimaxity and admissibility, and showed that the corresponding maximum likelihood estimates inherit both properties by being asymptotically Bayes. In addition, he proved that the superefficiency points form a Lebesgue null set in the parameter space. Soon afterwards (1955), Le Cam published his extension of Wald's theory of statistical decision functions, and went on to define the Locally Asymptotically Normal Condition, the notions of contiguity, distance (deficiency) between experiments, approximate sufficiency, limits of experiments, and much more. With the hindsight afforded by a perspective of over 50 years, there seems little doubt that among the most lasting achievements of the Berkeley school in this period are those of Le Cam, who was, in the words[13] of Grace Yang (PhD 1966), "the principal architect of the modern asymptotic theory of statistics."

Other noteworthy contributions of the founders from among many include Hodges and Fix's pioneering 1951 report on nonparametric density estimation,

[13] Yang (1999), p. 223.

Scheffé's 1953 discovery of what became known as the S-method of multiple comparisons, Lehmann and Stein's deep 1948 paper "Most powerful tests of composite hypotheses", the Hodges and Lehmann body of research on nonparametric estimators and tests, and Barankin's many papers on the theory of sufficient statistics and exponential families.

Statistical Theory: Non-founders

The next generation of statistical theorists began with Peter Bickel (PhD 1963) and Kjell Doksum (PhD 1965), both students of Erich Lehmann, who joined the faculty as assistant professors in 1963 and 1966, respectively. Both have broad interests, and they collaborated on the analysis of transformations, revisiting the famous work of Box and Cox on this topic. Bickel wrote on robust estimation and was a participant in the Princeton Robustness year (1970–1971). He also made extensive contributions to distribution-free methods, minimax, Bayes and later adaptive estimation, semiparametric models, bootstrap asymptotics, higher order expansions, and much more. More recently, topics he has worked on include hidden Markov models, regularization methods, and particle filters. Doksum's best known work was on tail-free and neutral random probabilities, which later linked up with work of Berkeley probabilists, and he also contributed to reliability theory, survival analysis, and local correlation.

The year 1970 saw the appointment of Rudolf Beran, who worked in many of the areas listed above for Bickel, but continued working on different aspects of the bootstrap and adaptive estimation for much longer, collaborating with Millar on minimum distance estimation and several other topics. In 1977, Ching-Shui Cheng started in the department as an assistant professor. He had written his Cornell PhD thesis on optimal experimental design under Jack Kiefer, the person who almost single-handedly created that field, and in 1979 Kiefer himself joined the department but his time in Berkeley was sadly cut short by a sudden, fatal heart attack in August 1981. Cheng has remained the department's sole person in experimental design, contributing extensively to several different aspects of this topic.

In 1980, Leo Breiman was appointed (see next section) and Charles Stone shortly afterwards. Their joint work on classification and regression trees (CART) and Stone's on nonparametric function estimation and splines took the department into new directions, and this movement continued with the appointment of David Donoho in 1984. Major interests of Donoho at that time were sparse modeling, signal recovery, and inverse problems, particularly in geophysics, and he worked on these topics together with a postdoctoral fellow, Philip Stark, who was to join the faculty in 1988. Donoho's later departure for Stanford was a great loss for Berkeley.

Later appointments in the area of statistical theory include Deborah Nolan (1987), who originally worked on U-processes and Bin Yu (1993), whose Berkeley thesis was on information theory and empirical processes. Both of them have moved away from pure theory, in Nolan's case toward statistics teaching, including writing books and developing new courses in this area, and working toward gender equity in the mathematical sciences. Yu remains interested in theory related to the applied problems she attacks, and her current theoretical interests include sparse modeling and machine learning. These last interests are shared with Bickel and Michael Jordan (appointed 1998), Peter Bartlett (2002), and Martin Wainwright (2004), the last three being joint with Electrical Engineering/Computer Science (EE/CS). Our most recent appointments in statistical theory are Noureddine El Karoui (2005), with interests in random matrices and mathematical finance, and Aditya Guntuboyina (2012), who works on applications of convexity to statistics, minimax lower bounds, and a variety of other topics.

For many years Berkeley was generally unsympathetic to the Bayesian approach to statistics, with the notable exception of Blackwell. Freedman had long been interested in consistency, inconsistency, and pathologies of Bayes estimators, and in the last few years, Michael Jordan and Cari Kaufman (appointed 2008) have made many positive contributions to Bayes theory and methods.

In the 1950s and early 1960s Blackwell, Breiman, and Thomasian made several key contributions to information theory, perhaps the most notable being Breiman's 1957 proof of the individual ergodic theorem of information theory. More recently, Yu and Wainwright have both worked on different aspects of coding theory.

Applied Statistics at Berkeley

Jerzy Neyman came to Berkeley in 1938 with considerable experience applying statistics and closely linking theory and applications. Neyman also regarded applications as an essential part of the statistician's education, commenting at the 1945 Berkeley Symposium.[14]

> The future mathematical statistician needs early contacts with experimental sciences. He needs them because, at this stage of the development of statistics, the experimental sciences are sources of theoretical problems. Also, he needs them because in almost any imaginable job which he may get after graduation he will be called upon to apply his theory to experimental or observational problems.

We see in his *First Course* how he viewed applications of theorems on probability (pp. 69–70):

> The first difficulty is that no practical problem is concerned with mathematical concepts... Therefore, the practical problem must be translated into probabilistic terms before its solution is attempted...In fact, this translation amounts to building up a mathematical model of the practical problem. ...In certain cases, the adequacy of the particular model can be tested empirically.... Naturally, the solution of the problem based on a given model applies to the model itself and not necessarily to the phenomena for which it was intended.

[14] Neyman (1949), p. 27.

> ....The second ...difficulty ...is that most observable phenomena are rather complicated. ....In many cases, ...we are forced to revise the model, sacrificing its adequacy in order to gain simplicity.

This view permeated all his applied work, the sheer volume of which is dizzying. Not long after arriving in Berkeley Neyman made contact with the California Forest and Range Experiment Station, and the Department of Entomology. In due course Neyman together with other members of the Stat Lab, especially Scott and Fix, extended their applied research to many other areas, including bacteriology, fisheries, public health, demography, accident proneness, astronomy (over 30 papers), weather modification (over 30 papers), carcinogenesis, molecular evolution, and ecology. For some discussion of some of this enormous body of work, and a complete list of papers, we refer to Neyman's obituary[15] and to the published Neyman Lecture, Brillinger (2008) for an overview of three examples.

Of the founders, it was principally Neyman, Scott, and Fix who conducted applied research, though in the late 1970s and early 1980s, Le Cam became involved in cancer research. David Brillinger's appointment in 1970 immediately expanded the intellectual scope of the department, although his work on point processes related to that of Neyman and Scott, and there was a connection to harmonic analysis (Loève). He came with a background in time-series analysis and experience working with his advisor John W. Tukey and others at Bell Labs. Soon after his arrival in Berkeley, Brillinger and his students were collaborating with a neurophysiologist, a seismologist, and with scientists studying the abundance and movements of wild animals such as whales. Over the years his applied interests broadened to include imaging, spectroscopy, risk assessment, demography, and meteorology.

The next appointee with applied interests was Leo Breiman, who joined the department in 1980. Coming from a second career as a private consultant, following a distinguished academic career at UCLA, he was interested in applications, but unlike Neyman and Brillinger, he was not particularly interested in getting closely involved with scientific collaborations. He was even less interested in using probability models as the basis for his statistical analyses, a view articulated in Breiman (2001). Leo's forte was creative ideas for dealing with standard statistical questions, typically with large data sets, widely used methods that are now part of the literature of data mining and machine learning. He was an expert on bootstrapping, cross-validation, and other methods used for variable selection in regression, and played a central role in the development of many novel techniques, including classification and regression trees (CART), alternating conditional expectations (ACE), archetype analysis, bagging, and random forests.

David Freedman joined the department in 1961, ostensibly as a probabilist, but one with a strong and critical interest in statistics. His extensive contributions to applied statistics were stimulated by his undergraduate teaching and his consulting. Freedman's 1971 consulting with the Bank of Canada on econometric models

[15] Kendall et al. (1982).

began his career as a trenchant critic of much statistical practice in the social sciences, and this was to be a preoccupation for the rest of his life. For over 35 years, he consulted widely in the public and private sectors, appeared as an expert witness on many occasions, and gave congressional testimony twice. In most cases he would carry out all the analyses himself, and if not, he would very closely supervise his assistants. He was a critic of the energy modeling of the early 1980s, census adjustment, much econometric modeling, the use of path analysis and other methods claiming to elucidate causal relations in the social sciences, and much more, and his position was invariably reached through critical reanalysis of key data sets.

Donoho's and Stark's work on inverse problems has been mentioned under statistical theory, while another appointment in this general area was Finbar O'Sullivan, now at University College, Cork, Ireland. Stark's later research, includes confidence sets, election auditing, public policy issues, and online statistical education. Terry Speed was a senior appointment in 1987, and while here he began a research focus on statistical genetics and genomics which continues in the department today through the work of Haiyan Huang (appointed 2003), and Elizabeth Purdom (2009), and Sandrine Dudoit in Biostatistics. John Rice was appointed in 1990, further strengthening the department's commitment to applied work. At that time he had worked on nonparametric function estimation and was collaborating with neurophysiologists. He later moved into interdisciplinary work on transportation and astronomy and contributed to the methodology of functional data analysis. Bin Yu's interest in the statistical theory of sparse approximation has been complemented by applied work in information technology, remote sensing, and neuroscience, while both Yu and Jordan carry out research on text mining. Yun Song and Rasmus Nielsen were both appointed to the department in 2007 via a campus initiative in computational biology, Song jointly, with CS and Nielsen jointly with Integrative Biology. Cari Kaufman shares the interests of Brillinger and Yu in interdisciplinary research in spatial and environmental statistics, as well as working on Bayesian methods in applications.

Consulting, Calculating, and Computing at Berkeley

Statistical Consulting

Neyman viewed engaging in consulting with faculty in other departments as an essential part of the role of a university statistician. In this regard, he was completely in agreement with views Hotelling expressed at the 1940 IMS meeting in Hanover, NH that Neyman attended[16]:

[16] Neyman (1949), p. 48.

> A specialist in statistics on a university faculty has a threefold function. In addition to the usual duties of teaching and research, there is a need for him to advise his colleagues, and other research workers, regarding the statistical methods appropriate to their various investigations. The advisory function is a highly important one for the activities of the university as a whole, …

Consulting was thus part of the Stat Lab experience from the outset. Writing on personal events taking place toward the end of the 1951–1952 academic year, Lehmann relates that,[17]

> An important activity of the (Stat) lab was the consulting service that provided help for faculty members in other departments with their statistical problems. Neyman handled some of these himself; others he passed on to members of the staff. He considered it one of our obligations to accept such assignments.

Juliet Shaffer, who joined the department as a lecturer in 1977, organized the Statistical Consulting Service and the course "Statistical Consulting" for over a decade, before it became shared more widely across faculty with applied interests. The pattern for the service and course has not really changed much since its inception in fall 1977. Graduate students work in pairs and are assigned times at which they are available in the consulting room to all comers. Once a week the class meets as a group with the course instructor, to discuss the problems that have come to them, and these class meetings are frequently augmented by individual discussions between students and the responsible faculty member. There is no obligation on the consultants to do the analyses for the client, but nevertheless, this sometimes happens. These experiences have been a formative part of the education of our students and have resulted in many lasting scientific interactions, as well as the passing on of much good advice on design and analysis of experiments and studies.

Calculating and Computing

The department has always had a symbiotic relationship with calculating and computing machinery. Writing about Evelyn Fix, Lehmann says[18] that "Her first big task came in 1942, when Neyman received a contract for bombing research from the National Defense Research Council. The work was computationally very intensive; it was carried out on desk calculators and consumed much time and effort. Evelyn was in charge of the calculations and presided over a hastily recruited, ragtag group of students (of which I was one for a short while) and faculty wives." In 1945 Neyman felt able to state[19] that in the course of its contract

---

[17] Lehmann (2008), p. 72.

[18] Lehmann (2008), p. 34.

[19] Neyman (1945), p. 83.

war research, "the Laboratory acquired an efficient set of computing machines and other equipment."

In the late 1960s, the department purchased a Wang electronic calculator. This was followed in 1974 by a DEC PDP-11/45 running Unix, shared by Mathematics, Statistics, and Computer Science. Thus the department had access to one of the first machines running Unix and S outside Bell Labs. By the late 1970s, teletype terminals were installed in the lab rooms to broaden access to the centralized computers and, later, terminals were installed in every faculty and graduate student office.

In the 1980s and 1990s our computational environment developed under the energetic leadership of Leo Breiman, who observed shortly after his arrival in 1981 that[20] "for the department to ever get into any kind of relationship with data, they had to have a decent computing facility." In 1986, the Statistical Computing Facility (SCF) was formed with Breiman as director of a staff consisting of a systems manager, an applications manager/software consultant, and an administrative manager. The department obtained its first VAX, an 11/750, in 1982, then Sun workstations and servers, and most recently Linux workstations, compute servers, network, and communication services, and iMac desktop computers located in undergraduate computing laboratories and faculty and student offices. We now believe that we have an efficient set of computing machines and other equipment.

The SCF also supports the instructional needs of the department. Phil Spector, who came to us from SAS in 1987 as the first applications manager, started a two-course graduate sequence in statistical computing. By the mid-1980s computing was integrated into most of the undergraduate major courses. Statistics students now routinely use personal laptops running R, which was developed by one of our ex-students, Ross Ihaka (PhD 1985), and his collaborator, Robert Gentleman.

Where Are We Now, and Where Are We Going?

The department grew slowly but steadily between the early 1960s and the late 1980s and early 1990s. New faculty were hired, our courses continued to evolve, the "sausage machine" continued to produce PhD graduates in large numbers, and we seemed to be adapting to the changing world, which became increasingly driven by computing and an emphasis on applications. The founders aged and began to retire in this period, but the next generation—students of the founders, or their counterparts elsewhere—came into the department and flourished. However, in 2000 we faced the daunting prospect of ten faculty members retiring in the period 2007–2014. In that year the Chair, John Rice, set up a "next decade" committee to examine how the department should respond, over the coming 10

[20] Olshen (2001), p. 191.

years, to the ongoing changes in the discipline of Statistics. This committee began their October 2000 report noting that:

> Today's conventional wisdom... is that the future of Statistics consists of interdisciplinary applied research together with some new synthesis of what constitutes core theory

Envisaging "new faculty hires as the main engine of change," they recommended hiring on average one person per year for the coming decade, and the introduction of the option of "PhD in Statistics with Emphasis in X" for several interdisciplinary subjects X. The report also recommended that the department should place greater value on joint appointments. We digress briefly to comment on this issue, before resuming our narrative.

Statistics has frequently reached out to people located in other campus departments. Joint seminars, co-supervision, and cross-listing of courses are some of the ways these interactions have been fostered. Sometimes, individuals seek a more formal relationship, one permitting them to be the primary advisor of Statistics graduate students, or to teach courses listed wholly within Statistics. In Berkeley there are two ways of responding to an interest of this kind: a split appointment in two departments or a 0% appointment in a second department.

Examples of both kinds of arrangements go back to the mid-1950s, when the agricultural and resource economist George Kuznets (1909–1986) was the first 0% appointment in Statistics. Others from the past include Roy Radner, a statistician within the Department of Economics, Jacob Feldman, an ergodic theorist from the Department of Mathematics, Richard Barlow from Industrial Engineering and Operations Research, a Bayesian statistician interested in reliability, and Paul Holland, who was a statistician within our School of Education before returning to the Educational Testing Service. Currently we have eight 0% appointments: Kenneth Wachter from Demography, Leo Goodman from Sociology, Nicholas Jewell, Mark van der Laan and Sandrine Dudoit from Biostatistics, Alistair Sinclair from Computer Science (CS), Bernd Sturmfels from Mathematics, and Jasjeet Sekhon from Political Science. Albert Bowker (1919–2008), founder of the Stanford Statistics Department, became Berkeley Chancellor in 1971 with a 0% appointment in Statistics. In 1980 he moved office to the Statistics Department itself and interacted steadily with students and faculty alike. The participation of 0% faculty in our department's activities has greatly enriched the experience of our graduate students, as well as fostering closer collaborations among faculty. Conversely, several of our current faculty have 0% appointments in other departments, including Mathematics, CS, and EE.

More challenging administratively, but potentially more effective at achieving the same aims, are formal joint appointments, usually 50:50. Our first such was Thomasian, who did his PhD with Blackwell on information theory, and later held a joint appointment between Statistics and EE. In the mid-1990s Breiman urged the department to recognize the explosion of a kind of theoretical and applied statistics, growing within Computer Science departments, which contrasted dramatically with the view of the subject held in (most) Statistics departments. He argued persuasively that we should expand our view of the discipline and narrow the gap between these two approaches by making suitable full or joint appointments of people with this background. After considerable discussion between Statistics and CS and quite an extensive search, Michael Jordan was appointed in 1998 to do just this, and his arrival on the scene was quickly perceived to be a success. There are now four members of department with joint appointments of this kind: Jordan (joint with CS), with interests in machine learning, computational biology, Bayesian nonparametrics, graphical models, and natural language processing, Peter Bartlett (CS), with interests in machine learning, control and robotics, Martin Wainwright (EE), signal processing, coding, graphical models, optimization, sparsity in high-dimensional statistical inference, and nonparametric statistics, and Yun Song (CS), computational biology and population genetics. We also have two faculty joint with Mathematics, Michael Klass, and Steven Evans, while Rasmus Nielsen is joint with Integrative Biology, with interests in evolutional theory and genetics.

Returning to developments initiated around the year 2000, it is interesting to note that between 2003 and 2012, we made 12 new appointments including two joint with CS, one joint with Electrical Engineering, three in applied and two in theoretical statistics, three in probability, and one joint with Integrative Biology. In the same period, several 0% appointments and adjunct appointments were made.

Developments over this decade overseen by three different chairs seem to have followed the 2000 vision for "departmental renewal" surprisingly closely.

Today

The Department of Statistics at UC Berkeley currently has 43 faculty: 13 with 100% Statistics appointments, seven joint appointments spread across Mathematics (2), Electrical Engineering (1), Computer Science (3) and Integrative Biology (1), three with 100% teaching appointments (2 senior lecturers, 1 continued lecturer), four adjunct appointees in probability, statistics, and finance, eight 0% appointments from within Biostatistics (3), Demography (1), Sociology (1), Computer Science (1), Mathematics (1), and Political Science (1), and eight emeriti, three of whom are active within the Graduate School. In round numbers, we teach 2,000 students from other departments each semester; we have 250 students taking a BA majoring in Statistics, 10 students in our MA program, and 50 in our PhD program.[21] Furthermore, several of our PhD students specialize in either Communication, Computation, and Statistics, or Computational and Genomic Biology, these being two of the campus' cross-disciplinary PhD Designated Emphases.

[21] The approximately 500 Berkeley Statistics PhD graduates are all listed at http://www.stat.berkeley.edu/people/alumni

We greatly value our interdisciplinary links and one of our goals is to maintain and extend these initiatives. Our graduate students are increasingly trained in a cross-disciplinary manner through collaborations with domain scientists and participation in the designated emphases mentioned above, and we aim to expand these efforts. We are also expanding our Master's program and updating its curriculum, and seeking to improve our interactions with business, industry, and government, in an effort to meet the huge and growing demand for data analysis skills in these areas.

Appendix

1. Books by Neyman and the Founders

For brevity we give first editions only. Many of these books have multiple editions, and translations into other languages. Also, we omit edited volumes, such as the *Proceedings of the Berkeley Symposia on Probability and Mathematical Statistics,* which are all readily available in PROJECT euclid.

David Blackwell and M. A. Girshick. (1954). *Theory of games and statistical decisions.* John Wiley and Sons, Inc., New York, xi + 355.

David Blackwell. (1970). *Basic Statistics.* McGraw-Hill, New York.

J. L. Hodges and Richard S. Crutchfield and David Krech. (1993). *StatLab : an Empirical Introduction to Statistics.* McGraw-Hill Ryerson, Limited, 376.

J. L. Hodges, Jr. and E. L. Lehmann. (1965). *Elements of finite probability.* Holden-Day Inc., San Francisco, Calif., vi + 230.

J. L. Hodges, Jr. and E. L. Lehmann. (1964). *Basic concepts of probability and statistics.* Holden-Day Inc. San Francisco, Calif., xiii + 375.

Lucien Le Cam. (1986). *Asymptotic methods in statistical decision theory.* Springer-Verlag, New York, xxvi + 742.

Lucien Le Cam and Grace Lo Yang. (1990). *Asymptotics in statistics.* Some basic concepts Springer-Verlag, New York, viii + 180.

E. L. Lehmann. (1959). *Testing statistical hypotheses.* John Wiley & Sons Inc., New York, xiii + 369.

E. L. Lehmann. (1975). *Nonparametrics: statistical methods based on ranks.* With the special assistance of H. J. M. d'Abrera, Holden-Day Series in Probability and Statistics Holden-Day Inc., San Francisco, Calif., xvi + 457.

Erich Leo Lehmann. (1983). *Theory of point estimation.* A Wiley Publication in Mathematical Statistics John Wiley & Sons Inc., New York, xii + 506.

E. L. Lehmann. (2008). *Reminiscences of a statistician: The company I kept.* Springer, New York, xii + 309.

E. L. Lehmann (2011). *Fisher, Neyman, and the Creation of Classical Statistics.* New York: Springer-Verlag.

Michel Loève. (1955). *Probability theory. Foundations. Random sequences.* D. Van Nostrand Company, Inc., Toronto-New York–London, xv + 515.
J. Neyman (1938). *Lectures and Conferences on mathematical statistics.* Graduate School, US Department of Agriculture. Washington D.C. 1938. 160.
J. Neyman. (1950). *First Course in Probability and Statistics.* Henry Holt and Co., New York, N. Y., ix + 350.
J. Neyman. (1967). *A selection of early statistical papers of J. Neyman.* University of California Press, ix + 429.
J. Neyman and E. S. Pearson. (1967). *Joint statistical papers of J. Neyman and E. S. Pearson.* University of California Press, iv + 299.
Henry Scheffé. (1959). *The analysis of variance.* John Wiley & Sons Inc., New York, xvi + 477.
Charles Stein. (1986). *Approximate computation of expectations.* Institute of Mathematical Statistics, Hayward, CA, iv + 164.

2. Selected Books by Non-founders

- The books below are representative of a much larger number written by faculty in Berkeley Statistics. Again, we restrict to first editions only, and omit reference to translations, conference proceedings, or collections of other kinds.

- David J. Aldous. (1985). *Exchangeability and related topics.* École d'Été de probabilites de Saint-Flour, XIII, 1–198, Lecture Notes in Math., 1117. Springer, Berlin, 1–198.
- Martin Anthony and Peter Bartlett (1999) *Neural Network Learning: Theoretical foundations.* Cambridge University Press. xiv + 389.
- Peter J Bickel and Kjell A. Doksum. (1976). *Mathematical statistics. Basic ideas and selected topics*, Holden-Day Series in Probability and Statistics Holden-Day Inc., San Francisco, Calif., xiv + 493.
- Leo Breiman, Jerome H. Friedman, Richard A. Olshen, and Charles J. Stone. (1983) *Classification and Regression Trees.* Wadsworth Advanced Books and Software, Belmont, CA.
- David R. Brillinger. (1975). *Time series. Data analysis and theory*, International Series in Decision Processes Holt, Rinehart and Winston, Inc., New York, xii + 500.
- Ching-Shui Cheng (2013) *Theory of Factorial Design.* Chapman and Hall.
- Lester E. Dubins and L. Jimmy Savage (1965) *How to Gamble if You Must: Inequalities for stochastic processes.* McGraw-Hill Book Co., New York-Toronto-London-Sydney. xiv + 249.
- Steven N. Evans. (2008). *Probability and real trees.* Lectures from the 35th Summer School on Probability Theory held in Saint-Flour, July 6–23, 2005 Springer, Berlin, xii + 193.

Fig. 1 From left to right. 1. E. L. Crow. 2. E. S. Keeping. 3. V. Lenzen. 4. D. G. Chapman. 5. C. Stein 6. E. Crow. 7. E. Fix. 8. S. W. Nash. 9. E. L. Scott. 10. D. Cruden Lowry. 11. J. Gurland. 12. I. Elvebach. 13. E. L. Lehmann 14. T. A. Jeeves. 15. E. Seiden. 16. E. A. Fay. 17. B. Epstein. 18. R. T. Birge. 19. I. Blumen. 20. Unknown. 21. B. M. Bennett. 22. J. L. Hodges, Jr. 23. T. Hodges. 24. H. Cramér. 25. G. R. Seth. 26. J. Neyman. 27. H. Hotelling. 28. Z. Szatrowski. (Stat. Lab. Univ. of Calif. Summer Session 1947)

- David A. Freedman. (2005). *Statistical models: theory and practice.* Cambridge University Press, Cambridge. x + 414.
- David A. Freedman. (2010). *Statistical models and causal inference.* A dialogue with the social sciences, Edited by David Collier, Jasjeet S. Sekhon and Philip B. Stark Cambridge University Press, Cambridge, xvi + 399.
- David Freedman and Robert Pisani and Roger Purves (1978). *Statistics.* W. W. Norton & Co.
- Deborah Nolan and Terry Speed (2000) *Stat Labs: Mathematical Statistics through Applications*, Springer. xviii + 300.
- Jim Pitman. (2006). *Combinatorial stochastic processes.* Lectures from the 32nd Summer School on Probability Theory held in Saint-Flour, July 7–24, 2002. Springer-Verlag, Berlin, x + 256.
- Jim Pitman. (1993). *Probability.* Springer, xi + 559.
- John A. Rice. (2006). *Mathematical statistics and data analysis.* Cengage Learning, 666.

- P. B. Stark (1997). SticiGui: Statistics Tools for Internet and Classroom Instruction with a Graphical User Interface. http://statistics.berkeley.edu/~stark/SticiGui
- Charles Stone (1995) *A Course in Probability and Statistics*. Duxbury, 838.
- Aram J. Thomasian. (1969). *The structure of probability theory with applications*. McGraw-Hill, 746.
- Martin J. Wainwright and Michael I. Jordan (2008). Graphical Models, Exponential Families, and Variational Inference. *Foundations and Trends in Machine Learning*. Vol. 1: No 1–2, pp 1–305.

3. Photographs

See Figs. 1, 2, and 3. For more photos, including L. Le Cam, M Loève, H. Scheffé and E Barankin, and others, please go to http://www.stat.berkeley.edu/photos

Fig. 2 From left to right. F. N. David, E. L. Scott, D. Blackwell and E. Fix. (Photo taken about 1960)

Fig. 3 From left to right. Front row: Deb Nolan, Philip Stark, Peter Bickel, Elizabeth Purdom, Roger Purves, Julie Shaffer, Martin Wainwright, Michael Klass. *Middle row*: Ani Adhikari, Cari Kaufman, Haiyan Huang, Jim Pitman, Chuck Stone, Jon McAuliffe, Sandrine Dudoit, Bin Yu, Michael Jordan. *Back row*: Jas Sekhon, Aditya Guntuboyina, Allan Sly, Peter Bartlett, Nick Jewell, Steve Evans, Elchanan Mossel, Terry Speed. (Faculty in February 2012)

References

Breiman L (2001) Statistical modeling: the two cultures. Stat Sci 16:199–231

Brillinger D (2008) The 2005 Neyman lecture: dynamic indeterminism in science. Stat Sci 23:48–64

Deming WE, Birge RT (1934) On the statistical theory of errors. Rev Modern Phys 6:119–161

Freedman D, Pisani R, Purves R (1978) Statistics. W.W. Norton & Co., New York

Kendall DG, Bartlett MS, Page TL (1982) Jerzy Neyman. Biogr Memoirs Fellows R Soc 28:379–412

Lehmann EL (1996) The creation and early history of the Berkeley Statistics Department. In: Ferguson TS, Shapley LS, MacQueen JB (eds) Statistics, probability and game theory: papers in honor of David Blackwell. Institute of Mathematical Statistics, Hayward, pp 139–146

Lehmann EL (2008) Reminiscences of a statistician. The company I kept. Springer, New York, pp xii + 309

Moore CC (2007) Mathematics at Berkeley. AK Peters, Wellesley, 341 pp

Neyman J (1945) Teaching and research at the Statistical Laboratory, University of California. Biom Bull 1:83–84

Neyman J (1949) Proceedings of the Berkeley symposium on mathematical statistics and probability. University of California Press, Berkeley, 501 pp

Olshen R (2001) A conversation with Leo Breiman. Stat Sci 16:184–198

Reid C (1982) Neyman—from life. Springer, New York, pp ii + 298

Yang GL (1999) A conversation with Lucien Le Cam. Stat Sci 14:223–241

University of Chicago Department of Statistics

Stephen M. Stigler

The University of Chicago's Department of Statistics was one of several born out of the growing appreciation of the power of statistical methodology following World War II. During the war there had been a great assembly of statistical talent in what was called the Statistical Research Group (SRG), with its base at Columbia University. Warren Weaver had been instrumental in the events leading to the founding of SRG in July 1942, with W. Allen Wallis (then at Stanford) and W. Edwards Deming (then at the Bureau of Commerce) also playing instigating roles. Weaver convinced the military that such an organization would be invaluable, and at the suggestion of Sam Wilks he recruited Harold Hotelling, then at Columbia, to head it. Hotelling in turn recruited W. Allen Wallis and Jack Wolfowitz as the initial staff, with Wallis to serve as operational head. Together they assembled a set of people who would constitute a who's who of the statistical generation that followed (Wallis 1980; Olkin 1991).

Following the war, the SRG members dispersed and either returned to their home institutions, or founded programs or departments at other institutions: Allen Wallis and Jimmie Savage to Chicago to start what would become a department, Abraham Wald and Jack Wolfowitz returned to Columbia (department formed in 1946 with Wald as Chair), Harold Hotelling moved to the University of North Carolina at Chapel Hill (department formed 1946 with Hotelling as Chair), Fred Mosteller to Harvard (department formed 1957 with Mosteller as Chair), Al Bowker and Abe Girschick to Stanford (department formed 1948 with Bowker as Chair), Harold Freeman to MIT, Ken Arnold to Michigan, Milton Friedman to Minnesota and later to Chicago, Herb Solomon to ONR and then Stanford,

The University of Chicago,
Department of Statistics,
5734 S. University Avenue,
Chicago, IL 60637, USA
S. M. Stigler (✉)
e-mail: stigler@uchicago.edu

A. Agresti and X.-L. Meng (eds.), *Strength in Numbers: The Rising of Academic Statistics Departments in the U.S.*, DOI: 10.1007/978-1-4614-3649-2_24,

Churchill Eisenhart to NBS. At least nine SRG members later served as chair of a Department of Statistics. Other statisticians who worked at least briefly or consulted with SRG included John Tukey, Sam Wilks, Jerzy Neyman, and I. Richard Savage.

Wallis was recruited by Chicago explicitly to start a program in statistics. He had begun his graduate career as a PhD student in Economics at Chicago in 1933. He subsequently studied with Hotelling at Columbia, and he taught at Stanford (with a visit at Yale 1937–1938) before Hotelling recruited him to SRG in 1942. He had by then already established himself as an excellent researcher and administrator despite the fact that he never received a PhD. Delays in receiving a degree were common in those days; Wallis's infinite delay was in part due to a Chicago policy of not awarding degrees to members of their own faculty.

Wallis first returned to Stanford for spring and summer 1946, but by then he and Jimmie Savage had received offers from Chicago. Both moved there in fall 1946. Wallis's initial appointment was as Professor of Statistics in the Business School; Savage went as a Rockefeller Foundation Fellow to the University's Institute of Radiobiology and Biophysics for the first year, becoming a Research Associate in the Institute in 1947. Both taught a variety of statistics courses from that first fall, including Wallis on Quality Control and Sampling Inspection, based upon materials developed at SRG.

Wallis and Savage had formed a high mutual respect for each other at SRG, and they soon began working on courses that would later become a core for a new statistics curriculum: Wallis on Statistics 301-302-303, a sequence entitled Statistical Inference that used only a little calculus but carried the subject to what was then a fairly advanced level on a par with much of the 1943 M. G. Kendall treatise, and Savage on Statistics 351-352-353, called Mathematical Statistics, treating measure theoretic probability and statistics to the level of Harald Cramér's 1946 book. Neither sequence used a text; both relied upon dittoed notes that were soon polished to stand essentially as texts. Wallis's notes were in fact an outgrowth of a text he had begun writing with Milton Friedman in 1940 under the title *Analytical Statistics*, only to be interrupted by the war.

Meanwhile the path toward the official formation of a department at Chicago was begun in May 1947, when a committee was appointed consisting of Tjalling Koopmans (Economics), Marshall Stone (Mathematics), and Allen Wallis, "to explore the need for statistics at the University and to consider both the organizational form that it should take and the possible personnel that should be added to the staff." When Stone was away on various leaves he was replaced on the committee, first by Paul Halmos and later by Saunders Mac Lane. Savage was added to the committee in November 1948. The outcome of their deliberations was predictable: They recommended the establishment of the Committee on Statistics to begin operations in fall 1949. They were initially granted a first year budget of \$36,000 (the committee had recommended \$75,000), to be added to the accounts of the Physical Sciences Division, where they would be housed in the same building as the Mathematics Department, Eckhart Hall.

Eckhart Hall was constructed in the late 1920s as the home of the Mathematical Sciences, with portraits of Newton and Gauss flanking the front door and the names of such other scientists as Laplace and Bernoulli, decorating other entrances. The mathematicians were pushed out in the early 1940s when the space was taken over by Enrico Fermi as the headquarters of the "Manhattan Project", but when Fermi moved to Los Alamos after the first sustainable chain reaction in December 1942, the mathematicians returned.

The University of Chicago's use of the academic unit of "Committee" is unusual. It is a department-like unit that can offer courses, admit graduate students, has a unit staff (often only one person), but usually lacks appointive power, so appointments all require a separate academic home in an established department. However, from the outset the Committee on Statistics was granted appointive power, becoming effectively a department, but the sailing was not easy. Even before fall 1949, $10,000 had been transferred away from the Committee budget to another department after strong pressure was applied. And by fall 1951, Wallis found that the name "Committee" caused difficulties in faculty recruitment, and in 1952 he requested a change in name to "Department" to recognize the status that they essentially held already. He encountered unexpected resistance: Vice President and Dean of the Faculty Pat Harrison was unenthusiastic, and he forwarded the request to Chancellor Larry Kimpton with a handwritten note, "I am reluctant to create a department. Wallis is both ambitious and competent. With a department back of him I am afraid that we are in for problems (whether we say "no" often or "yes" often)." Despite this, Kimpton would allow the change under two conditions, namely that the change in status was initially approved by the Physical Sciences Division, and that there would be no increase in funds tied to the change in name.

Under those conditions the matter was taken to the Divisional Policy Committee where narrowly provincial interests held sway. Notwithstanding that the Statistics budget was added to the Division explicitly for that and no other purpose, the Policy Committee would not approve the change without weighing this use of these funds against other divisional uses. Faced with this resistance, Wallis withdrew his request. The change in name would finally be approved on March 1, 1957, just as Wallis agreed to become Dean of the Business School, and Jimmie Savage (then serving as Acting Chair) was that day appointed Chairman of the new Department of Statistics. A new Policy Committee had approved the change in a sunny budgetary period, and the Division faculty had voted in favor, 71 to 8. In 1960, when again the university faced a budgetary squeeze, another Policy Committee threatened to withdraw support from the department, but cooler heads prevailed then, and since.

Appointments

The initial members of the committee were Wallis, Savage, Tjalling Koopmans, and Saunders Mac Lane. The brief minutes of their first four meetings survive and show they acted with dispatch. 21 November 1949: recommended the appointments of Charles M. Stein and Albert H. Bowker at the tenured level of Associate Professor. 5 December 1949: settled on the degree requirements for the Master's Degree (a formidable curriculum of 27 courses over 3 years). 16 December 1949: recommended appointment of Fred Mosteller as Full Professor, jointly with Sociology or Psychology. 20 February 1950: approved invitations at junior levels to Melvin Peisakoff, William Kruskal, T. E. Harris, and Raj Bahadur. Not all of these appointments were approved as recommended, but Bahadur joined the faculty in spring 1950, and Stein (as an Assistant Professor) and Kruskal (as an Instructor) arrived in fall 1950. Already in fall 1949 they had the instructional help of John Gurland who was an Assistant Professor with the Cowles Commission (then based at the university), and he remained in that role until he left Cowles in 1952.

In fall 1950, Leo Goodman and Murray Rosenblatt also joined the staff as Assistant Professors of Statistics. Goodman's appointment was from the beginning joint with Sociology. Even before Alex Brownlee arrived in fall 1951, Wallis could list 11 faculty as on the committee (including Milton Friedman and Harry Roberts) and 11 more as associated with the committee (including Paul Halmos and Sewall Wright) (Wallis, 1951). Within that first decade three other key appointments were made, David Wallace (1954), Paul Meier (1957), and Pat Billingsley (1958). These together with Bahadur, Goodman, and Kruskal became the core faculty for the next two decades when in the early 1960s Allen Wallis and Jimmie Savage left the university, Wallis to the Presidency of the University of Rochester and Savage to the University of Michigan and then to Yale. In the program's first decade the faculty was augmented by a parade of visitors: Joseph L. Hodges, Jr., Ulf Grenander, Esther Seiden, Henry Daniels, Dennis Lindley, Fred Mosteller, Donald Darling, Ingram Olkin, John Pratt, Kai Lai Chung, Bruno de Finetti, Jack Youden, Georg Rasch, and Frank Anscombe.

The late 1960s were a time of expansion in American higher education generally, and in statistics in particular. Between 1964 and 1969 Chicago hired seven excellent junior faculty. But this was also a time of heightened competition between institutions for academic talent, and by 1972 all but one of these had left the university, several despite determined efforts at retention: Stephen E. Fienberg, Morris L. Eaton, and Christopher Bingham all went to build programs at the University of Minnesota. The one who remained from the 1968 cohort was Michael J. Wichura, and he remained until his retirement in 2010. Subsequent new hires in the early 1970s included Shelby Haberman, Michael Perlman, Ronald Thisted, and Sandy Zabell. By the 1980 and 1990s the demand and budget were growing again, and the need to plan for the future was apparent. Several new faculty were hired at a tenured level (Stephen Stigler, Donald Rubin, Peter McCullagh, Steven Lalley), others hired as Assistant Professors were rapidly promoted to tenure (Wing Hung Wong, Michael Stein, Yali Amit, Augustine Kong, Per Mykland, Mary Sara McPeek, Xiao-Li Meng). Since 2000 Matthew Stephens, Wei Biao Wu, Mathias Drton, Keith Worsley, and John Reinitz have joined the list of tenured faculty with principal appointments in Statistics. Sadly, Worsley died within months of his arrival. In 2007 the department began a new expansion with a charge to grow in directions of scientific computation and applied mathematics. To date, Nina Hinrichs, the late Keith Worsley, John Reinitz, and Lek-Heng Lim have been hired within that initiative. We have also benefitted from the collegial and instructional support of several Part-Time Professors: Kirk Wolter (NORC), Mihai Anitescu (Argonne Labs), Paul Fischer (Argonne Labs).

A full list of the remarkable scholars who have held professorial appointments formerly, and those with current appointments on the department faculty, is appended. In addition, the program has hosted a large number of visiting scholars, many of whom have been engaged with the department for long periods following their visit. For example, Fred Mosteller chaired a visiting committee (which also included David Blackwell and Frank Anscombe) in 1970 and subsequently published their report (Mosteller 1971).

Instruction, Courses

As Allen Wallis wrote in 1952, "The character of the statistics program at the University of Chicago is in large measure determined by the character of the University itself." (Wallis 1952) This has entailed a linked dedication to both research and teaching, where these are not treated as separate enterprises. The early emphasis in teaching was on graduate instruction, at both the Master's and PhD levels, and this continues to the present day. The curriculum was constructed locally—it then and today differs from other graduate programs in flavor, but always with keen attention to national and international developments. The first decade's courses and their developing schedule reflect the evolving interests of a young and changing faculty, with both attention to basic sequences (301–303 and 351–353) and explorations to new areas in theory and methodology. Here, we had Savage's movement from "Statistical Theory of Decision Making" in 1950 to "Subjectivistic Statistics" in 1959, Charles Stein's "Theory of Minimum Risk" in 1950, David Wallace's "Series Expansions for Distribution Functions" in 1954, as well as visiting courses by Grenander, Lindley, Olkin, Chung, Youden, and Rasch, and much more.

The courses have often been accompanied by extensive notes; sometimes circulated to other universities and sometimes developed into books. Bill Kruskal's coordinate-free analysis of variance made it to the "notes" stage; Michael Wichura's extension and expansion of that topic achieved hard covers.

The earliest elementary course was offered by Harry Roberts in Winter Quarter 1955, Statistics 200, based upon the book *Statistics: A New Approach*, published in 1956 and jointly authored by Wallis and Roberts. Since then, the undergraduate college has played an increasingly important role in our instruction, with the innovative introduction of the use of computers (from the use of David Wallace's own system SNAP in the early 1970s, to Minitab in the 1980s, to the current use of R), and with the introduction of research level methods (such as categorical data analysis or principles of design) at an undergraduate level. In the 1990s the undergraduate program and the Master's programs had grown to the point that we added two positions of Senior Lecturer, one charged with helping to organize and maintain the undergraduate curriculum, the other with the supervision of the Master's program. We have been fortunate in attracting extremely capable people to fill both roles, the first by Michael Larsen and then Linda Collins and the second by Mei Wang.

Interdisciplinary Research

From its beginning in 1949 the program has put heavy emphasis upon collaborative research. Initially some of this was with outside agencies. Early on, the committee had two contracts with the U.S. Navy's ONR, one to support basic research and the other to consult on classified military research. The latter work grew out of contacts made during the war at SRG. This research has long since been declassified, and it involved such topics as the design and analysis of experiments involving brittle materials, munitions, and body armor. Within the university there were close contacts with social science through Leo Goodman's joint appointment in 1950 with the Sociology Department, and through Bill Kruskal and Jimmie Savage's insatiable interests in all manner of scientific investigation. Kruskal was heavily involved in particular with weather modification experiments ("rain making") in the 1950s and 1960s, and served as Dean of the Social Sciences Division 1974–1984. Don Rubin's appointment was joint with the Department of Education. With Paul Meier's arrival as a joint appointment with Biological Sciences, we began a long and intensive involvement with both medical and basic biological science research. This has taken many forms over the years, and at present there are two joint appointments with Human Genetics (Mary Sara McPeek and Matthew Stephens), one with Medical Genetics (Dan Nicolae), one with Molecular Genetics and Cell Biology and Ecology and Evolution (John Reinitz), and one member of our faculty, Ron Thisted, who chairs the Department of Health Studies.

Other significant centers of strength in statistics at the university without joint appointments but cooperating closely with the department have been the Schools of Business (including Arnold Zellner, George Tiao, Henri Theil, Al Madansky, Ruey Tsay, and Nick Polson) and Education (Darrell Bock, Ben Wright).

Influence Upon Other Programs

The University of Chicago has had important roles to play in other institutions' programs. Had Allen Wallis remained at Stanford in 1946 he would doubtless have become their first chairman; it was at his specific suggestion that Al Bowker was appointed Stanford's founding chair in 1948 (Stigler, 1999). Chicago's 1956 offer to Fred Mosteller was an essential catalyst to Harvard's decision to form a statistics department the following year with Fred as chair (Stigler 2008). Several Chicago faculty have gone on to form new departments at other institutions: Allen Wallis again (at Rochester, as President), Jimmie Savage (at Yale), Michael Perlman (at the University of Washington), Stephen Fienberg (at the University of Minnesota). Others have left to lead revitalization elsewhere: Paul Meier (Columbia), Leo Goodman (Berkeley), Donald Rubin (Harvard), Peter Donnelly (Oxford), Xiao-Li Meng (Harvard). Our students too have frequently gone on to play such roles. For example, the second student to received a Chicago PhD, Morrie DeGroot, was the founding chair of Statistics at Carnegie Mellon, and a later Chicago PhD, Robert Kass, has served as chair of that same department.

Published Works

The Chicago Statistics Department has been remarkably productive since its inception. The first decades saw the appearance of the papers that introduced the Kruskal–Wallis test, Goodman and Kruskal's work on association measures, Goodman's work on latent class models and loglinear models (with Shelby Haberman's later theoretical treatment as support and extension), Jimmie Savage's pathbreaking book *The Foundations of Statistics*, Wallace and Mosteller's study of the authorship of the Federalist papers, and the Kaplan–Meier estimate of a survival function, for many years the most cited paper in statistics. The 1957 book *The Statistical Analysis of Stationary Times Series* by Ulf Grenander and Murray Rosenblatt was conceived and begun when both were at Chicago in 1952–1953. One of Charles Stein's famous unpublished papers ("The Comparison of Experiments") was written at Chicago. Bahadur's series of papers on sufficiency and the theory of estimation have helped form the current state of those fields and are signaled by several eponyms including Bahadur Efficiency, the Bahadur example of the inconsistency of maximum likelihood, and the Bahadur Representation of Sample Quantiles.

Per Mykland's work, first on martingale models and methods, and more recently upon statistical finance has helped fuel a new school of attention to the role of statistics in dealing with uncertainty in modern finance. Wei Biao Wu continues the tradition of strong theoretical work on asymptotics in models of dependence. While at Chicago, Wing Hung Wong published deep and influential work on partial likelihood and (with Martin Tanner) on data augmentation, and Donald Rubin and Xiao-Li Meng made major progress in conceptualizing and operationalizing their models and approaches to modern sampling and missing data problems.

Alex Brownlee's textbook *Statistical Theory and Methodology* (1960, 1965) set a new standard for upper divisions texts for that time, thanks in part to the extensive suggestions and criticism from his departmental colleagues. Pat Billingsley's books, particularly *The Convergence of Probability Measures* (1968, 1999) and *Probability and Measure* (1979, 1986, 1995) remain classical treatments of their subjects. Peter McCullagh's *Generalized Linear Models* (with John Nelder) (1983, 1989) was the key text/research monograph leading to the establishment of one of the most fruitful areas in modern statistical modeling. His *Tensor Methods in Statistics* (1987) remains the definitive work on that subject, as are Michael Stein's *Interpolation of Spatial Data: Some Theory for Kriging* (1999) and Yali Amit's *2D Object Detection and Recognition: Models, Algorithms, and Networks* (2002) on theirs. Ronald Thisted's text *Elements of Statistical Computing* (1988) helped give impetus to that rapidly developing subject. Michael Wichura's *The Coordinate-Free Approach to Linear Models* (2006) is but the latest book to emerge from the department's long tradition of dedicated and innovative teaching, being based on lecture notes from a course which had influenced generations of students and itself been inspired by an earlier course by Bill Kruskal, who never published his book on the subject. Other books by current faculty include Mathias Drton's *Lectures on Algebraic Statistics* (2009) (with Bernd Sturmfels and Seth Sullivant), Kirk Wolter's *Introduction to Variance Estimation* (2nd Edition, 2007), Lars Peter Hansen's *Robustness* (2008) (with Thomas Sargent), Greg Lawler's *Random Walk: A Modern Introduction* (2010) (with Vlada Limic), and Stephen Stigler's *History of Statistics* (1986) and *Statistics on the Table* (1999).

Our faculty has also been heavily involved in professional service. This has included major editorial responsibility: Wallis, Wallace, and Stigler have edited *JASA*, Kruskal edited the *Annals of Mathematical Statistics*, Billingsley, Lalley, and Lawler have edited the *Annals of Probability*, McCullagh has edited *Bernoulli*. Six members of the faculty (Savage, Kruskal, Bahadur, Billingsley, Meier, and Stigler) have served as President of the IMS; Wallis (soon after he left Chicago) and Kruskal have served as Presidents of the ASA; Stigler has served as President of the ISI.

Degrees

The department has through the summer of 2010 graduated 166 students with PhDs and over 232 with Master's Degrees (we lack a count of Master's prior to 1962). The growth in the program is reflected in the following table:

Decade	Masters degrees	PhD degrees
1951–1960	NA	5
1961–1970	46	21
1971–1980	27	23
1981–1990	33	20
1991–2000	60	46
2001–2010	66	51
Totals	232	166

Fig. 1 The Jerzy Neyman honorary degree, June 12, 1959. L to R: Ghurye, Wallace, Goodman, Brownlee, Neyman, Billingsley, Meier, Kruskal

Fig. 2 At the John W. Tukey honorary degree, June 13, 1969. Kruskal (L) and Tukey

We are immensely proud of their achievements. The full list of these graduates can be found on our department Web page (People > Alumni).

The first degree awarded by the program was an honorary doctorate, awarded to Ronald A. Fisher in June 1952. It is the University of Chicago's practice to only award honorary degrees for extraordinary scholarship. The process is difficult (see Stigler 2009, for an example), but the department regards this as a way to publicly display our intellectual values. To date the recipients have been Ronald A. Fisher (1952), Harold Hotelling (1955), Jerzy Neyman (1959), Maurice S. Bartlett (1966), John W. Tukey (1969), Fred Mosteller (1973), Erich L. Lehmann (1991), Charles M. Stein (1992), Ulf Grenander (1994), Bradley Efron (1995), David Aldous (2000), Persi Diaconis (2003), Grace Wahba (2007), and David L. Donoho (2009). Citations may be found on our Web page (People > Honorary Degrees). The figures show pictures from five of these occasions (Figs. 1, 2, 3, 4, 5).

Fig. 3 The Fred Mosteller honorary degree, October 26, 1973. L to R: Perlman, Goodman, Meier, Mosteller, Wichura, Wallace, Kruskal, Billingsley, Haberman, Glick

Fig. 4 The Charles M. Stein honorary degree, June 12, 1992. L to R: Wallace, Kruskal, Bahadur, Thisted, C. Stein, Stigler, M. Stein

Fig. 5 The Grace Wahba honorary degree, June 8, 2007. L to R: McCullagh, Wichura, Stigler, Mykland, Wahba, Drton (rear), Lawler, McPeek, Nicolai (rear), Amit, Collins, Thisted

Appendices

I. Former faculty, in order of their first appointment	
1949–1950	W. Allen Wallis
	Jimmie Savage
	John Gurland
	Raj Bahadur
1950–1951	William H. Kruskal
	Charles M. Stein
	Murray Rosenblatt
	Leo A. Goodman
	Harry V. Roberts
1951–1952	K. Alex Brownlee
1954–1955	David L. Wallace
1956–1957	Sudhish Ghurye
1957–1958	Paul Meier
1958–1959	Patrick Billingsley
	Gopinath Kallianpur
1959–1960	Melvin Katz
1963–1964	Barry W. Brown
1964–1965	Gus Haggstrom
1966–1967	Christopher Bingham
	Morris L. Eaton
1968–1969	Stephen E. Fienberg
	Michael J. Wichura
	Frederick Glick
	Gordon Sande
1970–1971	Shelby Haberman
1973–1974	Joe Whittaker
1973–1974	Michael D. Perlman
1974–1975	Sandy Zabell
	Rod Little
1975–1976	Alan Izenman
1976–1977	Joan Sander Chmiel
1979–1980	Paul Sampson
1980–1981	Wing Hung Wong
1982–1983	Donald Rubin
	David Draper
1984–1985	Alan Welsh
1987–1988	Augustine Kong
1988–1989	Kung-Sik Chan
1991–1992	Xiao-Li Meng

(continued)

(continued)

1994–1995	Peter Donnelly Alejandro Murua Eric Kolaczyk
1998–1999	Zhiyi Chi
1999–2000	Michael Larsen
2000–2001	Partha Niyogi
2002–2003	Marc Coram
2003–2004	Linda Collins
2008–2009	Keith Worsley David Degras

II. Department Chairs

W. Allen Wallis 9/49-3/56
Jimmie Savage 3/56 -6/60
Paul Meier 7/60-6/66, 7/73-9/74, 10/83-6/86
William H. Kruskal 7/66-6/73
Michael D. Perlman 10/74-9/77
David L. Wallace 10/77-9/80
Patrick Billingsley 10/80-9/83
Stephen M. Stigler 7/86- 6/92, 7/05- 6/10
Peter McCullagh 7/92- 6/98
Michael L. Stein 7//98-12/01
Steven Lalley 1/02-6/05
Yali Amit 7/10-6/13

III. Current Faculty (Fall 2011)

Yali Amit
Mathias Drton
Lars Peter Hansen
Nina Hinrichs
Risi Kondor
John Lafferty
Steven Lalley
Greg Lawler
Lek-Heng Lim
Peter McCullagh
Mary Sara McPeek
Debashis Mondal
Per Mykland
Dan Nicolae
John Reinitz
Michael L. Stein
Matthew Stephens
Stephen M. Stigler
Ronald Thisted
Mei Wang
Kirk Wolter
Wei Biao Wu

References

Cramér H (1946) Mathematical methods of statistics. Princeton University Press, Princeton
Kendall MG (1943) The advanced theory of statistics. Griffin, London
Mosteller F (1971) Report of the evaluation committee on the University of Chicago Department of Statistics. Am Stat 25:17–24
Olkin I (1991) A conversation with W. Allen Wallis. Stat Sci 6:121–140
Stigler SM (1999) The foundations of statistics at Stanford. Am Stat 53:263–266
Stigler SM (2008) Statistics at Harvard? Am Stat 62:246–250
Stigler SM (2009) The honorable Erich L. Lehmann. In: Institute of Mathematical Statistics: lecture notes—monograph series, vol 57 (third Lehmann symposium volume), pp 8–10
Wallis WA (1951) Statistics at Chicago. Am Stat 5(April–May):16–17
Wallis WA (1952) Teaching methods and current statistical curricula at the University of Chicago. Estadistica J Inter Am Stat Inst 10:461–462
Wallis WA (1980) The Statistical Research Group, 1942–1945. J Am Stat Assoc 75:320–335

University of Connecticut Department of Statistics

Dipak K. Dey, Nitis Mukhopadhyay, Lynn Kuo and Ming-Hui Chen

The Department of Statistics at the University of Connecticut was founded in 1962. However, long before the official formation of this department, on September 16, 1950, Geoffrey Beall was appointed professor of statistics. As one of the major statistics departments in New England, it provides outstanding preparation for careers in academia, industry, or government. With a core faculty of 15 members whose teaching and research expertise span virtually all major specializations in statistical science, our department has both national and international reputation in undergraduate and graduate education, research, and service to the profession. The department offers BA/BS majors, BA/BS majors in mathematics and statistics jointly as well as Masters and PhD degrees in statistics.

Over the years, more than 117 PhD and 255 MS degrees have been awarded. Our graduates have found attractive positions in academia, industry, government, and other institutions including medical centers of repute.

The university's main campus is at Storrs, Connecticut where the department is housed on the third floor of CLAS Building. The department plays a key role in

---

University of Connecticut,
Department of Statistics,
215 Glenbrook Road U-4120,
Storrs, CT 06269-4120, USA
D. K. Dey (✉) · N. Mukhopadhyay · L. Kuo · M.-H. Chen
e-mail: dipak.dey@uconn.edu

N. Mukhopadhyay
e-mail: nitis.mukhopadhyay@uconn.edu

L. Kuo
e-mail: lynn.kuo@uconn.edu

M.-H. Chen
e-mail: ming-hui.chen@uconn.edu

A. Agresti and X.-L. Meng (eds.), *Strength in Numbers: The Rising of Academic Statistics Departments in the U.S.*, DOI: 10.1007/978-1-4614-3649-2_25,

hiring and nurturing statistics faculty members, including lecturers, who offer statistics courses in other campuses at Avery Point, Hartford, Stamford, Torrington, and Waterbury.

A History of the Department

Robert H. Riffenburgh, PhD Virginia Polytechnic Institute 1957, was hired in 1962 to design and develop the statistics program and headed the department until 1968. He resigned from the department in 1970 and moved to the Naval Ocean Systems Center and San Diego State University, California. Gottfried E. Noether, who received his PhD from Columbia University, was the second head of the department during 1968–1982. Uwe Koehn, who received his PhD from the University of Illinois and joined the University of Connecticut in 1968, served his first term as the head of the department from September 1982 to August 1987 and then served a second term from April 1991 to December 1996. Nitis Mukhopadhyay was the head of the department from September 1987 to October 1990. Dipak K. Dey served as the department head from January 1997 to June 2011. Joseph Glaz became the department head in July 2011.

During Riffenburgh's headship, the MS and PhD programs in statistics were developed and students recruited, an NIH biostatistics training grant was obtained, and the departmental colloquia series began with many illustrious speakers, including Frank Anscombe, George Barnard, M. G. Kendall, Tjalling Koopmans, P. C. Mahalanobis, C. R. Rao, Nicholas Rashevsky, John Tukey, Jimmy Savage, and Colin White. Also under Riffenburgh, the department expanded greatly, bringing on board Earl J. Bell (lecturer), Marilynn Dueker (lecturer), Friedrich Gebhardt, Alan E. Gelfand, Bruce McK. Johnson, Uwe Koehn, Gottfried E. Noether, Harry O. Posten, Dolly Smith (lecturer), Hugh Fairfield Smith, H. Jean Thiebaux, Derrick S. Tracy, and Sam Zahl. Timothy J. Killeen and Joseph Glaz were hired during Noether's headship. The department continued to grow during Koehn's term when Dipak K. Dey, Nitis Mukhopadhyay, Lynn Kuo, and Richard Vitale were hired. During Mukhopadhyay's term, Nalini Ravishanker, Suman Mazumdar, Igor Perisic, and Sandy Addelstein (lecturer) were hired. During Dey's term, Yazhen Wang, Vladimir Pozdnyakov, Ming-Hui Chen, Zhiyi Chi, Enrique Alvarez, Cyr M'Lan, Ofer Harel, Jun Yan, Sangwook Kang, Alejandro Vilagran, and Kathleen McLaughlin (lecturer) were hired. To get a glimpse at some of the faculty members, one may see Figures 1, 2, 3 and 4.

At the department's founding in 1962, statistical computation was conducted on electric calculators, which Riffenburgh replaced with electronic calculators. Computer access for large programs was limited to the University's IBM mainframe computer. By 1985, a single computer terminal and a card-punching machine were located in a corner of our mailroom. Until early 1987, every faculty member, student, and staff had a mainframe account with small annual allocation

Fig. 1 From *top* clockwise: Robert H. Riffenburgh, Hugh Fairfield Smith, Gottfried E. Noether, Samuel Zahl, H. Samuelson, Earl J. Bell, H. Jean Thiebaux, and Derrick S. Tracy

Fig. 2 From *left* to *right*: Harry O. Posten, David Salsburg, and Timothy J. Killeen

each. When Mukhopadhyay became the head in September 1987, he negotiated with the dean to buy two free-standing computers with connections to the IBM mainframe. This was a big move forward at the time which eventually energized faculty members to win several NSF awards to build our own high-power computer labs in the years that followed.

From the initial stages to the late 1980s, the administrative staff included Jeanne R. Young and Alice Kristoff. Since the early 1990s, Cathy Brown and Tracy Burke have been the mainstay in our central office.

Fig. 3 From *left* to *right*: **a** Nitis Mukhopadhyay, Alan E. Gelfand, and D. R. Cox during Pfizer colloquium 1994; **b** Nitis Mukhopadhyay, Uwe Koehn, Dipak K. Dey, and Harry O. Posten; and **c** Rick Vitale

Fig. 4 From *left* to *right*, *Top* row: Ming-Hui Chen, Cyr M'Lan, Zhiyi Chi, Vladimir Pozdnyakov, Alejandro Villagran, Jun Yan, Ofer Harel, and Sangwook Kang; *Bottom* row: Nalini Ravishanker, Joseph Glaz, Dipak K. Dey, Lynn Kuo, and Nitis Mukhopadhyay. Not shown are Suman Majumdar and Rick Vitale. Spring 2011

Pfizer Colloquium Series and Films for the Archive of ASA

The founding of Pfizer colloquium series began nearly 35 years ago in this department under Harry O. Posten and David S. Salsburg. Salsburg is this department's first PhD (1966) degree recipient and he held a high-level position at Pfizer Global Research and Development in Groton, Connecticut. The purpose of the colloquium was to make films of distinguished statisticians for the archive of the American Statistical Association (ASA). This joint enterprise has been continually supported by funding from this department, Pfizer Global Research and Development in Groton, Connecticut, and ASA.

Colloquium director Posten videotaped J. Neyman (1979), H. Cramer (1980), C. R. Rao (1981), G. E. Noether (1983), M. Hansen (1985), Herman Chernoff (1989), and many more. Timothy J. Killeen helped Posten with finer details of filming. From 1985, Mukhopadhyay contributed to the project. In a number of these wonderful films, Frank J. Anscombe from Yale University did a splendid job introducing the distinguished statistical scientist and the lecture.

Posten also worked with colleagues at other institutions to videotape distinguished statisticians for ASA's archive, for example, Ed Deming (1981), George Box (1982), R. C. Bose (1983), Oscar Kempthorne (1991), Marvin Zelen (1997), and many more.

Since Posten's passing, Mukhopadhyay has been directing this project, filming Brad Efron (2001), David R. Brillinger (2004), Chris Heyde (2005), Emanuel Parzen (2006), Barbara Bailar (2007), and Stephen E. Fienberg (2009).

This lengthy series of historically invaluable films are preserved in the archives of the ASA and distributed by the ASA.

New England Statistics Symposium

The New England Statistics Symposium (NESS) began at the University of Connecticut in 1987 under the inspiration of Herman Chernoff from Harvard University. Its purpose was to bring together New England statisticians to share research, discuss emerging issues, and network with colleagues. The annual NESS falls on a Saturday after April 15, hosted in odd years by the Department of Statistics at the University of Connecticut and in even years at another New England institution. Since 1988, other host institutions have included University of Massachusetts-Amherst (1988), University of Lowell (1990), Bentley College (1992), University of Rhode Island (1994), Worcester Polytechnic Institute (1996, 2006), Massachusetts Institute of Technology (1998), Brown University (2000), Yale University (2002), Harvard University (2004, 2010), Suffolk University (2008), and Boston University (2012). While only 52 participants attended the first NESS, current attendance exceeds 150.

The year 2011 marked the twenty-fifth anniversary of NESS. In the early years, the NESS was a one-day affair. Since 2009, NESS has expanded to 2 days by offering short courses on the first day and technical sessions on the second. Each year, NESS features two plenary speakers, special theme sessions, and contributed papers covering all aspects of statistics and probability. Plenary speakers, originally only from New England, in recent years have come from both far and near.

The impressive list of plenary speakers included John Hartigan, Donald B. Rubin, and Gottfried Noether (1987); Herman Chernoff and Andrew Rukhin (1989); Yali

Amit and Persi Diaconis (1990); Nan Laird and David Pollard (1991); Alan Gelfand and Robert Devaney (1992); Carl Morris and Stuart Geman (1993); Herman Chernoff and David Salsburg (1994); Andrew Barron and Cyrus Mehta (1995); Gerald Hahn and John Pratt (1996); Constantine Gatsonis and Richard Dudley (1997); James Stock and L. J. Wei (1998); Louise Ryan and Donald Geman (1999); David Mumford and Dipak K. Dey (2000); Joseph G. Ibrahim and Joseph Horowitz (2001); Leon Gleser and Bruce Levin (2002); Xiao-Li Meng and J. Michael Steele (2003); George Cobb and Andrew Lo (2004); Evarist Gine and David Harrington (2005); Lawrence H. Cox and Soren Bisgaard (2006); Paul Dupuis and Nitis Mukhopadhyay (2007); Andrew Lo and Martin Wells (2008); James O. Berger and Richard A. Davis (2009); Iain Johnstone and Jennifer Tour Chayes (2010); Brad Carlin and Jun S. Liu (2011); and Rick Durrett and Robert Kass (2012).

Since 2005, IBM T. J. Watson Research Center has sponsored the IBM student award given when UConn Department of Statistics hosts NESS. Since 2009, Smith Hanley Associates LLC has provided financial support for graduate students to attend NESS. In 2010, Microsoft and Google also sponsored the NESS student award.

Other Major Events and Activities in the Department

Joint UConn-UMass Colloquium Series

The joint UConn-UMass colloquium series was conceived in 1988 by Andrew Rukhin, University of Massachusetts-Amherst. In the fall semester, it is hosted by UMass-Amherst with a presentation by a UConn faculty and, in the spring semester it is hosted by us with a presentation by a UMass faculty. These colloquia are heavily attended by students and faculty alike from both institutions. This series has survived 24 years with no break, quite some feat by any standard.

International Workshop in Applied Probability 2006

The International Workshop in Applied Probability (IWAP) was originally orchestrated by Joe Glaz and some of his very close associates. The first and second IWAP were held at Caracas, Venezuela (2002) and Piraeus, Greece (2004), respectively. The third IWAP, was hosted by our department during May 2006. Bernoulli Society, IBM Research, Institute of Mathematical Statistics, Office of Naval Research, Taylor & Francis Group and the University of Connecticut sponsored the workshop. Nearly 200 participants from 26 countries took part in IWAP 2006. During the conference 160 invited talks were presented arranged in 40 parallel sessions, and 8 plenary lectures were delivered by Louis H. Y. Chen, Robert J. Elliott, Alan E. Gelfand, Steven Haberman, Nikolaos Limnios, Hosam

M. Mahmoud, Servet Martinez, and Gennady Samorodnitsky. Glaz was the chair of the local organizing committee. Faculty members from this department served extensively on the scientific and local organizing committees, organizing invited sessions, and presenting invited talks.

International Chinese Statistical Association Applied Statistics Symposium 2006

The International Chinese Statistical Association (ICSA) Applied Statistics Symposium was hosted by our department during June 2006. This annual statistics symposium featured keynote lectures by Xiao-Li Meng, James O. Berger, and Terry P. Speed, and plenary talks by Kung-Yee Liang and Jun S. Liu. There were also eight sets of nine concurrent oral presentation sessions and one poster session. More than 318 participants came from many countries around the world. This symposium received strong support from our university community and 11 companies, including Bristol-Myers-Squibb, Boehringer-Ingelheim Pharmaceuticals, Pfizer Global R&D, GSK, Organon, Amgen, Merck, IBM, Sanofi-Aventis, Johnson & Johnson, and Eisai Medical Research.

International Indian Statistical Association Conference 2008

The International Indian Statistical Association (IISA) conference was hosted by this department during May 2008, sponsored by us, the ASA, and partners from business and industry. It was attended by nearly 180 participants from academia (including many students), government, and industry from many parts of the world, including Bangladesh, Brazil, Brunei, Canada, Egypt, France, Germany, India, Italy, Japan, Russia, Spain, Sweden, and the USA and was the most highly attended IISA conference held in North America at the time. The program included plenary lectures by Jayaram Sethuraman and Marvin Zelen. A special feature was a series of lectures named after R. R. Bahadur, D. Basu, V. S. Huzurbazar, P. R. Krishnaiah, and P. V. Sukhatme and presented by Evarist Gine, Glen Meeden, L. J. Wei, Barry Arnold, and Sanat Sarkar, respectively. Dey, Mukhopadhyay (Chair), and Ravishanker served in the local organizing committee.

Statistical Consulting Services

Riffenburgh started the department's consulting service to other university departments in 1963. Koehn continued and expanded the consulting services in the 1980s and 1990s. Currently, the Statistical Consulting Services (SCS), primarily supported by this department, serves as a statistical consulting resource for external clients from business, government, and industry and provides statistical advice for faculty and graduate student research. It generates collaborative research projects and serves as a hands-on applied statistics training ground for our graduate students. The SCS also conducts short courses when appropriate. In the recent past, SCS has provided consulting services not only to various departments within the Storrs campus but also to the Law School, the University of Connecticut Health Center, and some local companies.

Department's Colloquia Series and Students' Seminar Series

This department retains its vigorous colloquium series throughout the year, hosting 12–15 colloquia each semester presented by active researchers in statistical science. Recently, colloquia series were run by Chi, Glaz, Harel, Yan, and Kang.

Additionally, we host a separate Students' Seminar Series, also held every week. Our graduate students make presentations accessible to all students and are encouraged to post their presentations on our website as student journals. This extremely important activity started in the late 1980s and is still going strong, currently coordinated by Kuo.

Pfizer Global Research and Development Student Fellowship Program

In December 2007, this department and Global Research & Development of Pfizer, Inc. jointly agreed to institute a Pfizer Global Research & Development Student Fellowship program. Under this agreement, a student fellow is selected from the graduate students in the department. This student fellow works at Pfizer for 10 hours per week during regular semesters and 20 hours per week in the summer. The student fellows included Wangang Xie (2008), Miaomiao Ge (2009–2011), and Ouyang Guang (2011–2012).

Traveler's Research & Development Student Fellowship Program

In 2000, this department and Travelers, Inc. jointly agreed to institute a Traveler's Research & Development Student Fellowship program, under the leadership of our alum Keith Holler. Under this agreement, a student fellow was selected from the graduate students in the department. This student fellow worked at Travelers for 10 hours per week during regular semesters and 20 hours per week in the summer.

The student fellows included Hai Xu and Yuchen Gu, both of whom later took permanent positions at Travelers. Currently, this program is being continued under the leadership of our alum, Patrick Wang from Travelers.

Founding and Organizing International Conferences Elsewhere

Glaz has been instrumental in inspiring and then continues organizing the IWAP internationally every 2 years. Mukhopadhyay inspired and founded the International Workshop in Sequential Methodologies (IWSM) and continues to organize the IWSM internationally every 2 years. Mukhopadhyay was instrumental in founding the first International Sri Lankan Statistical Conference that was hosted by University of Peradeniya, Kandy, Sri Lanka during December 28–30, 2004. De Silva and Mukhopadhyay published the jointly edited *Proceedings of the International Sri Lankan Statistical Conference: Visions of Futuristic Methodologies*. Mukhopadhyay was a Vice-Chairman for the sixth Calcutta Triennial Statistics Symposium that was held during December 2006 in Calcutta, India. He coedited its proceedings jointly with Manisha Pal as a special issue of the *Calcutta Statistical Association Bulletin*, October 2009.

Kudos to the Department

The department includes a large number of high-profile colleagues who are elected Fellows and Members of some of the internationally leading learned societies. Chen, Dey, Glaz, Mukhopadhyay, and Vitale are elected Fellows of the Institute of Mathematical Statistics. Chen, Dey, Glaz, Kuo, Mukhopadhyay, and Ravishanker are elected Fellows of the ASA. Chen, Dey, Mukhopadhyay, and Vitale are elected Ordinary Members of the International Statistical Institute.

Dey and Glaz are both elected members of Connecticut Academy of Arts and Sciences. Dey received the Outstanding Alumni Award (2007) from Purdue University. Starting July 1, 2011, Dey is serving as an Associate Dean in the College of Liberal Arts and Sciences, after serving this department as its head during the past 14 years. He has become a Fellow of AAAS in 2011.

Glaz (2006), Mukhopadhyay (2008), and Pozdnyakov (2006) received the Abraham Wald Prize in Sequential Analysis. Dey (2008) received the Board of Trustees Distinguished Professorship from the University of Connecticut. Harel received a 5-year career award supported by National Institute of Health (2010–2015) for missing data methodologies in HIV prevention trials.

A large number of us have provided outstanding support to our profession by serving in the editorial board of numerous journals of international repute. In particular: Chen has been a co-editor of *Sankhyā* since 2004, and one of the editors of *Bayesian Analysis* since 2010. Dey was a co-editor of *Sankhyā* during

1999–2001 and editor of *IMS Bulletin* during 1998–2001. Glaz founded *Methodology and Computing in Applied Probability* in 1997 and continues as editor-in-chief. Mukhopadhyay co-edited *Sequential Analysis* in 2003 and has been its editor-in-chief since 2004. Ravishanker has been an editor of *Applied Stochastic Models in Business and Industry - Theory & Methods* since 2008.

Research Profile

The areas of expertise of the faculty recognized internationally include: *applied probability* (Chi, Glaz, Mukhopadhyay, Pozdnyakov, Vitale), *Bayesian phylogenetics* (Chen, Kuo), *Bayesian statistical methodology and computation* (Chen, Dey, Harel, Kuo, M'lan, Ravishanker, Villagran, Yan), *categorical data analysis* (Chen, Dey), *causal inference* (Harel), *convex-geometric methods in probability and statistics* (Vitale), *copulas* (Yan), *decision theory* (Chen, Dey, Kuo, Mukhopadhyay), *design and analysis of epidemiological studies* (Kang), *diagnostic testing* (Harel), d*istribution theory and methods* (Dey, Mukhopadhyay), *econometrics* (Dey, Yan, Mukhopadhyay), *environmental engineering and transportation engineering* (Ravishanker), *environmental sampling* (Mukhopadhyay), *estimating equations* (Yan), *geometrical probability* (Glaz), *large deviations* (Chi), *limit theorems and approximations* (Mukhopadhyay, Pozdnyakov), *longitudinal data analysis* (Chen, Dey, Harel, Kuo, Yan), *mathematical finance* (Dey, Pozdnyakov), *microarray data analysis* (Chen, Kuo, M'Lan), *missing data analysis* (Chen, Harel, Mukhopadhyay, Yan), *multiple comparisons* (Chi, Glaz, Mukhopadhyay), *multivariate analyses* (Dey, Mukhopadhyay, Ravishanker), *nonparametrics* (Mukhopadhyay), *nonparametric Bayesian statistics* (Dey, Kuo), *occurrence of patterns* (Pozdnyakov), *probability approximations* (Glaz, Mukhopadhyay), *probability inequalities* (Glaz, Mukhopadhyay, Vitale), *reliability* (Dey, Mukhopadhyay), *scan statistics* (Glaz), *selection and ranking* (Mukhopadhyay), *sequential analysis* (Glaz, Harel, Mukhopadhyay, Pozdnyakov, Yan), *simultaneous inference* (Dey, Glaz, Mukhopadhyay, Ravishanker), *spatial statistics* (Dey, Yan), *statistical genetics* (Dey), *statistical inference* (Dey, Majumdar, Mukhopadhyay), *statistical methods in actuarial science and marketing* (Ravishanker), *statistical shape analysis* (Dey), *stochastic geometry* (Vitale), *stochastic processes* (Chi), *survey sampling* (Harel, Kuo, Mukhopadhyay), *survival data analysis* (Chen, Dey, Kang, Kuo, Mukhopadhyay, Ravishanker, Yan), *statistical image analysis* (Dey), and *time series analysis* (Dey, Mukhopadhyay, Ravishanker, Villagran).

This research profile may be broadly classified into three clusters supplemented with a list of selected books and papers in the reference section.

Statistical Methodology

Chen and Dey developed a new theory for skew link functions useful for categorical and longitudinal data analysis. They also made contributions to model determination, variable selection, prior elicitation, and Bayesian meta-analysis. Chen has advanced Bayesian statistical computing by providing new algorithms along with theory and software. Chi has developed major new results in multiple hypothesis testing. Glaz has developed probability inequalities and approximations for both discrete and continuous scan statistics. Dey developed unified theory, methodology, and application of skew-elliptical distribution.

Gelfand retired from here in 2002 and immediately started on a new career path. We appreciate the contributions that he had made during the years from 1969 to 2002. His blockbuster paper with Adrian Smith on sampling-based approaches to calculating marginal densities was published in 1990.

Harel has come up with a novel approach and coined the phrase "outfluence" to evaluate the importance of missing observations in data analysis. He has also provided more insight for the understanding of partial and latent ignorability of missing data and multiple imputations. Kuo formulated a stochastic search procedure using indicator functions for the Bayesian variable selection. She was the originator of sampling-based procedures for nonparametric Bayesian statistics, survival analysis, software reliability, and quantal bioassay. M'Lan has developed Bayesian sample size determination methods for cohort studies.

Mukhopadhyay has published intensely on key sequential, two-stage, and multistage sampling designs from perspectives of point estimation, hypothesis testing, ranking and selection, clinical trials, and nonparametrics. He also published a series of papers on sufficiency, ancillarity, minimum variance unbiased estimation, UMP tests, maximal invariants, and UMPI tests. Ravishanker has introduced Bayesian sampling-based approaches to ARMA processes and multivariate survival analysis with positive stable frailties. Yan has proposed novel methods for copulas, temporal process regression, dynamic survival models, and spatial statistics.

Applied Probability and Stochastic Processes

Chi has made contributions to large deviation and stochastic processes. Majumdar has worked on probability metrics and the central limit theorem. Glaz has made theoretical contributions in scan statistics, probability inequalities for multivariate distributions with dependence structures, approximations and inequalities for order statistics for dependent observations, and combinatorial and geometrical probability. Mukhopadhyay has produced probability distributions of stopping times, bounds, and approximations for t- and F-distributions, independence, correlations, nonassociation, multivariate t-distributions, and optimality of Jensen's inequality with unusual applications. Pozdnyakov has investigated patterns of waiting times in gambling, martingale approaches to scan statistics, and future prices. Vitale has discovered a strong law of large numbers for random compact sets and has focused on the relationship between convex geometry and questions in probability and statistics, including novel stochastic and correlation inequalities, bounds, and moments.

Interdisciplinary Research

This is one of the hallmarks in this department. Most faculty members are proactively involved in interdisciplinary research with other scientists. It not only helps to shape the research directions for us, but also helps to foster visibility within our university and scientific community.

Chen works with Anthony V. D'Amico at Harvard University on various issues arising in prostate cancer research. He also works with other scientists on Bayesian phylogenetics, clinical trials, meta-analysis, missing data, microarray, and structural equations. Chi has collaborated with neuroscientists on projects studying neural activity and its relation to sensory input and learning. Dey has collaborated with Kent Holsinger at UConn Ecology and Evolutionary Biology (EEB) on population genetics and ecology. He also collaborated with other scientists on cancer risk studies, data mining, proteomics, survival analysis, and statistical image processing.

Glaz collaborated with Peter Willett at UConn Electrical Engineering Department on applications of scan statistics. Harel has served as a biostatistical consultant, nationally and internationally, since 1997. Through his collaborative consulting, he has been involved with a variety of research areas, including Alzheimer, diabetes, nutrition, HIV/AIDS, and alcohol and drug abuse prevention.

Kuo collaborates with David Rowe at University of Connecticut Health Center and Dong-Guk Shin from Computer Science Engineering at UConn on microarray analysis, pathway, genetic network, and stem cell research. She also collaborates with Chen jointly with Lewis at UConn EEB on Bayesian phylogenetics. Mukhopadhyay has collaborated with Alex Tartakovsky and Aleksey Poluchenko from University of Southern California on network surveillance, change-point problems, and security monitoring.

Ravishanker works with J. N. Ivan at UConn on transportation engineering and with other scientists on actuarial science, environmental engineering, and marketing. Yan works with M. Willig at UConn Center for Environmental Sciences & Engineering on spatial–temporal data modeling and analysis. He also works with other scientists on biostatistical research.

Some Former PhDs

In the history of this department, we have been very fortunate that all its past graduates have made their own mark in the profession. We distinctly feel proud of all of them and we thank them for giving us the great opportunity to be a part of their lives during the formative years. Here, we present a short but representative list:

David Salsburg (PhD 1966) became one of the Vice-Presidents of Pfizer Global Research in Groton, Connecticut. Chris P. Tsokos (PhD 1968) has been the Director and a University Distinguished Professor at the University of South Florida-Tampa. Matthew Goldstein (PhD 1970) has been the Chancellor at CUNY. Salsburg (2008) and Tsokos (2010) won the first two Alumni Awards from this department.

Brad Carlin (PhD 1989) is the Head for the Division of Biostatistics and a University Chair Professor at the University of Minnesota. He who received the APHA Mortimer Spiegelman Award in 2000. Saibal Chattopadhyay (PhD 1993) became the Dean at the Institute of Management Calcutta, India. Bani Mallick (PhD 1994) became a University Distinguished Professor at Texas A&M in 2011. Sudipto Banerjee (PhD 2000), Division of Biostatistics, University of Minnesota, received the APHA Mortimer Spiegelman Award in 2011.

Summary and Outlook

In the 50 years of this department's history, it has made extensive and exemplary contributions to the profession of statistics. The vibrancy and relevance of its curriculum and research arguably have been second to none. The diversity, breadth, and depth this department has to offer could easily be the envy of many. But that does not mean that we are standing still and basking in our past glory. We will not do so. The world changes and presents daunting new demands and we pledge to change and adapt to live up to the future. We are continuing to train ourselves and our students to face the challenges that are yet to come.

Acknowledgment Our sincere thanks go to Robert H. Riffenburgh for providing helpful information. We also thank Keith Conrad for sharing the minutes from past UConn Board of Trustee meetings.

References: Selected Books and Papers

Selected Books and Edited Volumes

Chen M-H, Shao Q-M, Ibrahim JG (2000) Monte Carlo methods in Bayesian computation. Springer, New York

Dey DK, Müller P, Sinha D (eds) (1999) Practical nonparametric and semiparametric Bayesian statistics. Lecture notes series, vol 133. Springer, New York

Dey DK, Ghosh SK, Mallick BK (eds) (2001) Generalized linear models: a Bayesian perspective. Marcel Dekker, New York

Glaz J, Naus J, Wallenstein S (2001) Scan statistics. Springer, New York

Glaz J, Pozdnyakov V, Wallenstein S (eds) (2009) Scan statistics: methods and applications. Birkhauser, Boston

Harel O (2009) Strategies for data analysis with two types of missing values: from theory to application. Lambert Academic Publishing, Saarbrucken

Ibrahim JG, Chen M-H, Sinha D (2001) Bayesian survival analysis. Springer, New York

Mukhopadhyay N (2000) Probability and statistical inference. Marcel Dekker (Taylor & Francis Group), New York

Mukhopadhyay N, Solanky TKS (1994) Multistage selection and ranking procedures: second order asymptotics. Marcel Dekker, New York

Ravishanker N, Dey DK (2002) A first course in linear model theory. Chapman & Hall, CRC, Boca Raton

Selected Articles

Artstein Z, Vitale RA (1975) A strong law of large numbers for random compact sets. Ann Probab 3:879–882

Asgharian M, M'Lan CE, Wolfson DB (2002) Length-based sampling with right censoring. J Am Stat Assoc 97:201–209

Chen M-H (1994) Importance weighted marginal Bayesian posterior density estimation. J Am Stat Assoc 89:818–824

Chen M-H, Ibrahim JG, Sinha D (1999) A new Bayesian model for survival data with a surviving fraction. J Am Stat Assoc 94:909–919

Chi ZY, Geman S (1998) Estimation of probabilistic context-free grammars. Comput Linguist 24:299–305

Dey DK, Srinivasan C (1985) Estimation of a covariance matrix under Stein's loss. Ann Stat 13:1581–1591

Fine JP, Yan J, Kosorok MR (2004) Temporal process regression. Biometrika 91:683–703

Gelfand A, Dey DK (1994) Bayesian model choice: asymptotics and exact calculations. J R Stat Soc B 56:501–514

Gelfand AE, Kuo L (1991) Nonparametric Bayesian bioassay including ordered polytomous response. Biometrika 78:657–666

Ghosh M, Mukhopadhyay N (1981) Consistency and asymptotic efficiency of two stage and sequential estimation procedures. Sankhyā Ser A 43:220–227

Glaz J (1989) Approximations and bounds for the distribution of the scan statistic. J Am Stat Assoc 84:560–566

Glaz J, Naus J (1991) Tight bounds and approximations for scan statistic probabilities for discrete data. Ann Appl Probab 1:306–318

Harel O, Zhou XH (2006) Multiple imputation for correcting verification bias. Stat Med 25:3769–3786

Harel O, Zhou XH (2007) Multiple imputation review of theory implementation and software. Stat Med 26:3057–3077

Kang S, Cai J (2009) Marginal hazards regression for case-cohort studies with multiple disease outcomes. Biometrika 96:887–901

Kang S, Cai J (2009) Marginal hazards regression for retrospective studies within cohort with possibly correlated failure time data. Biometrics 65:405–414

Kundu S, Majumdar S, Mukherjee K (2000). Central limit theorems revisited. Stat Probab Lett 47:265–275

Kuo L, Yang T (1996) Bayesian computation for nonhomogeneous Poisson process in software reliability. J Am Stat Assoc 91:763–773

Majumdar S (1992) On topological support of Dirichlet prior. Stat Probab Lett 15:385–388

Marriott J, Ravishanker N, Gelfand A, Pai J (1996) Bayesian analysis of ARMA processes: complete sampling-based inference under exact likelihoods. In: Berry DA, Chaloner K, Geweke JK (eds) Bayesian analysis in statistics and econometrics: essays in honor of Arnold Zellner. Wiley, New York, pp 243–256

Mukhopadhyay N, Duggan WT (1997) Can a two-stage procedure enjoy second-order properties? Sankhyā Ser A 59:435–448

Pozdnyakov V, Steele JM (2004) On the martingale framework for futures prices. Stochastic Process Their Appl 109:69–77

Pozdnyakov V, Glaz J, Kulldorff M, Steele JM (2005) A martingale approach to scan statistics. Ann Inst Stat Math 57:21–37

Qiou Z, Ravishanker N, Dey DK (1999) Multivariate survival analysis with positive stable frailties. Biometrics 55:637–644

Stein ML, Chi ZY, Welty LJ (2004) Approximating likelihoods for large spatial data sets. J R Stat Soc B 66:275–296

Villagran A, Huerta G (2006) Bayesian inference on mixture-of-experts for estimation of stochastic volatility. Adv Econometr 20:277–296

Villagran A, Huerta G, Jackson C, Sen M (2008) Computational methods for parameter estimation in climate models. Bayesian Anal 4:823–850

Vitale RA (1979) Approximation of convex set-valued functions. J Approx Theory 26:301–316

Wolfson C, Wolfson D, Asgharian M, M'Lan CE, Ostbye T, Rockwood K, Hogan DB (2001). A reevaluation of the duration of survival after the onset of dementia. New Engl J Med 344:1111–1116

Yan J, Fine JP (2004) Estimating equations for association structures. Stat Med 23:859–874

University of Florida Department of Statistics

Alan Agresti, William Mendenhall III and Richard Scheaffer

Statistics at the University of Florida (UF) evolved from a small Statistical Laboratory in the 1950s to a department formed in 1962 to coordinate statistics teaching, collaborative research, and consulting around campus. It has grown to now encompass a highly rated graduate program with 17 faculty and 65 MS and PhD students. Throughout its history, the department has emphasized excellence in its teaching role, including the development of course textbooks, and about 7,500 students a year now take its courses.

Early Days: A UF Statistical Laboratory

The first course devoted to Statistics at University of Florida (UF) seems to have been given in 1926 in the Mathematics Department—"Introduction to Statistical Theory." This was followed in 1927 by "Elements of Statistics" in the General Business Department. In 1929 the graduate-level "Mathematical Statistics" was offered in the Mathematics Department and taught regularly from the book of the same name that had been published 2 years earlier by Henry Rietz of the University of Iowa. Also in 1929, "Agricultural Statistics" was introduced in the Agricultural Economics Department and was later sometimes taught by J. Wayne Reitz, who served as President of UF from 1955 to 1967.

University of Florida,
Department of Statistics,
102 Griffin-Floyd Hall,
Gainesville, FL 32611-8545, USA
A. Agresti (✉) · W. Mendenhall III · R. Scheaffer
e-mail: aa@stat.ufl.edu

R. Scheaffer
e-mail: scheaffe@stat.ufl.edu

A. Agresti and X.-L. Meng (eds.), *Strength in Numbers: The Rising of Academic Statistics Departments in the U.S.*, DOI: 10.1007/978-1-4614-3649-2_26,

The UF established a Statistical Laboratory in 1951, directed by Herbert A. Mayer of the Mathematics Department, to provide statistical consulting and computing services to students and faculty. Mayer had received his PhD at Iowa under Henry Rietz, and as director he reported directly to the dean of the Graduate School. Consultants at the Laboratory in the mid-1950s included David B. Duncan and Victor Chew, who taught a sequence of courses in experimental design and statistical analysis that had been initiated in 1949 in the Agronomy Department.

The Statistical Laboratory organized a symposium on Monte Carlo methods in March 1954, sponsored by the US Air Force, which was followed by a meeting of the Biometric Society (ENAR). Participants at these meetings included George Snedecor, Gertrude Cox, Maurice Kendall, and John Tukey. At ENAR, David Duncan presented his work on the multiple range test (Duncan 1955). The mid-1950s were productive years for the laboratory, and one might have expected a Department of Statistics to be formed, but this was not to be. Duncan left for the University of North Carolina in 1956 and Chew for North Carolina State in 1957. Their departure left a void in the teaching and consulting facilities, although Willard Ash arrived in 1957 to take up some of the slack, remaining at UF until 1967.

In 1958, President J. Wayne Reitz appointed Dr. Alva Esmond (A. E.) Brandt as "Statistician, Head of Statistical Section, Agricultural Experiment Station." Brandt had co-directed with George Snedecor, his advisor, the Mathematics Statistical Service at Iowa State when it was founded in 1927, and he was Snedecor's right-hand man when the Statistical Laboratory was founded there in 1933. Brandt's assignments included overseeing the formation of a statistics department. He was named head of a newly created Department of Statistics in the College of Agriculture in January, 1962. Brandt, who was due to retire in 1963, spent much of his last official year at UF working out budgetary details and interviewing prospective chairs for the department, hiring William Mendenhall III. By this time, course offerings in Statistics were scattered across at least eight departments on campus. A degree with an emphasis in Statistics could be obtained through the Department of Mathematics.

The Mendenhall Years, 1963–1977

Dr. William Mendenhall, a PhD graduate of the Statistics Department at North Carolina State University, came to UF from Bucknell University in 1963 to head the new Department of Statistics. He later noted that when he arrived for his interview in early 1963, there were no faculty members in the statistics department, no vacant position line items, no budget, no approved degrees, and no courses! On the positive side, three strong forces favored the new department. First, the College of Agriculture and the Agricultural Experiment Station were in desperate need of statisticians to serve as statistical consultants upon Dr. Brandt's retirement. Second, Leon Grinter, Dean of the Graduate School, had pushed for a statistics department as part of his goal of making the university a true research institution. Third, Robert Mautz, Vice President for Academic Affairs, was faced with a tight university budget and saw the opportunity of saving money by combining the many introductory statistics courses taught throughout the university into a few courses taught by the new department.

The immediate problem in spring 1963 was solving the staffing problem for the coming September. Mautz had suggested that departmental staff members could be drawn from departments in various colleges, but only one of these persons had a graduate degree in statistics, Willard Ash in Agronomy. Mendenhall recruited Larry Kupper, a chemical engineer by training whom he had met at a short course that he taught for the Bureau of Mines in Colorado in 1962, both to help teach the new introductory courses in statistics and to enter the new graduate program in statistics. In fall 1963, Larry Kupper and John Cornell were the first two graduate students in the department.

Roadblocks were still encountered. At first, the Mathematics Department refused to transfer their advanced statistics courses, and other departments refused to eliminate their competing courses until pushed to do so by Vice President Mautz. Receiving official approval for a Master's program in statistics was delayed, partly because of a letter from the President of Florida State University (FSU) to the President of UF noting that FSU had statistics faculty with strong theoretical interests and backgrounds, so the UF Statistics department should be restricted to developing an applied program. For fall 1964, Mendenhall was able to hire Dr. P. V. Rao, an assistant professor at the University of Georgia. Fred Barnett, a new PhD graduate from VPI (Virginia Tech), accepted the second teaching appointment, and Frank Martin was hired as a consultant in the Agricultural Experiment Station. New graduate students in 1964 included Robert Beaver and Richard Scheaffer, who had been students in the Mathematics Department at Bucknell University. For several years after that, the new classes of graduate students included students coming through the pipeline established from Bucknell by Mendenhall.

The research capabilities of the department improved significantly with the hiring from VPI in 1965 of John Saw, who had been a PhD student of F. N. David at University College, London. The Master's degree was approved in 1964, and the PhD program in 1968. The academic home of the department then moved to the College of Arts and Sciences, but strong ties with agriculture remained through a consulting unit, and a separate unit was established for consulting in the Health Sciences Center.

The remaining years of the Mendenhall chairmanship included the hiring of several young faculty who were to have long careers at UF (Fig. 1), including Richard Scheaffer, Ramon Littell, Jim McClave, Jonathan Shuster, Mark Yang, Alan Agresti, Dennis Wackerly, André Khuri, and Randy Carter and Ronald Marks in the Health Science Center consulting unit. Others who served several years in the department included James Boyett, David Hughes, and Lyman Ott. Also hired was Carol Rozear, who served as an administrative assistant in the

Fig. 1 Statistics faculty at the University of Florida in 1976. Row 1: Victor Chew and Frank Dietrich; row 2: Alan Agresti, Frank Martin, and John Saw; row 3: John Cornell, Robert Smidt, and James Boyett; row 4: Mark Yang, Randy Carter, Larry Peele, and William Mendenhall; row 5: Richard Scheaffer, P. V. Rao, Ronald Marks, Andre Khuri; row 6: Dennis Wackerly, Ramon Littell

department from 1970 until 2008 and who helped the program run smoothly for several chairs and for several generations of graduate students (Fig. 2).

In the early days, the department prided itself on being socially congenial. One tradition was an annual springtime "lasagna dinner," at which faculty would bring Italian food for an end-of-the-academic-year celebration at the chair's home. Many of the graduate students and young faculty were also good athletes, including a few who won All-American honors as undergraduates in various sports. As a consequence, for many years the department fielded very successful intramural sports teams in softball, basketball, and flag football.

The Scheaffer Years, 1977–1989

Dr. Richard Scheaffer, who had joined the faculty in 1967, noted that he chaired the department during its "teenage years," with all that is implied by that statement. By 1977, the department had clearly established strengths in teaching and applied research. Scheaffer made it a goal to strengthen the research capabilities. Since the faculty members had nearly all been hired as fresh PhD graduates, one

Fig. 2 Three chairs with the department's administrative assistant: Ron Randles, Carol Rozear, Bill Mendenhall, Dick Scheaffer

goal was to add senior faculty. Among these, two full professors were hired who already had strong international reputations for their research accomplishments—Ronald Randles from the University of Iowa in 1981 and Malay Ghosh from Iowa State University in 1982. Additional junior faculty hired who went on to have long careers at UF included Andrew Rosalsky, James Booth, Myron Chang, Jane Pendergast in the biostatistics consulting unit, and Ken Portier in the Institute of Food and Agricultural Sciences (IFAS) consulting unit. Others hired during this period who spent significant time at UF included Rocco Ballerini, Michael Conlon, James Kepner, Susan McGorray, Ralph O'Brien, and Geoff Vining, as well as Marilyn Saddler on the staff. During the early years of his administration, Scheaffer obtained funds to bring several noted statisticians, including C. R. Rao, William Cochran, and David Blackwell, to the department for extended visits.

By the twenty-fifth anniversary of the department in 1988, the department had 34 positions for statisticians, 28 of which were at the PhD level, and six at the Master's level, to serve in teaching, research, and consulting roles. About 50 graduate students were working on degrees and about 3,000 students were taking courses each semester from the department. Faculty research productivity increased, and seven faculty members were named Fellows of the ASA. But, in the early days the department had developed a reputation around campus for the excellence of its teaching, and faculty continued to pay careful attention to that role and regularly received campus-wide teaching awards. Richard Scheaffer himself had a major impact in the field of statistical education, helping to promote the teaching of statistics in secondary schools and initiate an Advanced Placement

Statistics course. With the growth of the department, having sufficient space was a continual problem, and the department moved twice to other buildings.

The first major research grant to the department came in 1979 when NIH funded the Pediatric Oncology Group Statistics Office. After that, grants to faculty members grew steadily, with awards coming through NIH, NSF, ONR, EPA, USDA, and numerous foundations and industrial groups. In addition, the faculty played a major collaborative role in grants that were awarded to other units of the university.

The Second Quarter Century

During its second quarter century, the department's objective continued to be to build a research program that balanced theory with applications while maintaining excellence in its service role. The success of the service role was reflected by continually increasing enrollments in basic statistics courses at the undergraduate and graduate levels.

The tradition of the statistics department having long-serving heads continued when Ronald Randles served as chair from 1989 to 2000. During this period, the financial status of the UF followed a roller coaster pattern that continues today, reflecting the state of Florida's often precarious financial situation. This often affected the department's ability to keep faculty and hire new ones. In 1992 the department moved to Griffin-Floyd Hall, one of the oldest buildings on campus, when it was renovated after being empty since 1978. This building had been the original home of the College of Agriculture and the location for its teaching of statistics more than 60 years previously. The department remains there today. After John Saw passed away in 1990, his mother made a donation to fund a departmental library in his name in the building.

During this period, the department offered several 2 1/2- day short courses at an Orlando hotel during the university's spring break each year, on topics in which the faculty had special expertise. These were well attended, largely by statisticians working in industry, and brought in significant discretionary funds for the department. A research initiative, starting in 1999 and continuing to the present day, was a Winter Workshop held in early January each year, focusing on a currently vibrant research area, and inviting worldwide leaders in the area to present talks (Fig. 3).

Several new hires during this period have had long careers in the department, including James Hobert, Brett Presnell, and the instructors Larry Winner and Maria Ripol. An external search for a chair to replace Randles led to the hiring of George Casella from Cornell University in 2000. Casella served as chair from 2000 until 2006, followed by interim chair Ramon Littell and then Mike Daniels, who is the current chair. New hires in these years included the senior hires, Linda Young, Mary Christman, Mike Daniels, and Hani Doss, the junior hires Pamela Ohman, Bhramar Mukherjee, Alex Trindade, Trevor Park, Kshitij Khare, and Xu Han, and the

Fig. 3 Attendees of the UF Winter Workshop in 2001 on Monte Carlo methods

Fig. 4 Statistics faculty and former faculty at the University of Florida in 2010. From left Bhramar Mukherjee, Linda Young, Trevor Park, Maria Ripol, Ramon Littell, George Casella, Mary Christman, Andre Khuri, Richard Scheaffer, John Cornell, Michael Daniels, Alan Agresti, Kshitij Khare, Hani Doss, James Hobert, Jon Shuster, Andy Rosalsky, Mark Yang, Robert Dorazio, Malay Ghosh, Larry Winner, Ron Randles, Jane Pendergast, Sam Wu

instructors Megan Mocko and Yasar Yesilcay who now together with Ripol and Winner handle most of the undergraduate service course teaching (Fig. 4).

A generous donation to the department by the Gill Foundation, created by Jack and Linda Challis Gill, led in 2001 to the annual Challis Distinguished Lecturer series. This has brought to the department visits by distinguished statisticians such as Persi Diaconis, Bradley Efron, Norman Breslow, and Peter Hall. Other donations by William Mendenhall, Myron Katzoff, and Kenneth and Janet Keene have funded fellowships and awards for graduate students.

In the new millennium, disagreements occurred between many members of the Biostatistics unit in the Health Center and the rest of the department about academic goals for the Biostatistics group. That group broke away from the Statistics Department to become a division in the Department of Epidemiology and Health Policy Research in the College of Medicine, with faculty including Mini Devidas, Wendy London, Jon Shuster, Sam Wu, and later Keith Muller. A separate Biostatistics group was established in the College of Public Health and Health Professions, with faculty including Babette Brumback and Mike Daniels. That group has maintained close ties with the Statistics Department, including a joint academic program. Currently, the Biostatistics groups are being merged while remaining in separate colleges.

As of 2012, at its fiftieth anniversary, the Statistics Department has about 65 graduate students in its Masters and PhD programs. The department's undergraduate major has been increasing in size, having about 50 majors and many more co-majors. During the last decade, in addition to its standard MS and PhD-level courses, the department developed several new applied courses that serve both undergraduate Statistics majors, as well as graduate students in other fields who want to take additional courses beyond the basic two-semester sequence in statistical methods that many departments require. In the 2011–2012 academic year, about 7,500 students took courses from the Statistics Department at UF. On the negative side, after many years of budget cuts at the University of Florida, the Statistics Department had shrunk from 25 tenured or tenure-track faculty in three units in 2000 to 12 such faculty in two units in 2011, as the service course teaching was increasingly handled by nontenure-track instructors.

Research Accomplishments and Textbooks

Areas in which UF Statistics faculty have made significant research contributions include Bayesian statistics (M. Ghosh), biostatistics (H. Doss, M. Daniels, J. Shuster), categorical data analysis (A. Agresti), decision theory (G. Casella, M. Ghosh), experimental design (J. Cornell), missing data (M. Daniels), mixed models (A. Khuri, R. Littell), multivariate statistics (R. Carter, J. Saw), nonparametric statistics (B. Presnell, R. Randles, P. V. Rao), Monte Carlo and computationally intensive methods (J. Booth, G. Casella, H. Doss, J. Hobert, K. Khare, B. Presnell), probability limit theory (A. Rosalsky), spatial statistics (M. Christman, L. Young), and statistical education (R. Scheaffer). The reference list at the end of this chapter shows some research articles written by UF Statistics faculty (while at UF) that have achieved more than 200 citations according to Google Scholar.

As the research achievements of the faculty increased over the years, the department received increasing recognition worldwide. In 2008, the department was ranked by the *U. S. News and World Report* as number nine in the Top Ten Statistics Departments. Special awards to the faculty include the ASA Founder's

Award to Linda Young, elected President of ASA and the ASA Founder's Award to Dick Scheaffer, Vice President of ASA and the Paul Minton Award by the Southern Regional Council on Statistics to Ron Randles, the Distinguished Achievement Medal from the ASA Section on Statistics and the Environment to Ramon Littell, long time editorship of *Sequential Analysis* by Malay Ghosh, the Youden Prize, Brumbaugh Award, and Shewhart Medal to John Cornell, elected Foreign Member of the Spanish Royal Academy of Sciences and editorship of *JASA*, *Statistical Science*, and *Journal of Royal Statistical Society Series B* for George Casella, and an honorary doctorate from De Montfort University (UK) and "Statistician of the Year" award from the Chicago chapter of ASA for Alan Agresti.

Several highly regarded textbooks evolved from the department's mission of teaching most of UF's introductory statistics courses. Bill Mendenhall's *Introduction to Probability and Statistics* was one of the first introductory statistics textbooks to achieve major sales, and several spin-off versions focused on various disciplines. *Mathematical Statistics with Applications*, which Mendenhall co-wrote with Dick Scheaffer and was later revised by Dennis Wackerly, continues to be one of the most popular undergraduate texts on this topic. Later textbooks achieved prominence for graduate level courses, such as Alan Agresti's *Categorical Data Analysis* and George Casella's *Statistical Inference* (with Roger Berger). The reference section at the end of this chapter lists several books written by members of the department.

MS and PhD Graduates

The department has through the summer of 2010 graduated 165 PhDs and 392 MS degrees. The growth in the program is reflected in this table:

Decade	MS Degrees	PhD degrees
1960–1970	28	3
1971–1980	81	24
1981–1990	81	32
1991–2000	94	49
2001–2010	108	57

The list of graduates and their date of degree can be found on the departmental web page, www.stat.ufl.edu/personnel/alumni.html. Over the years, many UF Statistics alumni have had distinguished careers in academia, industry, and government. The first graduate student, Larry Kupper, is Alumni Distinguished Professor of Biostatistics at the University of North Carolina, while his fellow student John Cornell had a long career in the IFAS statistics consulting unit at the University of

Florida. Of the early doctorates, James McClave followed a brief academic career by founding a highly successful consulting company (Info Tech) that now has more than 150 employees, Ken Stanley worked in Geneva for the World Health Organization as well as for the Departments of Biostatistics and Statistics at Harvard, Darryl Downing became Vice President of Statistical and Quantitative Sciences at GlaxoSmithKline, Ronald Marks served as head of the department's Biostatistics Unit for several years, William Louv was Senior Vice President at GlaxoSmithKline, Walter Offen served as Senior Research Fellow at Eli Lilly, Jim Schott at the University of Central Florida wrote the highly regarded book *Matrix Analysis for Statistics* and David Nickerson serves as chair of that department, and Samy Suissa was Director of Clinical Epidemiology at McGill University. Of the later graduates, among those having highly productive academic research careers were Parthansarasi Lahiri (University of Maryland), Dawn Peters (Oregon State University), Gauri Datta (University of Georgia), André Adler (Illinois Inst. Tech.), Joseph Lang (University of Iowa), I-Ming Liu (Victoria University of Wellington), Brent Coull (Harvard University), Ralitza Gueorguieva (Yale University), Galin Jones (University of Minnesota), Brian Caffo (Johns Hopkins University), Wolfgang Jank (University of Maryland), Bernhard Klingenberg (Williams College), and Samiran Sinha (Texas A&M University). MS graduates who went on to receive PhDs elsewhere and then have productive academic careers include Marcello Pagano (Harvard), David Stoffer (Pittsburgh), and Ruey Tsay (Chicago).

Acknowledgments The authors thank Victor Chew for information about the "Early Days" section, which was based on his remembrances of the twenty-fifth anniversary of the department in 1988, based partly on notes left by A. E. Brandt. The material in "The Mendenhall Years" and "The Scheaffer Years" is based on remembrances written by William Mendenhall and Richard Scheaffer for that event. For more details about the history of the department, see www.stat.ufl.edu/info/history.

Some Frequently-Cited Articles Written by UF Statistics Faculty

Agresti A (1992) A survey of exact inference for contingency tables. Stat Sci 7:131–177

Agresti A, Coull BA (1998) Approximate is better than exact for interval estimation of binomial parameters. Am Stat 52:119–126

Booth JG, Hobert JP (1999) Maximizing generalized linear mixed model likelihoods with an automated Monte Carlo EM algorithm. J R Stat Soc B 61:265–285

Duncan DB (1955) Multiple range and multiple *F* tests. Biometrics 11:1–42

Ghosh M, Rao JNK (1994) Small-area estimation: an appraisal. Stat Sci 9:55–76

Hobert J, Casella G (1996) The effect of improper priors on Gibbs sampling in hierarchical linear mixed models. J Am Stat Assoc 91:1461–1473

Littell RC, Pendergast J, Natarajan R (2000) Modelling covariance structure in the analysis of repeated measures data. Stat Med 19:1793–1819

Osborne MR, Presnell B, Turlach BA (2000) On the lasso and its dual. J Comput Graph Stat 9:319–337

Park T, Casella G (2008) The Bayesian lasso. J Am Stat Assoc 103:681–686

Some Textbooks Written by UF Statistics Faculty

Agresti A (2012) Categorical data analysis, 3rd ed. Wiley, New York

Casella G (2008) Statistical design. Springer, New York

Casella G, Berger R (2001) Statistical inference, 2nd edn. Brooks/Cole, Pacific Grove

Casella G, Robert C (2005) Monte Carlo statistical methods, 2nd edn. Springer, New York

Cornell JA (2002) Experiments with mixtures, 3rd edn. Wiley, New York

Daniels M, Hogan JW (2008) Missing data in longitudinal studies: strategies for Bayesian modeling and sensitivity analysis. Chapman & Hall, New York

Ghosh M, Mukhopadhyay N, Sen PK (1997) Sequential estimation. Wiley, New York

Khuri AI (2003) Advanced calculus with applications in statistics, 2nd edn. Wiley, Hoboken

Khuri AI (2009) Linear model methodology. Chapman & Hall, London

Khuri AI, Cornell JA (1996) Response surfaces: designs and analyses, 2nd edn. CRC Press, Boca Raton

Littell RC, Milliken GA, Stroup RD, Wolfinger RD (2006) SAS for mixed models, 2nd edn. SAS Institute, Cary

Randles R, Wolfe D (1979) Introduction to the theory of nonparametric statistics. Wiley, New York

Scheaffer RL, Mendenhall W, Ott RL (2005) Elementary survey sampling, 6th edn. Duxbury, London

Shuster J (1992) Practical handbook of sample size guidelines for clinical trials. CRC Press, Boca Raton

Wackerly D, Mendenhall W, Scheaffer RL (2007) Mathematical statistics with applications, 7th edn. Duxbury, Pacific Grove

University of Georgia Department of Statistics

John Stufken and Robert L. Taylor

Statistics at the University of Georgia (UGA) evolved from a few courses in several programs in the 1940s to an interdisciplinary Institute of Statistics in the 1950s and then to a formal Department of Statistics in 1964. While the Department of Statistics at UGA has seen many changes over the years, it has consistently housed faculty who have played leading roles in national and international statistical activities. Throughout its history, the Department has maintained a highly rated graduate program and an active undergraduate program which has awarded 152 PhD degrees, 544 Master's degrees, and 493 Bachelor's degrees from 1964 to 2010. It has taught the vast majority of statistics courses on campus, and has a long history of activity in collaborative research and statistical consulting. While its size has changed dramatically compared to the early years, more recent changes in the numbers of faculty and students have been modest, with the most notable change being the increased size of majors and co-majors in the undergraduate program. The department currently houses 18 tenure track faculty, six nontenure track faculty, about 70 graduate students and about 50 undergraduate majors. This article will trace the history of the Department over a 70-year period and will highlight selected developments during this period. Much of the early history was

University of Georgia,
Department of Statistics,
101 Cedar Street, Athens,
GA 30602, USA
J. Stufken (✉)
e-mail: jstufken@uga.edu

Clemson University,
Department of Mathematical Sciences,
Clemson, SC, USA
R. L. Taylor
e-mail: rtaylo2@clemson.edu

A. Agresti and X.-L. Meng (eds.), *Strength in Numbers: The Rising of Academic Statistics Departments in the U.S.*, DOI: 10.1007/978-1-4614-3649-2_27,

obtained from an extensive set of notes by A. Clifford Cohen, who was a faculty member at UGA from 1947 to 1978 and an active Professor Emeritus until his death in 1997. Program reviews and annual reports provided much of the departmental data for the seven decades of this history.

The Pre-Department History, 1941–1964

Offerings in statistics at UGA in 1941 consisted of two courses. One course was an elementary course in business and economic statistics which was taught in the College of Business Administration. The second course (Math 356) was a basic course in statistics which was taught in the Department of Mathematics of the College of Arts and Sciences. Some Statistics was also included in various courses in Psychology, Sociology, and Education. Very little progress in developing the statistics program at UGA occurred during the war years of 1941–1945.

When World War II ended, Dr. Tomlinson Fort was hired from Lehigh University to head the Department of Mathematics at UGA. Dr. Fort was charged with developing a graduate program in mathematics which would also embrace a program in statistics. Dr. Theodore Bancroft was recruited to head up the statistics program. Dr. Bancroft arrived in Athens in September 1946, but remained at UGA for only 1 year before accepting an appointment at Auburn University. During this year he taught the elementary statistics course, Math 356, and added a two-course sequence in statistical theory (Math 451–452) for advanced undergraduate and beginning graduate students. This sequence, under the name Stat 4510–4520, currently forms the required mathematical statistics sequence of the Bachelor's degree program. Also in 1946, the Agronomy Department of the College of Agriculture began offering a first course in statistical methods for agricultural workers with Professor Edwin James as the instructor.

Dr. A. Clifford Cohen (Fig. 1) joined the Department of Mathematics in June 1947 as a replacement for Dr. Bancroft. For several years, Dr. Cohen taught the elementary statistics course and the statistical theory courses and provided limited consulting service for graduate students and faculty members in Forestry, Agriculture, Education, Psychology, and other disciplines. In 1952, the Southern Regional Education Board (SREB), a public agency of 15 member states cooperating to improve education in the region, sponsored a Southern Regional Conference on Statistics which led to the formation of the SREB Southern Regional Committee on Statistics (SREB-COS). The Committee's purpose was "... to provide a means of joint planning and cooperative action by which the member institutions can coordinate and supplement their respective educational, research, and service programs in statistics to meet more effectively the present and future needs of the region and nation." The first organizational meeting of the SREB-COS was in Atlanta, Georgia, on October 5–6, 1955, and UGA was represented at

Fig. 1 A. Clifford Cohen around 1991

this meeting by E. Broadus Browne, a consulting statistician with the experimental research program at UGA. In accordance with a recommendation of the SREB-COS, a movement began at UGA to organize a centralized program in statistics, which ultimately led to the creation of an Institute of Statistics at UGA in 1958 with A. Clifford Cohen as Director.

The Institute of Statistics at UGA was charged with responsibility for conducting and encouraging statistical research, for providing computing and consulting services and for developing programs of instruction in statistics. Computing services were rendered through the fledgling computer center headed by Dr. James L. Carmon, who also had a joint appointment in the Institute and later in the Department of Statistics. Membership in the early days of the Institute also included James E. Fortson of the Computer Center, two faculty members in mathematics (Cohen & C. H. Kapadia), a faculty member from business, a faculty member from psychology, and a faculty member from forestry.

Pursuant to recommendations contained in a University self-study and with the encouragement from the SREB-COS, a Department of Statistics was established at UGA on July 1, 1964 with Dr. Cohen as Acting Head. Dr. Kapadia was the only other full-time faculty member of the department, but former members of the Institute of Statistics were included as adjunct members. For the first 2 years, the Department was housed in the Geography, Geology, and Mathematics Building of the Science Center. It later occupied the Lumpkin House (aka Rock House) for 2 years before moving to the Graduate Studies Building upon its completion in 1968. The name of the Department was at that time changed to Department of Statistics and Computer Science. In 1984, when it was again renamed as Department of Statistics, the Department moved to its present location, the Statistics & Computer Services Building.

Fig. 2 Rolf E. Bargmann around 1997

The Early Department History, 1964–1980

In September 1965, Dr. Carl F. Kossack was appointed as permanent Department Head and served in this capacity until 1979. Shortly thereafter, in 1965, Dr. Rolf E. Bargmann from IBM (Fig. 2) accepted an appointment as Professor, and Kanti Patel was appointed as Assistant Professor. The Department functioned with four faculty members and five adjunct members for 2 years. In the summer of 1966, it hosted the SREB Summer Session with financial support from NSF. Dr. James W. Walker of Georgia Tech augmented the Department's teaching staff for this session.

Similar to other universities, UGA experienced rapid enrollment increases in the 1960s. Five new assistant professors were added during the 1967–1968 academic year. The Department was authorized to offer the PhD degree in Statistics in 1967. Prior to this time, four doctorates in mathematics with dissertations in statistics had been awarded, with the first one going to Dr. P. V. Rao in 1963 with Dr. Cohen as his major professor. Also in 1967, Master's degrees and Bachelor's degrees in Statistics were approved. In 1968, the Department initiated a Master's program in Computer Science under the direction of Professor Bargmann.

Dr. L. R. Shenton was hired by the Computer Center in 1967 and had a joint appointment in the Department of Statistics. Accompanying Dr. Shenton to the Computer Center in 1967 was his PhD student, Kermit Hutcheson. Dr. Hutcheson joined the Department of Statistics in 1969 as an Assistant Professor, served as Assistant Head and Graduate Coordinator for several years, and retired in 2001.

Collaborative research and joint appointments were particularly prevalent during the early history of the Department. In addition to the aforementioned joint appointments, there were joint faculty appointments in forestry and education. In 1974, a joint appointment for an experiment station statistician and an Assistant Professor of Statistics and Computer Science was negotiated with the College of

Agriculture Experiment Station, and Dr. Glenn Ware was appointed to this position.

The Department enjoyed good successes in the late 1960s and 1970s with external funding via computing contracts, a large ONR Themis grant, a NASA contract and AFoSR funding. Consequently, the UGA statistics program had a strong computing flavor, with the exception of Cliff Cohen's work on the topic of reliability and life testing, and was closely tied to applications. During this time period, there was an unusually high rate of transition in the faculty, especially at the junior level, partly due to high demands for people with computational skills. During 1972–1976, the Department averaged awarding seven Master's degrees and four PhD degrees per year. In 1976 the Department's graduate program had approximately 50 students, about 30 of whom were at the Master's level. In 1977, a professional Masters in Applied Mathematical Sciences (MAMS) degree was established by the Department of Mathematics, the Department of Statistics, and Computer Science and the Management Science Program in the College of Business. Two to three MAMS degrees were awarded annually in statistics from 1977 to the early 2000s. In subsequent years, interest in the program faded, and MAMS was officially discontinued in the Department around 2005. Also in 1977, Bill McCormick joined the Department, making him the longest serving current faculty member. A national report on statistics programs by Ingram Olkin listed seven statistics faculty members at UGA in 1970, eight in 1975 and 10 in 1980. However, the size of the Department of Statistics and Computer Science was actually larger because it also contained several computer science faculty who were not included in Olkin's counts.

A New Beginning, 1980–1989

Dr. Kossack retired in 1980, and Professor Lynne Billard was recruited from Florida State's Department of Statistics to become Department Head. Under her leadership, the Department gradually changed from one that had mostly a computational and applied orientation to one with pillars of strengths in theory, applications, and computations. The broadening and strengthening of the graduate programs, combined with the hiring of many new faculty and a strong emphasis on research productivity, put the Department on an upward trajectory that not only raised its stature but that also provided the foundation for many later developments. Dr. Ralph A. Bradley retired from Florida State University in 1982 and was recruited to UGA as Research Professor of Statistics. Dr. Billard also recruited full professors, Bob Taylor and Ishwar Basawa in 1983 and 1986; respectively, to strengthen the mathematical statistics expertise of the Department and to fortify its senior leadership. Measure theory probability was added to the curriculum and the PhD exams in the early 1980s. Many Assistant Professors were also hired during this era, including Jonathan Arnold (jointly with Genetics), Jaxk Reeves, Marc

Sobel, Kooros Mahjoob, Ashim Mallik, Timothy Green, David Nickerson, John Stufken, TN Sriram, Somnath Datta, and Daphne Smith.

A graduate program review in 1987 reported the number of tenure track faculty in Statistics at 16, with 10 of them (seven Assistant Professors, two Full Professors, and one Research Professor) hired during Dr. Billard's period as Head. The Department of Statistics continued to build on national and international prominence during the 1980s with funded research projects, robust degree programs, and professional activities by its faculty members.

With rapid changes, these were in many ways very exciting and very formative years for the Department. But the changes were easier to embrace for those joining the Department as new members than for those who were continuing from the previous era. The standards and expectations were admirably high, as they should be for a department that aims at moving up, but also posed challenges, both for some of the continuing faculty and some of the new faculty.

In addition to laying the foundations for a more rigorous curriculum and a strong emphasis on research activity, senior faculty were during this period very active in the professional societies. Ralph Bradley served as President of ASA in 1981, and Lynne Billard served in the 1980s on numerous committees in professional societies, including as President of the Biometrics Society ENAR in 1985.

An Era of Expansion, 1989–1998

Professor Taylor became Department Head in 1989, and Professor Billard was named Associate to the Dean of Arts & Sciences. Christine Franklin joined the Department in 1989 as instructor. The Department continued its steady growth in faculty and graduate students with four new assistant professors being recruited in 1990 including Gauri Datta. John Stufken left UGA to join the faculty at Iowa State. Steady recruiting of faculty members continued through the early 1990s and compensated for other faculty departures as the Department grew to 20 faculty members. Among those recruited during this era, and currently still on the faculty, were Lynne Seymour and Dan Hall. Computerized instructional modules were added to the introductory statistics course, and the graduate program increased to 60 students. Graduate student offices were constructed in the Statistics Building along with a statistics lecture room and a computerized classroom. By 1997 enrollments in introductory statistics courses totaled more than 3,000 students per year with a total enrollment of 4,400 students in statistics courses. Kermit Hutcheson received the Honoratus award for being named Outstanding Honors Professor on five different occasions.

In October 1991, the Department celebrated the eightieth birthday of Professor Emeritus Cliff Cohen with professional activities which were attended by over 100 former students, colleagues, and friends. The Cohen festivities were highlighted by the dedication of the Cohen Reading Room in the Department which houses the professional journals and books. Research Professor Ralph Bradley retired in 1992, and retirement festivities were attended by over 60 former students and colleagues. The retirement served as a fund raising opportunity to establish an Annual Distinguished Lecture Series. In April 2011, Peter McCullagh was the twentieth invited speaker in this lecture series. To honor Professor Emeritus Rolf Bargmann and his pioneering work in computing at UGA, 27 of his former doctoral students spearheaded a drive to establish the Rolf Bargmann Computing Lab which was dedicated in May 1997.

UGA statistics faculty members continued to receive unusual recognition with professional fellowships and appointments during the 1990s. Professor Billard was named University Professor and received a UGA Creative Medal for Research in 1992. Professor Bradley received an ASA Founders Award in 1992, and Professor Taylor served as SCROS President in 1996. Professor Billard served as President of the International Biometrics Society in 1994–1995 and as ASA President in 1996.

Professors Basawa and Taylor published the *Selected Proceedings of the Sheffield Symposium on Applied Probability*, and set the stage for the Department to host several international symposia on probability and statistics at UGA. In 1996 the UGA Symposium on Estimating Functions was co-sponsored by IMS, and the selected proceedings were published by Basawa, Godambe, and Taylor. In 2000 the UGA Symposium on Inference for Stochastic Processes was co-sponsored by IMS, and the selected proceedings were published by Basawa, Heyde, and Taylor.

The Recent History and the Present, 1998–2011

Bill McCormick (1998–2000) and Ishwar Basawa (2000–2003) served as Interim Department Heads during a time that the Department attempted to recruit new leadership from the outside. While this was a period with formidable internal challenges, the Department maintained a remarkably high level of scholarly activity and productivity. These facts were already visible in an external review committee report in 1998. While the committee lavishly praised the quality of the curriculum, the research strength of both junior and senior faculty, and the leadership provided by Professor Taylor, it also expressed great concern about an apparent atmosphere of distrust among faculty. The hope was that new leadership could help to solve the problems, but as the committee noted in a word of caution, "... a head can only implement the collective wishes of the Department."

Multiple changes in the faculty line-up occurred during the period 1998–2003, leading to an overall decrease in faculty size. Faculty who were recruited during this period and are currently still in the Department include Xiangrong Yin (2000), Paul Schliekelman (2001), and Academic Professional Chandler Pike (2001). Some who were recruited during this period soon moved elsewhere, including Julie Berube and Jem Corcoran. Faculty who left for other positions or who retired during this period include Steve Rathbun (2001, Penn State University), Kermit Hutcheson (2001, retired), Bob Taylor (2002, Chair of the Department of Mathematical Sciences at Clemson University), Nancy Lyons (2002, retired), and Hubert Chen (2002, retired).

Active recruiting for a Department Head continued nationally, resulting in the hiring of John Stufken for fall 2003. Professor Stufken had been Program Director for Statistics at the NSF from 2000 to 2003, and was well acquainted with the Department since he had been a junior member of the faculty before leaving for a position in the Department of Statistics at Iowa State University.

The Department has changed dramatically since 2003, both in terms of personnel and in terms of research emphasis. The change in emphasis is partly driven by a changed modus operandi in scientific research. In an era in which more and increasingly complex data are collected in virtually all areas of science, opportunities for statisticians in interdisciplinary collaborations are almost limitless. While advances and needs in scientific research make this a golden era for statistics, mining the gold also comes with tremendous challenges. In an academic environment, it requires faculty who are strong in their statistical areas of specialization and who can make major novel contributions to these areas, but who are also interested in working on real problems as part of interdisciplinary research teams. With this model in mind, the Department recruited six new tenure track faculty between AY04 and AY07 (Nicole Lazar, Cheolwoo Park, Abhyuday Mandal, Jeongyoun Ahn, Yehua Li, and Lily Wang). Two additional tenure track faculty were recruited in AY11 (Jennifer Kaplan and Liang Liu), and are expected to work closely with collaborators in mathematics education and bioinformatics, respectively. During the same period, some faculty moved to other positions (Somnath Datta, University of Louisville; Robert Lund, Clemson University; Mary Meyer, Colorado State University; and Anand Vidyashankar, Cornell University) or retired (Ishwar Basawa in 2010; see also Fig. 3).

New developments at UGA also offer tremendous potential for new opportunities for Statistics. This includes the founding of a College of Public Health in 2005, the establishment of a new Medical Campus based on a 2008 partnership between UGA and the Georgia Health Sciences University, and approval from the Board of Regents in 2010 for a vast expansion of engineering degree programs. While the full impact of these developments on the Department is not yet clear, each will bring increased opportunities for teaching and research. At the time that this chapter is written, multiple teaching and research collaborations between Statistics and the new Department of Epidemiology and Biostatistics have already been developed.

The Department also houses a Statistical Consulting Center (SCC), which has existed since 1990, but saw some changes in 1998, and major changes in 2008. Since 2008 the personnel consists of a Director, an Associate Director, and around five graduate consulting assistants. The Director, Jaxk Reeves, a tenured faculty member, has been involved with the SCC since 1990. The position of Associate Director (AD), created in 2008, was held by Dr. Jien Chen until 2010 and is now held by Dr. Kim Love-Myers, a nontenure track faculty member. The SCC earns

Fig. 3 Graduate students at the railroad tracks in front of the department at the occasion of Professor Ishwar Basawa's retirement in 2010

income from the consulting projects that it handles, and also receives financial support from the Office of the Vice President for Research (OVPR), from the Graduate School, from the Franklin College of Arts and Sciences, and from small contributions from a few other colleges around campus. This support reflects the SCC's impact on both research and teaching. The SCC primarily serves graduate students and faculty, and has potential to initiate interdisciplinary research projects. It has also served an increasing number of outside clients during the last few years.

In AY12 the Department has 24 faculty (see Table 1), four adjunct appointees, four staff members, 73 graduate students, and 51 undergraduate students. The graduate students include 15 secondary students, often supported by their other department. The Department provides an assistantship (teaching, research, or consulting) for about 35–40 graduate students. A significant portion of the undergraduate students are double majors, and a few are even triple majors. Both the size and quality of the undergraduate program have made major strides over the last decade under the leadership of Undergraduate Coordinator Christine Franklin and with buy-in from a significant portion of the faculty. In AY10, the Department awarded 10 Bachelor's degrees, 17 Master's degrees, and six PhD degrees. Over the period 1964–2010, the Department has now awarded a total of 493 Bachelor's degrees, 544 Master's degrees, and 152 PhD degrees.

Table 1 Faculty in AY12 with title and years of service

Name	Current title	Years in department
Jeongyoun Ahn	Assistant Professor	2006–present
Lynne Billard	University Professor	1980–present
Gauri Datta	Professor	1990–present
Christine Franklin	Senior Lecturer and Undergraduate Coordinator	1989–present
Kim Gilbert	Lecturer	2008–present
Dan Hall	Associate Professor and Associate Head	1996–present
Jennifer Kaplan	Assistant Professor	2011–present
Nicole Lazar	Professor	2004–present
Yehua Li	Assistant Professor	2006–present
Liang Liu	Assistant Professor	2011–present
Kim Love–Myers	Academic Professional	2010–present
Abhyuday Mandal	Associate Professor	2005–present
Bill McCormick	Professor	1977–present
Jack Morse	Instructor	2006–present
Cheolwoo Park	Associate Professor	2005–present
Chandler Pike	Senior Academic Professional	2001–present
Jaxk Reeves	Associate Professor and Director SCC	1982–present
Paul Schliekelman	Associate Professor	2001–present
Lynne Seymour	Associate Professor and Graduate Coordinator	1994–present
T. N. Sriram	Professor	1987–present
John Stufken	Professor and Head	1986–1990; 2003–present
Lily Wang	Assistant Professor	2007–present
Mark Werner	Lecturer	2010–present
Xiangrong Yin	Professor	2000–present

Many of the current faculty are excellent teachers, and several have been recognized for this. In recent years, the Department has become very competitive for the Sandy Beaver Excellence in Teaching Award within the Franklin College of Arts and Sciences. While the Department had to wait for a first winner until 1999, by now five nominations for this award have been successful. Recipients of this award are Christine Franklin (1999), T. N. Sriram (2004), Chandler Pike (2005), Jill Smith (2008), and Nicole Lazar (2010). In 2008, Christine Franklin was a recipient of the Honoratus Award for winning the Lothar Tresp Outstanding Honors Professor Award for the fifth time. The Department also had an earlier success with the General Sandy Beaver Teaching Professorship, for which Rolf Bargmann was the sole winner in 1980.

In calendar year 2010, the Department taught a total of 18,397 credit hours. The number of credit hours stayed fairly constant over recent years, but increased significantly in AY11. A large introductory statistics course, STAT 2000, is the Department's staple. This 4-credit course attracts close to 3,000 students per year. In AY11, through an agreement with the Terry College of Business, the Department began teaching an introductory statistics course for business students, MSIT 3000. This 3-credit course could eventually draw close to 1,000 students per year.

To meet this increase, in 2010 the Department hired Lecturer Mark Werner, and converted temporary positions for Lecturer Kim Gilbert and Instructor Jack Morse into permanent positions.

Several new undergraduate and graduate courses have been introduced over the past years. Most notable at the undergraduate level is a two-semester capstone sequence, which culminates in a poster session near the end of the second semester in which students present the real projects that they worked on during that semester. At the graduate level, an extensive internal review of the graduate program was concluded in AY10, and resulted in, among other changes, the introduction of a sequence of courses aimed at exposing students earlier to research, writing, and sound professional practice.

In terms of research, a committee that conducted the most recent Departmental Review noted in its February 2005 report that "The UGA Statistics Department is a high profile research department with an active research faculty." The research profile of the Department has steadily increased since that time. Within the last couple of years, 14 of the then 16 tenure track faculty brought in external research funding as PI or co-PI from a variety of sources, including NSF, NSA, NIH, USDA, ARO, and BLS. The Department has also been increasingly competitive for the M. G. Michael Award for Excellence in Research in the Franklin College of Arts and Sciences. Established in 1944, this award was originally given to a single faculty member each year, but features currently up to three recipients per year. The Department has seen five successful nominations for this award: Clifford Cohen (1954), John Stufken (1988), Paul Schliekelman (2005), Nicole Lazar (2007), and Jeongyoun Ahn (2011).

The Department continued to organize and host a number of very successful research conferences. In 2004, it hosted the meeting of the International Indian Statistical Association, with a local organizing committee of Professors G. Datta, Sriram, Vidyashankar and Seymour. In 2009 Professors Basawa and Sriram organized an international Symposium on New Directions in Asymptotic Statistics. In addition, in 2009 the fMRI research group in the Department, led by Professor Lazar, started an annual one-day workshop for the Network of Greater Georgia Institutions for Neuroimaging and Statistics (NOGGINS). The Department also hosted the 2003 ASA sponsored TEAMS conference on teacher preparation in statistics, co-chaired by Christine Franklin and Bob Taylor.

Many faculty are active as Editors or Associate Editors for professional journals. Some have also written a textbook or research monograph. For the period covered in this section, this includes John Stufken (*Orthogonal Arrays: Theory and Applications*, Hedayat, Sloane and Stufken, Springer Verlag, 1999), Gauri Datta (*Probability Matching Priors: Higher Order Asymptotics*, Datta and Mukerjee, Springer Verlag, 2004), Lynne Billard (*Symbolic Data Analysis: Conceptual Statistics and Data Mining*, Billard and Diday, John Wiley & Sons, 2006), Nicole Lazar (*The Statistical Analysis of Functional MRI Data*, Lazar, Springer Verlag, 2008), and Christine Franklin (*Statistics: The Art and Science of Learning from Data*, 2nd Edition, Agresti and Franklin, Prentice Hall, 2008).

The current faculty includes five ASA Fellows (Billard, Datta, Franklin, Sriram, and Stufken) and three IMS Fellows (Billard, Datta, and Stufken). Many have other awards and honors to show for their efforts, and several are quite active within our professional organizations. While more details can be found on the Department's website at http://www.stat.uga.edu, we do want to mention the exceptional record of Lynne Billard. In addition to the aforementioned presidencies, Professor Billard's many awards during the period covered in this section include the ASA Wilks Award (1999), Elected Fellow of the American Association for the Advancement of Science (2000), the ASA Founders Award (2003), and the COPSS Elizabeth Scott Award (2008).

Looking Ahead

While it is by far easier, though not easy, to describe the past than to predict the future, looking ahead is critical if a department is to succeed. Unfortunately, as too many recent lean budget years have shown, any timetable for goals that require significant levels of funding is highly uncertain.

In AY12, the Department has 24 faculty, of which 18 are tenure track. In view of the growing number of students and credit hours taught by the Department, and in view of the increasing complexity of the rapidly growing collaborative research opportunities, a midsize statistics department like ours will need to grow in order to meet the on-campus demand for statistical training and research expertise. Such growth is more likely to occur, especially during leaner years, by working with deans and vice presidents to address priorities and problems for the College and University. In addition to the opportunities already mentioned in "The Recent History and the Present, 1998–2011", there are other areas of priority for the University, which include bioinformatics, computational science, environmental science (especially energy use and sustainability), and increasing the workforce in the STEM disciplines. These are all areas where Statistics can and must play a role, and where it can greatly benefit from doing so.

The Department's most recent strategic plan, which was produced in 2004, emphasizes five broad goals: (1) Increase visibility in the profession; (2) increase visibility and role on campus; (3) modernize curriculum and instructional technology; (4) increase activity in statistics education, with the goal of a statistics education graduate degree program; and (5) improve recruiting of undergraduate and graduate students. While the Department has made great progress on most of these goals, there remains ample room for further improvement. There is a relatively simple recipe towards success in most of these areas: Recruit talented, prolific faculty who are (1) outstanding researchers, whose work is driven by important problems in modern science; (2) outstanding teachers and advisors; and (3) willing to make sacrifices for the good of the Department.

The 2010 NRC rankings (based on data collected in 2005) show that the Department is one of the better Statistics departments in the southeast. However, the NRC overall interval rankings of 15–43 (R ranking) and 21–36 (S ranking) also indicate that the envisioned summit has not been reached yet. The continued development of junior faculty, their retention, and the opportunity to recruit and compete for the best available talent will be critical factors in determining whether the Department can continue its ascent.

Acknowledgments We are grateful to the staff in the Department for their help with collecting materials on which this brief history is based, and are especially indebted to Business Manager Tim Cheek for his tireless efforts. We are also grateful to faculty, referees, and the Editors for their helpful input on earlier drafts. Corrections and updates will be maintained on the Department's website.

University of Iowa Department of Statistics

Robert V. Hogg and Dale L. Zimmerman

1918–1942: The Rietz and Craig Years

The roots of the Department of Statistics and Actuarial Science at the University of Iowa can be traced back to the hiring of Henry L. Rietz as Head of the Department of Mathematics in 1918. Although statistics courses had been taught previously at the university—the first one, "The Method of Least Squares," was offered as early as 1895 and the second, "The Theory of Statistics" was introduced in 1906—no faculty member before Rietz could rightfully claim to be a statistician. In fact, Rietz himself was not originally a statistician, but an algebraist; his PhD thesis, written in 1902 at Cornell University, was on primitive groups (Crathorne 1944). But around 1905, while he was on the faculty of the Department of Mathematics at the University of Illinois, he apparently became interested in statistics. Between 1905 and 1910 he published several statistical papers through the Agricultural Experiment Station at Illinois, with titles like "Variability of Corn," "Principles of Breeding," and so on. In 1911, he published "On the Theory of Risk," the first of his many publications in actuarial science. In fact, it may have been his expertise in actuarial science as much as that in statistics or other areas of mathematics that got Rietz the job at Iowa, for by 1918 three actuarial courses and only one or two statistics courses were being taught every semester. In any case, by the time Rietz arrived at Iowa he had established himself as a leading scholar in both statistics

University of Iowa,
Department of Statistics
241 Schaeffer Hall,
Iowa City, IA 52242-1409, USA
D. L. Zimmerman (✉) · R. V. Hogg
e-mail: dzimmer@stat.uiowa.edu

R. V. Hogg
e-mail: rvhogg@gametrail.org

A. Agresti and X.-L. Meng (eds.), *Strength in Numbers: The Rising of Academic Statistics Departments in the U.S.*, DOI: 10.1007/978-1-4614-3649-2_28,

and actuarial science. He went on to become a vigorous promoter of both disciplines, writing 156 articles, reviews, or discussions, and 11 books over the course of his career. His 1927 book, *Mathematical Statistics*, was used as a text for many university courses in the years that followed.

Henry L. Rietz, 1943

For over a decade, Rietz was the only statistician on the Mathematics faculty. But being head of the department had its advantages, and he was able to attract many good students. One was Frank Weida, who went to George Washington University after graduating from Iowa and in 1935 established there the first Department of Statistics in a Liberal Arts and/or Science college. Another was Frank Satterthwaite, who is known for his eponymous approximation to the distribution of a linear combination of mean squares. But perhaps the best two of Rietz's students were Samuel Wilks, from Texas, and Allen Craig, from Florida, both of whom earned the PhD in 1931. Upon Wilks' graduation, he spent a year on a National Research Council (NRC) Fellowship at Columbia with Harold Hotelling (Mosteller 1964), then another year in England as an NRC International Research Fellow, where he worked with Karl Pearson and John Wishart. During these fellowship years he developed and published the multivariate test criterion now well known as "Wilks' lambda," which generalizes the F test of the univariate analysis of variance. In 1933, Wilks joined the Princeton faculty, where he remained for the rest of his professional life. Craig, on the other hand, was hired on at Iowa in 1931 and stayed until his retirement in 1970. So throughout the 1930s and early 1940s, mathematical statistics at the University of Iowa was taught by Rietz and Craig. Craig earned a reputation as a highly polished lecturer. It was said that he would begin writing his lecture in the upper left-hand corner of the blackboard, fill the board three or four times with beautifully crafted sentences, and just as the bell rang, put a period on his last sentence in the lower right corner with a flourish and walk out of the room, leaving his students in awe. Craig made many important research contributions as well. Perhaps the most important, which is now widely known as Craig's Theorem, gives necessary and sufficient conditions for the independence of quadratic forms in normal random vectors; see Driscoll and Gundberg (1986) for a history of this theorem.

No account of University of Iowa statisticians in those pre-war years is complete without a description of their major role in the formation and early leadership of the Institute of Mathematical Statistics (IMS). In the late 1920s, Harry Carver, a statistician at the University of Michigan, saw a need for a good publication outlet for mathematical statistics research. So in 1930, as Wilks and Craig were nearing completion of their studies at Iowa, Carver began publishing, at his own expense, the *Annals of Mathematical Statistics*. Now Rietz, at about the same time, thought that the mathematical statisticians in the U.S. ought to form an independent organization. In early 1935 he had Craig write to 100 or so statisticians about this, and in the summer of 1935, 60 or 70 came to a summer meeting of the American Mathematical Society in Ann Arbor and agreed to form the Institute of Mathematical Statistics. The group elected three officers: Henry Rietz as President, Walter Shewhart as Vice President, and Allen Craig as Secretary/Treasurer. In 1938, Carver turned the *Annals* over to IMS as their official journal. Sam Wilks was appointed its editor, serving through 1949, and Allen Craig and Jerzy Neyman were appointed its two associate editors. So the influence that Iowa statisticians had in shaping the IMS during its early development was quite remarkable.

1943–1965: The Craig and Hogg Years

In the late 1930s, Rietz's health began to fail. He retired in 1942 and died late the next year, just after he learned that the 1943 volume of the *Annals* would be published in his honor. (Incidentally, Rietz continues to be memorialized by the Rietz Lectures, which began in 1947 and are now given every third year at the annual IMS Meeting.) Meanwhile, with World War II going on, Craig had volunteered for the Navy and left to serve as an officer on a destroyer. At the war's end he returned to Iowa and picked up his research and teaching where he had left off. With Rietz gone, Craig necessarily carried the teaching load for statistics by himself. Overall, statistics and actuarial science course offerings in the department of mathematics numbered around ten per year during the late 1940s and 1950s, about two-thirds of them in statistics. In order to keep the actuarial science program going strong, the department had wisely hired Lloyd Knowler, another of Rietz's students, in 1939. In 1946, Knowler became Head of the Mathematics Department, a position he held until 1959. Besides teaching actuarial science, Knowler did several things that helped develop Statistics at Iowa. In the fall of 1944, he gave the first short course in Statistical Quality Control to industry people on campus. He eventually gave this course more than 40 times during his career, and he developed it into a course for students at the university as well. For this and his contributions to the founding of the American Society for Quality Control, he was awarded the Shewhart Medal in 1962. Also, in 1945, recognizing the importance of statistical methods in medicine and the health sciences, he began teaching courses in biostatistics for the College of Medicine. He taught these until the College of Medicine hired its own biostatisticians in the mid-1960s.

Allen T. Craig. 1965

During the early post-war period, Craig had several graduate students. One of them was Robert ("Bob") Hogg, a native of Hannibal, Missouri who came to Iowa in 1947 to study actuarial science. Hogg, like Craig, had served in the Navy during the war and then obtaind his Bachelor's degree in Mathematics from the University of Illinois. While pursuing his actuarial studies at Iowa, he took a class on statistical theory from Craig, and the two of them hit it off so well that Hogg decided to stay on and do a PhD under Craig's supervision, which he received in 1950. At that time the Mathematics faculty numbered 12 people, with Craig as the only statistician, and it was decided that they needed another. So they hired Hogg. Thus began a productive research, teaching, and writing partnership—and personal friendship—between Craig and Hogg that both men cherished (Randles 2007). Being single at the time, they often spent three or four nights a week at their offices working together on research on various topics, especially independence and sufficient statistics. Furthermore, Hogg and Craig were teaching a three-course sequence in probability and mathematical statistics at the senior/first-year graduate level. Finding themselves unsatisfied with existing textbooks at this level, they began to write up the material they were teaching. This project culminated in the publication of *Introduction to Mathematical Statistics*, a.k.a. "Hogg and Craig," in 1958. This book, which is now in its seventh edition, is undoubtedly one of the most popular mathematical statistics textbooks ever written. It has sold more than 125,000 copies, allowing Hogg to refer to himself, with a wink and a grin, as "the other famous author from Hannibal, Missouri."

Robert V. Hogg, 1971

1965–1983: The Formation and Expansion of the Department of Statistics

By the early 1960s, many other statistical programs were expanding rapidly and forming into departments, and in 1962, at Hogg's urging, the university created an "interdepartmental program in statistics" that could grant graduate degrees. The committee overseeing the program continued to make the case to university administrators for creating a department of statistics. Ted Bancroft, the Head of the Department of Statistics at Iowa State University, was quite helpful in this regard, as there was a question of duplication with that fine department to the west of Iowa City. But when the question of creating another statistics department in the state was put to him, Bancroft said he could not imagine a major research university without one. That laid the issue to rest, and the program was granted Department status in 1965, with five founding members: Hogg (as Chair); two other statisticians, Craig and John Birch; and two actuaries, Lloyd Knowler and James Hickman. Birch had been hired the previous year after receiving his PhD under David Blackwell. Hickman had earned his PhD in Actuarial Science at Iowa and joined the faculty in 1961. The actuarial science program thus moved with Statistics into the new department, where it has resided ever since.

The late 1960s were characterized by rapid growth of the young department. The National Defense Education Act had been passed by Congress in 1958, primarily due to the success of the Soviet Union's Sputnik, so money for science, engineering, and the mathematical sciences was flowing. By 1969 the department had more than doubled in size with the hiring of Tim Robertson, Jon Cryer, Fred Leone, Tim Wright, and Ron Randles, among others. Several outstanding PhD students graduated during this period, including Richard Dykstra, Thomas Hettmansperger, Douglas Wolfe, and Edward Wegman. All became prominent professors at leading institutions: Dykstra at the University of Missouri, Hettmansperger at Penn State, Wolfe at Ohio State, and Wegman at North Carolina. Expansion slowed considerably in the 1970s; notable hires in that decade were George Woodworth, James Broffitt, Stuart Klugman, Russell Lenth, and Johannes Ledolter. With these the department expanded its course offerings to include more applied statistics and broadened its research focus from mainly mathematical statistics to include applications of statistics in areas such as law and justice, engineering, and economics. Of course, along with new hires there were retirements and departures. Craig retired in 1970 and Knowler in 1976. Hickman left in 1970 for the University of Wisconsin College of Business, where he eventually became Dean; Leone left in 1973 to become Executive Director of the American Statistical Association, and Tim Wright and Ron Randles left for more southern climes (University of Missouri-Rolla and University of Florida in 1975 and 1981, respectively). In the early 1980s Dykstra returned to the department as Professor and Ralph Russo was hired, bringing the number of faculty to 13 in 1983. Research areas of distinction for the faculty during this period were robustness, nonparametric statistics, and order-restricted inference.

1983 to the Present

Hogg stepped down as chair in 1983, but he remained very active in the profession for a long time. He served as President of the American Statistical Association in 1988 and was a tireless promoter of statistical education and statistical quality management for many years thereafter. He finally retired in 2001. The Department Chairs that followed Hogg served comparatively short terms with the exception of James Broffitt, who chaired from 1993 to 2004 (see Appendix for a complete list) and skillfully directed the Department's move from MacLean Hall to Schaeffer Hall in 1997. From 1983 to 1994 the department again expanded considerably; in one year (1985), four new faculty were added, including Joseph Sedransk, who was the first chair to be hired externally. The number of regular faculty reached its all-time high of 20 in 1992. Current Statistics faculty who were hired during this period were Dale Zimmerman, Kung-Sik Chan, Joseph Lang, Jian Huang, and Osnat Stramer. With these and other hires the department added considerable strength in spatial statistics, time series analysis, categorical data analysis, survival analysis, and stochastic processes, and completed its transition from a department with a very strong emphasis on mathematical statistics to one with expertise throughout the full spectrum of modern applied and theoretical statistics. Elias Shiu joined the department in 1991 as the Principal Financial Group Professor and became the leader of the actuarial program.

Unfortunately, the growth that took place in the early 1990s proved to be unsustainable. As the university's budget was subjected to a series of cuts and reversions, so was the department's ability to retain faculty and replace those who retired or left for other reasons. Gradual attrition left the department with a faculty of 17 members in 2010. Among the faculty hired after 1994 were Kate Cowles, John Geweke (who left in 2010), Luke Tierney, N.D. Shyamalkumar, Qihe Tang, Rhonda DeCook, Jerome Pansera, Aixin Tan, and Joyee Ghosh. These people added particularly to the department's strength in actuarial science and Bayesian and computational statistics. Tierney, who was hired as the Ralph E. Wareham Professor in 2002, also served as chair of the department from 2004 to 2010. In 2010, Dale Zimmerman became chair and was named the inaugural Robert V. Hogg Professor.

Actuarial Science

The actuarial science program has been an integral part of mathematical sciences at the University of Iowa for over a century. In fact, of the 14 doctoral students supervised by Henry Rietz, eight wrote dissertations on actuarial science topics. This is no doubt due to Henry Rietz's enthusiasm for both actuarial science and statistics, and the passing down of this attitude to Allen Craig, Lloyd Knowler, and Robert Hogg that the actuarial science program allied itself with the Department of Statistics when it formed in 1965. In 1980, in order to formally recognize the importance of actuarial science to the department's overall mission, the department name was changed to Statistics and Actuarial Science. It is the only academic department in the United States with "Actuarial Science" in its name (there are others in Canada and elsewhere). More than 1,200 students have graduated from the actuarial science program at Iowa. Many of these have served as leaders of the actuarial profession as well as CEOs of their companies or organizations. Five former presidents of The Society of Actuaries are Iowa graduates—Victor E. Henningsen, Robert J. Myers, Edwin B. Lancaster, Charles L. Trowbridge, and Neil Parmenter. Myers was Chief Actuary of the Social Security Administration from 1947 to 1970, and holds the record for the number of times (175) that a person has testified before the U.S. Congress.

In 2009, on the basis of its job placement record, exam pass rates, and credentialed graduate rates among other factors, the actuarial program was designated by the Society of Actuaries as one of 21 Centers of Actuarial Excellence in the U.S. and Canada. Approximately 2.3% of new Fellows of the Society of Actuaries from 2000 to 2010 were University of Iowa alumni.

Other Statisticians at Iowa

The Department of Statistics and Actuarial Science has historically been the largest, but by no means the only academic unit at Iowa that had statisticians and taught statistics courses. Due largely to the efforts of E. F. Lindquist, a professor in the College of Eduction who founded American College Testing (ACT), the Education College developed a very strong Educational Measurement and Statistics program by the late 1960s. Some of its members were given secondary appointments in the new department of statistics and taught many of its applied course offerings in its early years of existence. These folk included Leonard Feldt, H. D. Hoover, Paul Blommers, Mel Novick, Bob Forsythe, Bob Brennan, and Michael Kolen. Another significant concentration of statisticians on campus resides in the Biostatistics Department within the College of Public Health. This unit formed initially as the Division of Biostatistics in the College of Medicine's Department of Preventive Medicine and Environmental Health in 1974, with Paul Leaverton as it first Director, and other original members Leon Burmeister, Judy Bean, William Clarke, and Robert (Skip) Woolson. Peter (Tony) Lachenbruch was recruited from the University of North Carolina in 1976 to become the second Director. When Lachenbruch left in 1985, Woolson became the third Director. The Biostatistics Division became the Department of Biostatistics in the newly created College of Public Health in 1999, with Woolson as its founding chair. Other founding members of this department were Leon Burmeister, Trudy Burns, William Clarke, Charles Davis, Jeffrey Dawson, Michael Jones, Jon Lemke, and Veronica Vieland. Kathryn Chaloner joined the faculty and became Chair of the Department in 2002. Current faculty members not already mentioned are Joseph Cavanaugh, Christopher Coffey, Dawei Liu, Jacob Olesen, Jane Pendergast, Brian Smith, Kai Wang, Gideon Zamba, Ying Zhang, and M. Bridget Zimmerman. Cooperation between the Statistics and Biostatistics Departments has been strong, especially in recent years. Several statistics faculty have secondary appointments in Biostatistics and vice versa, and graduate students in each department often take courses from the other.

Selected Books Authored by the Faculty

Kung-Sik Chan and Howell Tong, "Chaos: A Statistical Perspective" (Springer, 2001).
Jonathan D. Cryer, "Time Series Analysis" (Wadsworth, 1986).
R.L. Dykstra, Tim Robertson and F.T. Wright, "Order Restricted Statistical Inference" (Wiley, 1988).
Robert V. Hogg and Stuart A. Klugman, "Loss Distributions" (Wiley, 1984).
Robert V. Hogg, Joseph W. McKean and Allen T. Craig, "Introduction to Mathematical Statistics" 7th ed (Prentice Hall, 2005).
Robert V. Hogg and Johannes Ledolter, "Applied Statistics for Engineers and Physical Scientists" (Prentice-Hall, 1992)
Robert V. Hogg and Elliot A. Tanis, "Probability and Statistical Inference" 8th ed (Prentice Hall, 2005).
Luke Tierney, "LISP-STAT: An Object-Oriented Environment for Statistical Computing and Dynamic Graphics" (Wiley, 1990).
George G. Woodworth, "Biostatistics: A Bayesian Introduction" (Wiley, 2004).
Dale L. Zimmerman and Vicente Núñez-Antón, "Antedependence Models for Longitudinal Data" (Chapman & Hall/CRC Press, 2010).

Department Chairs

Bob Hogg, 1965–1983
John Birch, 1983–1986
Joe Sedransk, 1986–1989
Dick Dykstra, 1989–1992
Bob Hogg, 1992–1993
Jim Broffitt, 1993–2004
Luke Tierney, 2004–2010
Dale Zimmerman, 2010–2011

Departmental Faculty, 2003. Back row: Bob Hogg, Johannes Ledolter, Gordon Klein, George Woodworth, and Ralph Russo. Middle row: Joe Lang, Kung-Sik Chan, Luke Tierney, Russ Lenth, and Osnat Stramer. Front row: Elias Shiu, Grace Chan, Kate Cowles, Jim Broffitt, Jun Yan, Dale Zimmerman, Jian Huang, and Matt Bognar. Not pictured: Richard Dykstra, John Geweke, Tim Robertson.

Regular Faculty, in Order of Appointment

1965	Bob Hogg, Allen Craig, John Birch, Lloyd Knowler, James Hickman, Tim Robertson
1966	Jon Cryer
1967	Fred Leone, James Cole, Peter Wang
1968	Farroll T. Wright, Ronald Randles, John Ramberg
1970	James Broffitt
1971	George Woodworth
1974	Stuart Klugman
1975	Russell Lenth
1976	Richard Ziock
1978	Stephen Brier
1979	Johannes Ledolter
1981	George Runger
1982	Richard Dykstra, Jane-Ling Wang
1983	Ralph Russo
1984	Glenn Meyers

(continued)

(continued)

1985	James Calvin, Mark Conaway, James Sconing, Joseph Sedransk
1986	Dale Zimmerman
1988	Marianthi Markatou
1989	Jacques Carriere, Joseph Pignatiello, Louis Rizzo
1990	Martin Appel
1991	Kung-Sik Chan, Elias Shiu
1992	Bruce Jones, Joseph Lang, Jens Praestgaard
1994	Jian Huang, Osnat Stramer
1997	Mary Kathryn (Kate) Cowles, Sheldon Lin
1998	Grace Chan
1999	John Geweke
2002	Luke Tierney
2003	Jun Yan
2004	N. D. Shyamalkumar
2006	Qihe Tang
2007	Rhonda DeCook
2008	Jerome Pansera
2009	Aixin Tan
2010	Joyee Ghosh

Acknowledgments Much of the material for the history of Statistics at Iowa prior to 1983 was taken from remembrances written by Bob Hogg. We thank Jim Broffitt, Dick Dykstra, and Luke Tierney for their comments on the manuscript. Thanks also go to Carly Harryman at Pearson for information about sales of "Hogg and Craig"; to Michael Jones for historical information about the Department of Biostatistics; and to Tamara Siegel for locating many other useful sources of historical information.

References

Crathorne AR (1944) Henry Lewis Rietz—in memoriam. Ann Math Stat 15:102–108

Driscoll MF, Gundberg WR (1986) A history of the development of Craig's theorem. Am Stat 40:65–70

Mosteller F (1964) Samuel S. Wilks: statesman of statistics. Am Stat 18:11–17

Randles RH (2007) A conversation with Robert V. Hogg. Stat Sci 22:137–152

University of Michigan Department of Biostatistics

Trivellore Raghunathan

Introduction

I was extremely delighted when professors Alan Agresti and Xiao-Li Meng (my academic brother) informed me about this book project and asked me to write a chapter on the historical account of the Department of Biostatistics at the University of Michigan, for my childhood dream of being a historian is about to be fulfilled! I vividly recall the moment I went to my father and said, "I want to be a historian." Perhaps, seeing sparkles in my eyes, he said, "Good. Become whatever you want to be, but make sure that you try to do the best." (Looking toward my mother he added wryly "Don't expect me to retire" and, of course, I got the message). As far as writing this historical account, following his advice has been extremely difficult. At the core, the University of Michigan has, and had, very decentralized operations and has led to enormous success in interdisciplinary research but a nightmare for any amateur historian. However, many documents are stored at the Bentley Historical Library at the University of Michigan that provides the building blocks for tracing the history.

Before the creation of any formal department of biostatistics or statistics, the statistical activities at Michigan occurred in three distinct units: the Medical School, the Department of Mathematics, and the Institute for Social Research (ISR). The public health-oriented researchers in the Medical School played an active role in the formation of the Michigan State Department of Health in 1881 including setting up data systems for collecting and analyzing public health statistics. Eventual formation of the School of Public Health occurred in 1941 with Henry Vaughn as the Dean [School of Public Health Dean's Reports

University of Michigan,
Department of Biostatistics, School of Public Health,
Ann Arbor, MI 48109, USA
T. Raghunathan (✉)
e-mail: teraghu@umich.edu

A. Agresti and X.-L. Meng (eds.), *Strength in Numbers: The Rising of Academic Statistics Departments in the U.S.*, DOI: 10.1007/978-1-4614-3649-2_29,

(1941–current) and Francis papers (1917–1975)]. The Department of Public Health Statistics was a unit within the Department of Epidemiology, chaired by Thomas Francis. Dean Vaughn was the Commissioner of the Detroit Department of Public Health before joining the University of Michigan.

It is interesting to note that the tradition of housing Biostatistics in Epidemiology or some other unit continues to this day in newly formed Schools of Public Health such as at the University of Maryland and the University of Wisconsin at Milwaukee. However, all the successful and prominent schools of public health have a thriving and independent department of biostatistics. Often this recognition seems to elude the leaders who are forming the new schools of public health!

There was considerable statistical activity in the Department of Mathematics, where PhD dissertation topics included "Inverse Sampling" (Shen (1936)), "Fisher's Z-transformation" (Aroian (1939)), and an assortment of other related topics. R. A. Fisher came to Ann Arbor in August 1936 and gave three lectures on various topics. The notebook containing minutes of the Mathematics Club, which met regularly at 8 p.m., included many presentations on statistics theory and application. In August 1937, Neyman visited the department to give lectures on the theory of statistics. Several boxes of papers related to the Mathematics Department in the Bentley Historical Library at the University of Michigan (Department of Mathematics papers (1913–1981)) during this period, indicate very active research in statistics. Periodically, faculty members wrote historical accounts which are included in these boxes of papers. One detailed document was written by Wilfred Kaplan.

J. W. Glover, who chaired the Department of Mathematics from 1926 to 1934, was interested in applied statistics and actuarial mathematics and established the statistical laboratory with the help of H. C. Carver, who joined the department in 1916, and C. Craig who joined in 1922. Of course, H. C. Carver is well known for founding the "**Annals of Mathematical Statistics**," funded using his own personal resources to create an outlet for mathematically rigorous manuscripts. Those days, the **Journal of American Statistical Association** was much more focused on applications and descriptive statistics. Other notable events include Jimmy Savage obtaining his PhD in 1941 and Howard Raiffa in 1952.

Just after World War II, a small group of social science researchers, including Leslie Kish (who could be coined as a father of sampling techniques in practice) approached the University of Michigan for setting up a research program in the social sciences. Prestigious schools on both coasts apparently rebuffed similar requests from these researchers. The University of Michigan allowed them to form a research group under the condition that they should not expect any resources from the university. These founders struck a deal with the university that allowed them to retain their indirect costs on the grants and contracts (but pay for all building costs, utilities, etc.). Today, the result of the efforts of those founders and those following them resulted in the Institute for Social Research (ISR), with about $70 million in annual research expenditures!

Public health researchers in the Medical School and the State of Michigan Department of Health were deeply interested in poliomyelitis during this period and saw the origins of a large clinical trial, the Salk Vaccine Trial, which was eventually completed with the publication of a report in 1955 (Francis (1955)). The box containing Francis papers has interesting articles and memos related to planning the trial, where one group promoted the use of observational controls whereas others, most notably statisticians, researchers from ISR, researchers at the U.S. Public Health Service, and Francis himself advocated use of randomized study. Despite these arguments, the study began with observational controls consisting of the first and third graders as controls and the second graders providing consent receiving the vaccine.

Due to several issues that introduced bias in the study, perhaps the convincing arguments of the original advocates of randomization led to the introduction of a double blind randomized study component. Brownlee (1955) states in a review of the report on the trial notes, "It is a pity that explicit credit is not given to whomever was responsible for this change. However, only 41% of the trial was rescued and the remaining 59% blundered along its stupid and futile path." (p. 1007).

Interestingly, during this same time period, probability sampling methods were receiving considerable attention in political polling. In 1948, many political polls employing quota sampling predicted that Dewey would win the presidential election by defeating Truman by 5% points, whereas the Survey Research Center, using probability sampling methods, predicted a Truman victory (House et al. (2004)). The ensuing investigation led to a firm grounding for probability sampling methods for obtaining valid inferences about the population. Thus, random assignments of treatments and probability sampling methods for selecting subjects took a firm rooting based on the work conducted at Michigan.

Beginnings (1949–1960)

Due to the efforts of statisticians located at ISR, the Department of Mathematics and the School of Public Health, a separate department of Public Health Statistics was formed in August 1949 (8 years after being located in the Department of Epidemiology). The department consisted of Professor Clarence J. Velz as Chairman, Dr. Fay Hemphill as Resident Lecturer, Miss Helen L. Johnson as Instructor, Mrs. Mildred E. Harter as Resident Lecturer, and Mr. John J. Freysinger as Chief IBM Operator. The department could have been formed much earlier, but the faculty meeting minutes and the Dean's memos during the period 1944–1949 provide names of several individuals that Dean Vaughn and Francis, the Chair of Epidemiology, contacted but were unsuccessful in luring them to become the founding chair of the department. In June 1948, a conference on statistics for public health was organized under the joint planning of SPH faculty members and statisticians from local, state, and federal agencies, and was attended by more than 150 people.

The department's initial mission was teaching, research, and advisory service to the other departments of the School of Public Health, and related groups. The statistical method was seen as a tool in the various public health disciplines and

Clarence Velz, Founding Chair of the Department of Biostatistics (1949–1956)

practices, and not as an end in itself. Courses in public health statistics were taught, and research conducted on the prediction of poliomyelitis in 48 states, environmental analysis, public health economics, and demography. Professor Velz in the article "Vistas of Public Health Statistics" (Velz (1952)) articulates the role of public health statistician as an essential member of a public health team, setting the tone of the vision for the newly founded department.

The department's first doctoral thesis was completed in 1953. It was written by Deward Waggoner under the direction of Professor Velz, and titled "The Use of Vital Statistics in the Estimation of Postcensal Population Characteristics in the United States." The next few years saw the increased development of courses in public health statistics (to larger numbers of students), the creation of a sequence in survey methods, and increased statistical consulting. Many courses were developed in close collaboration with researchers at the Survey Research Center. In 1955 the department initiated a fellowship program for training public health statisticians, supported by a U.S. Public Health Service grant of $106,272. Five fellowships were awarded.

In 1956, Dr. Velz was appointed Chair of Environmental Health, and the following year Professor Felix Moore replaced him as the second chairman of the Department of Public Health Statistics. Felix Moore, who came from the U. S. Public Health Service, was a close collaborator of Jerome Cornfield and played a critical role in the Salk Vaccine Trial (Cornfield et al. (1956)). In that year eight MPH and two PhD students received degrees specializing in public health statistics. Richard Remington also produced a report comparing several methods for measuring blood pressure reactivity (Remington 1958).

Circa 1960–1970

In 1960, the department was renamed the Department of Biostatistics, as part of a major school reorganization. In 1961–1962, the department was involved in about 50 consulting projects inside and outside the school, and managed statistics for two

Felix Moore: The second Chair of the Department (1957–1971)

national groups engaged in the analysis of cancer chemotherapy. In 1962–1963 the faculty consisted of John Jacquez, Helen Johnson, Marcus Kjelsberg, Felix Moore, Mary Ellen Patno, Richard Remington, and Michael A. Schork (Instructor). In 1963–1964 the department had 11 graduate students. Later, Professor Remington became the Dean of the School of Public Health in 1974 and served in that capacity untill 1982.

The Dean's Report for 1966–1967 notes that "The Department of Biostatistics carries out its chief research function as counselor and guide to projects carried on elsewhere in the School and the University." The department in the late 1960s was relatively stable in terms of composition and activities.

Right from its founding, the Department of Biostatistics developed close ties with the Department of Mathematics in terms of courses and shared supervision of students. Meanwhile the Department of Mathematics had grown tremendously with three dominant areas listed in a memo from the Chair to the Dean as a part of the annual budget request: Pure Mathematics, Applied Mathematics (mostly engineering-oriented applications), and Computational Mathematics (actuarial mathematics and statistics). The statistical laboratory was mostly autonomous with a weak link to the department. The idea of the "statistical laboratory" focusing on consultation continues to this day as the Center for Statistical Consultation and Research (CSCAR), an autonomous unit under the Office of Vice President for Research.

In the late 1950s and early 1960s, several budget-related communications between the Chair of the Mathematics Department and the Dean of the School of Literature, Science & the Arts (LSA) pushed for faculty growth in Statistics. Various departmental memos and faculty meeting minutes also indicate struggles in operating a large department with diverse disciplines. In one memo, the Chair complains that "Statistics" has been neglected for the past 30 years. During this period, several statisticians were hired including Ericson, Robbins, Berk, Hill, Woodroofe, Starr, Darling, etc. Robbins negotiated several additional faculty lines

Richard Cornell, Chair
1971–1983

for the statistics area. The number of hires eventually reached a critical mass and led to the founding of the Department of Statistics in 1969. The Department of Statistics maintained, and continues to maintain, close ties with the Department of Biostatistics with shared courses, doctoral thesis supervision, and faculty joint appointments.

Circa 1970–1980

In June 1971, Felix Moore was succeeded as Department Chairman by Richard Cornell, formerly Professor at Florida State University. Randomized designs that reduce the number of trial participants who receive an inferior treatment were initiated by Richard Cornell's work on the ECMO trial. The department grew steadily under Cornell's leadership; for example, in the 1971–1972 academic year the department had 20 students, and in 1982–1983 had grown to 55 students. Richard Cornell also brought methodological research focus to the department. Various memos and annual reports prepared by him and archived at the Bentley library show his vision for the department as a balance of methodological and collaborative research, with a series of hires to implement this vision. George Williams joined the department in 1972 and Richard Landis in 1975. Anant Kshirsagar and Robert Wolfe were recruited in the late 1970s, in part to meet the increasing demand for doctoral student supervision. Cornell also brought a sense of belonging to a community among the faculty. For example, he developed a practice of the faculty eating lunch together every day to gain feedback from the faculty, and with his gentle demeanor developed close ties among the faculty. Many former faculty members remarked that his style of leadership made faculty very open and developed a sense of ownership of the department.

Richard Remington, Dean from 1974–1982

During this time, Richard Remington and Anthony Schork wrote a classic text "Statistics with applications to the biological and health sciences" (Remington and Schork (1970)). Kshirsagar published a classic text "Multivariate Analysis" (Kshirsagar (1972)) while still at Texas A & M University but soon after moved to the University of Michigan. Richard Remington became the third Dean of the School of Public Health in 1974. Under his leadership, the school developed a solid research program on hypertension and blood pressure that became his passion for the rest of his career. In recognition of these achievements, the American Heart Association organizes a plenary of lecture on methodological issues at their annual meetings.

Circa 1980–1990

Morton Brown joined the department in 1980, and in 1983–1984 he succeeded Richard Cornell as the fourth Department Chair. The Dean's report noted that Biostatistics averaged five PhDs per year over a 6-year period, second in the country in Biostatistics Departments. Faculty members Richard Landis and Lee-Jen Wei won Speigelman Awards for their research in the 1980s. Kshirsagar published a book "A course in linear models" (Kshirsagar (1983)).

In 1972, the Department of Health Management Policy in the School of Public Health developed a nonresidential program, dubbed "On Job/On Campus" (OJOC) for the health administrators to obtain a Master's degree while maintaining their full-time job (Penchansky (1975)). Under the leadership of Morton Brown, Bob Wolfe, and Tony Schork the department began a similar program in 1985 for clinical researchers. Under this program, students attend one extended weekend (Thursday–Sunday) a month for approximately 18 sessions. During these

Morton Brown Chair:
1984–1987

sessions, the courses offered include various aspects of design (clinical trials, observational studies, and survey sampling), conduct and measurement, analysis methods (Basics of statistical inferences, regression methods, survival analysis, and repeated measures analysis). The students are taught how to write research grants and evaluations. This program continues to this day with the enrollment of a fifteenth cohort that began their classes in October 2011 (since, the program takes 18 months to complete, the new students are admitted during odd numbered years). At the successful completion of the program, the students receive a Master of Science degree in Clinical Research Design and Statistical Analysis (MSCRDSA). The average class size has been around 40.

In 1987, Graham Kalton succeeded Morton Brown as Department Chair, serving a 3-year term before accepting a senior appointment at Westat. In 1990, Richard Cornell became Interim Chair of the Department, serving for 3 years before being appointed Interim Dean of the School of Public Health in 1993. After stepping down as the chair of the department, Morton Brown in collaboration with the Medical School created the Biometrics Outcome Research Core, a unit specializing in the design and conduct of multicenter clinical trials. Michael Boehnke, who joined the department in 1984, added Statistical Genetics to the department's research activities.

1990–Current

Rod Little joined the department as Chair and Professor in 1993. Under his leadership, the department saw unprecedented growth in both methodological and collaborative research. The department was able to hire several faculty with diverse research interest. The department had both Bayesians and Frequentists, parametric and non-parametric approaches, all aspects of designs: clinical trials,

Graham Kalton: Chair from 1987–1990

observational studies and survey sampling. With joint appointments of Jim Lepkowski, Rod Little, and myself (I joined the University of Michigan in 1994), ties with ISR became strong. Jeremy Taylor joined the department to head the Biostatistics Unit at the Cancer Research Center.

The Biostatistics Department at Michigan operates like many other statistics departments, where instructional faculty members have 9-month tenure-track appointments and generate their summer support on research grants. However, the faculty members were successful in generating funds through grants and contracts that could be used to cover a fraction of their salaries. The department developed a "Growth Model" under the leadership of Rod Little and Morton Brown in which the number of tenure-track positions could be increased with the commitment to fund a fraction of the salaries on grants and contracts and without increasing the commitment from the school. For example, if the department had 12 tenure-track slots and commits to cover 25% of the 9-month salaries for these positions, the number of slots could be increased to 15. Of course, there is a risk to the School as the salary obligations for all 15 positions will have to be covered, if the department cannot fulfill its commitment due to lack of funding. Dean Noreen Clark accepted this growth model and the department has been operating under the growth model ever since.

The department also established two additional centers (in addition to BORC), the Center for Statistical Genetics (CSG) and the Kidney Epidemiology and Cost Center (KECC). Researchers at KECC are helping with fundamental changes in the U.S. system for determining organ donor allocations, saving lives of many individuals with end-stage heart, kidney, liver, and lung disease, whereas the CSG has played a leading role in development and application of statistical methods for human gene mapping; its identification of nine type 2 diabetes gene regions in a paper in *Science* magazine, was cited by *Time* magazine as one of the 10 medical breakthroughs of 2007.

The department also began hiring faculty in the research professorial track. These faculty are funded on research grants but also do teach, if the instructional faculty were to generate funds more than needed to cover the commitment under the growth model and buy out of teaching courses.

Jack Kalbfleisch joined the department as chair in 2002 and continued to grow the department through hiring under the growth model. Faculty with very intellectually diverse research interests populated all ranks. Kalbfleisch also instituted several reforms to manage departmental budgets and accounting processes. Rod Little served as chair, once again from 2007 to 2009. I took over as the eighth chair of the department in January 2010, indeed to fill the big shoes of my predecessors!

In terms of books, Kshirsagar published a book on Growth Curves (Kshirsagar and Smith (1995)) and the second edition of books by Kalbfleisch and Prentice (2002) and Little and Rubin (2002) were also published.

During recent years, the department has greatly expanded in terms of the number of faculty, staff, students, and research funding. Currently, the department has 24 tenure-track faculty and 6 research faculty, 70 staff members, 152 students, including 69 PhD students, 51 Master's, and 32 OJOC students. The research-funding base is about \$27 million, compared with about \$2.7 million in 1993. In 2009–2010, the department faculty and students authored or co-authored 172 refereed research articles, including 62 where they were lead or senior authors. Every year a sizeable number of UM Biostatistics students and faculty present at the major statistics research meetings, and many students receive travel awards for their research.

Conclusions

From modest beginnings, the department has become one of the major departments in the country, notable for its outstanding PhD and MS programs and its wide-ranging methodological and collaborative research. People receiving PhD and Masters degrees from the department have become entrepreneurs (such as Lora Schwab and Mark Becker who co-founded STATPROBE), leaders in Industry (such as Ed Vonesh at Baxter Health Corporation), and great researchers in academia [such as the late Tom Ten Have, Danyu Lin (UNC, Chapel Hill)]. Former faculty members have moved on to new leadership positions outside the University of Michigan, such as Richard Landis (University of Pennsylvania), Graham Kalton (Westat), and Mark Becker (President, Georgia State University), and to other institutions, such as L. J. Wei and Xihong Lin, both at Harvard.

The National Research Council recently ranked the Michigan Department of Biostatistics as the top department of Biostatistics in the country. Currently, department faculty are working in many areas of developing and implementing

The current and former chairs: The photo taken at the anniversary celebrations. (Richard Cornell, Jack Kalbfleisch, Rod Little, and Trivellore Raghunathan)

design and methods to address the influence of social, environmental, contextual, economic, biological, clinical, and genetic factors affecting the population health. In terms of methods, the faculty and students are making influential methodological contributions in many areas, including the analysis of imaging data, survival and event history analysis, statistical genetics and bioinformatics, clinical trials, Bayesian methods, statistical computing, longitudinal data analysis, methods for missing data, and survey research. The development of biostatistical methods for cancer research through numerous R-grants and a Genomics Institute Initiative that began in 2010 under the direction of Goncalo Abecasis are exciting recent developments.

Obviously, institutions are not just built by a few individuals listed in any article. Hundreds of current and former students, staff, faculty, and leaders in the school and the university contribute toward creating a fine department. There are many unnamed heroes and no article of 5,000 words can do justice to the collective contributions. People ever affiliated with the department reading this article will find omissions and inaccuracies, and for that I apologize. One thing is for sure, we are all proud of what we have achieved and we are not satisfied just with past laurels. Their contributions will inspire the current and future generations to propel the department to greater achievements in teaching, research, and service. Michigan Biostatistics is, indeed, an exciting place to be.

References

Aroian LA (1939) A study of R.A. Fisher's z-distribution and the related F distribution. Ph.D. dissertation, Department of Mathematics, University of Michigan stored under the title "Continued fractions for the incomplete beta function" at Bentley Historical Library, University of Michigan, Ann Arbor

Brownlee KA (1955) Statistics of the 1954 polio vaccine trials. J Am Stat Assoc 50:1005–1013

Cornfield J, Halperin M, Moore F (1956) Some statistical aspects of safety testing the Salk poliomyelitis vaccine. Public Health Rep 71:1045–1056

Department of Mathematics (University of Michigan) records (1913–1981) Bulk 1935–1981; call number 90204 Bimu C26 2 UBImum C26. Bentley Historical Library, University of Michigan, Ann Arbor

Francis T, Korn R et al (1955) An evaluation of the 1954 poliomyelitis vaccine trials. Am J Public Health 45, 50 pp supplement with a 63 pp appendix

House JS, Juster FT, Kahn RL, Schuman H, Singer E (eds) (2004) A telescope on society: survey research and social science at the University of Michigan and beyond. University of Michigan Press, Ann Arbor

Kalbfleisch JD, Prentice RS (2002) The statistical analysis of failure time data. Wiley, New York

Kshirsagar AM (1972) Multivariate analysis. Marcel-Dekker, New York

Kshirsagar AM (1983) Course in linear models. Marcel-Dekker, New York

Kshirsagar AM, Smith WB (1995) Growth curves. CRC Press, London

Little RJA, Rubin DB (2002) Statistical analysis with missing data, 2nd edn. Wiley, New York

Penchansky R (1975) The OJ/OC program: a non-resident MPH program in medical care administration. Health Educ Behav 3(3):292–305

Remington RD (1958) Comparison of several tests of blood pressure reactivity, University of Michigan Press

Remington RD, Schork A (1970) Statistics with applications to the biological and health sciences. Prentice Hall, Englewood Cliffs

School of Public Health (University of Michigan) Publications: Dean's Report and Other Documents (1941) Call number 9713 Bimu C281 2. Bentley Historical Library, University of Michigan, Ann Arbor

Shen C (1936) Fundamentals of the theory of inverse sampling. Ph.D. dissertation, Department of Mathematics, University of Michigan stored at Bentley Historical Library, University of Michigan, Ann Arbor

Thomas Francis Papers (1917–1975) Bulk 1941–1969; Call number: 851285 Aa 2 Ac. Bentley Historical Library, University of Michigan, Ann Arbor

Velz CJ (1952) Vistas in public health statistics. Public Health Rep 67:725–732

University of Minnesota School of Statistics

Glen Meeden and Martha Coventry

The University of Minnesota sits on the banks of the Mississippi River, not far from where the river begins. A land-grant university, it was founded in 1851 to serve the "sons and daughters" of the state and further Minnesota's progress as it made the most of its natural resources—trees, minerals, rich soil, and the Mississippi itself.

In 2010, nearly 68,000 students attended the university and its branches around the state. The main campus in the Twin Cities is located in both Minneapolis and St. Paul, a division that plays a major part in the statistics story at the University of Minnesota.

Minneapolis has, among other things, the Academic Health Center, the law and business schools, the liberal arts, and engineering, as well as the administration. The St. Paul part of the campus, about a 20-min shuttle bus ride away, was built on a hill with farms flanking it to the west. Farmland still lies within its borders. This location made it a perfect place for agricultural research, which it still conducts today. St. Paul also houses such disciplines as genomics, natural resource studies, and veterinary and food sciences.

Beginnings

In the 1930s and 1940s, Minnesota was not so different from most other universities when it came to statistics. Many departments at Minnesota, like educational psychology, economics, mathematics, public health, and business, engaged in some statistics teaching and research.

University of Minnesota, School of Statistics,
313 Ford Hall, 224 Church Street S.E,
Minneapolis, MN 55455, USA
G. Meeden (✉) · M. Coventry
e-mail: glen@stat.umn.edu

A. Agresti and X.-L. Meng (eds.), *Strength in Numbers: The Rising of Academic Statistics Departments in the U.S.*, DOI: 10.1007/978-1-4614-3649-2_30,

The university's Graduate School even offered a Statistics PhD if a student could cobble together the right courses to get the plan approved. But there was no school or department dedicated to the field.

Nationwide during this time, statistics was beginning to be recognized as a discipline of its own and not just a tool to serve other fields such as medicine or psychology. To the south, Iowa State University had been building a department of future renown since 1934.

As the 1940s gave way to the 1950s, faculty members who worked in statistics began lobbying for a department of their own. In 1958, the College of Liberal Arts (CLA), located in Minneapolis, finally established the Department of Statistics. Minnesota's department was one of several that were founded around that time and went on to become leaders in the field, like Berkeley (1955), Harvard (1957), and Rutgers (1959).

The department hired several strong faculty members during the 1960s and early 1970s. Ingram Olkin was at the department from 1960 to 1961 and Richard Savage from 1961 to 1963. Three people who spent a significant amount of time in the department were Milton Sobel, who finished up his career at UC Santa Barbara; Professor Emeritus Bernard Lindgren who came from the Minnesota mathematics department—then and still a departmental ally—in 1963; and Robert Buehler who came from Iowa State, also in 1963. Somesh Das Gupta joined in 1967, William Sudderth in 1969, Donald Berry in 1970, and Morris Eaton in 1971.

The chairs of the still-small department were often recruited from within the university, rotating on loan from other disciplines. Palmer Johnson, the first chair, was in educational psychology and Leonid Hurwicz, future Nobel Prize winner, came from economics. Bernard Lindgren became chair in 1963.

The University Forms a School

In 1969, the split-campus plan led to a unique arrangement that set Minnesota apart from its peers. The School of Agricultural and the Agricultural Extension Service in St. Paul pressured the administration for their own statistics department to serve their consulting needs, and they got it. The first three hires for the new department were Frank Martin, who began in Biometry, Dennis Cook, and Kinley Larntz. Stephen Fienberg came on as chair and built a substantial group. He stayed at Minnesota until 1980. The administration decided that the new statistical presence in St. Paul should be allied with Minneapolis, so it disbanded the Department of Statistics and established the School of Statistics. The School then consisted of the Department of Theoretical Statistics in Minneapolis and the Department of Applied Statistics in St. Paul.

Dividing the new School of Statistics by discipline and geography, although not ideal in practice, worked fairly well. Professors tended to spend most of their time either in Minneapolis or St. Paul and seminars and faculty meetings would alternate between the two locations. The arrangement was probably most bothersome for the graduate students. At least 3 days a week they would have to travel to both locations. The free shuttle buses between them ran fairly often, but the ride was long. Students' memories of those days always include racing for the bus or waiting in the snow for the next one to arrive.

The Role of Consulting

The professors in the applied department had a different workload than their colleagues in Minneapolis—it was divided 50/50 between teaching and consulting. That meant teaching one course per quarter and then being available to any faculty at the university for consulting help the rest of the time.

This close relationship of the applied department with other fields was beneficial not only to their clients, but to members of the department. For example, Dennis Cook's development of Cook's distance was stimulated by a consulting problem brought to him from the Animal Science Department.

"It was a wonderful set up," says Christopher "Kit" Bingham, professor emeritus. "Over the years, I got involved in all sorts of great research with medical doctors, economists, musicians, food scientists, soil scientists, conservation biologists, and veterinary biologists."

The College of Liberal Arts paid the applied department salaries for 9 months; the Agricultural Experiment Station picked up the remaining 2 months. But that support slowly eroded. As time went on, new hires in the applied department were not given summer funding. In the 1990s, the CLA dean decided that the college could no longer subsidize consulting for any faculty. And, as more and more of the original hires retired or left, the Agricultural Experiment Station ended its support for consulting.

Currently, the School continues its commitment to consulting in many ways. The Statistical Consulting Service, led by Aaron Rendahl, a recent PhD graduate of the School, continues to offer consulting in St. Paul and around the university, but for a fee. The College of Liberal Arts continues to fund consulting for researchers in liberal arts. Many of the School's current faculty members continue to be involved in long-term consulting projects. For example, Birgit Grund has worked for many years on a large-scale AIDS research project funded through the university's School of Public Health.

Consulting is also alive and well among graduate students, and consulting experience is a requirement for all second-year PhD students. Today, they must complete a consulting project during their second summer. Most work on a project with researchers from different disciplines on campus, but some take internships away from the university. For those who remain on campus, and those with unfunded internships, the School supports them during the summer. The following fall semester, all the students must write up a short report and give a seminar-style talk on their consulting project.

Fig. 1 Seymour Geisser

Geisser Comes to Minnesota

When the university administration was setting up the applied and theoretical departments, it realized that the School would need a director to coordinate its activities. In 1971, it chose Seymour Geisser to fill this position. Although the two departments had their own chairs, all-important decisions, like tenure, were made by the School as a whole, with Geisser having the final authority (Fig. 1).

Geisser came from the University of Buffalo and was making his name as a brilliant and independent thinker. He believed statistical work and methods should be judged by their practical usefulness. A Renaissance man of prodigious curiosity, Geisser dropped Latin into a conversation, enjoyed reading history, and became an expert on DNA statistical analysis. He had coauthored, with Sam Greenhouse, the citation classic, "On methods in the analysis of profile data," published in *Psychometrika* in 1959.

"Seymour was an able administrator and a great scholar," says Glen Meeden, who succeeded Geisser as chair. "He quickly got the School up to speed and kept it there for 30 years."

Geisser ran up against the constraints of a public university budget, which waxed and waned with fluctuating state support. Known as a fighter, he would push the liberal arts deans as hard as he could to get what he needed for the School of Statistics. Despite financial limitations, he hired an impressive array of faculty with a broad range of specialties and philosophies during his tenure. He created an environment conducive to research and he left people alone so they could do their work.

Even as Geisser practiced a hands-off approach with faculty members, he drove them to excel and was a strong leader. One young professor, upon returning from a sabbatical, stopped in to see him. Geisser asked, "Well, how'd it go?" The professor replied, "I worked hard and kept busy." Geisser paused a second, then asked, "But did you accomplish anything worthwhile?"

Coming Back Together

After nearly 30 years of traveling between Minneapolis and St. Paul, many faculty members began hoping for a reunited School of Statistics. They felt graduate students were spending far too much time shuttling between locations for classes, seminars, and consultations with dissertation advisors. Most undergraduate classes were in Minneapolis. And bringing St. Paul faculty to Minneapolis could result in improved class size distribution and course offerings. Perhaps, they said, coming together could lead to closer collaboration among faculty.

During the late 1990s, both Twin Cities' locations undertook badly needed renovations. One building that benefitted was Ford Hall, an Art Deco edifice on the Minneapolis campus mall. The administration gave the School of Statistics the third floor and part of the fourth floor to house its faculty, staff, and classrooms. When the School moved into Ford Hall, the administration formally disbanded the theoretical and applied departments. All faculty were now just part of the School of Statistics and under one roof.

Faculty members chose to retain the designation "school" for the reunited group. It was a sentimental gesture, but they also liked the caché of the name at a university full of "departments."

A Well-Rounded Research Focus

Geisser was a leading Bayesian who emphasized the prediction of observables rather than estimation of unobservable parameters. His standard question to the speaker at the end of a School of Statistics seminar was, "Well, what does this have to say about prediction?" He had arrived in Minnesota when Bayesian decision theory was out-of-fashion in the United States. Today most statisticians take Bayesian theory for granted, but at that time the American statistics community was not very welcoming of the approach.

There were Bayesians on the faculty when Geisser arrived, including Robert Buehler, William Sudderth, Donald Berry, and Morris Eaton. Over the years, Geisser hired additional Bayesians David Lane, Luke Tierney, Kathryn Chaloner, Jim Dickey, and Glen Meeden.

But although Geisser was a committed Bayesian, he further added to an early group of non-Bayesians, which included Dennis Cook, Kinely Larntz, Sanford Weisberg, Frank Martin, and Kit Bingham. New hires David Hinkley, Gary Oehlert, Douglas Hawkins, and Charles Geyer were frequentists who flourished in the School.

By 1975, the School had an impressive array of younger faculty who covered most of statistics and a good piece of probability. In addition to Bayesian statistics, particular strengths included theory and applications of linear models, categorical data analysis, gambling theory, multivariate analysis, foundations of inference, and classical mathematical statistics, including statistical decision theory.

For more than 30 years, Cook and Weisberg have been leaders in the theory and application of regression analysis and graphics. Hawkins has been at the front of diagnostics and quality control. Geyer is an expert in Markov chain Monte Carlo methods. Recently, he has developed several R packages; the latest is "Aster" which is meant for biologists wishing to do life history analysis.

Geisser was interested in the foundations of probability and statistics, and both the frequentist and Bayesian approaches to statistical inference have been well represented in the School. Indeed the interaction between and reconciliation of the two approaches have been important themes in some of the faculty's research. Robert Buehler's foundational work, Meeden's Bayesian approach to finite population inference, the formal Bayes/decision theory work of Galin Jones and Morris Eaton, and Bill Sudderth's finitely additive approach to Bayesian inference have all contributed to a better understanding of this foundational relationship. The faculty continues to make important contributions to this basic area of research.

Computing at Minnesota

In 1972, the state of Minnesota had the largest educational computing network in the world. Called the MERIT system, it stretched from Moorhead in the far northwest to Rochester in the southeast. It was available to high schools, colleges, and universities and transmitted ten characters per second on teletype machines, which were used as command-line interfaces with the mainframes. The university's computer lab in Ford Hall had five or six teletype machines that tapped into MERIT and the School of Statistics began to integrate computing into its courses.

Soon the university bought its own central computers, primarily CDC Cyber mainframes. In addition to the standard statistical packages, the School faculty also wrote several innovative, interactive packages. This software implemented and brought into standard usage regression diagnostics and other ideas. Kit Bingham, for example, created an early interactive program called MATTER. Gary Oehlert joined the department in 1984 and soon after, he and Bingham began to develop MacAnova, a free open source statistics package.

Other faculty members who were interested in developing statistical software packages were Luke Tierney, Dennis Cook, and Sandy Weisberg. They brought object-oriented programming, modern interactive graphics, and interactive modeling into the mainstream of statistical practice with the programs they developed—Lisp-Stat and Arc, respectively. Moreover, Tierney's knowledge of computing was much deeper than most statisticians and quite unusual for then, and now. His expertise was crucial in developing the School's strength in computing.

After the mid-1980s as computers became more sophisticated, computing at the School of Statistics split into two major streams and one minor stream. Research and graduate computing moved to Unix, first on a VAX 11-750 in 1985, then on to a succession of workstations through the late 1980s and 1990s (Sun, then DEC, then HP), and most recently running Linux on high-end, multi-core Intel boxes.

Undergraduate and service teaching—teaching students from other departments—moved onto personal computers (both Macintosh and Windows) using a mixture of commercial software and locally written software.

The final computing stream is supercomputing. Minnesota has been at the forefront of academic supercomputing for three decades, and School of Statistics faculty can apply for time on the university supercomputers when they need large-scale number crunching.

Recently, statistical software development has turned more and more to R packages hosted on CRAN rather than stand-alone software. The School of Statistics faculty and students have contributed 16 CRAN packages in the past few years. Charles Geyer has been a leader in teaching students and faculty members the intricacies of R and how to write software packages. A few years ago when there was a problem with the computers and Geyer was not around, one of the graduate students said, "How can the School expect us to get any work done when Charlie is not here to make sure everything is working?"

Teaching as a Discipline

From the beginning, the courses the School offered were of two types: service courses for students from other departments and courses for its own graduate students. For years, the School has taught three introductory courses for undergraduates. The first is a non-mathematical course for liberal arts students. The second is a course for engineering students who know calculus. The third lies somewhere in between these two. It does not assume calculus, but assumes more mathematical sophistication than required for the first class. In this course, the students are expected to learn how to use the computer package R to solve their homework problems.

Statistic majors typically get interested in statistics after taking either the second or third introductory course. They then take a follow-up course, which builds their computing skills and introduces more statistical ideas. At this point, they are ready to take a theory course and to start taking some of the applied service courses.

For the most part, the School developed these service courses for graduate students from other departments. The topics included the usual suspects: regression, design, categorical, sample survey, nonparametric methods, multivariate methods, and time series along with some more recent additions: environmental statistics, mixed linear models, and model selection. Over the years, some faculty and graduate students have benefitted from interactions with Joan Garfield's group in the Department of Educational Psychology as she and her colleagues have explored ways to improve the teaching of statistics.

Like most U.S. statistics departments, Minnesota has never had a large number of undergraduate majors. It tended to average only 40 or so each year. Recently, however, the number has grown to around 80, in part because of an influx of international students.

Since most of the university's students live off-campus, the School has a Web server with R Web installed. This allows students with Internet access to use R over the Web. In addition, the School has developed an online Web-based system for grading homework. It allows each student to get his or her own version of a particular problem, which is then automatically graded. Although the system can only handle routine problems, it is a way to provide timely student feedback in an era of declining resources.

The School has a good working relationship with the biostatistics department. Biostatistics requires all its PhD students to take the School's PhD theory sequence, and most statistics students take one or two biostatistics courses as electives.

More Recent Hires

The School has hired more than half its faculty members in the last dozen years. The following are some brief highlights of their research.

Peihua Qiu's work has been in two main areas. The first is in jump regression analysis, which concerns regression modeling when the regression function has jumps or discontinuities. The second is in quality control where Qiu, along with Doug Hawkins and others, have developed changepoint methods for quality control problems.

Tiefeng Jiang's research focuses on two things: searching for similarities among two or multiple protein or DNA structures in three-dimensional spaces, and exploring random matrices and their relationship to statistics, mathematics, and statistical physics.

Galin Jones is interested in Markov chain Monte Carlo methodology with an emphasis on studying convergence rates of Markov chains. He is also interested in Markov chains in decision theory, hierarchical models and applications of statistical methodology in biological, agricultural, and environmental settings.

Some of Snigdhansu Chatterjee's work deals with climate change statistics and would be nearly unimaginable 20 years ago. He works with a large team consisting of climate scientists, hydrologists, and computer scientists.

For large and complicated problems arising from massive data sets, Lan Wang has developed lack-of-fit tests that can be used for model checking and some novel model selection methods. One important area she has worked in is quantile regression.

Xiaotong Shen and Hui Zou have been working on supervised, semisupervised, and unsupervised learning problems using regularization, where data with and/or

Fig. 2 The Minnesota faculty in 2011. From left to right, Adam Rothman, Xiaotong Shen, Hui Zou, Ansu Chatterjee, Glen Meeden, Dennis Cook, Galin Jones, Lan Wang, Doug Hawkins, Yuhong Yang, Birgit Grund, Bill Sudderth, Aaron Rendahl, Charlie Geyer, Tiefeng Jiang, Sandy Weisberg, and Peihua Qiu. Missing is Gary Oehlert

without labels have been processed through classification, clustering, and classification/clustering models with prediction as a goal.

Yuhong Yang has been working on theories and applications of model selection and combining statistical procedures, especially when a large number of models are involved. He has also studied optimal data splitting ratio in cross validation and has contributed to understanding the possibilities and limitations of statistical learning.

Adam Rothman is the School's only assistant professor and joined the faculty in the fall of 2010. His interests are in theory, methodology, and computational algorithms for statistical problems involving high-dimensional data. Specific topics include multivariate analysis, machine learning, and efficient optimization.

Taking Stock

Over the past 35 years, the School has maintained its international reputation as a center for teaching and research. This is evidenced not only by faculty publications and the School's PhD graduates, but also by the professional recognition afforded the faculty. Such recognition includes faculty visits to other institutions in the United States and abroad, countless presentations at professional meetings and other universities, and service contributions such as refereeing, associate editorships, and editorships. Stephen Fienberg was coordinating editor and applications editor of the *Journal of the American Statistical Association* from 1977 to 1979, David Hinkley served as editor of the *Annals of Statistics* from 1980 to 1982, and Morris Eaton was co-editor of the *Annals of Statistics* from 2004–2006. Nearly all of the current and former faculty of the School are or were fellows of the American Statistical Association or the Institute of Mathematical Statistics.

There are a variety of noteworthy faculty activities that have added to Minnesota's reputation for excellence and innovation. Certainly one is the publication of advanced level books. These include a multivariate analysis book by Morris Eaton; a prediction book by Seymour Geisser; a number of regression and graphics books written jointly and separately by Dennis Cook and Sandy Weisberg; a book on quality control by Douglas Hawkins and David Olwell; a finite population sampling book by Malay Ghosh and Glen Meeden; a book on gambling theory by Ashok Maitra and William Sudderth; an experimental design book by Gary Oehlert; and a book on jump regression by Peihua Qiu.

At the School's 25-year celebration in 1996, past students were the only speakers. In May 2011, the School commemorated its fortieth year (Fig. 2), and again only invited former students to come and speak about their work. With the strong nucleus of younger faculty, the School should remain a vibrant center for the practice and development of statistics, with the expectation that future celebrations will continue to feature former students who will bring perspective to the past and insights into the future of the field.

University of Missouri Department of Statistics

Nancy Flournoy and Melody Galen

The first course at the University of Missouri (MU) to mention statistics in its title was "Mathematical Theory of Probability, with Applications to Life Insurance and Statistics," offered by the mathematics department from 1906 to 1910, followed by "Probability and Statistics" starting in 1911. A committee was formed in 1961 to consider whether or not to form a new department in the College of Arts and Science (A&S) that was devoted solely to statistics. At this time, mathematics majors could have an emphasis area in "mathematics and statistics," and there was an accounting and statistics department in the School of Business and Public Administration (B&PA). Because B&PA was to have a new dean in fall 1962, the committee decided to wait for his arrival before making a recommendation. The dean of faculties called together a committee after the dean of B&PA had arrived. The idea of a new statistics department was a popular one, and it was established in 1963.

On the occasion of the department's 25th anniversary, which was celebrated in 1989, founding chair Dan Brunk shared correspondence he had retained since the committee's deliberation. Paul Burcham, chair of the math department in 1962, had written to Brunk, "It seemed the unanimous opinion of everyone there that a strong Department of Statistics in Arts and Science that had both a training and consulting function was most desirable." William Francis English, dean of A&S, was also supportive, recognizing the need for a solid graduate program in Statistics and pointing out that there were a number of departments to be served. He expressed the desire that within two years most statistics courses would be taught by the new department, though no department would be denied the right to a limited number of specialized courses.

University of Missouri,
Department of Statistics,
146 Middlebush Hall, Columbia,
MO 65211, USA
N. Flournoy (✉) · M. Galen
e-mail: FlournoyN@Missouri.edu

A. Agresti and X.-L. Meng (eds.), *Strength in Numbers: The Rising of Academic Statistics Departments in the U.S.*, DOI: 10.1007/978-1-4614-3649-2_31,

The Founding Four

The visionaries of the department, James Holstein and Frederick Williams, came from accounting and statistics in B&PA. According to Albert Winemiller, MA '65, Holstein and Williams comprised the domain of statistical expertise at MU prior to the formation of the department. They brought with them courses in regression, ANOVA, operations research, and sampling. They did not bring elementary statistics because Link Cobb, who stayed in B&PA, taught that class; after Cobb left the university a few years later, the course was transferred to the Statistics department. They also introduced a course in quality control, primarily for engineers. Williams' expertise was in sampling, operations research, and applied probability.

Burcham actively recruited Hugh Daniel "Dan" Brunk from the University of California, Riverside (UCR) to chair the fledgling department. Brunk, who had been part of the Department of Mathematics at MU from 1952 to 1961, came back to MU in 1963 and launched a successful program in theoretical statistics. He made significant research contributions in probability, Bayesian analysis, and mathematical statistics, including pioneering work on isotonic regression. Before it came to be known as isotonic regression, statisticians dubbed the process "Brunkizing" estimates.

Brunk had tried to recruit David L. Hanson to UCR, but Hanson was leery of the impending review for tenure he would soon face. Because he had been at IBM and Sandia Corporation for three years, he felt uncertain of entering the academic system and declined. However, in 1963, Brunk succeeded in recruiting Hanson as an associate professor in MU's new statistics department with no deadline for tenure. Hanson was an expert in rates of convergence and was interested in probability as it relates to statistics.

During his first semester, Hanson still was a rather inexperienced teacher, and he remembered four students in his measure theory class, three of whom were doctoral candidates in math. "One student would nod his head at everything I put on the board—so it appeared either that they already knew the stuff or that it was really easy for them," he said. "Of course, I just went faster and faster until the first homework assignment was turned in. It turns out that he was nodding all the time in hopes it would keep me from calling on him."

Brunk's student, Timothy Robertson, PhD '66, was the first granted a doctorate from the program; Hanson's student, Farroll Timothy Wright, PhD '68, was the third, with a dissertation on rates of convergence. After graduation, Hanson had some Department of Defense money to bring Wright back to work with his student Gordon Pledger, PhD '70, on convergence problems and order restricted inference (Hanson et al. 1973). The authoritative book *Order Restricted Statistical Inference* (1988) was written by Robertson, Wright, and Richard Dykstra, who joined the faculty in 1968. After spending time at the University of Missouri–Rolla and the University of Iowa, Wright was recruited back to MU in 1989 and served as chair 1995–2002.

When Brunk left for Oregon State University in 1969, Hanson took up advising Sharon LeDuc, PhD '71, who was deputy director of NOAA's National Climatic Data Center from 2001 until her retirement in 2011. LeDuc credits Hanson and Brunk, together with professors Jim McQuigg and Wayne Decker from atmospheric sciences, for determining the direction of her career.

Three of the founding four contributed to governing the department. Brunk was chair from 1963 to 1966, then Williams followed him for five years. Hanson took the reigns from 1971 to 1973, and Williams followed him from 1973 to 1976, plus a short stint as interim chair in 1980–1981.

Building a Department

During that first year, the four members of the statistics faculty weren't even housed together in the same building. Brunk and Hanson were in the engineering building with mathematics, while Holstein and Williams kept their offices in the accounting and statistics department in Middlebush Hall. The storeroom on the top floor of the Naval ROTC building (then at the corner of Fifth and Stewart, across from the power plant) was remodeled into offices, and the four moved into that space in the second year. 1964 was also the year that the department began offering an undergraduate degree in statistics.

Wallace E. Franck, a colleague of Hanson's at Sandia, was hired in 1964, followed by John Hewett, and William Bulgren in 1965. Gary Krause also joined them in 1965, but his appointment was in the Statistics Experimental Station in the College of Agriculture (now the College of Agriculture, Food and Natural Resources, or CAFNR) rather than A&S. This initiated an ongoing collaboration between CAFNR and statistics. Krause wrote early software to do least squares estimation, initially for a cross-breeding experiment in beef cattle.

With four additional people, their storeroom offices were getting cramped, and Brunk wanted a new building to house statistics and mathematics. A committee was formed to apply for a matching-funds grant from a government agency. After all was said and done, the computer science department was brought in to share the building. The Mathematical Sciences Building was completed in 1968, and statistics shared that space until 2002 when the then-new chair Nancy Flournoy moved it back to freshly renovated space in Middlebush Hall—where Holstein and Williams originally had their offices.

Prior to launching the department, there had not been a statistics degree at MU, but soon a few students from business came over to get their Masters degrees, and Robert Patton, PhD '67, Robertson, and Wright came from mathematics to get their doctorates. Patton became dean at Bowling Green State University and at the University of Southern Maine. Brunk's work attracted Raymond Bowman, PhD '66, and Charles Mann, PhD '69, early to the program. Charles R. Mann Associates, Inc., became hugely successful applying statistics to litigation and regulatory matters.

The department soon began offering introduction to mathematical statistics for engineering students. Some psychology, education, and agriculture students were required to take elementary statistics at undergraduate and graduate levels. The agricultural economics department helped the most to boost enrollments in higher-level statistics courses by requiring students to take regression and analysis of variance.

Gerald Chase, Keewhan Choi, George Denzel, and M. Vijaya Manon arrived in 1966, followed in 1968 by Dykstra, David Pierce, William A. Thompson, and Robert Tsutakawa. Thompson wrote *Applied Probability* in 1969. Tsutakawa passed away at the age of 81 on January 28, 2012. His first doctoral student, Darrel W. Parke, PhD '72, is a member of the Federal Reserve Board. Tsutakawa is known internationally for his work in statistical methods in bioassay, item response theory, and Bayesian hierarchical models. A review paper in *Statistical Science* (Chaloner and Verdinelli 1995) cited him for publishing one of the earliest papers (1972) putting design for nonlinear models in a Bayesian perspective. This work was motivated by discussions with David Wallace at the University of Chicago.

When James Higgins, PhD '70, was choosing graduate schools, Hanson called, and it sealed the deal for MU. "Geography and a personal call had as much to do with my decision as statistics. So much for decision theory!" he said.

Higgins took Monte Carlo methods from Bulgren and found the idea that one could solve extremely difficult computational problems with nothing more than the computer equivalent of tossing coins or rolling dice very appealing. It had a lasting impact upon his career. Classes with Holstein also made an impression. Though computers have markedly changed statistics courses, he still falls back on "old" ideas.

Shriniwas Katti, who joined the faculty in 1969, is known for his generosity to students. In 1987, he started the Katti Family Endowment Fund "to promote excellence in statistics and to promote scientific inquiry." To honor Katti's generosity and mentorship, and with a strong personal commitment to the next generations of statisticians, Ferrin Harrison, PhD '90, contributed substantially to this fund over a number of years before his death in 2009. Harrison spent his career at the FDA working with fellow student Sue-Jane Wang, MA '84. She is now associate director for Adaptive Design and Pharmacogenomics. Continuing the legacy, and honoring Harrison as a "passionate scientist and humanitarian," the FDA announced the Ferrin Harrison Internship starting in summer 2011.

Melvin Moeschberger arrived in 1970. At MU, he wrote *The Theory of Competing Risks* (1978) with H.A. David, and introductory calculus and statistics texts with Richard Madsen, who joined the faculty in 1971. Hewett claims bragging rights for matching his first doctoral student, John Spurrier, PhD '74, with Spurrier's future wife, Pamela, MA '74. John is distinguished professor emeritus of Statistics at University of South Carolina.

Asit Basu joined the department in 1974 when Williams was chair. During Basu's tenure as chair, from 1976 to 1983, the department was able to recruit Paul Speckman in 1982. Randall Eubanks (2006 correspondence) reported that

Fig. 1 Conference on Statistics: Present and Future, Columbia, Mo., April 14–15, 1989. Front row: James Higgins, Susan Diamond, Hari Mukerjee, Richard Madsen, Purushottam Laud, Dennis Weier, Shriniwas Katti, James Guffey, Cheryl Walgren. Second row: Frederick Williams, Mary Sue Beersman, Robert Tsutakawa, Pamela Spurrier, John Spurrier, Dipankar Bandyopadhyay, William Thompson, Cynthia Johnson, Carol Feltz, Chiranjit Mukhopadhyay. Third row: Nancy Bissey, Larry Lee, James Holstein, Tim Wright, John Hewett, Jon Maatta, Paul Speckman, David Hanson, Peter Zekert, Gary Shoop, Asit Basu, Hamid Fallahi. Back row: John Klein, Charles Champ, Richard Dykstra, Tim Robertson, Max Englehardt, James McHugh, Wallace Franck, Randy Hanson, Ben Buddy

> Speckman's unpublished 1981 technical report 'The Asymptotic Integrated Mean Square Error for Smoothing Splines,' laid the foundation for the modern asymptotic theory for smoothing splines. In 1985, Paul gives an exact, asymptotic bound for the best possible rate of convergence for the global mean squared error in nonparametric regression and develops a smoothing spline estimator variant that attains the bound. He (1988) derived exact, large sample expressions for the variance and bias of partial spline type estimators. Isolating the cause of the partial spline bias effect led to an estimator that converges at parametric rates with asymptotically negligible bias.

In 1989, Hewett and Basu organized a conference on the present and future of statistics in celebration of the department's 25th anniversary (Fig. 1).

The addition of Dongchu Sun, who had been Jim Berger's student, in 1992 while Tsutakawa was chair, brought a burst of energy and technical skills in Bayesian methods.

Expanding Our Scope

Tsutakawa earned his doctorate under Paul Meier at the University of Chicago in 1963. His work on bioassay design, hierarchical models, and Bayesian methods was a harbinger for major research thrusts in the department today. He also exemplified the modern statistician in finding motivation in applications. Tsutakawa shared three examples of applications that led him to Bayesian hierarchical models:

> In the early 1970s, an environmental biologist brought me a scatter plot of chemical readings on field mice plotted against days in captivity. The points appeared as if they had been chosen by tossing a dart at a rectangle. I learned there were three groups of mice trapped at three distances downwind of a lead smelter. When each point was identified by distance, animal, and time, there was a clear pattern of higher readings closer to the smelter and a steady decrease as days elapsed. The problem boiled down to three collections of random regression lines, one for each animal.
>
> In 1975, Art Dempster invited me to attend his seminar on variance components. When I asked him if my field mice problem had any relevance to the topic, he said, "I have just the algorithm," adding that he never understood variance components until he worked on the algorithm. Little did I know that he was referring to the now-famous EM algorithm, a general iterative procedure for getting maximum likelihood estimates with incomplete data. Eventually, I learned about many past attempts at solving the variance components problem. In particular, my colleague and close friend, Bill Thompson, was a pioneer in this area, as illustrated by his 1955 and 1962 papers.
>
> A few years after my sabbatical, Carl Marienfeld, from community and family medicine, was looking for a statistician to analyze the effects of public drinking water on cancer mortality in Missouri. Cancer records included classifications by year, age, sex, etc., matched with characteristics of each township's public drinking water, such as source, treatment, chemical analyses, etc. Previous conclusions were unreliable due to the small number of cases in most cells. Standard statistical approaches seemed quite inadequate.
>
> Several weeks later while driving home from a trout fishing trip, passing through many small rural Missouri towns, it suddenly crossed my mind that the mortality rates in these towns are random. Simultaneously, I thought about the EM algorithm.
>
> About this time, two graduate students from educational psychology insisted I attend a conference in Iowa City. During the three-hour drive, I learned that item responses and dose responses share the same models, the main difference being that ability parameters are unknown.

"Such stories" said Tsutakawa, "describe the motivation for my subsequent work on Bayesian hierarchical models. The development of Gibbs sampling and other Markov chain Monte Carlo procedures led us to produce full Bayesian solutions to, for example, the spatio-temporal disease mapping problem (Sun et al. 2000)."

Curricular Transformations and Faculty Ups and Downs

In 1979, a course on statistical consulting was added to the curriculum, giving students the opportunity to consult on real problems. Holstein, Franck, Thompson, and Williams retired in 1992, and Katti followed in 1995. Two new large lecture classes partially solved the problem of few faculty and offered the opportunity to employ international students as graduate teaching assistants—by creating positions in which students could grade and later be promoted to graduate instructors when they passed the state-mandated English proficiency test. Having a way in which international students could be supported before passing the English proficiency test, together with an NIH Biostatistics Graduate Training Grant for U.S. citizens since 2005, the number of statistics graduates has risen to over 50 since 2006.

Course requirements for doctoral students were few until about 2000 when requirements were reconfigured to include a new sequence on data analysis. Departmental offerings then remained relatively stable until around 2004 when increasing numbers of students who had not taken calculus were found in calculus-based classes. This was unsatisfactory for them, as they were unprepared for the assignments, and the education of statistics students was compromised with unqualified students in the classroom. Yet, their motivation demonstrated demand, so the department introduced a collection of applied courses that did not require calculus. "It was really a gamble whether the classes would attract enough students to run," said Flournoy. "But they have, so it was a win–win situation for everyone."

After Winemiller became president and CEO at PROS, the world leader in B2B pricing software utilizing real-time pricing decision science, he established two endowments as a way to help the department forge practical connections with other disciplines. Also the department recruited Flournoy as chair with the goal of expanding statistics' presence on campus. Nine new faculty members were subsequently added, but due to recent hiring freezes, the number of tenure-track faculty has again dropped to 12 in 2011—augmented by six teaching faculty: Jonathan Blankenship; Ryan Ellebrach; John Fresen; Leonard Hearne; Suhwon Lee, PhD '03 under Speckman; and Lawrence Ries, PhD '95 under Basu.

Spatio-Temporal Statistics

Prior to 1990, longitudinal data analysis and time series did not really include a spatial component. Noel Cressie's *Statistics for Spatial Data* (1991) popularized the discipline. While Tsutakawa and Sun provided the department with a foundation of expertise for Bayesian hierarchical modeling, the hiring in 1998 of Chris Wikle, who was Cressie's student, was a catalyst for the department's increasing prominence in spatio-temporal statistics. Wikle and Cressie (1999) contains the first full development of a space–time Kalman filter to handle the computational evolution of a space–time process. His statistical models incorporate application-specific partial differential equations that allow for dynamical space–time interactions.

Applying a hierarchical approach to meteorology, Wikle et al. (2001) developed a model of oceanic surface fluxes and wind stresses that has been put into production by the National Center for Atmospheric Research. Moving to the problem of predicting the behavior of complex processes in the immediate future, Wikle's motivations include tornado and wind behavior (Xu et al. 2005).

Zhuoqiong "Chong" He, who joined the department in 1999, became engaged in spatio-temporal sampling and modeling through her collaborations with the

Missouri Department of Conservation. Since arriving in 2001, Athanasios Micheas has migrated from the analysis of shapes (Micheas and Dey 2005) to the analysis of random sets (Micheas and Wikle 2009). More recent faculty whose work includes a spatial component are Scott Holan, Marco Ferreira, and Subaharup Guha.

Reliability and Biostatistics

Basu developed an internationally recognized research program in reliability. His NSF-sponsored workshop in 1978 began a series of international conferences on reliability, survival analysis, and quality control with sequels in 1984, 1986, 1988, and 1991. The 1984 conference was the first major one of its kind in North America. Among the participants were Ed Deming, Richard Barlow, Bruce Hoadley, Nancy Mann, Gary McDonald, Frank Proschan, and Nozer Singpurwalla. There was a new awareness of improving the quality of U.S.-made products. Deming was the de facto guru in the field after his phenomenal success in improving quality of Japanese products using the Deming philosophy. Deming wanted the quality of U.S.-made products to improve and was heard saying, "This is our chance. Let us not blow it." Basu's book with Rigdon (2000) was one of the first to consider Weibull models for analyzing data from repairable systems.

Moeschberger served on the doctoral committee of John Klein, PhD '80, a student of Basu. This had ramifications as Moeschberger and Klein later co-wrote a popular textbook *Survival Analysis: Techniques for Censored and Truncated Data (Statistics for Biology and Health)* (1977, 2003), which is used in the department's course on applied survival analysis. Klein heads the Department of Biostatistics at the Medical College of Wisconsin.

Expertise in survival analysis was strengthened with the recruitment of Jianguo (Tony) Sun in 1997. J. Sun is the world's leading expert on the analysis of interval censored failure time data (2006). Sun and Jack Kalbfleisch (1993) provide the first statistical procedure for comparing two groups using current status data. Sun et al. (2005) tackled informative observation times in semiparametric regression analysis of longitudinal data.

Joining the department in 2002, Flournoy closed a circle of interaction with Tsutakawa that began in the early 1980s when he visited the Fred Hutchinson Cancer Research Center in Seattle where she was director of clinical statistics. Having been an early developer of Bayesian experimental design (1972), Tsutakawa encouraged Flournoy to be the first to implement one for a clinical trial, which Flournoy described in one of the first papers on the elucidation of priors (1993). In this clinical trial, the experts' priors were far from subsequent experimental results, with lethal consequences, which made Flournoy leery of relying on expert opinion. She then focused on adaptive treatment allocation procedures with short memory and fast convergence properties (see Ivanova and Flournoy 2006).

Flournoy, with Durham and Li (1998), developed a procedure with binary responses that causes the next subject to be assigned, asymptotically, to the best treatment with probability one. Although generalized by others, inherent degeneracy of within-treatment sample sizes thwarted the development of statistical testing procedures until May and Flournoy (2009).

Flournoy organized workshops on adaptive design at The Fields Institute in 2003 and at ENAR in 2005. Min Yang joined the department in 2005, bringing new insights to crossover designs and optimal designs for nonlinear models. Flournoy and Yang organized the Sixth North American Conference on DAE: Design and Analysis of Experiments in 2009 at MU.

Jing Qiu and Sounak Chakraborty were hired in 2004 and 2005 to provide bioinformatics expertise. The department's reputation in biostatistics was recognized when it was one of only 12 recipients of an NIH training grant in 2007. It is the only biostatistics training award given to a department of statistics.

Additional Conferences

Speckman organized an NSF-Conference Board on Mathematical Sciences conference on longitudinal data analysis in 1997, which featured Nan Laird of Harvard. The First Albert Winemiller Symposium on Statistical Applications, held in 1999, was organized by Hewett, Speckman, Dave Larson (forestry), and Phil Wood (psychology). This series began as outreach encouraging interaction between faculty who practiced statistics across the MU campus. Wright was organizer in 2001; J. Sun in 2002; and He, Ries, and Wood in 2003. Then the series became international, each focusing on a special topic: natural resources in 2004 (He), social science methodology in 2006 (Thombs and Stanislav Kolenikov), and survival analysis in 2008 (J. Sun). In addition, D. Sun and Speckman organized the Fifth International Workshop of Objective Bayes in 2005, the same year Thombs organized the 12th Southern Regional Council on Statistics.

Consulting

An early consulting center that primarily served the business and medical schools has since branched out in many ways.

Biostatistics Group

Chase had a joint appointment in the Department of Community Health and Medical Practice in the School of Medicine. To cover the hole created when Chase left in 1974, Hewett began spending half of his time consulting in the medical and nursing schools. Then when Hewett became department chair in 1983, Madsen took over consulting for the School of Nursing, and Hewett, with the help of rheumatologists Gordon Sharp and Bernard Singsen, wrote the proposal that established the Biostatistics Group in the medical school. Hewett and Madsen comprised the original group, which is now organizationally aligned with MU's Office of Clinical Research.

Matthew Mayo, BA '86, MA '90, founding chair of the Department of Biostatistics at the University of Kansas Medical Center, said, "Holstein and Williams got me interested in statistics. Then, Madsen convinced me that I could be a graduate student and teach. Holstein, Speckman, Madsen, Franck, Basu, and Hewett were great teachers. Franck was great to work under as a teaching assistant."

Now, Hewett and Wade Davis direct the group. In addition to Hewett and Madsen, Davis, Gregory F. Petroski, and Youngju Pak make up the doctoral staff, and Bin Ge and Isabella Zaniletti are the master's staff. With the exception of Petroski and Pak, the five staff members are all alumni of the MU statistics department.

The quality of the group's work is reflected in Hewett's Lifetime Achievement Award from the Association of Rheumatology Health Professions.

CAFNR Statistics Experiment Station

Begun in 1965 with the hiring of Gary Krause by the College of Agriculture, the statistics experiment station (SES) was formed with the purpose of providing statistical consulting for agriculture faculty. Krause hired Jeannie Sebaugh in 1977, who left a few years later to start a consulting company in Columbia. Next came Mark Ellersieck in 1979.

When Krause retired in 2002, Flournoy joined the department as chair. The department was still on the second and third floors of the Mathematical Sciences Building, and SES was on the first floor. CAFNR appointed Flournoy as director of SES in addition to being chair of statistics, and she arranged for everyone to move together into Middlebush.

In the early days, regression and analysis of variance were the bread and butter of agricultural statistics. The appointment of Flournoy as head of SES reflected CAFNR administrators' recognition that a broader range of statistical tools was needed to support modern life sciences research. Krause's single salary line was split in 2004 to partially fund three statistics faculty members (Holan, Qiu, and Christie Spinka) each with different expertise relevant to CAFNR research thrusts. Spinka has since given up her full-time appointment, but Holan and Qiu have recently received tenure.

Fig. 2 MU Department of Statistics faculty in 2012, starting with back row: Associate Professor Scott Holan, Associate Professor Athanasios Micheas; row five: Associate Professor Sounak Chakraborty, Professor Paul Speckman, Associate Teaching Professor Larry Ries, Assistant Professor Michael W. Robbins, Instructor Ryan Ellebracht; row four: Professor Jianguo (Tony) Sun, Instructor Yolanda Kumar, Assistant Teaching Professor Leonard Hearne, Associate Professor Marco Ferreira; row three: Instructor Peggy Bryan, Professor Chris Wikle, Assistant Teaching Professor Suhwon Lee; row two: Assistant Professor Subharup Guha, Associate Professor Lori Thombs, Professor Zhuoqiong (Chong) He; front row: Assistant Teaching Professor John Fresen, Professor Dongchu Sun, Professor Nancy Flournoy

Life Sciences Center Consulting

Additional statistical support to faculty and students in the life sciences is provided by Leonard Hearne who is partially supported by the Life Sciences Center.

Social Science Statistics Center

Deriving from mission enhancement funds, the SSSC began in 2003 with the recruitment of Lori Thombs as director. The center is an interdisciplinary initiative between the MU Department of Statistics and the Department of Psychological Sciences that has provided an increasing number of statistical consultations. Key players in the formation of the center were Provost Lori Franz and faculty members Wright, Harris Cooper, Hewett, Speckman, and Phil Wood.

In 2003, two statistical software specialists employed with the university's information technology division, Ray Bacon and Margie Gurwit, were transferred to the new SSSC. Doug Steinley, a psychology faculty member who had earned his master's in MU's statistics department, was recruited as an affiliate faculty member of the SSSC. Typically, there are also two or three advanced graduate students working in the center. In 2008, Gurwit moved to the psychology department and was replaced by Lada Micheas.

Since August 2003, this group has provided gratis one-on-one consulting to MU faculty, staff, and graduate students; assistance with grant proposal writing; and statistical software training. Although it was formed with the purpose of helping researchers in the social sciences, the center never turns anyone away.

Going Forward

D. Sun assumes the chair in 2011. With a young, energetic faculty and a strong demand for statistics, the future of the department looks bright (Fig. 2).

References

Chaloner K, Verdinelli I (1995) Bayesian experimental design: a review. Stat Sci 10:273–304 (1995)

Cressie N (1991) Statistics for spatial data. Wiley, New York

David HA, Moeschberger ML (1978) The theory of competing risks. Griffin, London

Durham SD, Flournoy N, Li W (1998) A sequential design for maximizing the probability of a favorable response. J Can Stat 26:479–495

Flournoy N (1993) A clinical experiment in bone marrow transplantation: estimating a percentage point of a quantal response curve. In: Gatsonis C, Hodges JS, Kass RE, Singpurwala ND (eds) Case studies in Bayesian statistics. Springer, New York, pp 324–336

Hanson DL, Pledger G, Wright FT (1973) On consistency of monotonic regression. Ann Stat 1:401–421

Klein J, Moeschberger M (1977) Survival analysis: techniques for censored and truncated data (statistics for biology and health). Springer, New York

Klein J, Moeschberger M (2003) Survival analysis: techniques for censored and truncated data (statistics for biology and health), 2nd edn. Springer, New York

Ivanova A, Flournoy N (2006) Up-and-down designs in toxicity studies. In: Chevret S (ed) Statistical methods for dose-finding studies. Wiley, New York, pp 115–130

May C, Flournoy N (2009) Asymptotics in response-driven designs generated by a two-color, randomly reinforced urn. Ann Stat 37:1058–1078

Micheas AC, Dey DK (2005) Modeling shape distributions and inferences for assessing differences in shapes. J Mult Anal 92:257–280
Micheas AC, Wikle CK (2009) A Bayesian hierarchical non-overlapping random disc growth model. J Am Stat Assoc
Rigdon SE, Basu AP (2000) Statistical methods for the reliability of repairable systems. Wiley-Interscience, New York
Robertson T, Wright FT, Dykstra R (1988) Order restricted statistical inference. Wiley, Chichester
Speckman PL (1981) The asymptotic integrated mean square error for smoothing splines. Technical report
Speckman PL (1985) Spline smoothing and optimal rates of convergence in nonparametric regression models. Ann Stat 13(3):970–983
Speckman PL (1988) Kernal smoothing in partial linear models. J Roy Stat Soc Ser B 50:413–436
Sun J (2006) The statistical analysis of interval-censored failure time data. Springer, New York
Sun J, Kalbfleisch J (1993) The analysis of current status data on point processes. J Am Stat Assoc 88:1449–1454
Sun D, Tsutakawa RK, Kim H, He Z (2000) Spatio-temporal interaction with disease mapping. Stat Med 19:2015–2035
Sun J, Park D-H, Sun L, Zhao X (2005) Semiparametric regression analysis of longitudinal data with informative observation times. J Am Stat Assoc 100:882–889
Thompson W, Jr (1955) The ratio of variances in a variance components model. Ann Math Stat 26(2):325–329
Thompson W, Jr (1962) The problem of negative estimates of variance components. Ann Math Stat 33:325–329
Thompson W, Jr (1969) Applied probability. Holt, Rinehart, and Winston, Chicago
Tsutakawa RK (1972) Design of an experiment for bioassay. J Am Stat Assoc 67:584–590
Wikle CK, Cressie N (1999) A dimension-reduced approach to space-time Kalman filtering. Biometrika 86:815–829
Wikle CK, Milliff RF, Nychka D, Berliner LM (2001) Spatio-temporal hierarchical Bayesian modeling: tropical ocean surface winds. J Am Stat Assoc 96:382–397
Xu K, Wikle CK, Fox NI (2005) A kernel-based spatio-temporal dynamical model for nowcasting radar reflectivities. J Am Stat Assoc 100:1133–1144

University of North Carolina (Chapel Hill) Department of Statistics and Operations Research

Douglas G. Kelly

The First Two Decades

The Department of Statistics and Operations Research of the University of North Carolina in Chapel Hill (UNC) had its beginning at 1946 as the Department of Mathematical Statistics, under Harold Hotelling. Its founding was a major step in the expansion of the Institute of Statistics from its 1942 origin under Gertrude Cox in Raleigh. The Department of Mathematical Statistics at Chapel Hill complemented the existing Department of Experimental Statistics at North Carolina State College (now University). The Institute was further expanded in 1949 with the founding of the Department of Biostatistics at UNC-Chapel Hill under Bernard Greenberg. The document (Nourse et al. 1978) by Nourse et al. includes an extensive history of the Institute of Statistics to 1976, including details on all its constituent parts.

The founding faculty of the department had five members: Harold Hotelling (chair), P. L. Hsu, William G. Madow, Edward Paulson, and Herbert Robbins. Maurice G. Bartlett was a visiting professor in that first year. By 1948 Hsu, Madow, and Paulson had left Chapel Hill and had been replaced by Raj Chandra Bose, Samarenda Nath Roy, and Wassily Hoeffding. The department awarded its first PhD degrees in 1948; one was to George E. Nicholson, Jr., who joined the faculty at that time.

Hotelling stepped down as chair in 1952 and was replaced by George Nicholson, who served until 1971. The department became the Department of Statistics in 1952, as did the department at Raleigh around the same time (Figs. 1, 2).

Department of Statistics and Operations Research,
University of North Carolina, 336 Hanes Hall, CB#3260,
Chapel Hill, NC 27599-3260, USA
D. G. Kelly (✉)
e-mail: dgkelly@email.unc.edu

A. Agresti and X.-L. Meng (eds.), *Strength in Numbers: The Rising of Academic Statistics Departments in the U.S.*, DOI: 10.1007/978-1-4614-3649-2_32,

Fig. 1 The Department of Statistics in the late 1950s. Back row, left to right: D. B. Duncan, W. J. Hall, W. L. Smith, G. Zyskind (visitor), J. Ogawa (visitor), R. Gabriel (visitor), S. S. Shrikhande (visitor). Front row, left to right: G. Nicholson, S. N. Roy, H. Hotelling, R. C. Bose, W. Hoeffding

The department's faculty remained unchanged until 1953, when Robbins left for Columbia. Walter L. Smith joined the faculty in 1954. David Duncan and W. J. Hall were added within a year or so, and the department's size remained at eight until 1962. Between 1962 and 1965 four others joined: Norman L. Johnson, Indra M. Chakravarti, Robert L. Bohrer, and M. Ross Leadbetter (who remains an active member of the department at this writing). Bohrer left after 2 years to head the statistics group at the Research Triangle Institute, and the department strength was 11 until the expansion of the late 1960s.

Although it taught courses and granted PhDs, the department was not a regular unit of UNC's College of Arts and Sciences until 1966. Before then it was one of the research arms of the Institute of Statistics, supported largely by major research contracts from the Office of Naval Research (ONR), the Air Force Office of Scientific Research (AFOSR), and the Army Research Office (ARO).

Expansion, 1967–1971

The department added an astonishing 12 members to its faculty during these 4 academic years. Most of this was made possible by the efforts of George Nicholson in securing a Science Development ("Center of Excellence") Grant from the National Science

Fig. 2 The professors of the Department of Statistics in the early 1960s. Back row, left to right: H. Hotelling, G. E. Nicholson, N. L. Johnson. Front row, left to right: W. Hoeffding, R. C. Bose, S. N. Roy

Foundation (NSF), which provided funding from 1968 to 1975. Although university support from State revenues was insufficient to sustain the expansion at the level of 12 positions, 5 of the 12, all appointed in 1968 or 1969, remained with the department past 1975. They were Charles R. Baker, Stamatis Cambanis, Douglas G. Kelly (jointly with Mathematics until 1996), Gordon D. Simons, and Edward J. Wegman.

The department taught few undergraduate students until the early 1970s. Fifteen undergraduates were enrolled in statistics courses during the fall of 1969, and in the spring of 1970 there were 40. Beginning in the next year the department offered introductory courses for undergraduates and also encouraged undergraduate enrollment in some of its existing courses, and by 1975–1976 over a thousand undergraduates per year were enrolled in statistics courses. The number at this writing is over 2000. At the outset, the School of Business was the major supplier of undergraduates in statistics courses; now these courses enroll students from most departments and schools in the university. The number of undergraduate enrollments is more significant than at first sight, given that the university has no engineering school, and given that the Departments of Psychology, Sociology, and Economics have their own introductory statistics courses, which historically predate the Statistics Department's offerings.

The years 1969–1970 saw the culmination of considerable work in combinatorics, primarily under the leadership of R. C. Bose, who led efforts in the field until his retirement in 1971. (Colorado State University enjoyed Bose's presence for many years after 1971, benefiting from North Carolina's mandatory retirement age; Bose was one of the last UNC faculty members thus forced to retire.) In 1959 Bose, with S. S. Shrikhande and E. T. Parker, had famously settled, negatively, Euler's Conjecture on the nonexistence of Greco-Latin squares of order $4k + 2$ ($k = 1,2,\ldots$). Department activity in combinatorics included international conferences in France in 1965 (sponsored by NATO) and in Chapel Hill in 1967 (sponsored by AFOSR), and a Combinatorial Year (sponsored by AFOSR and NSF), which included some 20 visitors over the course of the 1969–1970 academic year and a week-long conference.

Activity in operations research also began during this period. George Nicholson's interest in the area led to the 1968 appointment of F. Jerome Gould, who instituted a program which in 1973 became the free-standing Curriculum in Operations Research. The history of this program and its eventual merger with statistics is described below.

A notable addition to the department in 1970 was June Maxwell, who became well known in the statistics community as the chief department administrator from 1970 until her retirement in 1998, and as a member of the editorial staff of the IMS Bulletin. The six department chairs who served during Maxwell's tenure all relied heavily on her to keep things running.

The 1970s and 1980s

In 1971 the department's faculty size was 15. In contrast to the 12 appointments made in the previous 4 years, there were only 9 between 1972 and 1990. Notable among these was Gopinath Kallianpur, who had received the PhD from the department in 1951. Kallianpur's arrival spurred the formation of the Center for Stochastic Processes, discussed below.

Others who joined the department during this period were Edward Carlstein, Raymond J. Carroll, Jianqing Fan, Chuanshu Ji (jointly with Mathematics until 1995), J. Stephen Marron, Donald St. P. Richards, and David Ruppert. Carroll, Richards, and Ruppert all left the department in 1987, and Fan left in 2003; the others remain active members of the faculty at this writing.

Another important addition was Pranab K. Sen, who had been a full-time member of the UNC's Department of Biostatistics since 1965. Sen accepted a joint appointment with Statistics in 1988.

By the end of the 1970s the department had a small undergraduate major program. The evolution of this program is described below.

The department's size in 1990 was essentially the same as it had been in 1970; but as was the case in most departments, those two decades had seen considerable change in the directions of applied work, the use of computers, and increased undergraduate Statistics education.

The Hotelling Lecture Series

In 1980, the department instituted its well-known annual lecture series in honor of founding chair Harold Hotelling. A distinguished guest speaker presents a series of talks, which are open to the public, and remains in residence at the department for several days. The 1980 Hotelling Lectures were given by David R. Cox, followed in subsequent years by other distinguished speakers: Herman Chernoff, Ole Barndorff-Nielsen, Frank Hampel, David Brillinger, David G. Kendall, Persi Diaconis, Pal Revesz, Willem van Zwet, C. R. Rao, Bradley Efron, Lucien LeCam, Peter Bickel, Ulf Grenander, Larry Shepp, David Donoho, David Siegmund, Herbert Robbins, Lawrence D. Brown, Nancy Reid, S. R. S. Varadhan, Stuart Geman, Iain Johnstone, Peter Hall, Ruth J. Williams, Terry Speed, Thomas Kurtz, and Peter McCullagh.

The Center for Stochastic Processes

Prominent among the activities of the department beginning around 1980 was the Center for Stochastic Processes. This was the joint effort of Stamatis Cambanis, Ross Leadbetter, and Gopinath Kallianpur, who had been individually funded by three separate agencies of the Department of Defense. Through a major cooperative effort of AFOSR, ONR, and ARO, they formed the Center, which flourished and produced extraordinary activity in probability and stochastic processes for some 15 years.

The Center's activities included a weekly seminar, a technical report series (over 500 reports between 1980 and 2000), and most notably the presence of an amazing number of distinguished visitors—on average, 15 per year from 1985 to 1995—who were in the department for periods of a week or two up to a full year.

Cambanis's untimely death in 1995 and Kallianpur's retirement a few years later reduced the activities of the Center. But the continued activity of Leadbetter, the appointment in 2000 of Amarjit Budhiraja, a 1994 PhD student of Kallianpur, and later appointments maintained the presence of probability and stochastic processes.

The 1990s

The department made few new appointments during the years from 1990 to 2000, but those of Richard L. Smith in 1991 and Andrew Nobel in 1994 were important and notable. The department also enjoyed the half-time presence of Willem van Zwet (from 1990 to 1996, jointly with Leiden) and Robert Adler (from 1996 to 1999, jointly with Technion).

The department hosted the third World Congress of the Bernoulli Society for Mathematical Statistics and Probability in 1994, jointly with the Annual Meeting of the Institute of Mathematical Statistics. Leading in the organization and arrangements were Chakravarti, Kallianpur, Leadbetter, and van Zwet.

In 1996, the department celebrated its fiftieth anniversary with a special meeting of the Institute of Mathematical Statistics; Herbert Robbins was among the featured speakers.

NISS and SAMSI

An important achievement of the statistics community in the Research Triangle area of North Carolina was the establishment in 1990 of the National Institute of Statistical Sciences (NISS). The national statistical organizations, along with the three Triangle universities (UNC, NC State, and Duke) cooperated to form NISS, whose stated mission (National Institute for Statistical Sciences, http://www.niss.org/) is "to identify, catalyze and foster high-impact, cross-disciplinary and cross-sector research involving the statistical sciences." Stamatis Cambanis led the effort on UNC's behalf, with the cooperation of the offices of the Dean of the College and the Vice Chancellor for Research. The Director and Associate Director of NISS are members of one of the three Triangle Universities' Statistics Departments. Alan Karr, appointed to the UNC Statistics Department, was the first Associate Director, under Jerry Sacks of the Duke department. Karr became Director of NISS upon Sacks's retirement.

In 2002, the three universities and NISS, along with the William R. Kenan, Jr. Institute for Engineering, Technology, and Science, collaborated to form the Statistical and Applied Mathematical Sciences Institute (SAMSI), one of the Mathematical Sciences Institutes of the NSF. It operates within NISS's building in Research Triangle Park. Steve Marron served as one of the first Associate Directors of SAMSI, under James Berger of the Duke department. Richard Smith is the present Director of SAMSI. Department faculty and graduate students have been extensively involved in SAMSI programs from its inception. The SAMSI website (Statistical and Applied Mathematical Sciences Institute, http://www.samsi.info/) provides descriptions of its mission and activities.

Operations Research

As noted above, George Nicholson began the department's long and fruitful association with the operations research community, which intensified with F. Jerome Gould's appointment in 1968. Gould, along with Jon Tolle (then in the Department of Mathematics), David Rubin, and John P. Evans (both in the School of Business), Richard Shachtman (who had joined Statistics in 1968), and others, founded a program which in 1973 became the free-standing Curriculum in Operations Research. Later, this curriculum became a full-fledged department in the College of Arts and Sciences, with a faculty of five to six and several joint appointments with Mathematics and Statistics. George Fishman, appointed in 1976, was a major contributor to this department's independence and vitality. Also, among the members of the OR Department were Vidyadhar G. Kulkarni (appointed in 1981), J. Scott Provan (1982), and Shaler Stidham (1986).

In 2003, partly to cut administrative costs and partly in recognition of overlapping intellectual interests, the Dean of the College merged the Departments of Statistics and Operations Research into a single department (STOR) of about 20 members, with Kulkarni as chair. While not unanimously favored at the time, this merger has proved beneficial to both groups, through the increased flexibility afforded by larger faculty size and budget as well as increased enrollments, and through interactions between the disciplines. Faculty members appointed since the merger have not been identified officially as belonging to one discipline or the other.

The Years Since 2000

Nine people appointed between 2000 and the time of this writing are still in the department: Gabor Pataki (2000), Vladas Pipiras (2002), Haipeng Shen (2003), Serhan Ziya (2003), Yufeng Liu (2004, jointly with UNC's Center for Genome Sciences), Nilay T. Argon (2006), Shu Lu (2007), Jan Hannig (2008), and Shankar Bhamidi (2009). The department's size at the time of this writing is 21.

The years since 2000 have seen an explosion of interdisciplinary research in the department; there are active collaborations with colleagues from Medicine, Genetics, Biology, Biostatistics, Economics, Business, and Computer Science. It no longer suffices to say merely that the department does "both theory and applications;" there is active research by department faculty members in theory, methodology, modeling, optimization, computation, data analysis, and visualization.

The Undergraduate Program

As mentioned earlier, the department began offering undergraduate Statistics courses around 1970, and had a small undergraduate major program by the late 1970s. This program was part of the Curriculum in Mathematical Sciences (CMS), begun under the leadership of W. Robert Mann in the Department of Mathematics with cooperation and support from the Departments of Statistics and Computer Science and the Curriculum in Operations Research, all of which were primarily graduate departments without their own undergraduate major programs. The CMS offered the BS degree and had five 'tracks,' in Applied Mathematics, Statistics, Operations Research, Computer Science, and Actuarial Science. It has graduated some 15–25 students per year since its beginning; around half of the graduates, especially in the early days, were in the Computer Science track. Jon Tolle led this program from the late 1990s until 2006; Douglas Kelly was the director from 2006 until 2011, when he was replaced by Vladas Pipiras.

In 2002, the Departments of Mathematics and Computer Science took over two of the tracks as formal undergraduate majors. The remaining tracks—Statistics, Operations Research, and Actuarial Science—were jointly administered, by the Departments of Statistics and Operations Research, as the curriculum in Mathematical Decision Sciences (MDS). When those two departments merged in 2003, MDS became the official undergraduate major program of the STOR Department. The three tracks no longer exist separately; students take a basic core including courses in Statistics and Operations Research, and then choose from a selection of upper level courses within and outside the department. The offerings in Actuarial Science, highlighted by two courses taught by Charles Dunn, an actuary who is an adjunct member of the department, attract an increasing body of students.

The Graduate Program

Graduate education and research has been at the core of the department's mission since its founding. It has produced over 275 PhDs in Statistics, an average of over 4 per year. The Operations Research group, as a curriculum and then a department, and continuing since the merger, has averaged two to three PhDs per year—remarkable since the OR faculty has rarely exceeded five in number.

Department alumni include numerous internationally recognized scholars, industry leaders, and influential professionals in both the Statistics and Operations Research communities. Lists of all the department's PhDs in Statistics and of recent PhDs in OR can be found on its website (Department of Statistics and Operations Research, University of North Carolina, Chapel Hill, http://stat-or.unc.edu).

The department has maintained separate graduate programs in Statistics and Operations Research and has recently instituted a program in Interdisciplinary Statistics and Operations Research (INSTORE). The three programs have different though overlapping core curricula, and each has its own comprehensive written examinations.

The MS program in Operations Research has consistently produced a large number of graduates—about 6–10 per year. These graduates have had almost universally good job placements, the more so since the department instituted the OR Projects course, through which MS students have provided technical expertise to a large number of university and community organizations and companies throughout the Research Triangle area.

Department Quarters

The department was housed in Phillips Hall, along with the Department of Mathematics and the Department of Physics and Astronomy, from its inception until 1998, when expansion of the Department of Mathematics required a move across the street to New West Building. After about 1970, the department's graduate students were housed in Mary Anne Smith Building, also across the street from Phillips, and in New West. Meanwhile, Operations Research was quartered first in Phillips Annex (a tiny building that also housed the university's typewriter repair shop), and then in Smith Building. After the two departments merged in 2003, the entire faculty moved into Smith, and the graduate students were temporarily located in Howell Hall, a 5-minute walk across campus. Finally, in 2008, all the faculty and graduate students of the Department of Statistics and Operations Research moved into newly remodeled Hanes Hall, where it occupies three floors.

Appendix A: Department Faculty

Name	Year appointed
Robbins, Herbert E.	1945
Hotelling, Harold; Hsu, P. L.; Madow, William G.; Paulson, Edward	1946
Bose, Raj Chandra; Hoeffding, Wassily	1947
Nicholson, George E., Jr.; Roy, Samarendra Nath	1948
Smith, Walter L.	1954
Duncan, David B.	1955
Hall, William J.	1957
Johnson, Norman L.	1962
Chakravarti, Indra M.	1964
Bohrer, Robert E.; Sen, Pranab K.; Leadbetter, Malcolm R.	1965
Dowling, Thomas A.; Tolle, Jon W.	1967
Baker, Charles R.; David, Herbert A.; Gould, Floyd J.; Kelly, Douglas G.; Shachtman, Richard H.; Simons, Gordon D.; Smith, Kempton J. C.; Wegman, Edward J.	1968
Cambanis, Stamatis; Cleveland, William S.	1969
Smith, Woollcott K.	1970
Rubin, David	1971
Carroll, Raymond J.	1974
Fishman, George	1976
Ruppert, David	1977
Begun, Janet; Kallianpur, Gopinath	1979
Kulkarni, Vidyadhar G.; Richards, Donald	1981
Marron, J. Stephen; Provan, J. Scott	1982
Carlstein, Edward	1984
Stidham, Shaler	1986
Ji, Chuanshu	1988
Fan, Jianqing	1989
Van Zwet, Willem	1990
Smith, Richard L.	1991
Karr, Alan	1993

(continued)

(continued)

Name	Year appointed
Nobel, Andrew	1994
Adler, Robert	1996
Ickstadt, Katja	
Budhiraja, Amarjit	2000
Pataki, Gabor	
Pipiras, Vladas; Tekin, Eylem; Zhu, Zhengyuan	2002
Shen, Haipeng; Ziya, Serhan	2003
Liu, Yufeng	2004
Argon, Nilay T.	2006
Lu, Shu	2007
Hannig, Jan	2008
Bhamidi, Shankar	2009

Appendix B: Department Chairs

Name	Years
Hotelling, Harold	1946–1952
Nicholson, George E.	1952–1971
Johnson, Norman L.	1971–1976
Simons, Gordon D.	1976–1981
Smith, Walter L.	1981–1986
Cambanis, Stamatis	1986–1993
Simons, Gordon D.	1993–1995
Kelly, Douglas G.	1995–2000
Smith, Richard L.	2000–2002
Carlstein, Edward	2002–2003
Kulkarni, Vidyadhar G.	2003–2009
Carlstein, Edward	2009–

References

Nourse ES, Greenberg BG, Johnson NL et al (1978) Statistical training and research: the University of North Carolina system. Int Stat Rev 46:171–207

Website, Department of Statistics and Operations Research, University of North Carolina, Chapel Hill. http://stat-or.unc.edu

Website, National Institute for Statistical Sciences. http://www.niss.org/

Website, Statistical and Applied Mathematical Sciences Institute. http://www.samsi.info/

The University of North Carolina at Chapel Hill Department of Biostatistics

Jane H. Monaco, Jianwen Cai, Lisa M. LaVange and Michael R. Kosorok

Introduction

The Department of Biostatistics in the Gillings School of Global Public Health has a distinguished past, a celebrated present, and bright future. With a commitment to groundbreaking research, excellence in training students, and public health service, the department has established its role as a leader in the biostatistics community. The department is committed to its mission "to forge dramatic advances in health science research that benefit human health in North Carolina, the US and globally, through the development of profound and paradigm-shifting innovations in biostatistical technology and the thoughtful implementation of biostatistical methodology to solve public health problems" (The University of North Carolina at Chapel Hill Department of Biostatistics 2011).

Greenberg Years (1949–1972)

The Department of Biostatistics was established in 1949 through the Institute of Statistics as a joint venture between the Departments of Statistics at the University of North Carolina (later renamed The University of North Carolina at Chapel Hill (UNC-CH)) and North Carolina State College at Raleigh (later renamed North Carolina State University (NCSU)). The original mission of the department was to offer public health statistics courses in the UNC-CH School of Public Health

The University of North Carolina at Chapel Hill,
Department of Biostatistics, 3101 McGavran—Greenberg,
Campus Box 7420, Chapel Hill, NC 27599-7420, USA
J. H. Monaco · J. Cai · L. M. LaVange · M. R. Kosorok (✉)
e-mail: kosorok@unc.edu

A. Agresti and X.-L. Meng (eds.), *Strength in Numbers: The Rising of Academic Statistics Departments in the U.S.*, DOI: 10.1007/978-1-4614-3649-2_33,

(SPH), to provide consulting services and to perform research in public health statistics. Drs. Gertrude Cox and Harold Hotelling were instrumental in the formation of the Department of Biostatistics and selected Dr. Bernard Greenberg to lead the new department (Nourse 1978). Greenberg, a recent graduate from North Carolina State College with a doctorate in experimental statistics, was only 29 years old when he was appointed as chairman.

Although degree programs were not initially part of the department's mission, the role was soon expanded to include graduate degrees. The first student graduated with a Master of Public Health (MPH) in 1951, and the first Master of Science in Public Health (MSPH) soon followed. Prior to 1965, doctoral students fulfilled their coursework requirements at either the Department of Statistics at North Carolina State College or the Department of Statistics at UNC-CH. The dissertation of a biostatistics student could be directed by faculty members in any of the three departments. These first PhD degrees were conferred beginning in 1953 as joint degrees. Dr. Harry Smith was the first student to receive a joint PhD degree. Smith subsequently had a distinguished career in the department before becoming Chairman of the Department of Biomathematical Sciences at Mt. Sinai School of Medicine. The SPH awarded PhD degrees in Public Health with an emphasis on Biostatistics during the 1950s and 1960s. Dr. Ahmed Sarhan, coauthor of the monograph Contributions to Order Statistics, was the first graduate from this program in 1955. In 1965, the department was authorized to grant the PhD in Biostatistics degree. The Master of Science (MS) degree in Biostatistics was initiated in 1969, followed by the DrPH degree in Biostatistics in 1970.

In its first two decades, the department pioneered efforts to establish the practice of cooperative multicenter trials. The National Cancer Institute awarded the department a contract in 1955 to become the statistical coordinating center for the Southeastern Cooperative Cancer Chemotherapy Study Group (SCCCS). Faculty from ten medical schools and the coordinating center, under the direction of Greenberg and Dr. James Grizzle, collaborated to investigate the effectiveness of chemotherapy agents. Research at the SCCCS led to the examination of the two-period crossover design and showed the limitations of its use in clinical trials (Grizzle 1965). A subsequent Veterans Administration grant supported research that introduced randomization of surgical procedures in the treatment of duodenal ulcers and led to important findings in the treatment of the condition. Greenberg published the well-known, first article regarding the design and conduct of collaborative clinical trials in the *American Statistician* in 1959. In 1967, Greenberg chaired the committee commissioned by the National Heart Institute of the National Institutes of Health (NIH) to develop procedures for conducting large, multicenter, clinical trials. The Greenberg Report (1967) is widely considered the landmark paper in this area. Another highly influential clinical trials paper (Freiman et al. 1978) exposed the widespread problem of under-powered trials.

The department grew rapidly in the 1950s with respect to faculty, students, space, and funding. However, Greenberg commented, "as additional burdens and responsibilities have mounted geometrically, there has been only an arithmetic increase in funds available for personnel" (Korstad 1990, p. 87). In 1953, the

National Heart Institute of the NIH awarded one of the first training grants to the department to train biostatisticians as health consultants. As part of this grant and others, the department was able to attract distinguished visiting professors such as David Cox, J.O. Irwin, David Duncan, David Newell, Ahmed Sarhan, Herbert David, Robert Elston, and Ruben Gabriel, some of whom remained at UNC-CH as faculty.

Greenberg was a visionary with respect to human rights as well as racial and gender equality. In the mid-1960s, as racial demonstrations and student unrest were common, Greenberg communicated the department's commitment to racial equality by encouraging students to patronize only integrated establishments. He recruited talented female biostatisticians including Regina Elandt-Johnson, Elizabeth Coulter, and Mindel Sheps.

Two special sessions in the summers of 1963 and 1964 were extremely successful in training biostatistics students. Biostatistics courses were taught by prestigious faculty, including Byron Brown, Herbert David, Daniel Horvitz, Robert Elston, Carl Erhardt, and David Newell. More than 100 students participated in these programs each summer including Drs. Clarence (Ed) Davis (future Department Chair), Gary Koch, Daniel Solomon (future Chair of the Department of Statistics and, later, Dean of School of Physical and Mathematical Sciences at NCSU), and O. Dale Williams.

Dr. Pranab K. Sen accepted an offer to join the faculty at UNC-CH in 1965. Sen remarked, "Eventually, I realized that UNC was one of the best places for statistics in America, if not the world, and by being here I could not only strengthen my background but develop additional ties with Indian schools." (Ghosh and Shell 2008, pp. 555–556). With more than 600 publications in the biostatistics and statistics literature, Sen is an internationally celebrated author of books and articles in nonparametric statistics, large sample methodology, and sequential analyses. He was named the Cary C. Boshamer Distinguished Professor of Biostatistics in 1982. In 2010, Sen received the Samuel S. Wilks award, one of the ASA's most prestigious awards, for his pioneering contributions to statistics, including exceptional mentoring of several generations of doctoral students. Figure 1 shows the faculty shortly after Sen's hiring.

The NIH awarded the department, jointly with the Department of Statistics at UNC-CH, a grant to develop methodology related to multivariate analysis in the mid-1960s. The grant provided support leading to many important papers in crossover designs, dose response curves, and categorical data by researchers Grizzle, Koch, and Sen within the department and other colleagues outside the department (V.P. Bhapkar, S.N. Roy, C.F. Starmer). The development of statistical software for multivariate analysis of variance, categorical data analysis, and multivariate nonparametric methods also resulted from this grant.

Two training grants, in Environmental Statistics and Demography, were awarded to the department in 1971. Dr. Lawrence L. Kupper, who joined the faculty in 1970, played a prominent role in obtaining the Environmental Training Grant, which has been continuously funded by the National Institute of Environmental Health Sciences for more than 38 years. From 1972 to 2006, Kupper served

Fig. 1 UNC-CH Department of Biostatistics, 1960s. Front Row: Robert Elston, Roy Kuebler, Elizabeth Coulter, Bernard Greenberg, Regina Elandt-Johnson, Forrest Linder, Dana Quade. Back Row: Herbert David, James Grizzle, Pranab K. Sen, Jay Glasser, Tony Lachenbruch, Harry Smith, Ruben Gabriel, Bradley Wells, Tom Donnelly

as the Director of the training grant which has supported hundreds of students with as many as 20 predoctoral and 8 postdoctoral students per year receiving support. The Demography Training Grant, directed by Dr. Chirayath Suchindran, was continuously funded for more than 30 years supporting many highly qualified trainees in population studies. Other training grants awarded during this period included the mental health training grant and the health services research grant.

Faculty members hired under Greenberg included Drs. Roy Kuebler and Ronald Helms. Kuebler, a beloved instructor, strengthened educational aspects in the program and was recognized with the first McGavran Teaching Award in 1975. During his academic career which spanned more than 30 years, Helms made important contributions in statistical methods and multiple collaborative areas such as sickle cell disease, asthma, and mental health. He cofounded Rho, a well-respected contract research organization, which continues to expand, and provide training opportunities for current graduate students.

Also hired during this period and making significant contributions were: Drs. Edmund Gehan, Thomas Donnelly, Bradley Wells, Bernard Pasternack, Dana Quade, Herbert David, James Abernathy, Peter Lachenbruch, Forrest Linder, Donna Brogan, Michael Symons, Lawrence Kupper, and David G. Kleinbaum.

When Greenberg went on to serve as the dean of the SPH at UNC-CH (from 1972 to 1982), the department had approximately 30 full-time faculty. In 1983,

Greenberg received the O. Max Gardner Award, awarded to one faculty member across all UNC campuses, for contributions to the welfare of the human race.

Grizzle Years (1972–1987)

In 1972, when Dr. James Grizzle was appointed chair, the department "was by far the largest of its kind in the world, both in the size of its faculty and the number of students" (The Body Politic, Jan. 1973, p. 6). Grizzle, who joined the faculty in 1960, became well known for his research in randomized controlled trials (Ruffin et al. 1969), and in weighted least squares analysis of categorical data (Grizzle et al. 1969).

Prior to becoming chair, Grizzle was instrumental in securing an NHLBI award for the department to serve as the coordinating center for the Lipid Research Clinics Program (LRC; 1971–1990). This award strengthened the department's leadership role in setting the standards for the coordination of large, multicenter studies, particularly in the areas of study design, data management, and statistical analysis. Grizzle was principal investigator (PI) of the original LRC coordinating center. He was succeeded by Drs. O. Dale Williams (1976–1990), Clarence Edward (Ed) Davis (1991–1997), Lloyd (Woody) Chambless (1997–2005), and Lisa LaVange (2005–present). Notable among the early LRC studies coordinated by the center is the Coronary Primary Prevention Trial (CPPT), which was the first major randomized controlled clinical trial to show that lowering cholesterol reduced mortality (JAMA 1984). The LRC Coordinating Center, renamed the Collaborative Studies Coordinating Center (CSCC) in 1986, is the oldest, continuously funded NIH Coordinating Center in the US, and has coordinated numerous studies involving hundreds of clinical centers throughout the US and the world. The longest-running such study is the NHLBI-funded Atherosclerosis Risk in Communities Study (ARIC), begun in 1986 and ongoing today. A cohort of over 15,000 persons, aged 45–64 years, living in four community areas across the US have been followed in ARIC to determine the incidence and risk factors of cardiovascular disease. Two other major studies administered through the CSCC with considerable public health impact include SOLVD (Studies of Left Ventricular Dysfunction, which examined the role of ACE inhibitors in reducing mortality in heart failure patients) and ACAS (Asymptomatic Carotid Atherosclerosis Study, which evaluated the efficacy of carotid endarterectomy in the reduction of stroke among asymptomatic atherosclerosis patients). In addition to innovations in statistical methods for multicenter studies, the CSCC has been an innovator in the field of data management, implementing remote data entry (1986) for the first time in an NIH multicenter study and developing Internet-based data management and tracking systems (2001).

The department is believed to be the first department of biostatistics to offer an undergraduate degree. Dr. Maura Stokes was the first recipient of this undergraduate degree in 1978 and went on to earn her doctoral degree in the department.

The Bachelor of Science in Public Health (BSPH) program, directed by Dr. Craig Turnbull until 2006, has graduated more than 200 talented students many of whom have pursued medical degrees or advanced degrees in biostatistics.

The field of demography became a departmental strength with significant contributions by Drs. Mindel Sheps, Forrest Linder, Bradley Wells, James Abernathy, and Chirayath Suchindran. The Laboratories for Population Statistics (POPLAB), begun in the late 1960s under the direction of Dr. Forrest Linder and funded by the Agency for International Development, conducted projects in the Philippines, Turkey, Ecuador, Kenya, Morocco, Indonesia, Mexico, Somalia, and Columbia. This cooperative project between the SPH and Carolina Population Center implemented new field survey methods for measuring birth rates and population growth in regions of the world where conducting a census of the population was not feasible.

In 1974, Dr. Gary Koch moved from Rosenau Hall to join Dr. Dennis Gillings and other young faculty in a temporary location, known as "Trailer 39." The friendship and collaborative relationship that Gillings and Koch established in that environment drew students eager to work with these remarkable faculty members. As a former graduate student recalls, "the trailer, as everyone called it, wasn't just a place to work—it felt like family." (Spivey 2008, p. 25).

Koch joined the faculty of the Department of Biostatistics in 1967 even before completing his doctorate. "He was sort of a phenomenon as a graduate student because he was writing papers that were appearing in *Biometrics*," (Spivey 2008, p. 24) according to Grizzle who hired Koch. Koch quickly established himself as an expert in the methodology of multivariate categorical data. For example, the research paper regarding the kappa statistic has been referenced more than 13,000 times (Landis and Koch 1977). In 1974, Koch received the Spiegelman Award given each year to an outstanding biostatistician under age 40. The Biometric Consulting Lab (BCL), directed by Koch since the mid-1980s, has provided biostatistical consulting services for investigators and training opportunities for biostatistics students. With more than 400 peer-reviewed publications, his legacy could easily be his contributions to the statistical literature; however, in addition to his exceptional research record, he has also had extraordinary impact as a mentor which was recognized in 2007 with the SPH Larsh Award for Mentorship. After more than 40 years of legendary mentoring, the Koch "student family" of former graduate students reaches around the world.

Gillings was recruited in 1971 for his contributions in Health Services Research. Together with Koch and the students of "Trailer 39," Gillings consulted on the design and analysis of clinical trials for pharmaceutical companies and developed statistical methods for drug development. The consulting work that began in the cramped environment of the trailer led to the creation of Quintiles, which was incorporated in 1982. Gillings left the faculty in 1988 to lead Quintiles full-time. Quintiles Transnational is currently the world's largest contract research organization with an annual revenue in excess of $2 billion and more than 20,000 employees worldwide.

Fig. 2 UNC-CH Department of Biostatistics, 1980s. (left to right) Marjolein Smith, Paul Stewart, Michael Symons, Dick Bilsborrow, Ed Davis, Dick Shachtman, Larry Kupper, Gary Koch, Pranab Sen, Dennis Gillings, Muhammed Habib, Bill Kalsbeek, David Kleinbaum, Kinh Troung, Shrikant Bangdiwala, Jim Knoke, Elizabeth Coulter, Jim Abernathy, Jim Grizzle, Chirayath Suchindran, Jim Hosking, Joe Janis, David Christiansen

Internationally recognized genetics expert, Dr. Robert Elston served on the faculty from 1960 to 1979. Known for the Elston–Stewart algorithm, he has coauthored several books on genetic epidemiology and biostatistics as well as trained many of the present-day leading statistical geneticists. Elston proceeded to become the chairman at Louisiana State University and then Case Western Reserve University. His departure left a void in the department's genetics program not addressed for nearly 20 years.

Dr. Lawrence Kupper, Alumni Distinguished Professor of Biostatistics, coauthored several books including Applied Regression Analysis and Other Multivariable Methods which has been adopted for use by approximately 100 universities. In addition to serving as the Director of the Environmental Training grant, the largest and most successful training program of its kind in the world, his many contributions included authoring or coauthoring more than 160 peer-reviewed articles. His research included statistical applications in epidemiology, environmental, occupational, and women's health. Kupper earned many awards for teaching and mentoring including two university-wide awards: the UNC-CH Distinguished Teaching Award for Post-Baccalaureate Instruction (1996) and the UNC-CH Mentor Award for Lifetime Achievement in Teaching and Mentoring (2007).

Other faculty members hired in this period included Drs. Clarence (Ed) Davis, Chirayath Suchindran, Richard Bilsborrow, Richard Shachtman, O. Dale Williams,

William Kalsbeek, Shrikant Bangdiwala, James Hosking, Lloyd (Woody) Chambless, Keith Muller, Paul Stewart, and Young Kinh–Nhue Truong (Fig. 2).

This period saw the beginning of the annual back-to-school picnics held at the Grizzle farm that are fondly remembered for good food, hayrides, music, and volleyball games. These social events, attended each year by approximately 150 students, faculty, staff and their families, continued for the 15 years that Grizzle served as chair.

Margolin Years (1987–1997)

After leading the statistical methodology section at the National Institute of Environment Health Sciences, Dr. Barry Margolin was hired as professor and chair in 1987. Margolin also served as the Director of the biostatistics facility of the UNC-CH Lineberger Cancer Center beginning in 1989. Margolin was an energetic student advocate, successfully lobbying for more office space, and better computing resources for the biostatistics graduate students. He placed a high priority on student funding for promising graduate students, strengthening the funding structure. Margolin was also a strong faculty advocate, successfully nominating several faculty members for ASA Fellow awards.

The Survey Research Unit (SRU) was established in 1990 under the direction of Dr. William Kalsbeek with the primary emphasis on specialized sampling plans and data collection. Since its inception, the SRU has been involved in approximately 300 projects ranging from the evaluation of triage procedures in emergency rooms to the investigation of teen dating behaviors. Kalsbeek has led the unit in applying state-of-the-art principles in conducting population-based research and in training students in the design, data collection, and analysis of survey studies.

Linear models research was strengthened during Dr. Keith Muller's more than 25 years of service to the department. He made valuable contributions in multiple areas including first authorship of two books, Regression and ANOVA: An Integrated Approach using SAS Software (Muller and Fetterman 2002) and Linear Model Theory: Univariate, Multivariate and Mixed Models (Muller and Stewart 2006). A respected mentor and teacher, much of his research focused on linear and nonlinear repeated measures models, medical imaging, and the development of software for power analysis.

The department has long endeavored to hire and graduate underrepresented minority biostatisticians. One of the first African-American female biostatistics faculty members, Mildred Francis, was hired in the early 1970s. Successful summer programs for minorities were instituted in late 1970s and early 1980s. Notable African-American graduates of the department include Delton Atkinson, George Bonney, Dubois Bowman, Randy Davis, Lloyd Edwards, Alula Hadgu, and Jean Orelien.

In 1990, Dr. Lloyd Edwards joined the department and made contributions in mixed model theory and applications, longitudinal data analysis, and clinical trials.

Dr. Bahjat Qaqish also joined the faculty in 1990, strengthening the department's research areas of generalized linear models and statistical computing. Increased emphasis on environmental statistics was accomplished in recruiting Dr. Haibo Zhou in 1997. With research contributions in missing data, respiratory diseases, and fertility modeling, Zhou serves as the Director of the Biostatistics Core for the Center for Environmental Medicine, Asthma and Lung Biology at UNC-CH and is a five-term associate editor for *Biometrics.*

By recruiting Dr. Jianwen Cai in 1992, the department's emphasis on survival analysis research was enhanced. A prolific researcher in multivariate failure time and recurrent event data, she has served as associate editor for *Biometrics*, *Lifetime Data Analysis,* and *Statistics in Biosciences*. Cai has been recognized not only for her research and service, but also for teaching through the 2004 McGavran Award for Excellence in Teaching. In 2006, Cai served as interim chair of the department and currently serves as Associate Chair.

Davis Years (1997–2005)

Dr. Clarence (Ed) Davis had been recruited to the department in 1972 by Grizzle. Davis became well known for research in clinical trials and cardiovascular disease epidemiology. He taught courses in clinical trials and public health research in more than 15 countries including an 11-year collaboration between the SPH and the University of Chile at Santiago. He served as the Director of the CSCC from 1991 to 1997 before being named the fourth chair of the department in 1997.

Under the direction of Davis, the faculty research productivity increased dramatically, including significant growth in the number of publications in the most prestigious methodological, as well as collaborative journals. Davis was also instrumental in recruiting talented women faculty members, increasing the number of female faculty from two to nine.

Recruiting several talented genetics faculty during this period reestablished the department's position in statistical genetics. Dr. Fred Wright joined the department in 2002 and directs the Carolina Environmental Bioinformatics Center. His research focus includes gene expression and controlling errors in multiple testing. In 2001, the department recruited Dr. Fei Zou, an accomplished researcher in statistical genetics with expertise in linkage analysis and genomics.

Davis was successful in procuring the department's first endowed chair and recruited Dr. Danyu Lin as the first Dennis Gillings Distinguished Professor of Biostatistics in 2001. The position was endowed by former faculty member, Dr. Dennis Gillings. With more than 120 peer-reviewed publications, Lin is an internationally recognized expert in survival analysis as well as statistical methods and software for genetic studies. Lin was the 1999 recipient of the Spiegelman Gold Medal. Some of his methods are used in several commercially available statistical software packages including SAS, S-Plus, and Stata. Lin has served as an associate editor of *Biometrika* since 1999 and been named in the Thomson ISI (Institute for Scientific Information) list of Highly Cited Researchers in Mathematics.

Research emphasis in Bayesian statistics within the department was significantly strengthened with the recruitment of Dr. Joseph Ibrahim, in 2002. Ibrahim, Alumni Distinguished Professor of Biostatistics, has authored or co-authored two books, Bayesian Survival Analysis and Monte Carlo Methods in Bayesian Computation, and more than 160 peer-reviewed articles, most in the premier statistical journals. In addition to interests in Bayesian statistics, he has also expanded the department's focus on genomics and missing data. He directs the Biostatistics and Data Management Core at UNC'S Lineberger Comprehensive Cancer Center and has served as the Director of Graduate Studies since 2003.

Many faculty additions (including Drs. Anastasia Ivanova (1999), Amy Herring (2000), Donglin Zeng (2001), and Lisa LaVange (2005)) during this period have impacted the direction of the department. Clinical trials research was expanded by Ivanova, whose particular focus is on adaptive and dose-finding designs. She has worked with researchers at the UNC-CH Lineberger Comprehensive Cancer Center to improve efficiency of oncology clinical trials. Herring has made research contributions in longitudinal analysis, missing data, and Bayesian methods as well as in collaborative areas of environmental and reproductive health. She was the 2011 President of the Eastern North American Region of the International Biometric Society (ENAR), following in the footsteps of Grizzle (1973), Koch (1979), and LaVange (2007). Zeng's research contributions include semiparametric inference for censored data, medical imaging, and empirical processes. In 2008, Zeng was awarded the American Statistical Association's Noether Young Scholar Award presented for contributions in nonparametric statistics to an accomplished statistician under age 35. LaVange joined the department as the first Professor of the Practice of Biostatistics. With more than 25 years' experience, she previously held leadership positions in several nonprofit and pharmaceutical industries, including serving as Vice President of Biostatistics at Quintiles. Additional faculty members hired in this period included Drs. Diane Catellier, David Couper, John Preisser, Todd Schwartz, Jane Monaco, and Michael Hudgens.

Kosorok Years (2006–Present)

Dr. Michael Kosorok joined the department as professor and chair in 2006. Prior to joining the department, Kosorok was a faculty member at the University of Wisconsin–Madison. He established a new research area within the department in theoretical and applied methods in empirical processes. He is the author of Introduction to Empirical Processes and Semiparametric Inference with other research endeavors in semiparametric methods with applications to clinical trials, personalized medicine, cystic fibrosis, and cancer. He was a coauthor on a publication (Farrell et al., 1997) which was instrumental in changing public health policy that resulted in neonatal screening for cystic fibrosis in all 50 states.

Fig. 3 UNC-CH Department of Biostatistics, Wrightsville Beach Retreat 2008. *Left column*: Hongtu Zhu, Lisa LaVange, Amy Herring, Jianwen Cai, Kathy Roggenkamp, Richard Bilsborrow, Shrikant Bangdiwala, Paul Stewart, Joseph Ibrahim, David Couper. *Center column*: Haibo Zhou, Fred Wright, Fei Zou, Rosalie Dominik, Wei Sun, Diane Catellier, William Kalsbeek, Jason Fine. *Right column*: Michael Kosorok, Chirayath Suchindran, Michael Hudgens, Haitao Chu, Pranab Sen, Kinh Truong, Danyu Lin, Jane Monaco, Todd Schwartz. On the Rail: Anastasia Ivanova, Bahjat Qaqish

Expansion of the genomics, statistical genetics, and imaging research areas has continued under Kosorok's leadership with the hiring of multiple tenured and tenure-track faculty members, including Drs. Jason Fine, Wei Sun, Hongtu Zhu, and Michael Wu. Fine and Zhu both have over 100 publications and have made internationally recognized contributions in biostatistics. Fine's contributions are in areas of survival analysis, diagnostic imaging, and epidemiological methods; Zhu's contributions are in brain imaging and other biostatistical methods (Fig. 3).

Preparing students for leadership positions in academia, private industry, and government continues to be a primary focus of the department. The department offers five degree programs: BSPH, MPH, MS, DrPH, and PhD. The Environmental Training Grant has been continuously funded for almost 40 years and is a crucial program in supporting students. The Genomics Training Grant provides support and training opportunities for highly qualified doctoral students in statistical genomics with an emphasis on cancer genomics. The Initiative for Maximizing

Student Diversity (IMSD) targets underrepresented students in an effort to increase diversity among doctoral level biostatisticians. As part of this initiative, Dr. Lloyd Edwards, co-principal investigator, mentors underrepresented PhD students and facilitates collaboration between biomedical and biostatistics graduate students.

The BSPH Program continues to be a unique, notable program which graduates a small number of highly qualified students. Dr. Jane Monaco has been the Director of the undergraduate program since 2006. A recent (2008) external departmental review noted, "...the BSPH is also a treasure in its own right.... both programs [BSPH and DrPH] are in fact small jewels in the Department's crown."

The Center for Innovative Clinical Trials (CICT) was established in 2007 as part of a gift by Joan and Dennis Gillings of $50 million to the Gillings School of Global Public Health. The interdisciplinary focus of the center, directed by Dr. Joseph Ibrahim, provides a mechanism to accomplish methodological and applied research in the design and analysis of clinical trials for faculty in several departments at UNC-CH in addition to researchers from the industry. By providing predoctoral and postdoctoral support, students have opportunities to study and to advance statistical methods in clinical trials. The five components of the CICT are methodological research, applied interdisciplinary research, education, outreach to industry, and evaluation of clinical trials methods.

Important departmental units include the Biometric Consulting Lab (BCL), the Collaborative Studies Coordinating Center (CSCC), and the Survey Research Unit (SRU). The BCL continues to provide consultation to investigators while offering students training and funding opportunities. The CSCC currently coordinates more than a dozen large multicenter studies and employs approximately 90 faculty, staff, and students. With annual expenditures over $12 million, the center serves as a valuable environment for research, teaching, and public service. The long-running ARIC (Atherosclerosis Risk in Communities Study) alone has produced more than 1,000 publications to date under the leadership of Dr. Chambless. The Hispanic Community Health Study, awarded in 2006 and recently completing enrollment of 16,000 US Hispanics/Latinos, is the largest study funded to date of this important and fast-growing minority population. Led by Dr. LaVange, this study seeks to evaluate risk factors and determine disease prevalence among US Hispanics/ Latinos. The SRU uses state-of-art technology and methodology to conduct surveys at local, state, and national levels. Dozens of population-based studies are conducted each year at the SRU by survey specialists and student research assistants who contribute in the aspects of sampling design, data collection, and complex survey analysis.

Current and former faculty, students, staff, and administrators from across the world gathered to celebrate the sixtieth anniversary of the Department of Biostatistics in October, 2009. The 5-day celebration included a "Festschrift" in honor of Dr. Gary Koch and was attended by more than 300 people.

In 2009, the National Cancer Institute awarded a $12.5 million, 5-year grant to design more effective clinical trials for cancer patients. Led by principal investigators, Kosorok, Dr. Marie Davidian at NCSU, and Dr. Stephen George at Duke University, "Statistical Methods for Cancer Clinical Trials" will develop new techniques for improving design and analysis of clinical trials and discovering and evaluating personalized cancer therapies. This grant has been successful in increasing cooperation and goodwill among statistical and biostatistical researchers at UNC-CH, NCSU, and Duke.

In 2010, the department comprised 36 primary faculty members, 9 joint faculty, 15 adjunct faculty, 68 staff, and 177 students (24 BSPH, 11 MPH, 24 MS, 29 DrPH, and 89 PhD). *U.S. News & World Report* ranked the Department of Biostatistics tied at tenth among doctoral programs in statistics in 2010. Only three departments of biostatistics were ranked higher. Among the current primary faculty members, 15 have been elected ASA fellows, and 7 have been elected IMS fellows.

A particular strength of the department is its physical location near the Research Triangle Park (RTP) within several miles of NCSU and Duke University, numerous pharmaceutical companies, government agencies, nonprofit agencies, and contract research organizations. The roots of many of these units can be traced to early efforts by Cox (NCSU and RTI), Greenberg, and Hotelling (UNC-CH Statistics) and the collaborative relationships across the three respective departments. Examples of nonacademic organizations in RTP include SAS, Research Triangle Institute, GlaxoSmithKline, Quintiles, Rho, Research Triangle Institute, PPD, NIEHS, EPA, FHI, National Institute of Statistical Science (NISS), and the Statistical and Applied Mathematical Science Institute (SAMSI). This proximity provides abundant opportunities for collaborative efforts for faculty and employment potential for students.

The current direction of the department builds on the successes achieved during a rich history and intensifies the commitment to lead in research, teaching, and service. The department continues to emphasize the synergy of collaborative and methodological research. Areas of current strength are recognized in statistical genetics, clinical trials, Bayesian statistics, imaging, longitudinal data, epidemiological methods, survival analysis, high-dimensional data, and semiparametric modeling. Growth is anticipated in statistical computing, personalized medicine, high dimensional data analysis, and environmental and spatio-temporal modeling. The department embraces its role as a training ground for talented students with an expanded curriculum and plentiful compelling applied research questions and methodological problems. The department stands poised to expand its role as a world leader in high-impact biostatistics.

Selection of Frequently Cited Articles by UNC-CH Biostatistics Faculty

(1984). The Lipid Research Clinic Coronary Primary Prevention Trial Results: I. Reduction of Incidence of Coronary Heart Disease. *Journal of the American Medical Association, 251,* 351–364.

Cai, J. W., Prentice, R. L. (1995). Estimating equations for hazard ratio parameters based on correlated failure time data. *Biometrika, 82,* 151–164.

Elston, R. C., Stewart, J. (1971). A general model for the genetic analysis of pedigree data. *Human Heredity, 21*(6), 523–542.

Greenberg, B. G. (1959). Conduct of cooperative field and clinical trials. *American Statistician, 13,* 13–17.

Grizzle, J. E., Starmer, C. F., Koch, G. G. (1969). Analysis of categorical data by linear models. *Biometrics, 25 (3),* 489–504.

Ibrahim, J. G., Chen, M. H., Lipsitz, S. R., Herring, A. H. (2005). Missing-data methods for generalized linear models: A comparative review. *Journal of the American Statistical Association, 100,* 332–346.

Landis, J. R., Koch, G. G. (1977). The measurement of observer agreement for categorical data. *Biometrics, 33,* 159–174.

Lin, D. Y., Zeng, D. (2006). Likelihood-based inference on haplotype effects in genetic association studies (with discussion). *Journal of the American Statistical Association, 101,* 89–104.

Sen, P. K. (1968). Estimates of the regression coefficient based on Kendall's tau. *Journal of the American Statistical Association, 63(324),* 1379–1389.

Truong, Y. K., Stone, C. J. (1992). Nonparametric function estimation involving time-series. *Annals of Statistics, 20,* 77–97.

Selection of Books by UNC-CH Biostatistics Faculty

Draper, N. R., Smith, H. (1998, 3rd edition). *Applied Regression Analysis* Wiley.

Ghosh, B. K., Sen, P. K. (1991). *Handbook of Sequential Analysis* CRC.

Ibrahim J. G., Chen M. H., Sinha, D. (2001). *Bayesian Survival Analysis* Springer-Verlag.

Kleinbaum, D. G., Kupper, L. L., Muller, K. E. (2007, 4th edition). *Applied Regression Analysis and Other Multivariable Methods* Duxbury Pr.

Kleinbaum, D. G., Kupper, L. L., Morgenstern, H. (1982). *Epidemiologic Research: Principles and Quantitative Methods* Wiley.

Kosorok, M. (2008). *Introduction to Empirical Processes and SemiParametric Inference* Springer.

Muller, K. E., Stewart, P. W. (2006). *Linear Model Theory: Univariate, Multivariate and Mixed Models* New York: Wiley.

Sarhan, A. E., Greenberg, B. G. (1962). *Contributions to Order Statistics* New York: John Wiley and Sons, Inc.

Stokes, M. E., Davis, C. S., Koch, G. G. (2000, 2nd edition). *Categorical Data Analysis Using the SAS System* SAS Institute, Inc.

Sen, P. K., da Motta Singer, J. (1993). *Large Sample Methods in Statistics: An Introduction with Applications* Chapman & Hall/CRC.

Distinguished Graduates

Distinguished graduates of the department include: Ronald Forthofer, Carl Henley, Peter Imrey, Robert Woolson, George Williams, Dennis Tolley, Yosef Hochberg, Richard Landis, Kerry Lee, Frank Harrell, Elizabeth Delong, Eric (Rocky) Feuer, Lisa Lavange, A. John Bailer, Edward Stanek, Maura Stokes, George Howard, Lloyd Edwards, Alula Hadgu, F. DuBois Bowman, John Preisser, Stuart Gansky, Robert Lyles, Catherine Tangen, Limin Clegg, Diane Catellier, Christopher Coffey, Antonio Sanhueza, Douglas Taylor, Douglas Schaubel, Guosheng Yin, and Jean Orelien.

Acknowledgments The authors are grateful for helpful conversations with Drs. Clarence (Ed) Davis, James Grizzle, Gary Koch, Pranab Sen, and Chirayath Suchindran. Information was also obtained from video recordings of Drs. Bernard Greenberg, Ed Davis, and James Grizzle, including the fiftieth anniversary celebration. The Carolina Public Health magazine (Fall 2008, Volume 1 Number 5) provided valuable information. The authors appreciate helpful suggestions from the reviewers which led to substantial improvement.

References

(1973) The body politic, newsletter of the School of Public Health University of North Carolina at Chapel Hill 1(1):6

(1984) The lipid research clinic coronary primary prevention trial results: I. Reduction of incidence of coronary heart disease. J Am Med Assoc 251:351–364

(1988) Organization, review, and administration of cooperative studies (Greenberg Report): a report from the Heart Special Project Committee to the National Advisory Heart Council. May 1967. Controll Clin Trials 9:137–148

Chen MH, Shao QM, Ibrahim JG (2000) Monte Carlo methods in Bayesian computation. Springer, New York

Farrell PM, Kosorok MR, Laxova A, Shen G, Koscik RE, Bruns WT, Splaingard M, Mischler EH (1997) Nutritional benefits of neonatal screening for cystic fibrosis. New Engl J Med 337(14):963–969

Freiman JA, Chalmers TC, Smith H Jr, Kuebler RR (1978) The importance of beta, the type II error and sample size in the design and interpretation of the randomized control trial. New Engl J Med 299(13):690–694

Ghosh M, Schell MJ (2008) A conversation with Pranab Kumar Sen. Stat Sci 23(4):548–564

Greenberg BG (1959) Conduct of cooperative field and clinical trials. Am Stat 13:13–17

Grizzle JE (1965) The two-period change-over design and its uses in clinical trials. Biometrics 21:467–480

Grizzle JE, Starmer CF, Koch GG (1969) Analysis of categorical data by linear models. Biometrics 25 (3):489–504

Graduate Program Review, Department of Biostatistics, November 2008.

Ibrahim JG, Chen MH, Sinha D (2001) Bayesian survival analysis. Springer, New York

Korstad RR (1990) Dreaming of a time: the School of Public Health, 1939–1989. UNC Printing, Chapel Hill

Kosorok M (2008) Introduction to empirical processes and semiparametric inference. Springer, New York

Landis JR, Koch GG (1977) The measurement of observer agreement for categorical data. Biometrics 33:159–174

Muller KE, Fetterman BA (2002) Regression and ANOVA: an integrated approach using SAS software. SAS Institute, Cary

Muller KE, Stewart PW (2006) Linear model theory: univariate, multivariate and mixed models. Wiley, New York

Nourse ES, Greenberg BG, Cox GM, Mason DD, Grizzle JE, Johnson NL, Jones LV, Monroe J, Simons GD Jr (1978) Statistical training and research: the University of North Carolina system. Int Stat Rev/Revue Internationale De Statistique 46(2):171–207

Ruffin JM, Grizzle JE, Hightower NC, McHardy G, Shull H, Kirsner JB (1969) A co-operative double-blind evaluation of gastric "freezing" in the treatment of duodenal ulcer. New Engl J Med 281(1):16–19

Spivey A (2008) A mentor for life. Carolina Public Health 1(5):24–25

The University of North Carolina at Chapel Hill, Department of Biostatistics (2011) http://www.sph.unc.edu/bios

University of Pennsylvania Department of Statistics

Edward I. George, Abba M. Krieger, Donald F. Morrison and Paul Shaman

It was one of the very first statistics departments in the United States. Established in 1931 as a spin-off from the Department of Economics, it was called the Department of Economic and Social Statistics, a name that remained until 1964. Along with the social science departments at the University of Pennsylvania, the Departments of Economics, Sociology, and Political Science, its home was in the Wharton School of Finance and Commerce. But unlike those departments, which eventually moved to the School of Arts and Sciences, it remained in Wharton and never moved.

Both the research and teaching missions of the department evolved considerably over its long history. During its first 25 years, excellent service teaching was the primary mission, and research focused mainly on the study of economic and social time series. Moreover, in these early years, the department looked inward for talent by hiring its own graduates, a practice which continued through the 1950s. This began to change in the 1950s as faculty and doctoral students gradually turned their research attention to statistical theory and methodology. By the next decade, all faculty hires were coming from outside the university.

In 1964, the department was renamed the Department of Statistics and Operations Research when an operations research group left the Case Institute of Technology for the University of Pennsylvania and merged with the department. At that point, the mission of the graduate program was revised to firmly incorporate theoretical statistics, which became the department's principal research focus by the 1970s. But the alliance with the operations researchers was not to last, as they soon left to form another department. A new curriculum for the Statistics

University of Pennsylvania,
Department of Statistics,
The Wharton School,
Philadelphia, PA 19104-6340, USA
E. I. George (✉) · A. M. Krieger · D. F. Morrison · P. Shaman
e-mail: edgeorge@wharton.upenn.edu

A. Agresti and X.-L. Meng (eds.), *Strength in Numbers: The Rising of Academic Statistics Departments in the U.S.*, DOI: 10.1007/978-1-4614-3649-2_34,

Doctoral Program, one that stressed theoretical statistics, was approved by the University, and shortly thereafter, in 1975, the department was finally renamed the Department of Statistics. With the steady incorporation of theoretical statistics during the 1970 and 1980s, the foundations were now set for the department that it was to become. The subsequent 20 years saw the assembly of a thriving diversified modern statistics department, with research spanning the development of theory and methodology, with increasing attention to applications and interdisciplinary work, with continuing emphasis upon innovative high quality teaching at all levels, and with a rigorous full-fledged doctoral program.

Early History

The initial faculty members in 1931 were three: Stuart A. Rice, Simon Kuznets, and J. Parker Bursk. Rice had been Professor of Sociology from 1926 to 1930, and Professor of Sociology and Statistics from 1930 to 1940, but took a succession of leaves during 1931–1936, when he was appointed Chairman of the United States Central Statistical Board. In 1933, Rice served as president of the American Statistical Association, and from 1947 to 1953, he served as president of the International Statistical Institute. Kuznets had earned a doctorate at Columbia in 1926, studying economic patterns in prices while a Research Fellow at the Social Science Research Council. This work led to publication in 1930 of his book Secular Movements in Production and Prices. A part-time member of the department from 1930 to 1936, Kuznets became Professor of Statistics from 1936 to 1954. Kuznets served as president of the American Statistical Association in 1949, as president of the American Economic Association in 1954, and was elected as a member of the National Academy of Sciences, the American Academy of Arts and Sciences, the American Philosophical Society, and the British Academy. Bursk graduated with the Wharton undergraduate class of 1923, received his PhD in 1931 as a dissertation student of Kuznets, and became an Assistant Professor at the creation. Bursk became Chairman in 1932 and would hold this position for 31 years. E. Douglass Burdick received his PhD under Kuznets in 1935, and joined the department faculty, rising through the ranks to Professor until his sudden death in January 1961. Donald S. Murray, also a Kuznets dissertation student, finished his PhD studies in 1944, and became a faculty member. Figure 1 shows early members of the department. He taught in the department for many years before moving to the central university administration. Morris Hamburg, also a Kuznets student, received his doctorate in 1952, becoming a faculty member in 1953. Hamburg avoided formal administrative roles, but served in leadership roles in the most distinguished way. He continued until his retirement in 1992.

Kuznets was the department's clear scholarly presence (Fig. 2). During his tenure, he directed 12 dissertations while conducting research on national income data, work that would be recognized by his selection in 1971 as a Nobel laureate, the third year of existence of the Economic Sciences Prize. Bursk, conversely, was a charismatic teacher with eclectic interests and a keen concern for the practical

Fig. 1 The four early members of the department. From left to right: Doug Burdick, Don Murray, Simon Kuznets, and Parker Bursk

Fig. 2 A small group of students discusses formation of national income with a world authority, Simon Kuznets

application of statistical methods to economic and social data. An excellent instructor, he expected the same from all his colleagues and always received it, from the teaching fellows to the standing faculty. It was this excellence of teaching that distinguished the department across the university in the earlier decades when research contributions to statistical theory and methods were scant.

The Middle Years

In the 1950s an action at the university occurred that would greatly affect the department, if not immediately, at least a generation later. President Gaylord Harnwell commissioned the Educational Survey of the University of Pennsylvania, an intensive examination of every academic department and program. The Survey employed both internal committees and eminent scholars from other institutions to make recommendations for the future direction of each department and program. The outside authority for Survey XIX, Statistics and Statistical Services, was Tjalling C. Koopmans, the Yale econometrician and 1975 Nobel laureate in Economics. The committee described the existing offerings in statistics as "decentralized" by substantive disciplines, but with the greatest concentration in the Department of Economic and Social Statistics. In their report of June 30, 1958, they recommended that this organization be retained. At the same time they strongly recommended that the university seek out and recruit a leading scholar in mathematical statistics who would chair a separate department in that discipline within the Graduate School. The incumbent would be expected to hire two additional faculty in mathematical statistics as soon as possible. The new department would offer courses leading to masters and doctoral degrees in theoretical statistics. Its faculty would serve as consultants in theoretical statistics throughout Penn, and would interact with the Wharton department and any others at the university. The committee charged with forming the proposed Department of Mathematical Statistics interviewed a number of candidates, but no one accepted an offer.

Perhaps as a consequence of the Educational Survey, the Wharton undergraduate program was drastically revised in 1959 and 1960. A required year-long mathematics course was added, one that included calculus and linear algebra. Indeed, a new two-semester introductory statistics course was designed with that as a prerequisite. The use of integral calculus clearly set the course apart from traditional and mundane treatments of basic statistical methods. Ultimately, the MBA statistics course would also have a calculus prerequisite and a higher mathematical level.

Bursk's health had been declining, and he died suddenly in April 1963. Murray assumed the role of Acting Chairman. With only one other full-rank faculty member actually involved, the department was in a weak position, and perhaps faced absorption into another, larger, department. In 1964 the operations research group at Case Institute of Technology indicated its intention to leave, and negotiations began to combine that group with the department. The Case faculty and graduate students came on July 1, 1964. Shortly after his arrival, the new chairman, Russell L. Ackoff, asked certain of the department faculty to prepare a document describing a graduate program in theoretical statistics. He took that proposal to the Provost as evidence that the department would be moving in that direction.

In addition to serving as department chairman, Ackoff was the director of the Management Science Center of The Wharton School. Members of the statistics faculty were encouraged to participate in the consulting program of the Center as buy-outs from their teaching loads. The center also provided support for graduate students in the Operations Research and Statistics Doctoral Programs.

The demands of a busy management consulting center and an academic department were heavy, the marriage of the statistics and operations research groups was stormy, and Ackoff turned the chair position over to Richard C. Clelland in January 1966. Clelland served as chair through June 1971, when he was appointed acting dean of The Wharton School for the coming academic year.

In a short time, Ackoff's interests shifted from operations research to the discipline he called social systems science. The original OR faculty had largely departed, and Ackoff assembled a small cadre of like-minded colleagues into the Social Systems Science Department within Wharton. Ultimately, the group left Penn for a new home in the University City Science Center, several blocks from the university. On Ackoff's passing in 2010 he was warmly remembered for his contributions as a pragmatic management guru.

In 1972 the new Wharton dean, Donald C. Carroll, arrived from the Sloan School at M. I. T. He came with a strong background in computing and the mathematical sciences. He immediately formed a committee to reconstitute the Graduate Group in Business and Applied Economics as the Wharton Doctoral Programs. Each program designed its new curriculum leading to the PhD, and the Wharton faculty unanimously approved the new doctoral group. The statistics program was not restricted to business, economics, or any other substantive discipline and for that reason the department name was changed to the Department of Statistics on July 1, 1975. The Provost declared the department to be the statistics department for the entire university. The stage was now set for it to become a full-fledged modern statistics department, a department that would further distinguish Wharton from all other business schools.

The Later Development of the Department

The trend of the department toward mathematical statistics in both research and teaching was understandable, given the backgrounds of some of its members. Clelland came from graduate studies in pure mathematics. The theoretician Max Woodbury was briefly in the department during 1952–1954. John S. de Cani taught courses in nonlinear optimization and dealt with probabilistic and statistical models in a variety of applications. Much later the shift was evident in new faculty appointments. Jean Gibbons Fielden and Donald F. Morrison joined the department in the fall of 1963. Both had been dissertation students of Herbert A. David at Virginia Polytechnic Institute and State University.

David K. Hildebrand joined the department in 1965 after graduate study at Carnegie Institute of Technology, and a decade later began to produce introductory texts. Hildebrand cared deeply about improving the quality of business school statistical teaching, and was one of the founders of the annual conference Making Statistics More Effective in Schools of Business. James Pickands III came in 1969.

Having written his dissertation under Simeon Berman at Columbia, he was a highly regarded mathematical statistician who made seminal contributions in the area of extreme values, and as such, served as a mentor to junior faculty and PhD students with a more theoretical bent. In the 1970s and 1980s strong faculty hires included Monique Guignard–Spielberg (1972), a researcher in linear and nonlinear optimization; Abba M. Krieger (1974), with wide interests in both theory and applications; Paul Shaman (1977), a time series analyst; Robert A. Stine (1983), whose research covers credit scoring, model selection, classification and time series; and Paul R. Rosenbaum (1986), who works in the design and analysis of observational studies and experiments, and in health outcomes research.

In the late 1970s, an analysis center was established under the leadership of Lawrence S. Mayer, who came from Princeton. Well-funded, the center focused on statistical applications relating to energy, a response to the energy crises of the times. Noteworthy was the stream of high-level young researchers who spent time in the center. They included Yoav Benjamini, Richard D. DeVeaux, Stine, and Scott L. Zeger.

The department's faculty size was very small in the early years, but by the 1970s it had grown to fluctuate around 14. The size continued to increase slowly over the next 20 years, as the overall number of Wharton faculty grew substantially. The currently authorized size of the standing faculty is 21, with an additional 11 secondary appointments for faculty from other departments.

The years since 1990 have seen a dramatic transformation of the department's faculty (Fig. 3). Guignard–Spielberg transferred to the Department of Operations and Information Management in 1987, as the department turned its focus solely to statistics. The remaining departmental faculty who had started prior to 1970 retired and left throughout the 1990s, and Hildebrand died prematurely in 1999. In the first major move and turning point of the 1990s, the department hired J. Michael Steele in 1990. Steele's work in applications of probability and mathematical finance gave the department strong new directions that blended nicely with the interests of faculty and doctoral students at Wharton and across the university. With the arrival of Mark G. Low in 1991, the department began its acquisition of a very strong core of researchers in statistical inference, nonparametric function estimation, and decision theory. Dean P. Foster was hired in 1992, giving the department a strong presence in variable selection and in topics relating to economics, including inference for stochastic processes and game theory. The department's ascendancy in mathematical statistics was established when Lawrence D. Brown and Linda H. Zhao joined the department in 1994. Brown came with exceptional achievements in statistical theory, and with multiple honors, including membership in the National Academy of Sciences. Zhao brought further strength in Bayesian analysis and decision theory. In 1998 Abraham J. Wyner joined the department, providing expertise in probabilistic modeling, information theory, and data compression.

Moving to the 2000s, T. Tony Cai came to the department in 2000, further cementing the preeminence of mathematical statistics. Cai's wide-ranging interests in statistical inference included high-dimensional inference, functional data analysis,

Fig. 3 Some department faculty in the 1990s. (Top row, left to right) David Hildebrand, Don Morrison, Mike Steele, Abba Krieger. (Bottom row, left to right) Bob Stine, Paul Rosenbaum, Dean Foster, Mark Low (holding Robyn Low)

large-scale multiple testing, nonparametric function estimation, and methodology and applications with wavelets. Next came Edward I. George in 2001, a senior hire who had already made significant contributions to Bayesian and frequentist decision theory, shrinkage estimation, tree modeling, and variable selection. He was quickly followed in 2002 by another senior hire, Andreas Buja, a preeminent researcher in modern multivariate analysis, statistical computing, and data visualization.

Six of the current faculty came as junior appointments in the next succession of hires: Dylan S. Small in 2002, with expertise in the analysis of observational studies, and in applications of statistics to questions arising in public health, medicine, and economics; Shane T. Jensen in 2004, with expertise in bioinformatics, hierarchical modeling, and MCMC methods; Alexander Rakhlin in 2008, with expertise in machine learning, sequential decision making, statistical learning theory, and applied probability; Sham Kakade in 2009, with expertise in machine learning, probabilistic artificial intelligence, algorithmic statistics, and game theory; Zongming Ma in 2010, with expertise in high dimensional statistical inference, nonparametric statistics, random matrix theory, and graphical modeling; and Emily B. Fox in 2011, with expertise in Bayesian and nonparametric Bayesian approaches to time series and longitudinal data analysis.

Three other recent appointments have served to add luster and breadth to the department. After a long and distinguished career at Bell Laboratories, spanning 34 years, followed by 13 years at Rutgers, Lawrence Shepp joined the department in 2010. A researcher with significant contributions in many areas of probability, statistics, and image reconstruction, he has received multiple honors, including membership in the National Academy of Sciences, the Institute of Medicine, and

Fig. 4 The faculty of the department for 2011-2012. From left to right: (1st row) Emily Fox, Linda Zhao, Nancy Zhang, Richard Waterman, Jean Lemaire, Mike Steele; (2nd row) Abba Krieger, Dylan Small, Ed George, Sasha Rakhlin, Paul Rosenbaum; (3rd row) Bob Stine, Paul Shaman, Larry Brown, Don Morrison, Dean Foster, Zongming Ma, Richard Berk; (4th row) Tony Cai, Shane Jensen, Adi Wyner, Andreas Buja. (Absent: Mark Low, Sham Kakade, Larry Shepp)

the American Academy of Arts and Sciences. Coming to the department from the faculty at Stanford, Nancy Zhang joined the department with tenure in 2011. Her research has focused on the development of new methodology for the statistical analysis of high-throughput biological experiments, adding substantial strength to the department's expertise in bioinformatics. Finally, Jean Lemaire, a distinguished professor of the Insurance and Risk Management Department and Director of the Actuarial Science Program at Wharton since 1987, migrated his primary appointment to the department in 2011. This appointment establishes a departmental research and teaching presence in the allied statistical areas of insurance and actuarial science. Figure 4 shows the current department.

Teaching

Throughout its history, the department has always taught undergraduates, master's students, and doctoral students. Though the bulk of its teaching has taken place in Wharton, the department has also functioned as a teaching department for much of the entire university, a responsibility that has gradually increased over time.

The first 11 doctoral degrees awarded by the department were all completed under the direction of Simon Kuznets. The first three recipients, Bursk (1931),

Burdick (1935), and Murray (1944), and later Hamburg (1952), all became faculty members, as previously noted. Two more graduates of the department, Robert C. Jones and de Cani, were students of the Wharton financial economist Irwin Friend, and joined the faculty in 1958. They retired in 1994 and 1995, respectively. The pace of PhD production in the department picked up in the 1950s, with 17 awarded. All of the early dissertations dealt with the statistical analysis of economic data. For instance, Bursk's dissertation was entitled Seasonal Variations in Employment in Manufacturing Industries: A Statistical Study Based on Census Data, and the title of Hamburg's thesis was The Income Elasticity of Food Purchases, 1929–1948. It was not until 1968 that a departmental graduate wrote a dissertation focused on statistical methodology. This marked a turning point, as all departmental dissertations since the late 1970s have addressed statistical theory and/or methodology. In the first 30 years, the department graduated 22 PhDs with five becoming ASA fellows, including Hamburg, and de Cani. The following table shows the volume of PhD production by decade.

Decade	Number of PhDs
1930–1939	2
1940–1949	1
1950–1959	17
1960–1969	10
1970–1979	15
1980–1989	16
1990–1999	21
2000–2009	44

For a long time the department matriculated terminal master's students, as well as doctoral students. This practice stopped in the late 1980s when Wharton eliminated matriculation into terminal master's programs except for the MBA program. The department's doctoral program matriculation counts ranged from 2 to 6 per year in the 1980s and 1990s, becoming more stable after 2000, ranging from 4 to 6 per year. The percentage of matriculants completing the doctorate has also increased sharply, with essentially all now finishing.

Moves to change the department's curriculum began to emerge in the 1960s. Partly in response to the Educational Survey, five of the department faculty collaborated on a text for the two-semester statistics course required of all Wharton undergraduates (Clelland, et al., 1966, 1973).

During 1959–1960 Hamburg spent the academic year at the Harvard Business School in a Ford Foundation program introducing Bayesian statistical decision theory to business school faculty. On his return to Penn, Hamburg incorporated this material into the basic courses of the Wharton MBA Operations Research Option, and he directed several doctoral dissertations in decision theory. One of his outstanding students, Paul E. Green, became a mainstay and much-cited member of the Wharton Marketing Department.

In the graduate program Fielden introduced a course in nonparametric inference and began work on a text (Gibbons, 1970) for it. Morrison developed a two-semester course in multivariate statistical theory and methods, and produced a book for the methods component (Morrison, 1967). Concurrently, Hamburg prepared a text for the MBA statistics course with an extensive treatment of statistical decision theory and Bayesian inference (Hamburg, 1970). And Hildebrand developed a two-semester course in statistics for undergraduate psychology and biological basis of behavior majors that extended into complex experimental designs and the analysis of variance for cognitive investigations. Hildebrand went on to write a text for the course, and also texts for the introductory courses taught to Wharton undergraduate students and to MBA students. All of his texts reflected updated methods of teaching.

At the start of the 1990s, Wharton instituted a major overhaul of the MBA core curriculum. In response to this, Hildebrand produced notes that completely redesigned the required statistics course in the MBA core. The new treatment stressed interpretation and, with computer usage, avoided the details of statistical computation. Many real data sets were utilized to illustrate the use of methodology and provide interpretive examples. Later these notes were thoroughly revised and expanded by Foster, Stine, and Richard P. Waterman, a faculty member from 1993 to 1999. This led to publication of two casebooks (Foster, Stine and Waterman, 1998a, 1998b). Later, Foster and Stine authored an introductory statistics text for both the undergraduate and MBA students (Stine and Foster, 2011).

Research

In various combinations with each other and with other researchers, Larry Brown, Tony Cai, Mark Low, and Linda Zhao have collaborated to investigate a variety of statistical issues with a special focus on nonparametric function inference. An important substream began with a paper by Brown and Low describing the asymptotic equivalence of two forms of nonparametric function estimation problems, nonparametric regression, and the white-noise signal estimation problem. Cai, with partial collaboration of Brown, has developed a long stream of innovations in the practice and theory of wavelet estimation. Cai and Low have devoted extensive attention to research into general adaptivity properties. Recently, Brown, Low, Zhao, and other coauthors have joined together these research streams, using wavelets, and ideas from the adaptivity investigations to propose improved nonparametric density function estimators.

A good deal of Michael Steele's research has focused on those parts of probability theory that share a connection with combinatorial optimization. The classic problems of this field are those associated with the traveling salesman problem, minimal spanning trees, and optimal matching. The central unifying object of this field is the subadditive Euclidean functional introduced by Steele in 1981. Steele has also studied discrete time processes, worked on interest rates, and written a text on stochastic calculus with applications to finance (Steele, 2001). Another theme throughout his research work has been the development of inequalities, and he is the author of a monograph on this subject (Steele, 2004).

Abba Krieger's research has included the development of methodology for grouped data, categorical data, sample surveys, and observational studies. He also developed stochastic models for applications, and obtained worst-case bounds for operations research problems. In the 1980s with Paul Green, he engaged a wide variety of problems in marketing research, with a prominent focus on conjoint models. Krieger's recent applied work is in neuroscience, studying seizure propagation as a result of epilepsy and addressing various clinical questions such as whether seizures are predictive and whether they are localized.

Paul Rosenbaum's research has focused on design and analysis of observational studies and experiments, and on health outcomes research. With Donald B. Rubin, Rosenbaum developed the propensity score. He is the author of two books on observational studies (Rosenbaum, 2002, 2010). Dylan Small has studied causal inference, longitudinal data analysis, and applications of statistics to health studies. He has particularly contributed to methods for using instrumental variables, developing methods of sensitivity analysis, nonparametric estimation, building a stronger instrument and correction for measurement error, as well as applications of instrumental variables in neonatology comparative effectiveness studies. Rosenbaum, Small, Krieger, and Shane Jensen have collaborated in various combinations on research on observational studies.

In the early 1990s, Dean Foster and Ed George started a push for regression with large data sets by introducing the risk inflation criterion (RIC), which used an oracle argument to gauge variable selection uncertainty. Later in the decade, they developed empirical Bayes variable selection criteria with adaptive penalties. In the early 2000s, Foster and Robert Stine pushed large stepwise regressions to new limits, using a variety of techniques to improve the accuracy, and guarantee theoretical properties of such faster methods. Foster and Stine also reconsidered model selection from an information theory point of view, and developed new methods for testing multiple hypotheses. In another stream of work, Stine and Paul Shaman analyzed bias for several methods of autoregressive parameter and spectral estimation.

Beyond his work with Foster, George's research with other co-authors has included the methodological development of Bayesian variable selection, Bayesian CART, Bayesian treed modeling and Bayesian additive regression tree (BART) modeling, as well as the development of theoretical minimaxity and admissibility results for multivariate predictive density estimation.

A primary line of Andreas Buja's research is in data visualization, pushing the reach of what can be "seen" in high-dimensional data. He has also contributed to multidimensional scaling methodology, providing visualization of objects as varied as social networks, and manifolds in high-dimensional spaces. In recent years he has introduced statistical inference to data visualization, to help avoid reading "tea leaves," or random patterns in data. Buja's other lines of work include function fitting, machine learning, and, most recently in collaboration with Richard Berk, Brown, George, Zhao and others, the problem of "post-selection inference."

Abraham Wyner has contributed research in a variety of areas, including information theory, machine learning, and climate science. His work has provided fundamental insight into the workings of the Lempel–Ziv data compression algorithm as well as into the limitations of boosting algorithms. He has recently focused on the statistical analysis of climate field reconstructions, demonstrating the large uncertainties underlying conventional wisdom about global warming.

Shane Jensen has worked extensively on statistical modeling in molecular biology and genetics. He has developed novel hierarchical models and semi-parametric approaches to address the rapid advances in DNA sequence analysis, regulatory network inference, gene expression analysis, and genome evolution. He has also collaborated with Wyner on the use of statistical models of player performance in major league baseball, with a focus on the quantitative analysis of fielding ability.

Alexander Rakhlin, working with co-authors, has formulated a minimax theory for sequential prediction problems. Analogs of many classical results from statistical learning theory can be proved in the setting of worst-case sequential prediction. A further generalization of these results gives a unified treatment to such problems as calibration of forecasters, Blackwell's approachability, regret minimization, and more.

Sham Kakade's research is in machine learning and artificial intelligence, with the objective of contributing to the statistical tools needed to make progress on core AI problems. The focus is on the core challenges facing machine learning: representational learning, large-scale learning and algorithmic statistics, and decision making in high-dimensional action spaces.

Zongming Ma's research focuses on high-dimensional statistical inference. Of particular interest are problems where only a small number of the unknown parameters under a suitable model are pertinent to statistical inference. Key questions include selection and estimation of the relevant parameters.

Emily Fox's research focuses on developing Bayesian and Bayesian nonparametric methodology for multivariate time series analysis. She has developed regime-switching and heteroscedastic models with application to a wide variety of domains including analysis of human motion, speech data, EEG, and stochastic volatility of stock indices.

Lawrence Shepp's research interests are diverse, including probabilistic, combinatorial, and statistical analysis models for problems arising in physics, engineering, and communications. They include computer tomography, automatic pattern recognition, probabilistic models for phase transitions, connectedness of random graphs, the mathematics of finance, and genetics.

Nancy Zhang's research interests lie at the intersection of statistics and biology. Her contributions involve innovative modeling methods for genomic profiling including new approaches for variable selection, data subsampling, false discovery rate control, and importance sampling.

A major contribution of Jean Lemaire's research has entailed the application of Markov chains to bonus-malus systems for automobile insurance (Lemaire 1995). He has also studied the adverse effects of prohibiting insurers from seeking genetic testing information, employed multiple decrement theory to estimate the life expectancy reduction from gunshot wounds, and used panel data techniques to improve prediction of insurance sales.

Service

Departmental members have served as editors and officers for the professional societies, and in administrative positions at the University of Pennsylvania. Morris Hamburg was editor of The American Statistician from the April 1956 issue through that of February 1972. Under his stewardship the magazine matured from news and notices to one carrying more technical and scholarly articles. Donald F. Morrison assumed the editorship with the April 1972 issue, and continued that trend, with the encouragement of expository papers, until the end of his term in December 1975. Larry Brown and Tony Cai have both served as co-Editor of The Annals of Statistics, Brown during 1995–1997, and Cai during 2010–2012. Michael Steele was the first Editor of The Annals of Applied Probability, 1991–1993, Andreas Buja was Managing Editor of the Journal of Computational and Graphical Statistics, 1997–2000, and Ed George was Executive Editor of Statistical Science, 2005–2007. Further, both Brown and Steele have held the position of IMS President, Brown in 1992–1993, and Steele in 2009–2010. George was President of the International Society for Bayesian Analysis, 2003, and Nancy L. Geller, a departmental member during the 1970s, was the 2011 ASA President. Brown is Chair of the Committee on National Statistics for the term 2010–2013. Paul Shaman was IMS Managing Editor during 1986–1989, and IMS Managing Editor for Statistics, 2002–2007.

At the University of Pennsylvania, Richard Clelland was Acting Dean of The Wharton School during 1971–1972 and a Wharton Associate Dean from 1975 to 1981. He became the first Deputy Provost of the university when the position was created in 1982, and continued until his retirement in 1992.

Honors and Awards

Simon Kuznets' Nobel Prize (1971), Larry Brown's membership in the National Academy of Sciences (1990), and Larry Shepp's membership in three academies, the National Academy of Sciences (1989), the Institute of Medicine (1992), and the American Academy of Arts and Sciences (1993), have already been mentioned. Both Brown and Shepp have been awarded honorary degrees, Brown from Purdue University (1993) and Shepp from New York Polytechnic University (2004).

Fig. 5 Larry Brown celebrates his 70th birthday at a Wharton conference in his honor with Linda Zhao and three of his former PhD students, Jim Berger, Iain Johnstone and Tony Cai. From left to right: Zhao, Berger, Johnstone, Cai and Brown

Brown (Fig. 5) received the Wilks Award from ASA (2002) and the C. R. and B. Rao Prize (2007). Shepp received the Paul Lévy Prize (1966) and the IEEE Distinguished Scientist Award (1979). Jean Lemaire received the International INA-Lincei Prize from the Italian Academy of Sciences (1988). Mike Steele received the Frank Wilcoxon Prize from ASQC and ASA (1990). Paul Shaman was awarded the Harry C. Carver Medal by IMS (2004). Paul Rosenbaum was the recipient of the COPSS George W. Snedecor Award (2003), and Tony Cai the COPSS Presidents' Award (2008). Alexander Rakhlin was given an NSF CAREER Award for 2010–2012.

Brown delivered the IMS Wald Lectures in 1985 and an invited World Mathematics Conference Lecture in 2002. IMS Medallion Lectures have been given by Brown (2006), Andreas Buja (2002), Cai (2009), Ed George (2010), and Mark Low (2008).

Six faculty members, six secondary faculty members, one adjunct faculty member, and 12 former faculty members have been named ASA Fellows. As noted, five graduates of the doctoral program, two of whom were faculty members, are ASA Fellows. Nine faculty members, one secondary faculty member, and four former faculty members are IMS Fellows.

Eleven members of the department have been appointed to named chairs in The Wharton School. They are Steele (1991), Brown (1994), Krieger (1996), Rosenbaum (2001), George (2002), Buja (2003), Lemaire (2005), Cai (2007), Foster (2007), Low (2007), and Shepp (2010). The department has a long list of teaching awards. John de Cani (1964), Abba Krieger (1978), Robert Jones (1979), and Morris Hamburg received Lindback Awards for Distinguished Teaching. Krieger

(1977), Richard Waterman (1997) and Robert Stine (2011) were recipients of the Helen Kardon Moss Anvil Award, given once a year for excellent teaching to MBA students. The David W. Hauck Award for Excellence in Undergraduate Teaching, given twice a year, was awarded to Krieger (1996), Stine (2001), Lemaire (2008), and Jensen (2009). The Award for Excellence in Undergraduate teaching was given to Lemaire (2004, 2006, 2007, 2008, 2011), Steele (2010) and Low (2011), and the Provost's Award for Distinguished PhD Teaching and Mentoring was given to Brown (2011). In addition to these, George, Krieger, Stine, and Waterman (who has continued to teach MBA classes in an adjunct capacity) have been the recipients of multiple teaching awards.

Appendix

Department Names	
Department of Economic and Social Statistics	1931–1964
Department of Statistics and Operations Research	1964–1975
Department of Statistics	1975–

Department Chairs	
J. Parker Bursk	1932–April 1963
Donald S. Murray (Acting Chair)	May 1963–June 1964
Russell L. Ackoff	July1964–December 1965
Richard C. Clelland	January 1966–June 1971
John S. de Cani	July 1971–June 1976
Donald F. Morrison (Acting Chair)	July 1976–December 1976
John S. de Cani	January 1977–June 1978
Donald F. Morrison	July 1978–June 1985
David K. Hildebrand	July 1985–June 1990
Paul Shaman	July 1990–June 2002
Abba M. Krieger	July 2002–June 2008
Edward I. George	July 2008–

Standing Faculty Roster in 2011–2012, with year of appointment

Lawrence D. Brown, 1994; Andreas Buja, 2002; T. Tony Cai, 2000; Dean P. Foster, 1992; Emily B. Fox, 2010; Edward I. George, 2001; Shane T. Jensen, 2004; Sham M. Kakade, 2010; Abba M. Krieger, 1974; Jean Lemaire, 2011; Mark G. Low, 1991; Zongming Ma, 2010; Alexander Rakhlin, 2009; Paul R. Rosenbaum, 1986; Lawrence Shepp, 2010; Dylan S. Small, 2002; J. Michael Steele, 1990; Robert A. Stine, 1983; Abraham J. Wyner, 1998; Nancy Zhang, 2011; Linda H. Zhao, 1994.

Secondary Appointments in Department of Statistics, Academic Year 2011–2012
Richard A. Berk, Criminology; Robert F. Boruch, Graduate School of Education; Eric T. Bradlow, Marketing; Francis X. Diebold, Economics; Daniel F. Heitjan, Biostatistics and Epidemiology; Michael Kearns, Computer and Information Science; J. Richard Landis, Biostatistics and Epidemiology; Hongzhe Li, Biostatistics and Epidemiology; Max Mintz, Computer and Information Science; Robin Pemantle, Mathematics; Ben Taskar, Computer and Information Science.

References

Clelland RC, de Cani JS, Brown FE, Parker Bursk J, Murray DS (1966) Basic statistics with business applications. Wiley, New York

Clelland RC, de Cani JS, Brown FE (1973) Basic statistics with business applications, 2nd edn. Wiley, New York

Foster DP, Stine RA, Waterman RP (1998a) Basic business statistics. Springer, New York

Foster DP, Stine RA, Waterman RP (1998b) Business analysis using regression, Springer, New York

Gibbons JD (1970) Nonparametric statistical inference. McGraw-Hill, New York

Hamburg M (1970) Statistical analysis for decision making. Harcourt Brace Jovanovich, Orlando, Florida

Lemaire J (1995) Bonus-malus systems in automobile insurance. Kluwer, New York

Morrison DF (1967) multivariate statistical methods. McGraw-Hill, New York

Rosenbaum PR (2002) Observational studies, 2nd edn. Springer, New York

Rosenbaum PR (2010) Design of observational studies. Springer, New York

Steele JM (2001) Stochastic calculus and financial applications. Springer, New York

Steele JM (2004) The Cauchy–Schwarz master class. Cambridge University Press

Stine RA, Dean PF (2011) Statistics for business: decision making and analysis. Addison-Wesley, Boston

University of Pittsburgh Departments of Biostatistics and Statistics

Howard Rockette, Leon Gleser and Carol Redmond

The Early Years: 1949–1969

The Department of Biostatistics was founded in 1949 as one of five original departments of the Graduate School of Public Health (GSPH). The objectives of the Department were to:

1. Foster an understanding of statistical logic as it applies to the quantitative study of public health problems and human biology through formal courses and day-to-day contact with students and faculty of the School.
2. Apply and demonstrate the application of statistical methods to the study of new and pressing problems in public health through research and community service.

At first, the Department had two faculty members, Antonio Ciocco, Professor and Chair, and Paul Densen. Because of the demographics of the environment in which the school was located, emphasis was given to the study of industrial populations. In the early 1950s, the Department enlarged the number of faculty and course offerings. Two notable additions were the appointments of Donovan Thompson in 1952 and Ching Chun Li in 1954. The departmental progress reports indicate increased attention to both meeting the quantitative needs of other departments in the School and in the training of biostatisticians. Departmental faculty believed that training biostatisticians cannot be accomplished solely within a school of public health, but must "involve, on one hand, institutions which can provide instruction in basic statistical theory and on the other hand, organized health agencies which can provide actual experience in public health activities."

University of Pittsburgh,
2717 Cathedral of Learning,
Pittsburgh, PA 15260, USA
H. Rockette (✉) · L. Gleser · C. Redmond
e-mail: herbst@pitt.edu

A. Agresti and X.-L. Meng (eds.), *Strength in Numbers: The Rising of Academic Statistics Departments in the U.S.*, DOI: 10.1007/978-1-4614-3649-2_35,

During the period 1952–1956, three PhD degrees, three MS Hygiene degrees and four MPH degrees were granted in biostatistics.

In the late 1950s and early 1960s, the Department had three active NIH training grants and was also developing an externally funded research program. Many research grants were in areas that the Department would continue to develop in future years including the study of the health of occupational populations, investigations of the treatment and etiology of cancer, and genetic aspects of disease. GSPH continued to grow and in 1957 moved into a newly constructed building at the corner of Fifth Avenue and DeSoto Street, a location which it still occupies. In 1962, the School had 60 faculty members and eight were primary in the Department of Biostatistics. Students were typically individuals with previous work experience who took leave from their career to pursue an academic degree. A notable example was Sidney Cutler (National Cancer Institute) who graduated with a doctorate in biostatistics in 1961, a few years after publication of his classic paper with Ederer on actuarial life tables.

Growing interest in Human Genetics led to the decision to broaden the program, and in 1967, a University-wide doctoral program in Human Genetics was adopted. From that year until the creation of a separate department in 1989, the Department of Biostatistics incorporated into its mission, objectives, and strategic planning the two tracks of Public Health Statistics and Human Genetics. Concurrent with expansion of the program in Human Genetics was a major revision of the Public Health Statistics curriculum. The modified curriculum enabled accepting students with training in Mathematics but no experience in Statistics. The 1962–1967 progress report acknowledges that the policy of limiting training "to mature individuals with adequate preparation in statistical theory and demonstrated capacity for research" was, in part, because "the faculty of the Department could not offer courses in statistical theory and the University lacked a Department of Statistics". The Department of Biostatistics recruited faculty to teach additional statistics courses. One of these faculty members, B. R. Rao, who remained with the Department for more than 20 years, subsequently held a joint appointment with the Statistics Group in the Faculty of Arts and Sciences.

In conjunction with increased training in statistical methodology, the Department emphasized the importance of direct contact of students with ongoing field research. In addition, a two-semester course was added to introduce students to statistical methods for public health research. The course, taught by Visiting Adjunct Faculty Sidney Cutler, William Haenszel, Nathan Mantel, and Marvin Schneiderman (later including Mitchell Gail and David Byar), was offered every year for more than 2 decades. During this time period Jerome Cornfield also held a primary academic appointment in the Department (1968–1972). In his notable 1969 paper, he comprehensively addressed, as well as advocated, the use of Bayesian methods in applications. Other ongoing research included development of analytical techniques utilizing stochastic processes for quantitative description of human reproduction; investigations of the mortality patterns of steelworkers in Allegheny; a variety of cytogenetic investigations; and development of methods in population genetics. A particularly important area of research, Population Statistics, was directed by Mindel Sheps (1961–1965). In several seminal papers published during that time, she (in collaboration with Ed Perrin) developed stochastic models to describe the reproductive history of couples.

Selected Publications

Sheps MC, Perrin EB (1964) Human reproduction: A stochastic process. Biometrics 20(1):28–45
Li CC (1955) Population genetics. University of Chicago Press.
Sheps MC, Perrin E (1963) Changes in birth rates as a function of contraceptive effectiveness: some applications of a stochastic model. Am J Public Health 53:1031–1046
Li CC (1964) Introduction to experimental statistics. McGraw-Hill Book Co.
Li CC (1967) Genetic equilibrium under selection. Biometrics 23(3):397–484
Redmond CK, Smith EM, Lloyd JW, et al (1969) Long-term mortality study of steelworkers, III. Follow-up. J Occup Med 11:513–521
Cornfield J (1969) The Bayesian outlook and its application. Biometrics 25(4):617–657

Two Programs/One Department: 1969–1989

Directions established in the late 1960s laid the foundation for the academic programs for the next 20 years. The two main tracks, Public Health Statistics and Human Genetics, had undergone considerable revision in academic offerings had clearly stated objectives and were committed to expanding the academic diversity of the faculty. Since statistical methods were an essential component in Population Genetics, the coexistence of both groups within the same department was natural, and C. C. Li was an ideal candidate to head the Department (1969–1975). C. C. had authored seven textbooks including the first book on population genetics. "An Interview with Dr. Li" by Timonthy Chen and John Tai in *Statistical Science* (1998) provides insights about this remarkable individual. During his time as chair, considerable interaction continued between the faculty in the two tracks on student committees and in collaborative research. Two faculty trained in Mathematical Statistics (Sati Mazumdar 1971, and Howard Rockette 1972) joined the Department and there were now seven faculty in Public Health Statistics and three in Human Genetics. During this period, the term Public Health Statistics Program was replaced by Biostatistics Program.

From its inception, the Department of Biostatistics had interest in studying industrial populations. Ciocco, along with a doctoral student (Bill Lloyd) on leave from the National Cancer Institute, had initiated work investigating mortality

Fig. 1 Faculty of the Department of Biostatistics, September 2011

patterns of steelworkers in Allegheny County. Under the direction of Carol Redmond, who joined the faculty in 1967, this study continued for more than 30 years. It was not only a model for the conduct of retrospective mortality studies of that era, but also stimulated the interest of departmental faculty in the application of Biostatistics and Epidemiology to problems in occupational and environmental health. The epidemiological studies of coke oven workers formed the basis for the development of the Occupational Safety and Health Administration's 1977 standards for exposure to coke oven emissions.

The 1967 recruitment of Philip Enterline, Professor of Biostatistics, also contributed to the prominence given health studies of industrial populations within the Department. Enterline, who had previously identified the health risks of asbestos workers, continued to investigate health patterns of industrial populations in Pittsburgh, both as a faculty member and as Department Chair (1975–1982). A 1986 departmental progress report indicates that the Department had investigated the mortality patterns of 15 industrial cohorts, several industrywide, consisting of a total of more than 220,000 workers. Studies included steelworkers, asbestos workers, coal miners, autoworkers, nickel-exposed workers, aluminum workers, arsenic-exposed workers, and petrochemical workers. Accompanying the investigation of specific occupational groups, the Department contributed to the development of statistical methodology and software related to occupational and environmental problems. These included incorporation of time-dependent covariates into the multistage model, refinement of the application of time series to mortality studies of air pollution, development of methods in computational toxicology to identify potential carcinogens, development of the Occupational Cohort Mortality Analysis Program (OCMAP), and quantitative risk assessment models to predict risk at low exposure levels (Fig. 1).

In a 1977 agreement between the Department and Dr. Bernard Fisher, Professor of the School of Medicine and Chairman of the National Surgical Adjuvant Breast Project (NSABP), Carol Redmond became the Director of the NSABP Biostatistical Center. The original studies of the NSABP were randomized Phase III clinical trials of patients with Stage I and Stage II breast cancer designed to evaluate multi-modality treatments including differing surgeries, radiation, chemotherapy, immunotherapy, and hormonal therapy. Over time the NSABP mission expanded to include randomized trials for adjuvant colorectal cancer (1976) and prevention trials for both breast (1995) and colorectal cancer (2002). Trials conducted by the NSABP have had a major impact on the reduction in the extent of surgery required for initial treatment of early stage breast cancer, the use of adjuvant chemo and hormonal therapy to improve survival of patients with breast and colorectal cancer, and the use of tamoxifen (Breast Cancer Prevention Trial) and raloxifene (STAR Trial) to prevent breast cancer.

Redmond became chair in 1982, and applications of Biostatistics to occupational and environmental health and the clinical trials of the NSABP continued as a major component of the research effort of the Department. Of the nine faculty members in the biostatistics program of the Department in 1985, eight had substantial research in statistical methods and/or applications in occupational and environmental health and four were associated with the NSABP.

In 1980, the Department's research program became more diverse when collaboration was established with the Otitis Media Research Center (OMRC) at Children's Hospital of Pittsburgh. OMRC's research included 12 clinical trials to determine the most effective method of management of infants and children with acute, recurrent acute, and chronic otitis media with effusion. Important results from these trials include the lack of efficacy of decongestant–antihistamine in treating otitis media with effusion, quantifying the effectiveness of treating acute otitis media with antibiotics, the benefit of tonsillectomy in treating chronic ear disease, and the impact of delaying tube insertion on language development.

In the 1980s, collaborations were also initiated with the Department of Radiology. Originally, these collaborations focused on the design and analysis of investigations to improve the diagnostic accuracy of chest images and were particularly important in providing the transition from original film to picture archiving and communication systems. Collaborations with radiology have expanded over time to encompass research efforts with the PET facility, the MR Research Center, and development of statistical methodology to evaluate better diagnostic imaging systems. Faculty members presently involved in this collaboration include Rockette, Weissfeld, Bandos and Mazumdar.

In 1985, the University established the Pittsburgh Cancer Institute (PCI). The Department created a position for a Biostatistics Core Director and recruited faculty to provide clinical and basic science investigators with statistical and computer-related expertise in design, execution, analysis, and reporting of cancer-related research studies. PCI research ranges from basic science studies, Phase I and Phase II oncology clinical trials, epidemiological studies including those related to cancer prevention and awareness to investigations of behavioral and health sequelae of cancer treatment.

The Human Genetics Program now had six faculty members and had developed a track in genetic counseling (MS) which led to a major increase in matriculating students. In the period 1980–1985, the program in human genetics graduated 50 students with a masters degree and seven students with a doctorate compared to 23 and 13, respectively, in the Biostatistics program. Research productivity of the Human Genetics Program was greatly enhanced by the recruitment of Aravinda Chakravarti, a population geneticist, and Robert Ferrell, a biochemical geneticist interested in genetic epidemiology. Li retired in 1983, but continued to contribute to the literature on population genetics. The success of the program, its increasing expansion into modern genetic laboratory sciences, and the established recognition of the faculty beyond quantitative genetics led to creation of a separate Department of Human Genetics in 1989. This was the first Human Genetics Department in a School of Public Health.

Selected Publications

Lloyd JW (1971) Long-term mortality study of steelworkers. V. Respiratory cancer in coke plant workers. J Occup Med 13(2):53–68

Li CC (1975) Path analysis, a primer. Boxwood Press

Fisher B, Montague E, Redmond C, et al (1977) Comparison of radical mastectomy with alternative treatments for primary breast cancer. A first report of results from a prospective randomized clinical trial. Cancer 39(6 Suppl):2827–2839

Fisher B, Bauer M, Margolese R, et al (1985) Five-year results of a randomized clinical trial comparing total mastectomy and segmental mastectomy with or without radiation in the treatment of breast cancer. N Engl J Med 312(11):665–673

Mandel EM, Rockette HE, Bluestone CD et al (1989) Efficacy of amoxicillin with and without decongestant antihistamine for otitis media with effusion in children: Results of a double blind randomized trial. N Engl Med 320(8):479–484

Fisher B, Costantino J, Redmond C, et al (1989) A randomized clinical trial evaluating tamoxifen in the treatment of patients with node-negative breast cancer who have estrogen-receptor-positive tumors. N Engl J Med 320(8):479–484

1990–2000

The 1990s were a period of gradual progress in many of the areas that had already been initiated. Stewart Anderson, 1989; Roslyn Stone, 1990; and Lisa Weissfeld, 1990 joined the Department as core members of the teaching faculty. Subsequently the number of faculty members during this decade remained relatively stable at about 13. Rockette succeeded Redmond as Chair of Biostatistics in 1996. In 1995, Sam Wieand, Director of Colorectal Trials at the Mayo Clinic, succeeded Redmond as Director of the Biostatistical Center of the NSABP. The study of mortality patterns of industrial populations remained a strong component of the departmental research effort. Gary Marsh, a close collaborator of Enterline's, assumed responsibility for directing many of Enterline's ongoing studies. These included investigations of the health effects of fiberglass workers, chemical workers, and workers in the mineral wool industry.

New collaborations with the Medical School continued to be developed. Mazumdar became involved with multiple research projects in psychiatry. Most projects were related to depression in the elderly and included studies of the relationship of sleep deprivation and spousal bereavement on depression, randomized trials evaluating initial treatment and maintenance therapy, and investigations of factors related to recurrence of depression. In the 1990s, Redmond, Weissfeld, and Stone participated with the Department of General Medicine on a Patient Outcomes Research (PORT) grant to compare health outcomes and cost effectiveness for patients with community acquired pneumonia. This led to additional ongoing collaborations with clinical investigators to evaluate comparative health outcomes for alternative medical interventions. Weissfeld initiated collaborations with the Department of Critical Care Medicine within the Clinical Research, Investigation, and Systems Modeling of Acute Illness (CRISMA) laboratory, a group of adult and pediatric critical care physicians who specialize in clinical research of the acutely ill. Stone began collaborative efforts in health services research with the Veteran's Administration (VA) Healthcare System, the UPMC Healthcare System, local health insurance providers, and pediatric trauma centers. Many of these studies are now coordinated through the Center for Health Equity Research and Promotions at the Pittsburgh Veteran's Administration (VA).

All of the collaborations mentioned in this chapter have been ongoing for at least 2 decades and address important public health problems that have resulted in numerous peer-reviewed publications. In addition to providing a source of funding for faculty and graduate students, questions arising in the design or analysis of these research studies stimulated important methodological research and often provided the basis of topics for student dissertations. Noteworthy examples include (1) the Bryant-Day approach of incorporating toxicity and efficacy jointly into the design of Phase II clinical trials; (2) the Obuchowski-Rockette method of analyzing multi-reader ROC studies; (3) a series of methodological papers coauthored with Mitchell Gail and other NCI statisticians to extend and validate a model to predict individualized risks of women for developing breast cancer; and (4) survival analysis with extensions to multiple time to events and nonproportional hazards. Although the scope and diversity of faculty research increased markedly during the 1990s, the number of matriculating students remained relatively constant with six to seven students per year. A list of selected publications during this time period can be found in an expanded version of this chapter at www.biostat.pitt.edu.

A Period of Growth: 2000–Present

The components were in place for a period of growth. The three longstanding major sources of funding (NSABP, Biostatistical Core of PCI, and studies of industrial populations) that supported both faculty and students were well integrated into the Department. There were more than a half dozen additional ongoing research collaborations with the Medical Center in areas each of which could provide partial support for one or two faculty members and several students. The VA Center for Health Equity Research and Promotion (CHERP) was funded in 2001, and quickly became a focus for studies on the quality and equity of healthcare within and outside the VA system. There was also a large core of faculty with training to teach advanced statistical methodology and mentor doctoral students. The Human Genetics Department developed several strong statistical geneticists (Dan Weeks and Eleanor Feingold) and the relationship between the two departments remained close. Cooperation continued to increase with the newly formed Statistics Department (see appendix). Conditions outside the School of Public Health also were conducive to growth. There was an increasing demand for biostatistics, the job market was strong, international students "discovered" the field of Biostatistics as a possible discipline distinct from Statistics, and the University of Pittsburgh Medical Center was showing tremendous growth in NIH funding which brought an increasing need for biostatistical collaboration.

The doctoral program was upgraded and the website underwent major revisions to emphasize the strong job market, the balance of training in both applications and methodology, and the opportunity to obtain "hands on" experience in the many collaborative research projects within the Department. By 2003, the number of applications to the Department had tripled. From 1989 to 2001, the average number of graduates per year had been 6.5. For the 5-year period, 2002–2006, the average number was 10.5 and since 2007, the average number has been 20 per year. Several other factors contributed to this rapid growth. The Department was awarded four NIH training grants during this period, one from NCI to train cancer biostatisticians, one from NIMH to train psychiatric biostatisticians, a general training grant from NIGMS to train biostatisticians for work in new, evolving areas of biomedical research, and a T-15 grant to introduce quantitatively oriented undergraduates to the area of Biostatistics. In addition, a relationship was developed with the Clinical Translational Science Institute, based in the Medical School, which resulted in the funding of three to five doctoral students per year. Finally, an arrangement was made with Feingold and Weeks to mentor some of the doctoral students in Biostatistics who expressed an interest in statistical genetics.

In 2004, Jim Schlesselman was recruited and appointed as Director of the Biostatistical Core of the Pittsburgh Cancer Institute. Schlesselman had been Director of the Biostatistical Core for the Miami Cancer Institute. This same year John Bryant, a statistician in the University of Pittsburgh's Department of Statistics who had been working with the PCI and the NSABP, transferred to become Professor of Biostatistics. Five new junior faculty members (including Jong-Hyeon Jeong, Abdus Wahed and George Tseng) were added as core teaching faculty. During this period, there was a tripling of the annual number of methodological publications and the acquisition of five externally funded grants in statistical methodology. The Department also appointed six new faculty members who would spend the majority of their time on the ongoing collaborative research efforts. Much of the research investigating public health problems in industrial populations was now in the Center for Occupational Biostatistics and Epidemiology (COBE), founded in 2007 by Marsh.

Sadly, the Department lost two highly regarded colleagues and friends when both Sam Wieand and John Bryant, who was Wieand's successor as Director of the

NSABP Biostatistical Center, died of cancer within a short time of one another in 2006. Their loss had both a profound personal and professional impact on our faculty. Together they had led the NSABP for more than a decade.

In 2007, Joseph Costantino, who had been the primary biostatistician for many of the major clinical trials conducted by the NSABP for the last 2 decades, was appointed as Director of the Biostatistical Center. In 2009, Rockette stepped down as Department Chair and Marsh was appointed Interim Chair. During this time period, the Department implemented an agreement to be an active participant in a newly formed school-wide Public Health Dynamics Laboratory (PHDL). The Director of the Center, John Grefenstette, has an appointment in the Department of Biostatistics, as do two newly appointed assistant professors.

Departmental collaboration in PORT had established a core faculty group with expertise in evaluation of comparative effectiveness of alternative clinical interventions on health outcomes. The 2010 recruitment of Sally Morton, who is known for her contributions in the area of evidence based medicine, provides an opportunity to expand the Department's role and research programs in this area. Morton is the first externally recruited chair since the School's formation. Currently, the Department has 32 primary faculty, 22 secondary faculty, and more than 100 full- or part-time students. The major externally funded research projects in the Department are maintaining their viability in a period of restricted resources, and the breadth of research interests in the Department produces the capability to adapt to a changing academic environment. The academic program has undergone review and appropriate modifications are being made. The Department is planning on further increasing the size of the faculty and the School is in the process of major renovations.

A primary challenge facing the department, shared by most departments of Biostatistics, is how to achieve the proper balance between the development of new, innovative statistical methodology and the demands of performing relevant, high-impact collaborative research and concurrently incorporating this balance into the training of our students. The faculty is optimistic about meeting this challenge given the strengths of the department and the University of Pittsburgh. (See expanded chapter at www.biostat.pitt.edu for selected publications during this period.)

Appendix

Creation of a Department of Statistics: 1968–2011

There was little organized research activity in statistics prior to 1968 at the University of Pittsburgh, outside of the Department of Biostatistics. There were isolated individuals in various departments either doing statistical research or with an interest in statistical research, but no center for such activity existed. Non-calculus introductory statistics courses were taught at the undergraduate and graduate level in several departments as a preparation for courses in research methodology in their fields, and the Department of Mathematics taught two-term calculus-based probability and mathematical statistics courses for undergraduates and for graduates.

In other universities, mathematical statistical research usually started in departments of Mathematics. An Italian probabilist, Giuseppi Pompilj, spent a few years in the Department of Mathematics in the 1960s, but until Henry Block was hired as an assistant professor in Mathematics in 1968, there was no one in mathematics interested in statistical research. In 1971, Asit Basu came as an Associate Professor from Northwestern, followed in the same year by Tom Savits. Basu was successful in inviting senior statisticians with research interests in reliability to visit for short stays, notably Milton Sobel, Frank Proschan and John Gurland. J. K. Ghosh was also a visitor for several years. The new chair of the department, Gene Deskins, encouraged development of this statistical research group and also supported the change of the name of the department to the Department of Mathematics and Statistics. In 1976, P. R. Krishnaiah was hired to coordinate statistical research in the University and to provide consulting. Krishnaiah in turn brought in C. R. Rao as a University Professor. Several other statistics faculty members were hired. Soon after Allan Sampson joined the group as associate professor in 1978, he and Block set up a comprehensive statistics program with a Bachelors of Statistics and Masters in Applied Statistics. By 1982, the group had grown and prospered to the point that the Conference Board of Associated Research Councils ranked the group as the most improved statistics group in the country, also ranking it in the top third of statistics groups. In that year, David Stoffer and Satish Iyengar were hired by the statistics group and the Center for Multivariate Analysis was established. The Journal of Multivariate Analysis was published out of the Center for Multivariate Analysis.

The Center of Multivariate Analysis was intended to be a focal point for statistical research in the University of Pittsburgh. Its permanent members were Rao and Krishnaiah (who served as director). A list of 30 team members prepared at the time includes seven members of the Department of Mathematics and Statistics, seven members of the Department of Biostatistics, three from Psychology, two members each from the Departments of Economics and Sociology, and one member from each of the departments of Electrical Engineering, Industrial Engineering, Psychiatry, Medicine, Anthropology, Neurological Surgery, and Radiology. Completing the list were T. L. Saaty (Graduate School of Business) and N. H. Timm (Planning and Budget, and author of a respected textbook on multivariate statistical analysis).

Some of the faculty members in the statistics group wanted to create a separate Department of Statistics, but Rao and Krishnaiah were concerned about additional administrative responsibilities (including an undergraduate major). However, Krishnaiah's premature death in the summer of 1987 and Rao's departure for Penn State (taking the Center for Multivariate Analysis with him), changed matters greatly. In 1989, Leon Gleser came as a full professor with the understanding that the statistics group would work to separate from the mathematicians and create a department of their own. Although the University administration agreed in principle, it took 8 years of hard work to complete the job. A key step was the consolidation of statistics service courses in the School of Arts and Sciences under the control of an Applied Statistics Education Committee, for which the new Department of Statistics took a leadership role. In January 1997, the Department of Statistics was officially created, with Allan Sampson as its first chair. In 1999, Satish Iyengar became the chair of the department, and with the exception of two 1-year leaves of absence, when Henry Block acted as chair, has continued to serve as chair to the present day.

Although the Department of Statistics was planned as a small department, there was expectation of some modest growth as enrollments grew and new research grants were obtained. Unfortunately, soon after the establishment of the department, a freeze on tenure-stream appointments in the School of Arts and Sciences was imposed. The department did add two non-tenure-stream faculty members, justified by an explosive growth in their service course enrollments. The two new faculty members thus added, Senior Lecturers Carl Bodenschatz and Nancy Pfenning, have since won prestigious Arts and Sciences and University-wide teaching awards. Dr. Bodenschatz also was responsible for reviving and growing the undergraduate major, and helped devise a new 5-year BS/MA Applied Statistics Degree that has attracted some outstanding students. In 2008, Dr. Gleser (who has acted as Graduate Director throughout the department's existence) received the Provost's Award for Mentoring of graduate students.

Even while they were in the Department of Mathematics, the senior faculty in the department had close collaborative and consulting ties with the Department of Psychiatry. Their collaborations grew, enabling the department to justify and create two joint 50–50 tenure-stream positions with Psychiatry. These positions have attracted a series of bright and creative young faculty interested in working in a cross-disciplinary environment: Xin Ming Tu, Hernando Ombao, Wesley Thompson, and currently Yu Cheng. Some of the senior faculty members also maintain close research collaborations with faculty and research groups in the Department of Psychiatry.

The Department of Biostatistics and the Department of Statistics also have maintained close ties. Three faculty members from the Department of Statistics have secondary (joint) appointments with Biostatistics. Students from the two departments can take courses from either department, and a course on life testing is cross-listed between the two departments. Faculty members from one department often serve on the dissertation committees of the other department. Also, some faculty members in the Department of Biostatistics were previously faculty members (Bryant, Wieand) or students (Weissfeld) in the statistics group, before it became a department. Greg Yothers, a faculty member in the Department of Biostatistics, and Kaleab Abebe, who has a secondary appointment in Biostatistics, both received their PhDs from the Department of Statistics. The two departments were rated together in the recent ratings of departments of statistics done by the National Research Council.

In its 14 ½ years of existence (1997–2011), the Department of Statistics has graduated 50 PhDs, 99 masters degrees and 63 BS degrees in Statistics, and has made significant research contributions in time series, function analysis, reliability and life testing, statistical dependence and statistical inequalities, adaptive clinical trials, measurement error models and inference, statistical meta-analysis, and modeling and inference in neuroscience (see www.stat.pitt.edu for further details). One of its faculty members (Gleser) was Executive Editor of *Statistical Science* from 1998 to 2000, and another (Stoffer) recently completed a 2-year term as Program Director of the Statistics Division, Mathematical Sciences, of the National Science Foundation. The department is also well represented on editorial boards and FDA, NIH and NSF review panels. Thus, although the department remains small (eight tenure-stream faculty, two teaching faculty), its impact on research and training in the field of statistics is substantial.

University of Washington Department of Biostatistics

Norman E. Breslow

Origins

Between 1957 and 1974 the extramural grants program of the National Institutes of Health (NIH) grew 10-fold, from $100 million to $1 billion. These "golden years" of NIH expansion fuelled biomedical research throughout the country, not least at the University of Washington (UW) School of Medicine. When epidemiologist Tom Grayston became Chair of Preventive Medicine in 1960, he set out to develop research to the level of other medical departments. Realizing the need for an expanded Biostatistics Division, he recruited a recent Stanford statistics graduate then at Pittsburgh who had been recommended by Doug Chapman, a well-known statistician in the Mathematics Department.[1] Ed Perrin arrived in 1962 to start building the new program.

Chapman and Perrin organized the Biomathematics Group in 1963 with faculty from Biophysics, Chemistry, Fisheries, Genetics, Mathematics, Medicine, Preventive Medicine, and Zoology. Their proposal titled "Program Leading to Degrees of M.Sc. and Ph.D. in Biostatistics" was accepted in 1964 and the Group, an interdisciplinary program administered by the Graduate School, received degree granting authority. Chapman was the first Group Chair (1964–1969),

[1] See A History of the Statistics Department at the University of Washington, pp. 511–524, for Chapman's many important contributions.

University of Washington,
Department of Biostatistics,
PO Box 357232 Seattle,
WA 98195-7232, USA
N. E. Breslow (✉)
e-mail: norm@u.washington.edu

A. Agresti and X.-L. Meng (eds.), *Strength in Numbers: The Rising of Academic Statistics Departments in the U.S.*, DOI: 10.1007/978-1-4614-3649-2_36,

followed by Perrin (1969–1972). Enrollment increased from 3 students who transferred into the program at its start to 36 in 1972.

Both Chapman and Perrin had access to funds that allowed them to attract students and recruit faculty to their respective academic departments. Chapman, who held a joint appointment in Fisheries, created a Center for Quantitative Science (CQS) in Fisheries, Forestry, and Wildlife with support from the Ford foundation. He later became Dean of Fisheries (1971–1980). Chapman and Perrin together administered an NIH biometry training grant. Two major "pathways" leading to a Biomathematics degree evolved, one in Biostatistics and the other in Quantitative Ecology and Resource Management (QERM). This chapter concerns primarily the Biostatistics pathway and the Biostatistics Division (later Department) whose faculty played the largest role in its development.

The Biostatistics Division in the Department of Preventive Medicine grew rapidly. Dick Kronmal, a UCLA biostatistics graduate recruited in 1964, developed a departmental computer center that was invaluable for processing the large quantities of data being collected by Grayston, and other epidemiologists. This and his Conversational Computer Statistical System[2] (CCSS) for medical records management laid the foundation for later success in competing for major coordinating centers. In 1966, Perrin recruited from Pittsburgh his previous boss and mentor Donovan Thompson (Fig. 1), a sampling expert who had co-developed the inverse probability weighting method now known as Horvitz-Thompson estimation. Thompson's broad experience in public health and NIH funding sources was of great value to the entire department. Norman Breslow, another Stanford graduate, joined the faculty in 1968, succeeded Perrin as statistician for the Children's Cancer Study Group, and became a founding member of the National Wilms Tumor Study Group. Polly Feigl was recruited the next year from the University of Pennsylvania Medical School to consult with medical faculty.

Grayston led the transformation of Preventive Medicine into a School of Public Health in 1970, his faculty having grown from 4 to 40 during the decade. Perrin became Chair of the Department of Biostatistics and Paula Diehr, another UCLA graduate, was its first faculty addition. Diehr focused her research on Health Services and eventually held a joint appointment there. Pat Wahl (PhD, 1971, Kronmal) joined the department after graduation, established herself as an outstanding teacher and researcher in cardiovascular disease, and ultimately became Dean of Public Health (1997–2010). With Pat, Paula, and Polly on the faculty, the future role of women in the new department was well secured.

[2] Gary Anderson (PhD, 1969, Kronmal) further developed CCSS into the commercial Scientific Information Retrieval system.

Fig. 1 Donovan Thompson, 1982

Donovan Thompson and the Formative Years

The decade of the 1970s was a formative period for Biostatistics. When Perrin left in 1972 to head the National Center for Health Statistics, Feigl served as interim Chair of Biostatistics while Thompson served as interim Dean of Public Health. Thompson then became Biostatistics Department Chair (1973–1983) while Kronmal headed the Biomathematics Group (1973–1985). A beloved mentor who created a lasting spirit of openness and collegiality, Thompson was also a visionary and entrepreneur who recognized and stimulated the demand for biostatisticians. He soon found opportunities for former Mathematics faculty, recruiting Lloyd Fisher, who later joined Kronmal to found the landmark coordinating center for the Coronary Artery Surgery Study (CASS), in 1972 and Al Hallstrom, who likewise coordinated studies of Seattle's famed emergency medical services program Medic 1, in 1974. These coordinating centers were important as a mechanism for faculty participation in major biomedical research projects, as a stimulus to statistical methods research, as an avenue for training and support of graduate students and, through indirect cost recovery, as a means to support staff and infrastructure.[3]

Thompson worked with local surgeon Bill Hutchinson to build a Program in Epidemiology and Biostatistics, now the Division of Public Health Sciences (PHS), at the Fred Hutchinson Cancer Research Center (FHCRC). He recruited Ross Prentice (1974) from Waterloo to be the first Biostatistics faculty member employed by the center and assisted in the recruitment and mentoring of other FHCRC biostatisticians and epidemiologists. Prentice succeeded Thompson as Director of PHS (1983–2002; 2007-). Thompson recruited Gerald van Belle (1974) from Florida State, another outstanding teacher who later chaired the School's

[3] See Appendix D in http://www.biostat.washington.edu/sites/www/content/files/Department_History/HomeNav/About_Us:_History_of_the_Department/app_d_pdf_4de3fbcb57.pdf for a partial listing of coordinating centers.

Department of Environmental Health (1990–1998), Kathryn Davis (PhD, 1974, Kronmal) who succeeded Fisher at CASS, Tim Derouen (1975) from Tulane, who became Chair of Dental Public Health Sciences (1986–2003) and Art Peterson (1975) from Stanford, who later led the FHCRC program in youth smoking research. Breslow was recruited back (1974) from the International Agency for Research on Cancer (IARC). Besides their important academic contributions, these recruits all eventually served as directors of training grants or of major coordinating centers or both. Chuck Bell, the department's first African American, joined the faculty in 1977. Nancy Temkin took a joint appointment with Neurological Surgery and Biostatistics that same year.

The number of students enrolled in Biomathematics Group degree programs also grew quickly, increasing from 36 in 1972 to 80 in 1974. The core PhD curriculum included courses taught by Mathematics faculty in real variables, probability, and statistics; by Biostatistics faculty in medical biometry and applied statistics; and by Fisheries and Forestry faculty in mathematical modeling. The Biostatistics PhD pathway was unusual in its emphasis on a strong foundation in statistical theory and methodology as well as exposure to applications. Biostatistics electives reflected faculty interests and expertise: statistical computing; methods for survival and categorical data; design of medical studies. Qualifying exams were administered jointly by faculty in Mathematics, Biostatistics, and Fisheries. Graduates of the program were as varied and outstanding as their faculty mentors. They included renowned mathematical statisticians such as Jon Wellner (PhD, 1974, Shorack) and Bruce Lindsay (PhD, 1978, Breslow). Wellner, Professor of Statistics and Biostatistics at the UW since 1983, is a major figure in empirical processes and semiparametric inference. Lindsay, since 1992 Willaman Distinguished Professor of Statistics at Penn State, is an expert on likelihood inference. Early graduates also included outstanding biostatisticians such as John Crowley (PhD, 1973, Breslow), William Fairweather (PhD, 1973, Thompson), and Loveday Conquest (PhD, 1975, Kronmal). Crowley began a distinguished academic career at Wisconsin, then returned to the UW and FHCRC where he directed the Statistical Center for the Southwest Oncology Group (SWOG) and founded the nonprofit Cancer Research and Biostatistics (CRAB). Fairweather spent 25 years at the Food and Drug Administration (FDA), where he became Associate Director for Epidemiology and Biostatistics in the Center for Drug Evaluation and Research, garnering several FDA awards before founding his own consulting firm. Conquest joined the UW Fisheries faculty, won several teaching awards, and currently directs the QERM interdisciplinary graduate program.

Through his contacts at the National Academies, Thompson secured a contract to recruit US biostatisticians, and epidemiologists to work with Japanese counterparts at the Radiation Effects Research Foundation (RERF) in Hiroshima. Several department faculty and students spent time at RERF as did many others from around the country and abroad. With Thompson's encouragement, Prentice founded in 1979 the Statistics and Epidemiology Research Corporation (SERC)

Fig. 2 Department faculty in the 1979–1980 academic year. Bottom row (L to R): Chuck Bell, Mary Jo Gillespie (staff), Dick Kronmal, Nancy Temkin, Tim Derouen, Kathryn Davis, Polly Feigl. Middle row (L to R): Paula Diehr, Donovan Thompson, Pat Wahl, Ed Davis (visitor), Art Peterson, Al Hallstrom. Top row (L to R): Gerald van Belle, Vern Farewell, Lloyd Fisher, Norm Breslow, Joel Brodsky, Ross Prentice

whose shareholders were largely UW and FHCRC faculty. Initially established to garner federal research funds earmarked for small business, SERC also proved useful for faculty consulting, for software development, and for sponsorship of short courses. It later evolved into the contract research organization AXIO.

By the end of the decade the department was well recognized for its excellence in development of new statistical methods motivated by applications in clinical trials and epidemiology. These included failure time analyses related to the Cox model and (conditional) logistic regression for the analysis of (matched) case–control data. Breslow and Day's first IARC monograph, The Analysis of Case–Control Studies, was published in 1980 as was the first edition of Kalbfleisch and Prentice's Statistical Methods for Failure Time Data. To celebrate, the department organized through SERC a short course series featuring the two 1980 texts. An oral history of this remarkable period, containing interviews with many of the faculty founders, is available through the School's video archives.[4] Figure 2 shows the faculty in the 1979–1980 academic year.

[4] http://sph.washington.edu/history/videos.asp

Table 1 Biomathematics and biostatistics degrees awarded

Years	MS	PHD
Biomathematics graduates		
1965–1969	8	2
1970–1974	16	13
1975–1979	48	21
1980–1984	23	21
1985–1986	6	8
TOTALS	101	65
Biostatistics graduates		
1986–1989	25	13
1990–1994	33	31
1995–1999	24	26
2000–2004	37	40
2005–2009	36	31
Totals	155	141

Transfer of the Biomathematics Degree to the Biostatistics Department

With financial support from Public Health Dean Robert Day, who was later Director of FHCRC, Biostatistics became the *de facto* administrative home for the entire Biomathematics Group. Biostatistics faculty and students came to dominate Group activities. Some students who were initially attracted to Biomathematics by the prospect of working in quantitative ecology and related fields found that faculty resources and job opportunities were limited, and they eventually pursued a more traditional Biostatistics curriculum. This change in the Biomathematics Group's interdisciplinary character upset some faculty; Professor Joe Felsenstein of Genetics resigned in protest. Nonetheless, the reputation of the department for high quality methods research, and of the degree program for well-trained graduates, continued to grow. A 1982 National Research Council evaluation of 64 doctoral programs in Statistics/Biostatistics ranked the UW Biomathematics/Biostatistics program second highest for "improvement in program quality." Stanford and Chicago were rated first and second for overall scholarly quality of the faculty. By 1995, UW Biostatistics was ranked sixth nationally in overall quality, and second among biostatistics programs.

When appointed Chair in 1983, Breslow set out to regularize the situation for Biostatistics students so they obtained a Biostatistics degree. According to a 1985 review by the Association of Schools of Public Health, the UW by then produced 16 % of all Biostatistics PhD's in the country. While some faculty lamented the loss of the interdisciplinary Group structure, the majority felt it was time for the Department to award its own degrees as did similar departments in Schools of Public Health at Harvard, Hopkins, UCLA, and UNC Chapel Hill. This was accomplished in 1986 with the dissolution of the Biomathematics Group and transfer of its degree granting authority to Biostatistics (Table 1). QERM students received degrees through Biostatistics for an interim period until 1990, when a new interdisciplinary program was established within the Graduate School to serve their needs.

After formation of the Statistics Department in 1980,[5] Biostatistics and Statistics faculty continued the cooperation they had enjoyed within the Biomathematics Group structure by cross-listing courses, sharing in curriculum development, and jointly administering exams. This cooperation contributed to the continued emphasis in the Biostatistics PhD curriculum on solid grounding in statistical theory.[6] The unique and genuinely close relationship between the two departments, which expanded resources available to students in both, was further solidified via a series of joint appointments. Wellner and Elizabeth Thompson, who accepted a joint appointment in 1988, both mentored several Biostatistics students while fully funded by Statistics. Kronmal was Biostatistics' first joint appointee. Statistics, as part of a planned expansion through joint appointments, chose Biostatistics as partner for their first "bounty" position. Finbarr O′Sullivan was recruited from Wisconsin via Berkeley to be its first occupant; since 1999 Jon Wakefield, formerly at Imperial College London, has occupied the position.

Continued Expansion

Biostatistics continued to expand during the 1980s, with many faculty recruited into Research or Regular, without-tenure-by-reason-of-funding (WOT) positions to accommodate the continuing demand for statistical input into biomedical research. Jackie Benedetti (PhD 1974, Kronmal), for example, returned to the department from UCLA in 1980 with a joint appointment in medicine, where she provided statistical support for research on sexually transmitted disease (STD). Stephanie Green was recruited in 1986, primarily to work at SWOG. Benedetti joined Green and Crowley at SWOG in 1991 and eventually wrote with them Clinical Trials in Oncology (1997, 2002).

There was increasing demand for basic instruction in Biostatistics from graduate and professional students in Public Health and throughout the Health Sciences. A three quarter service course series in Medical Biometry was organized in which Wahl and van Belle played leading roles, for which van Belle received the university's prestigious Distinguished Teaching Award in 1985. Fisher and van Belle developed a text, Biostatistics: A Methodology for the Health Sciences, first published in 1993. Van Belle later wrote Statistical Rules of Thumb (2002). A two quarter sequel in methods for analysis of categorical and survival data in

---

[5] See A History of the Department of Statistics at the University of Washington, pp. 511–524.

[6] Indeed, the second year core course in statistical theory remains a requirement for the Biostatistics albeit not the Statistics PhD.

epidemiology was organized around the Breslow-Day IARC monographs, the second of which, *The Design and Analysis of Cohort Studies*, appeared in 1987. Barbara McKnight, a Wisconsin student of Crowley's recruited in 1982, taught in both series and received the School's Outstanding Teaching Award in 1988 and 2006. More recently, the categorical and survival analysis courses for second year MS Biostatistics and PhD Epidemiology students have used as primary texts two books co-authored by Dave Hosmer (PhD 1971, Fisher) and Stan Lemeshow, with the Breslow-Day books as backup. Susanne May, a student of Hosmer's recruited from UCLA as Associate Professor in 2008, joined Hosmer, and Lemeshow in writing the second edition of Applied Survival Analysis: Regression Modeling of Time to Event Data (2008).

Tom Fleming was recruited from the Mayo Clinic as Professor in 1984 and had a joint appointment in Statistics after 1987. He brought strengths in clinical trials, experience with FDA advisory committees, and important contacts with the pharmaceutical industry. He developed a highly popular course on study design that paired MDs and Biostatistics students in writing project proposals and conducting mock site visits. He led the department's research program in HIV/AIDS as PI on methodology grants, training grants, and coordinating centers. Fleming's 1991 text with Harrington, Counting Processes and Survival Analysis, served him and others well in teaching advanced courses. During his tenure as Department Chair (1993–2006) he oversaw the establishment of several endowments: for the Genentech Professorship; for the Prentice Professorship, designed to enhance teaching opportunities for FHCRC faculty; and for the Breslow Lectureship.[7] Together with 1990 faculty recruit Danyu Lin, he organized the first Seattle Symposium in Biostatistics: Survival Analysis (1995). This highly successful series, highlighting faculty strengths and interests, continued with Analysis of Correlated Data (2000), Statistical Genetics and Genomics (2005), and Clinical Trials (2010).

Additions to the faculty during the 1990s included: Brian Leroux from British Columbia in 1991; David Yanez from Arizona State, Jim Hughes (PhD 1993, Guttorp, Statistics), Mary Emond (PhD 1993, Self), and Lianne Sheppard (PhD 1992, Prentice) in 1993; Patrick Heagerty from Hopkins and Scott Emerson (PhD 1988, Fleming) from Arizona in 1995; Barbra Richardson from UCLA and Mary Lou Thompson from South Africa in 1996; and Thomas Lumley (PhD 1998, Heagerty) in 1998. Leroux, who took a joint appointment with Dental Public Health Sciences, is now Professor in both departments and an expert in multilevel and longitudinal data analyses with applications in oral health. Hughes, longtime instructor for the large, introductory service course, won the School's Outstanding Teaching Award in 2007. He and Richardson are lead biostatisticians for Seattle's STD research centers. Heagerty, another School Outstanding Teacher (2009), participated as co-author with Diggle, Liang and Zeger (his Hopkins mentor) in the

---

[7] See Appendix E in http://www.biostat.washington.edu/sites/www/content/files/Department_History/HomeNav/About_Us:_History_of_the_Department/app_e_pdf_4de3fbcb58.pdf for a listing of Genentech and Prentice Professors and Breslow Lecturers.

second edition of the influential advanced text *Analysis of Longitudinal Data* (2002), writing in particular the chapter on causal inference. His other research interests include marginally specified mixed models and ROC methods for survival data. Heagerty and Lumley joined Fisher and van Belle as co-authors of the second edition of the basic biostatistics text, while Lumley wrote Complex Surveys: *A Guide to Analysis Using R* (2010) based on his software development. Emerson succeeded McKnight as Graduate Program Advisor during Fleming's chairmanship; van Belle and Feigl had occupied this important position during Breslow's tenure as Chair, while Yanez has directed the graduate program since 2005. Sheppard, an expert on health effects of air pollution, is now Research Professor of Biostatistics and Environmental Health.[8]

Affiliate Institutions

The importance of the UW-FHCRC partnership to graduate student training and support, and to the enhancement of the research reputations of both institutions, would be hard to overemphasize. Several department graduates joined the Research or Regular WOT Biostatistics faculty ranks with primary appointments at FHCRC, usually after spending time elsewhere: Steve Self (PhD 1981, Prentice) in 1984; Margaret Pepe (PhD 1986, Fleming) in 1988; Michael Leblanc (PhD 1989, Crowley) in 1995 and Peter Gilbert (PhD 1996, Self) in 2002. Although seen only occasionally in the classroom, consistent with their funding source, Prentice, Self, Pepe, and Gilbert together with other FHCRC faculty played a key role as dissertation advisors and as employers and mentors to student research assistants. Since 2000, Self has directed the Statistical Center for HIV/AIDS Research and Prevention (SCHARP) and since 2007 served as co-Director of the FHCRC Vaccine and Infectious Disease Division (VIDD). Pepe has involved dozens of students and young faculty in her research programs, particularly on ROC methodology as summarized in her 2003 book *The Statistical Evaluation of Medical Tests for Classification and Prediction*. Gilbert, who works with Self at SCHARP, has already chaired five graduate committees. Pepe and Gilbert both held Prentice Professorships. Several former department faculty and postdocs have also gone on to FHCRC positions, and PhD biostatisticians at FHCRC now rival the department in number. In view of limited classroom teaching opportunities, and the administrative complexities of recruitment and promotion in the Research or Regular WOT ranks, most FHCRC PhD biostatisticians now hold university appointments as Affiliate Professors. This by no means precludes their participation in department activities. For example, Ying Qing Chen, Affiliate Associate Professor

---

[8] See Appendix A in http://www.biostat.washington.edu/sites/www/content/files/Department_History/HomeNav/About_Us:_History_of_the_Department/app_a_pdf_4de3fbcb4b.pdf for a listing of faculty in the Division or Department of Biostatistics.

recruited to FHCRC/SCHARP from Hopkins in 2004, often teaches the advanced course in survival analysis and chairs a PhD committee.[9]

Another important research affiliation was with the Center for Health Studies, now the Group Health Research Institute, at Seattle's Group Health Cooperative (GHC). Bill Barlow (PhD 1986, Prentice) joined the GHC staff as Assistant Scientific Investigator in 1989 with concurrent appointment as Affiliate Assistant Professor in Biostatistics. A frequent and gifted instructor of service courses, and a School Outstanding Teacher (1996), he moved to the Research faculty ranks in 1993 and became Senior Research Biostatistician at CRAB in 2003. Five other GHC biostatisticians, two of them department graduates, currently hold affiliate faculty appointments. Andrew Zhou, now Professor of Biostatistics and director of a training grant for biostatistics in mental health, was first recruited in 2002 as Affiliate Associate Professor with primary appointment as Director of Biostatistics at Seattle's VA Puget Sound Health Care System. He is the lead author of *Statistical Methods in Diagnostic Medicine* (2002, 2011).

Collaboration with Medical School Faculty

Biostatistics departments are often located in schools of public health for historical reasons, yet their expertise is needed throughout the health sciences for research consultation. At the UW this need was partly satisfied through a consulting class, run jointly by Statistics and Biostatistics, in which researchers (students or faculty) from throughout the university met with student statistical consultants and a faculty mentor in one-hour sessions. This service, adequate for dealing with smaller problems, was provided free of charge as a training program for students. At the other end of the spectrum, Biostatistics and Biomedical faculty collaborated on large, grant supported clinical trials and observational studies from which they both received support. Yet there remained a need for sustained statistical collaboration while developing research proposals and for faculty level statistical input into smaller, unfunded medical studies. Unfortunately, since the loss of capitation funds that supported Feigl as consultant to Medicine during her early years, UW Medical School deans had not responded favorably to repeated requests that they allocate a small quantity of general funds to support faculty level consultation. In 1986, the department provided seed money for a Biostatistics Consulting Unit (BCU) headed by a recent PhD graduate who set up a cost center, with hourly charges for his time as consultant and for staff time for data management. This failed within a year or two due to the lack of professional rewards for the consultant and the stresses of providing fast results just prior to grant submission

[9] See Appendix B in http://www.biostat.washington.edu/sites/www/content/files/Department_History/HomeNav/About_Us:_History_of_the_Department/app_b_pdf_4de3fbcb56.pdf for a listing of adjunct and affiliate faculty.

deadlines, and the BCU reverted to the data management service that Kronmal had earlier organized.

The problem remained unsolved until Fleming led a working group composed of Biostatistics and Medical faculty that in 2005 submitted a proposal for a Center for Biomedical Statistics (CBS) to the Deans of Public Health and Medicine. The proposal was necessarily vague about administrative structures and financial models, but strongly made the case that a mechanism was needed to promote biostatistical collaboration in both basic and clinical medical science. The CBS was established within the department in 2007 with Heagerty as Director and with Emond and a recent PhD graduate at Seattle Children's hospital as faculty consultants, all part-time, plus two PhD and three MS staff biostatisticians and two student research assistants. Core support was provided by an NIH funded Institute of Translational Health Sciences (ITHS) involving the UW, FHCRC, Seattle Children's, GHC, and other regional research institutions. Fees were charged for consultation beyond minimum limits which depended on whether or not a CBS biostatistician was included in a grant proposal. The CBS has succeeded in providing interesting, funded projects for student support and stimulus for methods developments by faculty. A survey of ITHS investigators after 2 years of operation gave the CBS highest marks for "satisfaction" among a dozen core units and the third highest rating overall, behind Administration and Research Education. The department is hopeful that it has finally found the proper mechanism to promote collaboration between Biostatistics faculty and biomedical researchers, to the ultimate benefit of both.

Path to the Future

Recruitment in 2006 of the department's current chair, the prominent statistical geneticist Bruce Weir, highlighted a broadening from its traditional strengths in statistical aspects of clinical trials and epidemiology toward new strengths in statistical genetics that had been building for some years. Soon after her 1987 arrival from Stanford, Ellen Wijsman, now Professor of Medical Genetics and Biostatistics, worked with Elizabeth Thompson on a popular Statistical Genetics seminar for students in Biostatistics and Statistics. She developed a course on statistical methods in medical genetics featuring linkage analysis and chaired a Biostatistics PhD committee within the framework of the traditional degree program. The recruitment of Stephanie Monks (1999) from NCSU, Katie Kerr (2001) from UCLA and John Storey (2003) from Stanford brought additional faculty resources in genetics and needed support for a separate Statistical Genetics Program, which began in 2000 with PhD tracks in Statistics and Biostatistics. Biostatistics faculty who currently participate in the program include Kerr, McKnight, Thompson, Wakefield, Weir, and Wijsman together with recent recruits Tim Thornton (2009) from Chicago via Berkeley and Sharon Browning (2010) from Auckland. Other faculty participants have primary appointments in Medical Genetics and in Genome Sciences, where Weir and Thompson hold joint

Fig. 3 Department faculty in the 2010–2011 academic year. Bottom row (L to R): Pat Wahl, Lyndia Brumback, Nancy Temkin, Jackie Benedetti, Bruce Weir, Lurdes Inoue, David Yanez, Jon Wakefield. Second row (L to R): Dick Kronmal, Daniela Witten, Ira Longini, Barbara McKnight, Robyn McClelland, Lianne Sheppard, Sharon Browning, Betz Halloran. Third row (L to R): Tom Fleming, Elizabeth Brown, Ellen Wijsman, Kathleen Kerr, Mary Lou Thompson, Ken Rice, Barbra Richardson. Top row (L to R): Tim Thornton, Norm Breslow, Jim Hughes, Susanne May, Art Peterson, Scott Emerson, Adam Szpiro, Patrick Heagerty

appointments. The Summer Institute in Statistical Genetics that Weir directed at his longtime base at NCSU, has since moved to Seattle and provides summertime teaching opportunities. Daniela Witten from Stanford and Ali Shojaie from Michigan, both recruited in 2010, bring new strengths in the analysis of high-dimensional data that should enhance the department's contributions to genome sciences in future years. Together with Thornton, Gary Chan (2008) from Hopkins and Adam Szpiro (2009) from MIT, they bring to five the number of Assistant Professors recruited during the past 3 years. Robyn McClelland (PhD 2000, Kronmal), Associate Director of the umbrella Collaborative Health Studies Coordinating Center, joined the faculty as Research Associate Professor in 2010.

Department strengths in Bayesian statistics, initiated with Wakefield's appointment, also developed markedly during the past decade with the recruitment of Elizabeth Brown from Harvard and Lurdes Inoue from Duke in 2002, Peter Hoff (joint, with primary in Statistics) from Wisconsin in 2003 and Ken Rice from Cambridge UK in 2004. Wakefield's work emphasizes applications in spatio-temporal epidemiology, Brown's the joint analysis of longitudinal-survival data, and Rice's foundational issues at the interface of frequentist and Bayesian inference. Hoff authored *A First Course in Bayesian Statistical Analysis* while Inoue and her mentor wrote *Decision Theory: Principles and Approaches*, both published in 2009. Departmental resources in the statistics of infectious disease took an equally dramatic leap in 2006 with recruitment from Emory of the high-profile team of Betz Halloran and Ira Longini, both as Professors of Biostatistics with primary appointments at FHCRC/VIDD. Frequently consulted by national and international authorities for their expertise in modeling disease outbreaks, they published *The Design and Analysis of Vaccine Studies* with Struchiner in 2009. Both have participated in classroom teaching via appointments as Prentice Professor and served as faculty on a new NIH funded Summer Institute for Statistics and Modeling in Infectious Diseases that Halloran directs. Figure 3 shows the faculty in the 2010–2011 academic year.

In summary, after 40 years of operation the UW Department of Biostatistics has an outstanding, diverse faculty, a large and vibrant graduate program, and continuing close relationships with UW Statistics, FHCRC, GHC, Puget Sound VA, and other Seattle institutions that provide excellent opportunities for research collaboration. It counts among its graduates many leaders in statistical science in this country and abroad: Professors at Harvard, Hopkins, Michigan, USC and in London, Milan, Prague and Stockholm, to mention but a few, including chairs at Emory, Nebraska, Penn State, and UNC Chapel Hill; Directors of two Biometry Branches at the National Institutes of Health; and President of a national medical university in Taiwan. Numerous visitors, including several eminent statisticians and epidemiologists, have chosen to spend time in the department.[10] In 2011, *US News and World Report* ranked the academic quality of department faculty 3–4 (tie) among the faculties of 67 biostatistics/statistics graduate programs throughout the country. With the recent recruitment of new faculty with new skills and interests, the department is well positioned to maintain its record of educational and scientific excellence.

[10] See Appendix C in http://www.biostat.washington.edu/sites/www/content/files/Department_History/HomeNav/About_Us:_History_of_the_Department/app_c_pdf_4de3fbcb57.pdf for a partial listing of visitors.

University of Washington Department of Statistics

Michael D. Perlman, Fritz W. Scholz and Galen R. Shorack

Pre-history (1939–1971)

The modern era of Statistics at the University of Washington (UW) began in 1939 with the arrival of Z. W. (Bill) Birnbaum in the Department of Mathematics. There was but a single course in descriptive statistics. Bill began the process that led to the development of one of the first comprehensive undergraduate statistics programs in the United States.

Born in Lwów, Austria–Hungary, in 1903, Birnbaum studied mathematics there under Steinhaus (his major professor for his 1929 PhD) and Banach and then continued his studies at Göttingen—still central to world mathematics—and enjoyed contact with many of its luminaries. After working as an actuary, Bill emigrated to the U.S. and accepted a research assistantship at New York University. Contact there with Seattle native Harold Hotelling of Columbia led to his appointment at UW Mathematics—with recommendation letters from Courant, Landau, and Albert Einstein. By 1948 he had founded the Laboratory of Mathematical Statistics to serve as a "focus of statistical activity within the university." In particular, the physical component of this lab would soon hold a room full of the then-state-of-the-art Friden calculators. Bill also instituted the Laboratory of Statistical Research in 1948, through its long association with the Office of Naval Research, funding for the graduate students and faculty components of the statistics program at Washington was strengthened and expanded. He also served as a consultant at Boeing, and subsequently at the Boeing Scientific Research Laboratories (BSRL) from its inception in 1962 until its demise in 1971. That

University of Washington,
Department of Statistics,
Box 354322, Seattle, WA 98195-4322, USA

M. D. Perlman (✉) · F. W. Scholz · G. R. Shorack
e-mail: michael@stat.washington.edu

A. Agresti and X.-L. Meng (eds.), *Strength in Numbers: The Rising of Academic Statistics Departments in the U.S.*, DOI: 10.1007/978-1-4614-3649-2_37,

association would include joint seminars and research papers involving Birnbaum and UW Mathematics probabilist Ron Pyke (Birnbaum PhD 1956) as well as UW Mathematics (statistics) graduates Sam Saunders (Birnbaum PhD 1956) and Al Marshall (Birnbaum PhD 1958) at BSRL. Bill's work included distribution-free statistics, reliability, and competing risks. He served as editor of the *Annals of Mathematical Statistics* during 1967–1970.

Douglas Chapman joined UW Mathematics in 1949. A Canadian fresh from a PhD at Berkeley (under Neyman), he was a key figure in the development of the program in statistics within Mathematics until his relocation to the UW College of Fisheries in 1968, where he served as dean from 1971 to 1980. Chapman's prime interests centered on wildlife statistics (especially estimating marine mammal populations) and population dynamics, with other papers on more traditional topics such as estimation within gamma models and distribution-free statistics. He chaired the International Whaling Commission's Scientific Committee from 1965 until 1974 during a period when new methods were developed to assign whale quotas. He warned early on that the numbers being taken by whalers in the 1960s were not sustainable. He also chaired the U.S. Marine Mammal Commission's Scientific Committee from 1973 to 1976 and the entire commission from 1976 to the early 1980s.

Significantly, Chapman and Ed Perrin organized the Biomathematics Group in 1963 (Doug served as its first chair, 1964–1969). It was this unit that would solidify the close cooperation between statisticians on the upper (arts and sciences) and lower (medical, fisheries/forestry) campuses. Early on, a majority of the students in graduate courses in statistics (within Mathematics) came from lower campus.

Ronald Pyke (another Canadian), after spending 2 years each at Stanford and Columbia, joined the UW mathematics faculty in 1960—thus creating a significant nucleus of high visibility. Pyke was a probabilist with broad interests at the interface of probability and statistics and provided a real sense of energy within the Statistics Group. Indeed, much of the day-to-day contact and organization of group efforts within the Statistics Group was directed and energized by Ron. His interests included distribution-free statistics, Markov renewal processes, spacings, empirical processes, and Brownian sheets. He served as the first editor of the Annals of Probability, 1972–1975.

The 1960s brought several significant additions (Table 1) to the Statistics Group in Mathematics. Don Ylvisaker (Stanford PhD under E. Parzen; time series, reproducing kernel Hilbert spaces) arrived in 1961. Galen Shorack (Stanford PhD under L. Moses) arrived in 1965. Pyke sat in on his first nonparametrics class, leading to their well-known joint 1968 and 1969 papers on applications of empirical processes to asymptotic properties of rank statistics. Other Shorack interests were linear combinations of functions of order statistics, robustness, and reliability; he began his 17-year work on empirical processes that later became the now classic monograph "Empirical Processes with Applications to Statistics" (1986) with Jon Wellner, a graduate of the UW Biomathematics program in 1975.

Table 1 Statistics Group and probabilists in Mathematics 1939–1971

Statistics Group	Probabilists
Z. W. Birnbaum (1939–1974)	Robert Blumenthal (1956–1997)
Douglas G. Chapman (1949–1968)	Ron Getoor (1956–1966)
Robert Tate (1953–1965)	David Newman (1961–1965)
Ronald Pyke (1960–1998)	John Woll (1961–1966)
Donald Ylvisaker (1961–1968)	Ramesh Gangolli (1962–1997)
Galen R. Shorack (1965–)	Robert Smythe (1969–1976)
John van Ness (1966–1970)	Y.K. Chan (1969–1977)
Lloyd Fisher (1966–1972)	

In 1966, both John van Ness (Brown PhD under M. Rosenblatt; time series) and Lloyd Fisher (Dartmouth PhD under J. Lamperti; probability theory) joined the Mathematics faculty. Both moved toward statistics, especially Fisher who sought to teach all available statistics courses.

Further developing the undergraduate program, this group also established both Master of Science (thesis) and Master of Arts (nonthesis) degrees in Statistics within UW Mathematics. Other Math and Biostatistics faculty that contributed to statistics and probability teaching included Norm Breslow, Chuck Bell, Al Hallstrom, Ramesh Gangolli, Doug Lind, Dan Wulbert, Steve Monk, John Westwater, Barney Glickfeld, Jim Jans, Robert Nunke, Kitty Baxter, and visitors Stan Sawyer and Aziz Laghrari.

Chapman was the first to see the need for a separate department of statistics. In 1967, he polled the Statistics Group to see if anyone else was ready. In the wake of a lukewarm response, he prepared for his exit in 1968, which would lead to the formation of the Center for Quantitative Science (CQS) in Forestry, Fisheries, and Wildlife. He established his new home in the College of Fisheries—a natural choice, based on his long relationship and collaboration with members of that college. He would serve as Dean of Fisheries 1971–1980 and as Director of CQS 1980–1982.

After Chapman's departure from Mathematics, the other members of the Statistics Group soon realized that it was indeed time to form a separate department for statistics. This endeavor was led by Ron Pyke. By November 1969, an Ad Hoc Committee on Statistics appointed by Phillip Cartwright, Dean of the College of Arts and Sciences (A&S), recommended that a Department of Statistics be established. The committee stated that the purpose of the new department would be "to provide a more readily identifiable focus of mathematical statistics and probability on campus." It would be charged to work closely with Mathematics, CQS, and Biostatistics.

On November 5, 1970, A&S Acting Dean William Phillips announced that the outgoing Dean Cartwright endorsed the Ad Hoc Committee's recommendations to form a Department of Statistics by July 1, 1971. Phillips's letter noted that Mathematics, CQS, and Biostatistics held the largest concentrations of statisticians but that Economics, Oceanography, Psychology, Sociology, Business Administration,

Engineering, and Fisheries and Forestry all contained interested faculty. Lloyd Fisher spent much of 1970–1971 working for the proposed department: writing up course descriptions, researching enrollments, writing proposals.

In the end, although the formation of the new department appeared to have gained approval at the highest university levels, it became a casualty of the economic downturn known locally as the Boeing Bust, which saw Boeing employment levels plummet from 108,900 to 38,690 by April 1971. This downturn not only doomed the formation of the department at that time but also spelled an end to the rich history of the Boeing Scientific Research Laboratories. The final rejection was delivered by outgoing Acting Dean Cartwright, though it was rumored that this was a courtesy to spare the incoming A&S Dean George Beckman (Dean 1971–1978, UW Provost 1978–1988).

Formation of the Department (1971–1979)

As consolation to the Statistics Group, the next available position in Mathematics had been committed to a statistician. In 1972, after some struggle within Mathematics to retain this commitment, Fritz Scholz (Berkeley PhD under Erich Lehmann, 1971) was appointed Assistant Professor. However, his position only replaced Lloyd Fisher's, who transferred to Biostatistics at the same time.

The status quo was maintained through much of the 1970s, but Mathematics shrank from 70 faculty in 1965 to 57 in 1978 and understandably sought to maintain its most prized areas of strength. Birnbaum retired in 1974, while Fisher, van Ness, Smythe, and Chan had left Mathematics between 1970 and 1977.[1] Computer Science had split from Mathematics in 1967 and was growing rapidly. Moreover, demand by Mathematics majors was shrinking, while demand for nontechnical mathematics courses for the social sciences was growing.

Much of the impetus for the final creation of a statistics department came from the Biostatistics Department, located in the School of Public Health and Community Medicine (SPHCM). Biostatistics, although a thriving and vigorous department with a strong graduate program, was funded primarily by federal grant support and had relatively few state-funded positions to support its graduate teaching. For this it relied heavily on the statisticians and probabilists within Mathematics, but their numbers had dwindled during the 1970s. In 1977, Biostatistics Chair Donovan Thompson invited Pyke, Shorack, and Scholz to discuss transferring all three of their positions from Mathematics to Biostatistics.

Meanwhile, the applied mathematicians within Mathematics were also feeling the stresses of the times. In 1977, they proposed a Division of Statistics and Applied Mathematics (DSAM). Al Goldstein, Herbert Brownell, Terry Rockafeller, Ron Pyke, and Victor Klee came to support the DSAM, though Pyke was

[1] Mathematics added probabilists Bruce Erickson in 1973 and Richard Bass in 1977.

simultaneously exploring other options for statistics. Shorack (on sabbatical at ANU in Canberra) wrote—recalling the “lost year” that Fisher had spent in the failed 1968–1971 attempt for a Department of Statistics—that he would again work hard for a Department of Statistics or would support the DSAM proposal, but only if the UW administration would first commit irrevocably to the formation of such a unit.

On October 26, 1978, the Ad Hoc Committee on Applied Mathematics chaired by Bill Richardson, Dean of SPHCM (which housed the Biostatistics Department), issued its report to the Provost, George Beckmann, who would prove very supportive of the creation and nurture of the new Department of Statistics. The Ad Hoc Committee strongly endorsed the formation of a Department of Statistics in the College of A&S.

The Provost accepted the Ad Hoc Committee’s recommendation and on January 24, 1979, Acting Dean Frank Miyamoto of A&S appointed a Planning Committee for Statistics chaired by Paul Hodge, Associate Dean for Sciences. Its mandate was “to prepare a plan whereby the College of A&S could best create a Department of Statistics that would benefit the students of the University, the statistics faculty in the various colleges, and the general reputation of the University in the field of statistics, building on the already outstanding programs in Biostatistics and Mathematics.” It was also charged with responding to “the increasingly disjointed nature of statistics at the University and the resulting inefficiency of efforts in both teaching and research” as well as “to the strong desire [of] the statistics faculty within the Department of Mathematics to have its discipline recognized, given visibility and autonomy, and tied in more closely to statistics programs elsewhere on campus.”

The Planning Committee reported as follows: “The Department of Statistics will satisfy … the considerable need for a focal point for the discipline of statistics, serving to coordinate and unify where possible the extensive teaching, research, and consulting activities in statistics on campus …. [It] should have 4 and 6 positions in 1979–1980 and 1980–1981, respectively, together with several joint positions with other departments. During 1981–1986, additional regular and joint positions should be added, including two joint appointments with the social sciences…. [Its] course offerings should include unified service courses as well as applied and theoretical core courses …. A Statistical Consulting Center should be established to coordinate consulting activities within the University ….”

These recommendations were accepted by Provost Beckmann, and the Department of Statistics was established by the Board of Regents on July 13, 1979. The department’s development over its first decade was well supported by Dean Ernest Henley and Associate Dean David Prins of the College of A&S and by Provost George Beckmann.

Statistics Faculty 1979–1989

The initial faculty members in 1979–1980 were Pyke[2] (Professor), Shorack (Professor), Michael Perlman (Chicago; chairman), and Piet Groeneboom (Visiting Assistant Professor). Z. W. Birnbaum, although already retired, was retroactively made Emeritus Professor of Statistics, Fritz Scholz (Boeing) was given an affiliate appointment, and Hubert (Tad) Blalock and Chapman were given 0% faculty appointments.

During 1980–1982, Doug Martin (100%), Cliff Lunneborg[3] (25%), and Ben King[4] (50%) transferred all or part of their appointments to Statistics from Electrical Engineering, Psychology, and the Business School, respectively. Joe Felsenstein (UW Genetics), Charles Nelson (UW Economics), and David Mason (Delaware) were given adjunct/affiliate appointments.

The first appointees initiated by the department arrived in 1980: Peter Guttorp (Assistant Professor) and June Morita (Lecturer), PhDs from Berkeley of David Brillinger and Kjell Doksum, respectively. In 1981, Paul Sampson (PhD Michigan; visiting assistant professor, Chicago) joined the department as Assistant Professor. Mary Lou Thompson (Princeton) was acting assistant professor in 1981–1983.

During 1981–1989 the following new appointments were made: Andreas Buja[5] (ETH Zurich), Judy Zeh (UW EE), Jon Wellner (UW, Rochester), Don Percival (UW Applied Physics Lab), Antonio Possolo[6] (Yale), Werner Stuetzle (ETH Zurich, Stanford), John McDonald (Stanford), Andrew Siegel (Princeton), Elizabeth Thompson (Cambridge UK), Adrian Raftery (Trinity College Dublin; joint w. Sociology), Finbarr O'Sullivan[7] (Wisconsin; joint w. Biostatistics), and Julian Besag[8] (Durham UK).

In its early years, the department made a concerted effort to gain national and international visibility by hosting summer and sabbatical visitors attracted by the appealing environment of Seattle and the Pacific Northwest. In a single year, 1981–1982, the department hosted 21 extended visitors, who contributed greatly to the academic discourse in the young department.

---

[2] Ron subsequently returned to Mathematics; he passed away in 2005.

[3] Tragically, Cliff drowned in Lake Washington in 2005.

[4] However, Ben soon moved to the Educational Testing Service in Princeton.

[5] Andreas is now at the Wharton School, University of Pennsylvania.

[6] Antonio moved to Boeing's Applied Statistics Group in 1989; he is now Division Chief for the Statistical Engineering Division of NIST.

[7] Finbarr moved to the University of Cork, Ireland, in 1996.

[8] Julian passed away in 2010.

Academic Courses and Degrees 1979–1989

The new department inherited several courses from Mathematics: the one-quarter undergraduate service course 311 (joint with Econ 281 and 580[9]), the undergraduate sequence 341-2 for statistics majors, the MS graduate theory sequence renumbered 511-2-3, and the PhD graduate theory sequence 581-2-3. An immediate issue was the disposition of the undergraduate (394-5-6, 491-2) and graduate (521-2-3) probability courses located in the Mathematics Department and taught by Pyke, Shorack, and probabilists Bob Blumenthal, Bruce Erickson, and Richard Bass. Agreement was reached for a 50–50 division of these courses between Mathematics and Statistics, averaged over a 2-year cycle; this agreement has remained remarkably stable.

Despite its limited graduate course offerings, statistics was able to implement a graduate program relatively quickly by utilizing biostatistics graduate courses covering topics such as linear models, applied regression analysis, categorical data, and survival analysis. Meanwhile, embracing the computer revolution in statistics, including the bootstrap, interactive computing and graphics, S, and the Princeton robustness study, Statistics began to develop its own graduate courses in robustness (Martin), time series (Martin, Percival), stochastic modeling (Guttorp), sample surveys (Morita), multivariate analysis (Perlman, Sampson), statistical computing (Buja, Stuetzle, McDonald), empirical processes (Wellner), statistical genetics (Thompson), Bayesian statistics (Raftery), and spatial statistics (Besag, Guttorp). In a related vein, Martin's relationship with Bell Labs led to his founding the statistical software company Stat Sci in Seattle in 1988, its prime product being S-Plus. A number of future statistics department graduates would be employed there.

The department's graduate program was established officially in 1982 after the mandatory 3-year review period, although several UW graduate students had already transferred unofficially to statistics. Its first PhD dissertation "The statistics of long-memory processes" was completed in 1983 by Don Percival under the supervision of Doug Martin.

At the same time, the department established an undergraduate statistics major but soon also entered into a collaboration with Mathematics, Computer Science, and Applied Mathematics on the Applied and Computational Mathematical Sciences undergraduate degree. This program provides a well-rounded foundation in the mathematical sciences with additional concentration in one of the participating fields, including a statistics pathway.

Upon their arrival in 1980, Guttorp and Morita introduced a new service course, Stat 220, based on the then-new textbook developed by Freedman, Pisani, and Purves at Berkeley. In 1981, Doug Martin convinced the Civil and Electrical Engineering Departments to include the newly created calculus-based introductory

---

[9] Charles Nelson, chairman of Economics, subsequently transferred Econ 281 and 580 to Statistics.

course Stat 390 in their curricula. Cliff Lunneborg introduced upper division undergraduate courses in applied multivariate analysis and nonparametric statistics.

Relationships with Biostatistics and Boeing

Statistics and biostatistics have maintained an extremely close and effective collaboration. The two departments share major core requirements, notably the MS and PhD statistical theory and the PhD statistical methodology sequences and the associated qualifying exams; consulting training for graduate students; the Statistical Genetics PhD pathway; and frequent joint seminars and social events. Graduate students in one department often conduct their dissertation research under a supervising faculty member in the other department. All this ensures that the faculty and graduate students of the two departments interact and know each other well, both personally and professionally.

When Scholz left Mathematics for the Applied Statistics Group (ASG) at Boeing in 1978, he reestablished the important connection to Boeing that had existed earlier via Boeing Scientific Research Laboratories (BSRL) in the 1950s and 1960s.

Recent Years (1990–2011)

Space does not allow a comprehensive listing of faculty and student accomplishments, so this section presents only some of the highlights. A more thorough listing can be found at the following websites:

Faculty research, publications, awards: http://www.stat.washington.edu/research/books/
PhD graduates research, publications, awards:
Current faculty and students: http://www.stat.washington.edu/www/people/
PhD theses: http://www.stat.washington.edu/people/students/thesis/
Technical reports: http://www.stat.washington.edu/research/reports/
Honors and awards 2001–2010: http://www.stat.washington.edu/news/recent/

Interdisciplinary Programs

National Research Center for Statistics and the Environment. NRCSE (1996–2002) was established at UW under a $5 million collaborative agreement with the EPA, with Peter Guttorp as director and Paul Sampson as associate director.

NRCSE conducted research in problems related to air and water pollution. The center had 33 members from seven schools and colleges at the university, three postdocs, and 29 graduate students. Over the years 229 visitors spent time at the center. NRCSE provided a platform for multidisciplinary interaction with respect to a wide range of problems related to statistical analysis of environmental concerns. The research was published in 6 books and 145 scientific papers.

Statistical Genetics. The StatGen program at the UW, initiated in 1999–2000, is a joint endeavor of Statistics and Biostatistics. It consists of PhD tracks in the two departments and an Interdisciplinary Graduate Certificate program. The current core faculty number 18, from Statistics, Biostatistics, Genome Sciences, Medical Genetics, Epidemiology, and FHCRC, and the program works in close partnership with the program in Computational Molecular Biology. The core course sequence developed in 1999–2000 continues to evolve and attracts students from a variety of graduate programs. The program has been led by Elizabeth Thompson since 1999 but gained a broader foundation with the arrival of Professor Bruce Weir to chair the Department of Biostatistics in 2005. His long-running NSF-funded Summer Institute in Statistical Genetics is now firmly established at UW and brings worldwide students and faculty to Seattle each June. Led by Bruce Weir, the UW StatGen program also secured an NIH Training grant which provides core support for outstanding students in the subdiscipline.

Center for Statistics and the Social Sciences. In the late 1990s, the department spearheaded the establishment of CSSS in 1999, to foster collaborative research between statistics and the social sciences and to provide a rich menu of statistical and quantitative methods courses for social science graduate students. The university provided six faculty positions for CSSS, most to be jointly appointed between statistics and a social science department. Contrary to the understanding of most Statistics faculty members, these were not joint appointments in the traditional sense; instead their budgets reside in a separate CSSS budget controlled by the CSSS director, who reports directly to the A&S dean.

Core CSSS faculty with appointments in statistics have included Mark Handcock, Martina Morris, Peter Hoff, Sibel Sirakaya, Elena Erosheva, Adrian Dobra, and Tyler McCormick. Adrian Raftery was the founding director; the center is now directed by Thomas Richardson. PhD tracks have been established in nine PhD programs around the university based on CSSS courses. CSSS has helped to stimulate the rapid development of research at the interface between statistics and the social sciences nationally over the past decade.

Computational Finance. The CompFin program was established in 2004 by Doug Martin and now includes faculty in Applied Mathematics, Economics, and Finance. It has the following integrated academic focal points: portfolio construction and asset management, risk management, and computational methods. The curriculum includes not only traditional quantitative finance theory and methods but also cutting-edge methods such as modeling and use of fat-tailed skewed asset returns distributions, downside risk management, nonlinear correlations, and robust factor models. The computational focus is on the use of the open source R programming language and modeling environment, with its many rapidly evolving open source packages for quantitative finance.

Faculty Research, Teaching, Publications, Honors (Partial List)

Julian Besag made seminal contributions to stochastic modeling, especially by modeling local specifications (for example via Markov random field models). This approach proved successful in many fields, including agricultural field trials, spatial epidemiology, and image analysis. Julian was elected a Fellow of the Royal Society in 2004.

Fred Bookstein's research has established the foundations of morphometrics, statistical methods that structure the space of measurements as rigorously as conventional methods structure the space of cases. Applications of this program, including the methodologies of partial least squares (from chemometrics) and shape coordinates (from morphometrics), have contributed to fields as diverse as human evolution, anatomical birth defects, capital punishment, and finite element analysis. He is coauthor of the textbook, "Virtual Anthropology" (2011) on the role of image analysis in general and morphometrics in particular across the sciences of human origins.

Adrian Dobra's research interests include graphical models, stochastic computing and multidimensional contingency tables, Bayesian statistics, applications in the social sciences, spatial epidemiology, genomics, and disclosure limitation.

Elena Erosheva studies the development and application of statistical methods to address issues in the social, behavioral, and health sciences, focusing on discrete data analysis, hierarchical, latent variable modeling, and Bayesian methods. Major contributions include developing the class of mixed membership models for multivariate data analysis that generalizes such special cases as the Latent Dirichlet Allocation model developed by computer scientists and the admixture model used in genetics.

Tilmann Gneiting has contributed to the theory of spatial and space-time covariance functions and developed (with Adrian Raftery) tools for assessment of probabilistic weather forecasts.

Peter Guttorp's research has focused on space–time models involving air pollution, climate, and hematopoietic stem cells. His work with Paul Sampson on nonstationary spatial covariance has been very influential. He has published two well-known monographs on stochastic processes: "Statistical Inference for Branching Processes" (1991) and "Stochastic Modeling of Scientific Data" (1995). Peter also contributed to two of the four assessment reports of The Intergovernmental Panel on Climate Change, established in 1988 by the United Nations Environment Programme and the World Meteorological Organization to provide assessments of the science of climate change. This panel received the Nobel Peace Prize in 2007.

Mark Handcock's work at UW was motivated largely by questions in the social sciences and demography. He has focused on the development of statistical models for the analysis of social network data, distributional comparisons, environmental statistics, spatial statistics, epidemiology, and labor economics. Recent applications have been to social relations networks with the objective of understanding the social determinants of HIV spread. He has also developed survey sampling methods for network sampling and for combining survey and population-level information. His book "Divergent Paths: Economic Mobility in the New American Labor Market" (2001), coauthored by Annette D. Bernhardt, Martina Morris, and Marc A. Scott, won the Lester Prize in Labor Economics.

Peter Hoff has developed a variety of statistical methods for multivariate data analysis, including methods for network analysis, covariance, and copula estimation. He has applied these methods to various problems in the biological and social sciences, in particular, international conflict and trade data. He has also worked on nonparametric Bayesian inference in the context of convex models and cluster analysis. His recent book "A First Course in Bayesian Statistical Methods" (2009) is in its second printing.

Cliff Lunneborg's three textbooks "Elementary Multivariate Analysis for the Behavioral Sciences: Applications of Basic Structure" (1983, Robert Abbott co-author), "Modeling Experimental and Observational Data" (1994, 2000), and "Data Analysis by Resampling: Concepts and Applications" (2000), formed the core of the Statistics Department's undergraduate upper division elective curriculum.

Doug Martin has worked extensively on robust methods for signal processing in time series analysis. In recent years he has applied these methods to problems in finance, including portfolio optimization and risk management, options and derivatives, and data mining. He has coauthored the textbooks "An Introduction to Modern Portfolio Optimization with S+NuOPT, S-Plus, and S+Bayes" (2005) with Bernd Scherer and "Robust Statistics: Theory and Methods" (2006) with Ricardo Maronna and Victor Yohai.

Marina Meila's research interests include graphical probability models, machine learning, computational algorithms, and data mining. Specific topics include intransitivity in classification and choice, gravimetric inversion with sparsity constraints, manifold learning, and proteomics (interpreting the very complex signature of an amino acid sequence that is subjected to collision-induced dissociation).

Vladimir Minin works on developing statistical methods for phylogenetic reconstruction and for inferring demographic histories of populations from genomic data. His methodological work revolves around inference for partially observed stochastic models. He uses these stochastic models to study genetic diversity of rapidly evolving pathogens. Vladimir has worked on elucidating spatial patterns of recombination in the HIV genome, inferring population dynamics of influenza, and genome-wide scans for positive selection in the *E. coli* genome.

June Morita, with Eisenhower Foundation funding (1994–1996), brought week-long Quantitative Literacy Workshops to local elementary school teachers. These workshops were developed through the American Statistical Association. During 1996–2004, she was co-PI on the NSF-funded projects "Creating a Community of Mathematics Learners" and "Expanding the Community of Mathematics Learners" working with six local school districts to improve the level of mathematics understanding and the mathematics teaching effectiveness of elementary and middle school teachers.

Martina Morris's research is interdisciplinary, intersecting with demography, economics, epidemiology and public health, and statistics. Her current projects include the study of partnership networks in the spread of HIV/AIDS, the impact of economic restructuring on inequality and mobility, and the development of Relative Distribution methods for statistical analysis.

Don Percival is active in time series analysis, with emphasis on application of spectral and wavelet analysis to a variety of problems, including characterization of inaccuracies in atomic clocks, investigation of climate variability in the Arctic, and more recently, a depth/time analysis of temperature variations in an Australian subtropic dam. He also works on the time series analysis of Martian atmospheric pressure, Antarctic ice cores, atmospheric aerothermal turbulence, discharge behavior of rat motoneurones, heterogeneity measurements in sand on beaches, and subtidal coastal sea level fluctuations. He has coauthored the books "Spectral Analysis for Physical Applications: Multitaper and Conventional Univariate Techniques" (1993) and "Wavelet Methods for Time Series Analysis" (2000), both with Andrew Walden.

Michael Perlman contributes to multivariate analysis, notably in collaboration with Steen Andersson and David Madigan on the theory of graphical Markov models. They characterized the essential graph for equivalence classes of Markov models determined by acyclic directed graphs (=Bayesian networks) and introduced the AMP (alternative Markov property) for the wider class of chain graphs, which include both directed and undirected edges to model both causal and associative relationships. He served as editor of the *Annals of Statistics* during 1983–1985.

Adrian Raftery's research has focused on Bayesian model averaging, model-based clustering, and the statistical analysis of deterministic models, with applications to the social and environmental sciences. His 1995 *JASA* Discussion Paper "Inference from a deterministic population dynamics model for bowhead whales" with Geof Givens (UW PhD 1993) and Judy Zeh received the 1996 ASA Outstanding Application Award. Adrian served as Coordinating and Applications Editor of *JASA* in 1998–2000, was identified by Thomson-ISI as the most cited researcher in mathematics in the world for the decade 1995–2005, and was elected to the National Academy of Sciences in 2009.

Thomas Richardson works at the interface between causal and graphical models. His main contribution has been the development of a nonparametric graphical theory of path diagrams (which include directed and bidirected edges). This has included the construction of parameterizations, fitting algorithms, and model selection procedures. His main collaborators have been Steffen Lauritzen (Oxford), James Robins (Harvard), and Peter Spirtes (CMU).

Paul Sampson's primary research topics have been spatial and spatio-temporal modeling in environmental statistics, Partial Least Squares methods in applied multivariate analysis, and geometric morphometrics, including the analysis of 3D MR brain images, in conjunction with his role as senior statistician for the Seattle Longitudinal Study on Alcohol and Pregnancy. He continues as the founding director of the department's Statistical Consulting Program.

Fritz Scholz retired from the Applied Statistics Group at Boeing in 2006 but continues teaching at UW. His research efforts include the theory of estimation, large sample theory, nonparametric statistics, software reliability, quantile estimation, risk analysis, bootstrapping, and tolerance analysis. He presented the Keynote Address "Statistics in Aviation: Celebrating 100 Years of Flight" at the 10th Spring Research Conference on Statistics in Industry and Technology at Dayton, Ohio.

Galen Shorack's research interests include empirical processes, asymptotic properties of rank statistics, linear combinations of functions of order statistics, robustness, and reliability. In addition to his book "Empirical Processes" (1986) with Wellner, he published the advanced graduate textbook "Probability for Statisticians" in 2000.

Werner Stuetzle has worked intensively on nonparametric methods in multivariate analysis, statistical applications of computer graphics, and scientific computing. He currently serves as Divisional Dean for the Sciences at UW.

Elizabeth Thompson continues to contribute to the development of methods for model-based likelihood inference from genetic data, particularly from data observed on large and complex pedigree structures both of humans and of other species, including inference of relationships among individuals and among populations. While at UW she has published two monographs on statistical genetics: "Pedigree Analysis in Human Genetics" (1986) and "Statistical Inferences from Genetic Data on Pedigrees" (2000). She was elected to the National Academy of Sciences in 2008.

Jon Wakefield has worked on hierarchical modeling in pharmacokinetics/pharmacodynamics and in spatial epidemiology. With respect to the latter, understanding ecological bias and overcoming this bias via outcome-dependent sampling is of particular interest. More recently, Jon has worked in genetic epidemiology and on various problems associated with high-throughput technologies.

Jon Wellner studies limit theories and inequalities for empirical processes, including preservation results for both Glivenko-Cantelli theorems and Donsker theorems. He has investigated semiparametric models with missing data and problems in shape-constrained estimation and has developed new test statistics for problems involving sparse normal means. In addition to "Empirical Processes" (1986) with Shorack, his books include "Information Bounds and Nonparametric Maximum Likelihood Estimation" (1992) with Piet Groeneboom, "Efficient and Adaptive Estimation for Semiparametric Models" (1993) with Peter Bickel, Chris Klaassen, and Ya'acov Ritov, and "Weak Convergence and Empirical Processes" (1996) with Aad van der Vaart.

Judy Zeh, following in Doug Chapman's footsteps, served as chairman of the Scientific Committee of the International Whaling Commission during 2000–2002.

Photographs of Z. W. Birnbaum, Ron Pyke, and Isobel and Doug Chapman appear below:

University of Wisconsin Department of Statistics

Norman Draper, Richard Johnson, Stephen Stigler, Kam-Wah Tsui and Brian Yandell

The University of Wisconsin Statistics Department commemorated its fiftieth anniversary in 2010. A gala celebration marked the event. However, the year when Statistics was first introduced in the University of Wisconsin is unclear. Perhaps it was the establishment of a meteorological station in the 1850s in a building that was soon nicknamed "Old Probabilities". Perhaps it was in the Religion Department in the 1870s when they discussed Cross-validation. It certainly predated 1890, when George Comstock, the Director of Washburn Observatory, published a text based upon his course on the Method of Least Squares, targeted to "students of physics, astronomy, and engineering." The first page of that book shows an estimated density function; the second chapter was about "The Distribution of Residuals."

The next few decades were relatively quiet. Probability was taught in the Mathematics Department, including a course offered by Warren Weaver (faculty 1920–1932). Henry Scheffé, who achieved later eminence at Berkeley with his classic book *The Analysis of Variance*, took Weaver's course. Scheffé received both BA and PhD degrees from Wisconsin and taught there from 1935 to 1937, but he taught only pure Mathematics, not Statistics. What statistics activity there was

University of Wisconsin,
Department of Statistics, 1220 Medical Sciences Center,
1300 University Avenue, Madison, WI 53706-1685, USA
N. Draper · R. Johnson · K.-W. Tsui (✉) · B. Yandell
e-mail: kwtsui@stat.wisc.edu

University of Chicago, Department of Statistics,
5734 S. University Avenue, Eckhart 102,
Chicago, IL 60637, USA
S. Stigler

A. Agresti and X.-L. Meng (eds.), *Strength in Numbers: The Rising of Academic Statistics Departments in the U.S.*, DOI: 10.1007/978-1-4614-3649-2_38,

at the time was diffuse. There was some active statistical interest in the agricultural area, particularly in Agronomy where James Torrie was hired in 1940 as a red clover and soybean breeder. He and R. G. D. Steel published their book *Principles and Procedures of Statistics* in 1960. Courses were also taught in Economics and Business. A loose *Division of Statistics* organization provided a cloak for such activities.

Why did little more happen until George Box arrived in 1959? We must look back to the curious events of 1940–1941, when a bold attempt to create a statistics program crashed and burned. In 1940, the University of Wisconsin invited Milton Friedman to visit the Economics Department with the specific charge of strengthening their statistical offerings. Friedman, who won a Nobel Prize in 1976 for his work on monetary theory and the consumption function, is now much better known as an economist. However, in 1940, he was more of a mathematical statistician. His PhD thesis advisor was the well-known Harold Hotelling, who later headed the Statistics Department at Chapel Hill, NC.

Friedman came to Wisconsin in fall 1940 as a visiting Lecturer. He produced a detailed 9-page report on the dismal situation faced by a Wisconsin student interested in studying statistics, and he suggested the creation of a new Department of Statistics. He wrote:

> A student cannot secure training at the University of Wisconsin sufficient to qualify him to teach advanced statistics or to do independent work in the field of statistical methods. Even if he takes all the work offered he will be but indifferently qualified to do research involving the application of modern statistics.

The administration was greatly impressed, and in the spring offered him a position as Associate Professor, charged to implement the program he had outlined. He indicated he would accept, but before the appointment could be approved by the Regents, a dispute broke out. A group of senior professors in Economics, led by Edwin Witte, demanded that the Regents cancel the offer, claiming that, as able as Friedman might be, he was too young for such an appointment. He was, after all, merely a visiting lecturer! The students in the Economics department backed Friedman. On June 3, 1941, Friedman decided he did not want to come where he was not wanted, and he withdrew to return to the National Bureau of Economic Research in New York and a subsequent career in economics.

Friedman emerged intact, but Wisconsin did not; it would be 18 years before a second attempt would be made. During the year Friedman was at Wisconsin, 18 faculty members joined to form a Division of Statistics. The initial intention was to coordinate courses in statistics throughout the university and to supervise a minor in Statistics. Among the initial members were A. B. Chapman, C. Eisenhart, J. H. Torre, F. A. Gaumnitz, and M. H. Ingraham. By the mid-1950s, when pressure for a broader set of statistical offerings was growing, Gaumnitz was Dean of Commerce and Ingraham was Dean of Letters and Science. Letters of support for a new department of statistics were gathered from several departments and colleges.

During the Spring of 1958, Stephen Kleene, Chair of Mathematics, approached Jerzy Neyman to see if he was interested in starting a new Statistics department;

Fig. 1 Department of Statistics, George Box discussing nonlinear estimation with attentive students (1960s)

nothing came of this. The administration then invited George E. P. Box, who was visiting at Princeton, where he headed the Statistical Techniques Research Group. With the support of the Dean of Engineering and the Dean of Letters and Science, Rudolf Langer, head of the Mathematics Research Center, offered George a 1-year visiting position. George came in 1959, was appointed Professor of Mathematics in January 1960 and formed the Department of Statistics in the fall. He was on the faculty 1960–1991, until his retirement.

George had years of experience in the practical benefits of statistical knowledge from his work with the British Army in World War II, and afterward in the Dyestuffs Division of Imperial Chemical Industries in Manchester, UK. He had worked with scientists of all types; he was adroitly skillful in negotiating with all manner of potential partners; and he could personally demonstrate the enormous value that statistical applications could bring to research work. These skills facilitated the building of connections across the university and, in particular, did not arouse any defensive reactions such as those shown by senior economists in 1941. On the contrary, he was welcomed by all, and especially by the Division of Statistics members. The new department would not only teach the *science* of statistics; it would demonstrate that statistics could be applied to *practical problems* to the advantage of all participants. Therefore, an important feature in building up the faculty would be to make a number of joint appointments, with the cooperation of willing partner departments. Each such new appointee would thus have to be doubly acceptable. By the mid 1960s, George had developed a strong research program with members of the Chemical Engineering Department; this fueled work on nonlinear estimation, including reaction models (Fig. 1).

George Box's early appointments set the tone: John Gurland (1960–1987) and Norman Draper (1961–1999) were already in Madison, visiting the Mathematics Research Center. To this core were added three joint appointees, whose theses George had mentored at Wisconsin: George Tiao (1962–1983), Business; W. G. (Bill) Hunter (1963–1986), Engineering; and S. M. (Sam) Wu (1965–1979), Mechanical Engineering. Subsequent hires in the 1960s, Jerome Klotz (1965–1999), Gouri Bhattacharyya (1966–1995), Richard Johnson (1966–2008), George Roussas (1966–1976), Stephen Stigler (1967–1978), Grace Wahba (1967–) and Bernie Harris (1967–2002) strengthened the theoretical component of the department and led to its reputation for both strength and balance. The attached chart (Fig. 6) shows how hiring progressed from there.

A consulting laboratory (StatLab) was created in 1966 with faculty member Don Watts (1965–1970) as the first head. The StatLab exposed our own students to the practical statistical problems encountered by faculty and students throughout the university and trained them to be useful consultants. This led to the idea of a visiting Statistician in Residence, an experienced person who would not only be able to help nonstatisticians with their statistical problems but also provide a variety of statistical know-how to our students. One-year appointments were made to J. Stuart Hunter (1967–1968), Graham Wilkinson (1968–1969), Don Behnken (1969–1970), G. Morris Southward (1970–1971), Harvey Arnold (1972–1973), and Svante Wold (1973–1974). Brian Joiner, cocreator of Minitab, joined the faculty in 1974 and ran the StatLab until he resigned in 1984 to form his own private consulting company. Asit Banerjee was hired after Joiner to run the StatLab, which closed around 1987.

The UW-Madison PhD dissertations in Statistics span a wide range of topics, from pure theory to development of applied methods, and in recent years there has been an increasing emphasis on statistical computing. Our PhD program has produced over 400 PhDs. Many of our graduates have held leading positions in academia, industry and government.

Around 1977–1978, George Box initiated a major change, separating the PhD and the Masters Degree requirements. The courses, examinations and other requirements are now totally separate. The masters program has evolved in requirement of a one-semester consulting course and a week-long written and oral exam driven by scientists with real data problems. This masters program is highly regarded and widely emulated and has yielded over 500 MS graduates.

Our undergraduate major in Statistics was for a long time a rather small affair. However, society recognizes the need of statistical reasoning and the jobs have emerged. Demand for BS degrees has followed, and our program has grown. Further, the recognition that Statistics is an information science (or data science) has led to new collaborations and a continually increasing interest in Statistics within our campus.

Research highlights from the early years include George Box's development of time series, first with Gwilym Jenkins and then with George Tiao. Box and Norman Draper made substantial advances in the theory of response surface designs and Draper and Harry Smith published one of the early regression texts that emphasized the study of residuals. Box and Tiao produced a classic work on Bayesian analysis that featured the analysis of real problems. Richard Johnson and Dean Wichern published their popular text on applied multivariate analysis and George Roussas authored a monograph that increased access to Le Cam's contiguity work. Grace Wahba began her series of publications on the statistical application of splines and Stephen Stigler emerged as an expert on the history of statistics with a series of articles, followed later by a book. The basic required Mathematical Statistics courses for graduate students, 709–710, were recast and improved by Gouri Bhattacharyya. Gouri later collaborated with Richard Johnson on a couple of introductory statistics texts.

The faculty's research work was published for many years in a series of red-covered technical reports, sent to a large mailing list. Most of them appeared subsequently in the peer-reviewed statistics Journals published by the major Statistical Societies. ASA and IMS meetings typically included several faculty speakers. The Department's early initial decades were thus an exceptional period of production and growth and it quickly became one of the top departments in the country. Greg Reinsel (1976–2004) became a leading researcher on time series and supported numerous students through his collaboration with George Tiao on decades-long analyses of stratospheric ozone pollution data and global warming temperature data. Greg's sudden death in 2004 while jogging shocked us all. Chien-Fu "Jeff" Wu (1977–1990) conducted a very active research program in the design and analysis of experiments until he moved on to increased responsibilities elsewhere.

Grace Wahba (1967–) a member of the National Academy of Sciences, has been a world leader in the study of ill-posed problems. Tom Leonard (1980–1995), Kam Tsui (1980–) and Michael Newton (1991–) enhanced the Bayesian component of the program both in teaching and research. Doug Bates (1980–2011) is an expert in nonlinear estimation, and has been active in the development of the R statistical software language since its very beginning. In the late 1980s, Wei-Yin Loh (1982–) turned his attention to the design of classification and regression tree algorithms, 10 years before the development of machine learning. Jed Frees (1984–1997) held a joint appointment with the School of Business and Tom Kurtz (1986–2008) shared an appointment with the Department of Mathematics. Jun Shao (1996–) published a book on Mathematical Statistics which has become a standard reference. Richard Johnson (1966–2008) was founding editor of Statistics and Probability Letters and served in that capacity for 25 years. Kjell Doksum (2002–2010) was a welcome addition for both our teaching and research program. Chunming Zhang (2000–) demonstrates wide expertise in model selection and functional analysis. Zhengjun Zhang (2005–) has been a pioneer in extreme value theory, with particular attention to finance applications. Yazhen Wang (2009–) focuses as well on financial statistics, with particular emphasis on long term memory processes. Zhiguang "Peter" Qian (2006–) works on design of experiments, computer models, the interface between statistics and optimization and statistical methods for high-technology. Bret Hanlon (2010–) studies variable selection and branching processes with a variety of applications.

Joint appointments with intimate connections to other scientific units of the campus were central to Box's vision for the Department of Statistics. Over the years, the joint appointment positions became concentrated in two programs, Biometry in the School of Agriculture and Life Sciences and Biostatistics in the Medical School. The Biometry program began in the late 1970s with Kim Andriano (1979–1981) and directed in its early years by Erik Nordheim (1977–). The program was strengthened by Brian Yandell (1982–) and Murray Clayton (1983–), and more recently by Bret Larget (2002–), Jun Zhu (2005–) and Cecile Ane (2006–). Murray Clayton was director of Biometry from 2005 to 2010, followed recently by Jun Zhu. The Biometry Masters program is unique, as students are coadvised by a Biometry faculty member and their Biology PhD advisor. Biometry faculty maintain active collaborations across the biology departments in agriculture, veterinary medicine and basic biological sciences, with considerable expertise in statistical genomics and spatial statistics.

John Van Ryzin (1969–1979) led a group including John Crowley (1973–1981) that strengthened research and consulting connections with medical faculty and scientists. The Biostatistics Program started from the UWCCC Biostatistics Shared Resource. Dave DeMets (1982–) was hired to oversee statistical activities in the Medical School, which evolved from division to center in 1986 and to department by 1991. Karl Broman (2007–), Rick Chappell (1990–), Jason Fine (1998–2008), Sunduz Keles (2004–), Christina Kendziorski (2001–), Michael Kosorok (1992–2006), Michael Newton (1991–), Barry Storer (1984–1996), and Sijian Wang (2008–) have enhanced and broadened the Biostatistics program. A name change to the Department of Biostatics & Medical Informatics (BMI) in 1996 reflected the changing demands on this group. Recent collaborations between biostatistics and informatics faculty have risen from this wise pairing of complementary fields.

Statistics and BMI maintain close relationships. Graduate students in Statistics and Biostatistics reside in the same program, creating a unique synergy to the benefit of both departments. Most Biostatistics faculties have joint appointments in Statistics, and several other Statistics faculty have affiliate appointments in BMI. One example of this blending is the leadership our campus plays in Statistical Genomics, with faculty from Statistics, BMI and Biometry building courses, training students and leading workshops in this emerging field.

Fifty plus years on from the department founding, we recall these and others who played vital roles in this exciting period. Some have died: John Gurland (1960–1987), Bernie Harris (1967–2002), Jim Hickman (1972–1993, joint with the Business School), Bill Hunter (1963–1986, joint with Engineering), Jerry Klotz (1965–1999), Greg Reinsel (1976–2004), Jerry Senturia (1972–1978), John Van Ryzin (1969–1979) and Sam Wu (1965–1979, joint with Mechanical Engineering). Retired faculties include Gouri Bhattacharyya (1966–1995), George Box (1960–1991), Norman Draper (1961–1999), Richard Johnson (1966–2008), Brian Joiner

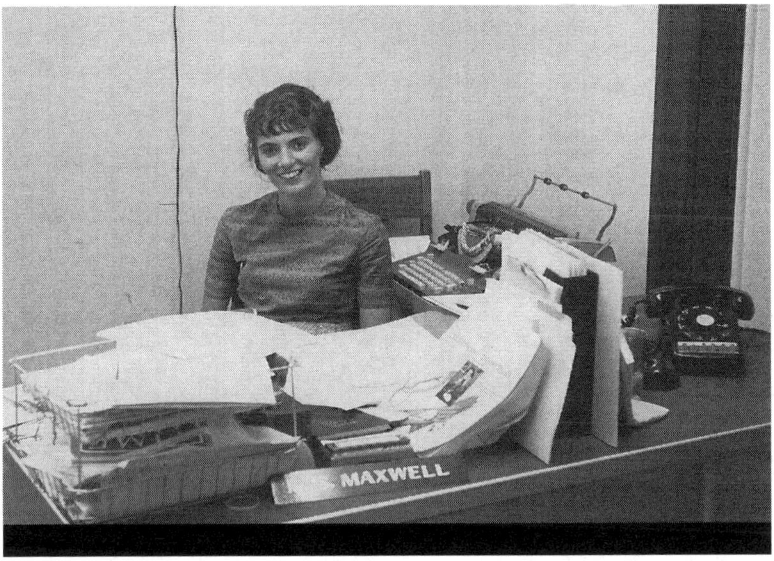

Fig. 2 Our first secretary, June Maxwell (1961–1963)

(1974–1984), Thomas Kurtz (1967–2008, joint with the Mathematics Department), Bob Miller (1968–2005, joint with the Business School) and Bob Wardrop (1974–2006). Others have scattered near and far: Jim Bondar (1966–1970), Don Watts (1965–1970), Irwin Guttman (1962–1970), Tom Leonard (1980–1995), George Roussas (1965–1976), Joe Sedransk (1969–1974), Stephen Stigler (1967–1979), George Tiao (1962–1983), and Chien-Fu Jeff Wu (1977–1990). Early visitors included J. Stuart Hunter (1960–1961), and Gwilym Jenkins (1964–1965). We fondly remember all of these colleagues.

Last, but certainly not least, we remember the office staff: June Maxwell (1961–1963), Mary Ann Clarke (1962–1994), Mary Esser (1963–1997), Wanda Gray (1967–1987), Gloria Scallisi (1988–2000), and Candy Smith (1970–2008) from past days, and all their successors, currently Nancy Brinkerhoff (2002–), Jude Grudzina (1997–) and Denise Roder (1997–).

We have had many stellar students over the years. Our 560 MS and 400 PhD degree graduates are scattered all over the world. We are deeply proud of them, even though we cannot claim credit for *all* their achievements. Many will remember the outrageously funny student-faculty Christmas skits, which were featured at George Box's annual Christmas party for many years. The students made fun of the faculty, and vice versa. For example, one such offering featured "an all star cast plus Brian Joiner". Another had the ambiguous line, "The students are revolting".

The Department has resided in four different physical locations over the years. Initially, we occupied a three-bedroom house on Johnson Street. George Box and Norman Draper each had a bedroom while the first four students occupied the master bedroom. June Maxwell (Fig. 2) ran the Statistics Department from the

Fig. 3 The Computer Science and Statistics Center, 1210 West Dayton Street (1967–2004)

ground floor as its Secretary for the first years, demonstrating great ingenuity in tackling the many teething problems that inevitably arose.

When our old house was demolished for a new building on the site, we moved to a rented set of small apartments above Tiedeman's drug store at 710 University Avenue at Lake Street. The building was owned by three very nice elderly sisters, two of whom were badly handicapped. The university tried to take over the building in those early days but the sisters (now dead) prevailed in court. Our second floor space, shared with another department consisting of one professor and his secretary, had many bathrooms and one large six-sided lecture room where George Box once jokingly confided a secret "just between these six walls". Within those same six walls, a student taking his final PhD orals responded impatiently to one of the examiners, "That's a *silly* question." The ensuing explosion and uproar required a 10-min break, with the candidate sent into the corridor. After offering an apology, he passed, and later had a successful career in another country.

Figure 3 shows (partly) the Computer Science and Statistics Center, our third building, which we shared with Computer Science. Initially the two departments were mixed together, but little cooperation was generated. As Computer Science expanded rapidly, stages 2 and 3 of the building were added, but when the site was filled entirely, Statistics was expelled to the former main city hospital at 1300 University Avenue in 2004. We now occupy a strange series of separate and/or nested rooms that remind us nostalgically of those above Tiedeman's drug store. Appropriately, we live in a veritable distribution of offices. No boring same-size offices for *us*.

Figure 1 shows our founder George Box in class in the 1960s. Note the suit and tie in more formal times. On the board is the famous A → B → C example of nonlinear estimation in reaction models. As we write this, George is in his 90s. Many of our former students have reminisced about George's "Monday Night Beer Sessions" held in the basement of his house, during which a speaker, most often from another field, would informally introduce a research problem he or she was studying, and would then be involved in a general discussion of how statistical methods could impact the research. "I learned more in those sessions than I did in regular class" is a sentiment often heard expressed by our alumni.

Wisconsin statistics faculty wrote many important books that had worldwide sales over many years. Among these were: *The Future of Statistics* (Watts ed. 1968); *Evolutionary Operation: A Statistical Method for Process Improvement* (Box and Draper 1969); *Contiguity of Probability Measures: Some Applications in Statistics* (Roussas 1972, digital version 2008); *A First Course in Mathematical Statistics* (Roussas 1973); *Statistical Concepts & Methods* (Bhattacharyya and Johnson 1977); *Statistics, Principles and Methods* (Bhattacharyya and Johnson 1984); *Nonlinear Regression Analysis and Its Applications* (Bates and Watts 1988); *Spline Models for Observational Data* (Wahba 1990); *Bayesian Inference in Statistical Analysis* (Box and Tiao 1992); *Time Series Analysis* (Box, Jenkins and Reinsel 3rd ed. 1994); *Statistics for Business: Data Analysis and Modelling* (Cryer and Miller 1994); *The Jackknife and Bootstrap* (Shao and Tu 1995); *Practical Data Analysis with Designed Experiments* (Yandell 1997); *Business Statistics—Decision Making with Data* (Johnson and Wichern 1997); *Applied Regression Analysis* (Draper and Smith 3rd ed. 1998); *Statistical Reasoning and Methods* (Johnson and Tsui 1998, alternate edition, 2003.); *Mixed Effects Models in S and S-Plus* (Pinheiro and Bates 2000); *Mathematical Statistics* (Shao 2nd ed. 2003); *Statistics for Experimenters* (Box, Hunter and Hunter 2nd ed. 2005); *Data Monitoring in Clinical Trials: A Case Studies Approach* (DeMets et al. 2005); *Mathematical Statistics: Exercises and Solutions* (Shao 2005); *Response Surfaces, Mixtures and Ridge Analyses* (Box and Draper 2nd ed. 2007); *Applied Multivariate Statistical Analysis* (Johnson and Wichern 6th ed. 2007); *Introduction to Statistical Methods for Clinical Trials* (Cook and DeMets 2008); *Fundamentals of Clinical Trials* (Friedman et al. 2010); *Probability and Statistics for Engineers* (Miller and Freund 8th ed. 2011 but revised by Johnson since the 4th ed).

Sadly, we don't have complete pictures of the department. Figures 4 and 5 show some faculty from the twenty-fifth and fiftieth anniversary gatherings, respectively. Figure 6 shows all faculties ever in the UW-Madison Statistics Department. See our web page (www.stat.wisc.edu) for current regular and affiliated faculty, and for a collection of historical pictures.

The succession of chairs of the Statistics Department is: George E. P. Box (1960–1969), Irwin Guttman (1965–1966), Norman Draper (1968–1973, 1994–1997), George Tiao (1973–1975), John van Ryzin (1975–1977), Gouri Bhattacharyya (1977–1979), John Gurland (1979–1981), Richard Johnson (1981–1984), Robert Miller (1987–1991), Douglas Bates (1991–1994), Greg Reinsel (1997–2001), Erik Nordheim (2001–2005), Jun Shao (2005–2009), Kam-Wah Tsui (2009–2011), Brian Yandell (2011–).

Fig. 4 Statistics Department twenty-fifth Anniversary. George Box is flanked by Claire Box and Dennis Cox. Rich Johnson and his wife Bobbie are behind, as are Bob Wardrop, Thomas Wehrly, Tom Leonard, Ian Hau, Connie Shapiro (Page), David DeMets, Kam Tsui, Kyungmann Kim, and Doug Nychka

Fig. 5 Fiftieth Anniversary Organizing Committee (June 3–4, 2010). From left, Kjell Doksum, Grace Wahba, Jun Shao, Kam Tsui, Rich Johnson, Brian Yandell, and Kyungmann Kim

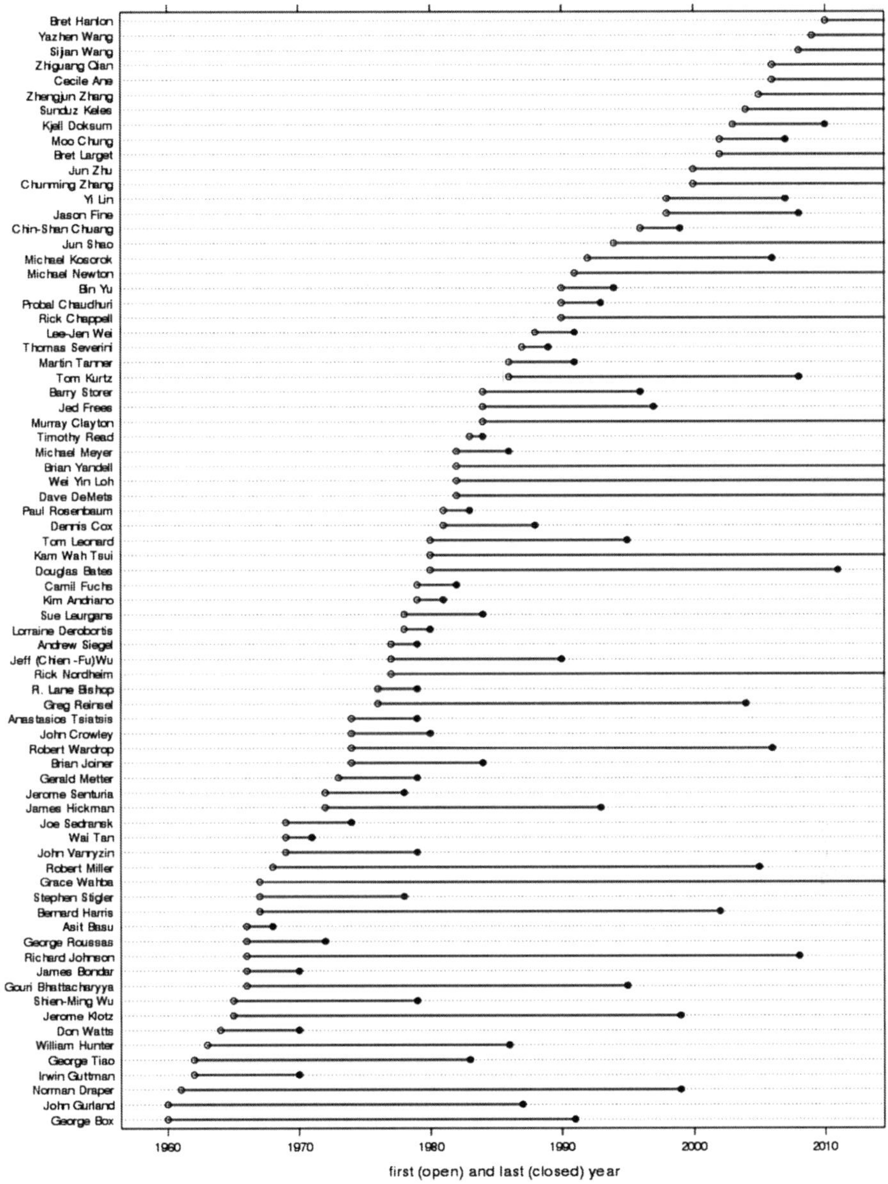

Fig. 6 History of statistics faculty at UW-Madison

While in Wisconsin, some of our faculties have been society presidents: George Box (ASA 1978; IMS 1980); Tom Kurtz (IMS 2006); Jun Shao (ICSA 2007). We have also had several editors of journals: Doug Bates (core development team of R, *Current Index to Statistics* 2001–2003); Richard Johnson (*Statistics &*

Probability Letters 1985–2010); Brian Joiner (*Current Index to Statistics*); Michael Newton (*Annals of Applied Statistics* 2006–2009); Jun Shao (*Journal of Multivariate Analysis* 2002–2005, *Sankhya* 2002–2007); Brian Yandell (*Amstat Online* 1999–2002)

Virginia Tech Department of Statistics

J. C. Arnold, K. Hinkelmann, G. G. Vining and E. P. Smith

The Department of Statistics at Virginia Tech celebrated 60 years of teaching, research, and service in 2009. Since the first Master's student, James Henry Davidson, graduated in 1945 we have granted over 507 B.S. degrees, 723 M.S. degrees, and 311 Ph.Ds. Our success is due to the foundation that was laid by Boyd Harshbarger starting in 1935 and the continued efforts of department heads and our faculty to maintain and enhance our programs. A brief history of the department is provided below. The departmental webpage (www.stat.vt.edu) contains other articles on the history as well as some interviews with selected faculty and graduates.

The Beginning of the Department at VPI

Boyd Harshbarger was one of the early pioneers in Statistics in the United States. He founded one of the earliest Departments of Statistics in the country at Virginia Polytechnic Institute, popularly known in those days as VPI. In 1935, he organized and taught the first courses in Statistics through the Department of Mathematics. He received a Rockefeller Fellowship and left to pursue a PhD degree in 1940. He wrote his doctoral dissertation under the direction of the distinguished Professor W.G. Cochran at Iowa State College and George Washington University and returned to VPI in 1942.

After returning to VPI, Dr. Harshbarger organized mathematical statistics courses for the Mathematics Department and applied statistics courses for the

Virginia Tech,
Department of Statistics,
406 A Hutcheson Hall,
Blacksburg, VA 24061-0439, USA
J. C. Arnold · K. Hinkelmann · G. G. Vining · E. P. Smith (✉)
e-mail: epsmith@vt.edu

A. Agresti and X.-L. Meng (eds.), *Strength in Numbers: The Rising of Academic Statistics Departments in the U.S.*, DOI: 10.1007/978-1-4614-3649-2_39,

Fig. 1 1947 Regional Summer Session

Agricultural Economics Department. A Statistics option was soon offered jointly by the two departments.

1947 Regional Summer Session

During those early years Dr. Harshbarger aggressively started working toward a degree-granting Department of Statistics. A major happening towards that goal was one of which he was very proud. It was the 1947 Summer Session in Statistics, a joint effort of VPI, The Iowa State College, The Survey Center of the University of Michigan, The University of North Carolina, and the US Department of Agriculture (Fig. 1). This was the first such session in Statistics in the United States. It was a huge success with over 100 students from 37 states and six foreign countries. There was a very distinguished faculty which included Raymond Jessen of Iowa State College, Maurice Kendall of the University of London, Renis Likert of the Survey Research Center, George Snedecor of Iowa State College, R.C. Bose of the Indian Statistical Institute, and of course, Boyd Harshbarger of Virginia Polytechnic Institute.

In addition to the distinguished faculty, the list of seminar speakers included Maurice Kendall, George Snedecor, Gertrude Cox, W. E. Deming, and Harold Hotelling. Although separated by great distances, these scientists supported the department by attending the summer sessions and supporting the department in other ways. One contribution was the establishment of the Kendall Library which holds many of the books and journals from Sir Maurice Kendall's personal library. Kendall had a long-standing close connection to the department which he visited several times.

1948 Statistics Section and Statistical Laboratory Forms a Separate Department of Statistics

After the successful Summer Session, momentum towards a formal department of statistics picked up dramatically. Boyd Harshbarger was able to convince the President of VPI that, although the Ivy League Schools did not yet have statistics departments, VPI should have one. A Statistics Section and Statistical Laboratory were organized in 1948 with a separate budget in the college's Agricultural Experiment Station. The laboratory was organized to offer statistical consulting for the Agricultural Experiment Station as well as the entire college. It still serves that purpose as the Laboratory for Interdisciplinary Statistical Analysis (LISA) for the entire university and also serves to train the department's graduate students.

Courses in Statistics were taught as far back as 1927 at Virginia Tech and M.S. degrees in statistical topics date back to 1945 (James Henry Davidson, The Estimation of a Missing Value in a Lattice Design Using Inter-and Intra-Block Information.). However, it was in 1949 that a separate Department of Statistics was approved, offering a Master of Science degree in Statistics. This was one of the first degree-granting statistics departments in the United States. The Department was housed administratively in the College of Arts and Sciences. Milton Terry was hired in 1949 and then in 1950 Ralph Bradley and David Duncan were hired, bringing the faculty size to four.

PhD Program Approved in 1951

The formal approval of the PhD program came in 1951. The faculty size grew to seven in 1953; it included Ralph Bradley, David Duncan, M.C.K. Tweedie, P. G. Sanders, C. Y. Kramer, Tom Russell, and Boyd Harshbarger. The faculty size grew rapidly to ten in 1955 as shown in Fig. 2.

All of these gentlemen had distinguished careers. Ralph Bradley later founded and developed the Department of Statistics at Florida State University. David Duncan was a prolific researcher and gained fame through the wide use of his multiple comparison test. Clyde Kramer extended Duncan's procedure and this new procedure was also extensively used worldwide. During the early years, there were many outstanding faculty members in the department at various times, including Paul Minton, who founded the Department of Statistics at Southern Methodist University, H. A. David, Rolf Bargmann, John Saw, and Leonard Shenton.

One of Harshbarger's significant hires was I. J. Good in 1967. Jack came to the Department of Statistics from Oxford University and was fond of telling us that he arrived on the 7th hour of the 7th day of the 7th month of year 7 in of the 7th decade (1967) and was put in apartment seven of block seven of Terrace View Apartments, all by chance. He served the department and greater community until his first retirement in 1987 at age 70. Following a conference to honor his retirement, the state

Fig. 2 1955 Statistics Department Faculty. Pictured are from left to right: Ralph Bradley, Lowell Wine, Hale Sweeny, Thomas Russell, William Thompson, John Freund, Kenneth Tweedie, Clyde Kramer, Paul Somerville, and Boyd Harshbarger

overturned the mandatory retirement, and Jack overturned his retirement request. He retired for a second time in 1994. Jack was fond of debating the value of Bayesian statistics and is pictured in Fig. 3 with Oscar Kempthorne in a quiet moment.

1957 B.S. Program Approved

The Bachelor of Science degree in Statistics was approved in 1957. The department at Virginia Tech was one of the earlier programs in Statistics to place an emphasis on the B.S. degree program. Undergraduate programs in Statistics were not typical in those days. The first undergraduate coordinator was Whitfield Cobb, who was followed by Professors Clyde Kramer and Walt Pirie.

Leadership of Department with SREB Committee on Statistics

The Statistics Department at Virginia Tech was instrumental in the development of Statistics in the South. A Committee on Statistics was coordinated through the Southern Region Education Board (SREB), and consisted of representatives from 14 southern states from Maryland to Texas. Their charge was to promote statistics graduate programs in southern colleges and universities. Boyd Harshbarger was the first chairman of this committee, and other members at that time included

Fig. 3 Jack Good (right) with Oscar Kempthorne

Gerrtude Cox (N. C. State), E. B. Brown (Georgia), G. E. Albert (Tennessee), Dana Card (Kentucky), T. W. Hildebrant (Oak Ridge Labs), Carl Marshall (Oklahoma A&M), Herbert Myer (Florida), George Nicholson (North Carolina), John Stockton (Texas), and Rutledge Vining (Virginia).

After the Boyd Harshbarger Era

Dr. Harshbarger resigned as Department Head in 1972 (and retired in 1976) and following a 1-year interim by Clyde Kramer, Jesse Arnold was elected as the second Department Head in 1973 (Fig. 4). During those almost 10 years of Arnold's tenure, the faculty size increased, and more structure was added to the graduate program. A major change was initiated in the governance of the department as faculty committees were utilized more than during the Harshbarger years. Departmental administration began to operate in the fashion of a chairmanship rather than a headship.

During those years, the Department greatly expanded the Statistical Consulting Center to offer statistical assistance to the entire university. Professor Raymond Myers took over the administration of the Center and changed it from a model based on a single faculty member with two to three clerks with desktop rotary calculators to a modern computer-based program with a statistical programmer and graduate student support. A major change that Myers implemented was the focus of the Consulting Center to emphasize the training of graduate students in statistical consulting and the support and involvement of the entire faculty in consulting activities. The photo of the faculty in 1976–1977 is given in Fig. 5.

In 1982, the reins of the department turned over to Klaus Hinkelmann. Klaus was widely recognized for his work on experimental design and continued the focus of the department on applied statistics.

Fig. 4 Maurice Kendall, Boyd Harshbarger, and Jesse Arnold

Fig. 5 1976–1977 Statistics Faculty. Standing is Clyde Kramer, Jerry Mann, Jack Good, Marvin Lentner, Larry Lee, Jesse Arnold, Marion Reynolds, Don Jensen, Ray Myers, Dick Krutchkoff, Klaus Hinkelmann and Ray Gaskins. Seated are Walt Pirie, A. Katiyar, Tom Bishop, S.K. Lee, Bob Schulman, and Eugene Seneta

The department kept a position open for visiting faculty who taught specialty courses for graduate students, presented several colloquia throughout their tenure, and interacted on research with faculty members and students. Among the visitors were: Eugene Seneta (Australian National University, 1976–1977), B.K. Ghosh (1978–1980), Oscar Kempthorne (Iowa State, 1983–1984), Sudhakar Dharmadikari (Northern Illinois University, 1984–1985), Donald Burdick (Duke University, 1985–1986), Charles McGilchrist (University of New South Wales, 1986–1987), Klaus Daniel (University of Bern, Summer 1988), Madan L. Puri (Indiana University, 1987–1988), Murari Singh (India, 1988–1989), Arijit Chaudhuri (Indian Statistical Institute, 1989–1990), Don Ramirez (University of Virginia, 1990), Steven Durham (University of South Carolina, 1990), Donald Michie (Turing Institute, Spring 1992), and Nozer Singpurwalla (George Washington University, Fall 1991). The visiting position was sadly lost to the budget cuts in the 1990s.

During these years, the number of graduate students varied from 50 and 70 per year, and the department awarded approximately 6 Ph.Ds and 20 M.S. degrees per year. This number is small compared to the number of PhD graduates in the 1960s and early 1970s when a large National Institute of Health (NIH) Training Grant supported most of our graduate students and the department graduated approximately 15 Ph.Ds per year. The course requirements for the graduate degrees were changed with the university wide change from the quarter to the semester system in 1988, but still focused on a solid program in experimental design and applied statistics.

Also, during these years, the conscientious effort to recruit undergraduates directly from high school or other quantitative programs at the university led the department to have one of the Clyde Kramer and largest undergraduate statistics programs in the country. Under the guidance of Walt Pirie, the number of undergraduate students grew from about 50 to between 80 and 90. This resulted in graduating between 12 and 25 students annually.

The Department, additionally, increased its course offerings (mainly service courses) at the Northern Virginia Graduate Center with regular and adjunct faculty. The plan was to offer a graduate program leading to the M.S. degree, but those plans never materialized when the budget cuts in the early 1990s took place, leading to a reduction of regular faculty there, and later to the demise of the program.

Although retired, Boyd Harshbarger was a common visitor to the department and frequently gave advice to faculty and the Department Head. His advice often indicated that the University of the 1940s had changed quite a bit (VPI had become Virginia Polytechnic Institute and State University, or VPI & SU, in 1970s). Klaus Hinkelmann recounted the following story: “Boyd Harshbarger came by the office occasionally to inquire about things. When I started to complain about budget cuts and reduction of funding, he invariably gave me the advice to ‘go to the president (of the university) and/or to the general at Ft. Lee.’ This is, of course, what he did—quite successfully—when he was department head. He did not understand or appreciate that this no longer worked, mainly because of a much more structured university (and military) bureaucracy with deans, vice presidents, provosts, etc.”

On the occasion of I. J. Good’s seventieth birthday, the Department organized and hosted a conference in 1987 on “Foundations and Philosophy of Probability and Statistics.” Speakers included Jim Berger, David Brillinger, Arthur Dempster, Persi Diaconis, I. J. Good, Joseph Kadane, Terry Seidenfeld, Oscar Kempthorne, David Kendall, Isaac Levi, Donald Michie, Donald Rubin, Mark Shervish, Jagdish Srivastava, Patrick Suppes, and Shelley Zacks. The proceedings were published as a special issue of the *Journal of Statistical Planning and Inference.* An interview with Jack was published in *Statistical Science* in 1996 (Banks 1996). The conference was originally intended to celebrate Jack’s retirement as well as his contributions. At the time there was a mandatory state retirement at age 70. A change in the law occurred after his announcement but before his retirement date, so he was able to have the celebration without the retirement (Jack officially retired in 1994).

Other colloquia also flourished: from 1984 to 1986 Yash Mittal organized, in addition to the weekly colloquia, the “Lecture Series in Probability and Its Applications” with national, and international speakers.

In the late 1980s, the Department began to develop a departmental computer laboratory and introduced the use of computer software, such as Minitab, and SAS in the classroom. This proved to be appreciated by the students as it made the subject of statistics a bit more interesting.

The Department currently has a very active Mu Sigma Rho chapter, which was initiated in 1976 by Oscar Kempthorne, one of the founders of the national organization. The chapter is run by student officers and sponsors two colloquia each year. Since 1976, its initiated membership has grown to 78 faculty and honorary members, 267 graduate students, and 148 undergraduate students.

Jack Good continued to be a keystone of the department during these years and in the 1990s there were numerous awards and videos commemorating his work. These include the video tapes "I. J. Good and Donald Michie in conversation with David Kahn and Karen Frenkel, Part 1: From codebreaking to computing: Remembrances of Bletchley Park fifty years later; Part 2: From wartime to peacetime, the years of development of computing and artificial intelligence" and "Good talking—I. J. Good conversation with Persi Diaconis" (for ASA film series of distinguished statisticians) were produced (see Banks, 1996 for more on Jack Good).

In 1993, the Department appointed Marvin Lentner as Head of the department. Six years later, the Department celebrated its fiftieth anniversary by holding a conference that brought in a number of alumni and speakers. The proceedings were published as special issues in the *Journal of Statistical Computation and Simulation* (a journal founded by one of our faculty members, R.G. Krutchkoff).

New Directions

Geoff Vining took over as department head in 1999 and was head until 2006. At the time of his hiring, the department had lost several faculty members to early retirements, and budget cuts created difficulties for the department. A major focus of his attention was the reinvigoration of the department and its continued emphasis on applied statistics, especially industrial statistics.

Under Vining, the department grew both in numbers and vision as a strong applied statistics program with an emphasis on industrial statistics. Two new senior hires, John Morgan and Bill Woodall strengthened our program in experimental design and industrial statistics. With Ray Myers, our program in industrial statistics was widely recognized as the strongest in the country as demonstrated by several important awards. Myers and Woodall have been honored by the American Society for Quality's (ASQ) Shewhart Medal, the society's top technical award. Marion Reynolds, Bill Woodall, and Ray Myers have twice received the ASQ Brumbaugh Award for the best paper published in a given year by an ASQ journal. Geoff Vining also has been awarded the Brumbaugh. In addition, Bill Woodall and Geoff Vining are past editors of the *Journal of Quality Technology*.

Also under Geoff the department initiated mentoring programs for new faculty. The department also improved courses for graduate and undergraduate students

Fig. 6 2009 faculty. *Bottom row*: Yili Hong, Leigh Harrell, Pang Du, Inyoung Kim, Ying Liu, Dong-Yun Kim, Marlow Lemons. Middle row: Leanna House, Feng Guo, Eric Smith, Golde Holtzman, Bill Woodall. Top row: Marion Reynolds, Scotland Leman, J.P. Morgan, Eric Vance, George Terrell, Jeff Birch. Not shown: Geoff Vining, and Ina Hoeschele

and developed mentoring programs for graduate students. Additionally, the Department developed a track system that allows students to focus on an application area as well as take a classical approach towards a degree. Current tracks include Biostatistics, Environmental Statistics, Computational Statistics, and Industrial Statistics. Another major contribution was the development of a Corporate Partners Program. These partners consists of statisticians from major companies that serve as an advisory group to the department and students, help students with financial support, and teach short courses.

The Department saw a decline in faculty numbers starting in 2004. This decline was due in part to retirements and a drop in the state budget which resulted in reductions at the university. Additionally, several faculty members left the department and new appointments were not made. In addition, the Statistical Consulting Center was closed due to budget issues.

Eric Smith replaced Geoff Vining in 2006. With the aid of a report by outside reviewers, a number of changes were made that led to several new faculty hires (the 2009 faculty are in Fig. 6). One of the major changes was a rebirth of the consulting program which was renamed as the LISA. The department was able to hire a director (Eric Vance) and an assistant director (Chris Franck) as well as support several graduate students. The program provides free support to staff, students and faculty throughout the University and is an excellent opportunity for students in statistics to develop communication skills and experience with real data.

The Department also made major changes to improve the quality of both the undergraduate and graduate programs. New graduate classes were added to provide students with a greater diversity of classes. The undergraduate program was revitalized by the addition of two instructors who have developed new courses and programs to enhance the undergraduate experience. Changes to the program include a new focus on computational statistics including Bayesian and non-Bayesian methods and simulation methods.

Summary and Outlook

Virginia Tech's Statistics department has a rich history and tradition that cannot be easily summarized in this short paper. The number of graduates continues to grow and graduates hold or held positions in prominent companies, universities, and government agencies. Our faculty and students have published numerous papers, received too many awards to mention, and have written many books. A few of the titles are given in the reference section.

The Department of Statistics has undergone many changes over its 60 years of history, and yet it has preserved its original core values in many ways. The department found its origins in collaboration and continues to this day to be a model of collaborative endeavor for scholars and their departments throughout the university. Our students receive the highest order of training to become professional collaborators while department research programs are predicated on the interface between scientific application and statistical theory. Students receive a sound foundation in theory, application, and collaboration and our faculty and students are not constrained to work in any single area. This was not only the philosophy of Boyd Harshbarger, but also that of Jack Good, Jesse Arnold, and so many other notables who followed in Harshbarger's footsteps. The Department of Statistics at Virginia Tech has a bright future, providing opportunities for every student, and faculty member to invent the future.

References and a Small Sampling of Relevant Books and Papers

Banks D (1996) A conversation with I. J. Good. Stat Sci 11:1–19

Gibbons JD, Chakraborti S (2011) Nonparametric statistical inference, 5th edn. CRC Press, Boca Raton (Jean Gibbons was the second PhD who graduated from our department)

Hinkelmann K, Kempthorne O (2008) Design and analysis of experiments. Introduction to experimental design, vol 1, 2nd edn. Wiley-Interscience, New York (a second volume was published in 2005)

Myers RH, Montgomery D (2002) Response surface methodology, 2nd edn. Wiley, New York

Reynolds MR Jr, Amin RW, Arnold JC, Nachlas JA (1988) X charts with variable sampling intervals. Technometrics 30:181–192

Vining GG, Kowalski S (2011) Statistical methods for engineers, 3rd edn. Brooks/Cole, Boston

Woodall WH (2000) Contradictions and controversies in statistical process control. J Quality Technol 32:341–350

Yale University Department of Statistics

John Hartigan

This history borrows extensively from a very complete and affectionate history of the Department in 1963–2000 written by Barbara (Amato) Kuslan available at stat.yale.edu.

Beginnings

From Stigler (1978), the first doctorate in Mathematical Statistics in the United States was awarded in 1876 by the Sheffield Scientific School at Yale to Mansfield Merriman for a thesis on least squares. A Yale graduate, Erasmus de Forest, did significant early research between 1870 and 1885 on the runs test for residuals, on Monte Carlo estimation of the variance of a statistic, and on the gamma distribution; he had no formal affiliation with the university though he lived in New Haven. (However, he is remembered at Yale for endowing the Erasmus de Forest professor of mathematics.)

The first statistics course at Yale was:

Economics 21: An elementary course in the use and abuse of statistics. The sources and reliability of statistical data are discussed, and the methods of distinguishing true and false inferences are pointed out.

Assistant Professor Irving Fisher, 1896–1897.

Professor Fisher, 1867–1947 was co-founder of the Econometrics Society, as well as the first PhD in Economics at Yale (under the direction of Willard Gibbs

Yale University,
Department of Statistics,
PO Box 208290
New Haven, CT 06520-8290, USA
J. Hartigan (✉)
e-mail: john.hartigan@yale.edu

A. Agresti and X.-L. Meng (eds.), *Strength in Numbers: The Rising of Academic Statistics Departments in the U.S.*, DOI: 10.1007/978-1-4614-3649-2_40,

and William Graham Sumner) and, as they say at Yale, true blue all the way through; undergraduate degree and graduate degree and entire academic career at Yale until his retirement in 1935.

The course Economics 21 (with evolving number, description, and content) has been taught in the Economics Department ever since.

In Public Health, the course Public Health 3, Principles of Vital Statistics, was first taught by Mr. Dublin, in 1918. Mathematics 28b, Elementary Statistics, was taught by Mr. Patton in 1922 and continued thereafter. A graduate course, Mathematics 190, Mathematical Statistics, was taught by Dr. Crum in 1921 and 1922, and by Professor Ore every second year in the 1930s, but was suspended during the war because of the diminished intake of graduate students. It was resumed in 1946 taught by Professor Votaw. A graduate course Psychology 104, Statistical Methods in Psychology, was taught for the first time in 1933 by Mr. Newhall. An undergraduate course, Psychology 22, Statistical Methods in Psychology, was taught for the first time in 1942 by Mr. Hovland.

In order to coordinate the teaching of statistics in many different departments, a university committee on Statistics, chaired by Professor E. Cuyler Hammond, Professor of Biometry in the Department of Public Health, was established in 1953, together with a Statistical Laboratory. The committee continued, in various forms, until the 1990s; it remains true that Statistics is taught in many departments. And the Statistical Laboratory has continued in various forms, and remains a valuable resource for the teaching of Data Analysis and Applied Statistics.

Founding of the Department

By 1963, there were nine undergraduate courses and 32 graduate courses in Statistics being taught in 10 different departments (five graduate courses in Economics and five in Epidemiology and Public Health). It was then that Yale University, which is not precipitate in these matters, decided that the time had come to set up a statistics department. On 1 July, Francis Anscombe, previously of Princeton University, became the founding chair, Alan James transferred in from the Yale Mathematics Department, and the biometrician Chester Bliss, who had been teaching a Biometrics course at Yale since 1942, became a lecturer. In the first year, the department offered courses at the undergraduate level: Introduction to Statistics, Mathematical Statistics, and Statistics Laboratory that had been previously taught by Alan James in the Mathematics Department. The first staff member was Barbara Amato, who was an ever gracious, generous, profoundly efficient, stalwart member of the department for 32 years. Readers are recommended to read her history of the department to see what the department was like from the inside during those years.

The initial focus of the new department was development of a graduate program. Frank Anscombe wished to ensure a substantial applied component. Thus, the 1964–1965 catalog read as:

> A distinctive feature of the program is a year of practical work under the supervision of a member of the department… it is believed that such practical work is highly effective in complementing the theoretical courses, and leading to real insight into the subject.

The new department was located in the Division of Social Sciences rather than Natural Sciences, since the majority of statistics courses being offered before its establishment were in the Social Sciences. This has had the effect of preserving a strong applied component in the department's graduate program; one of the requirements for the PhD is that students prepare a report on a statistical problem in collaboration with a faculty member from another department, under the direction of a Statistics faculty member. Three graduate students were admitted in 1963–1964, and one of these, Richard Olshen (now of Stanford University), was the recipient of the first PhD to be granted by the new department.

A generous 5-year grant from IBM assisted the setting up of the new department, helping to elegantly furnish Dana House, which became its home in 1965, and to allow for a number of distinguished visitors in the first years while the department was recruiting faculty. In 1964, Alan James returned to his home in Australia. The department secured the services of the brilliant polymath and eloquent Bayesian advocate Jimmie Savage, who became Eugene Higgins Professor of Statistics. In that year, the visiting professors were game theorist Robert Aumman (now Nobel Laureate), and Harald Cramer. The following year, the visiting professors were George Barnard and Cedric Smith.

Jimmie Savage

By 1970, the department was staffed by Professors Anscombe and Savage (Chair), Associate Professor Hartigan recently recruited from Princeton, Assistant Professors Feder, Margolin, and Muirhead, and Lecturers Hilary Seal (Accounting), and Chester Bliss (Biometry). Hartigan's interests included Bayes theory, data analysis, statistical computing, and classification. He was the founder of the Statistical Laboratory at Yale and initially developed Yale's computer-oriented data analysis courses. He was later president of the Classification Society, and one of the founding editors of Statistical Science.

In 1970, the graduate student enrollment was 13. There were graduate courses in probability, statistical theory, foundations of statistics, time series, linear models, experimental design, analysis of data matrices, stochastic theory of a risk business, and an introductory course for non-statistics graduate students. There were undergraduate courses in Probability, Statistics, and Data Analysis, descended from the original courses on those subjects that had been taught by Alan James in the Mathematics Department before the department was founded. Many more graduate statistics courses were offered in Administrative Sciences, Economics, Epidemiology and Public Health, Forestry, Political Science, Psychology, and Sociology. The graduate program had been established, with a significant Bayesian presence. We were making only a minor contribution to the undergraduate teaching of statistics, and to the demand for statistical teaching in the other disciplines.

Quoting from the Kuslan History:

> On November 1, 1971, our world was shattered when Jimmie Savage, just before his 54th birthday, died very suddenly of a heart attack at his home in New Haven. I think the words of his former student, Don Berry, (printed in The American Statistician, Volume 26, No. 1, February, 1972) say it best:
>
> On November 1, the statistical community lost one of its own. Whenever a man of the stature of Leonard J. Savage passes on, it is the custom of a periodical such as yours to cite what you deem to be his professional contributions. In the case of Jimmie Savage these can be nothing but impressive. Readers will lament that the profession has lost one of its most innovative members, perhaps the most innovative member. I am writing not to alter that impression but to augment it. No doubt you will not say in your tribute that he was a man of unlimited compassion, patience, and love. To his colleagues and students each moment with him was cherished—around him the atmosphere tingles, the tingles emanating not only from his sparkling intelligence and wit, but also from his empathy and warm-heartedness. He was ever willing to discuss a subject at any level, and ever capable of leading one's fallacious argument gracefully and elegantly over a nearby logical cliff. I cannot imagine the existence of another who combines raw intelligence and teaching ability as effectively as did Jimmie Savage.
>
> It is difficult for me at this moment to understand why the entire world is not grieving the loss of such a man. How unfortunate it is that so very few knew that he was in their midst. For me, and for most who knew him, and for all who loved him, there will remain an enormous gap in life. Speaking for myself, the person whom I most tried to impress is gone and no one could possibly take his place. I grieve as much for humanity as I do for his family.

It is painful, even at this distance in time, to reflect on the personal loss to the members of the department at Jimmie's sudden death. He was a great colleague, a great teacher, a great boss. His presence in the department, almost from its inception, gave immediate credibility to the department in the academic community and within the university. There was some question whether or not the department was viable without Jimmie. An informal contact at the University of Chicago suggested that the whole department (faculty and students) should be absorbed into the statistics department there. We decided quickly not to pursue this path: first of all, we rather liked where we were; second, our own administration was quick to show continuing support; and third, any prolonged consideration of abandoning ship is likely to cause abandoning ship! So, the immediate need of the department was to find a distinguished senior figure to begin rebuilding. Peter Huber agreed to visit for 6 months to help with our immediate needs, and we had a number of distinguished visitors over the next few years: Jim Press, Warren Hirsch, Willi Maurer, Ted Hannan, Leo Breiman, Beat Kleiner, Laszlo Engelman, Jose Bernardo. The visitors were valuable in adding to the intellectual resources of the department for the benefit of faculty and students alike, but of course the tasks of organizing teaching and mentoring students remained with the continuing faculty.

Faculty Positions

After a number of failed attempts at hiring senior faculty, the department was very pleased when Richard Savage, then at Florida State, joined the faculty in 1974. Richard had previously served as editor of the Journal of the American Statistical Association, and in 1974–1976 served as editor of the Annals of Mathematical Statistics. In 1984, he was president of the American Statistical Association. In 1974, the faculty consisted of Frank Anscombe, John Hartigan (Chair), Richard Savage, Barry Margolin, and Robb Muirhead. There were 16 PhD students enrolled.

In 1977, David Pollard was appointed Assistant Professor, and the faculty consisted of Frank Anscombe, John Hartigan, Mary Ann Maher, Robb Muirhead, David Pollard, Richard Savage (Chair). Eventually, David became the first Assistant Professor promoted to tenure, specializing in probability theory, empirical processes and LeCam theory.

Quoting from the Kuslan history:

> The University Committee on Statistics reported a need for additional coordination on the teaching of statistics within the University… In particular, the Committee recommended: (1) joint appointments between the Department of Statistics and the substantive departments; (2) cooperation between the departments in the teaching of basic courses at both undergraduate and graduate levels; and (3) the active participation of the administration to maintain a statistics program with efficiency and quality.

At that time, there were 19 graduate courses in the Department of Statistics, and 24 courses with substantial statistical content being offered by other departments in the Faculty of Arts and Sciences; and four undergraduate courses in the Department, with 10 undergraduate courses being offered outside the department. The Department of Epidemiology and Public Health offered 12 graduate courses.

As it always had been, more of the teaching of statistics was being done outside the department than by the department. Economics had a substantial distinguished group of econometricians, about the same size as our department; they had been teaching statistics 60 years longer than we had. The Division of Biostatistics within the School of Public Health was also about the same size. The other departments would typically have one or two members with an interest in statistics who would teach both elementary and advanced courses, and provide assistance in data analysis and experimental design to their colleagues. It is not such a bad model for these departments, but harmful to us because it keeps the size of our faculty small, and makes recruiting students and faculty more difficult. We have always argued that we should have principal responsibility for teaching the introductory level courses, which have much the same content in all the departments.

At Yale, departments used to have rigorously administered faculty counts which they clung to with the same determination as, say, Bosnia-Herzogovina clings to a border province. One senior faculty slot was equivalent to two junior slots, and if you promoted someone to tenure, you gave up two junior slots in exchange to preserve the total senior slot count. Retirements and resignations of tenured faculty nominally gave the provost half a slot, but in real life every such retirement and resignation resulted in a determined campaign to retain the full slot. The campaign was nearly always successful, being approved by the divisional committee composed of faculty members from different departments each of them totally averse to losing faculty slots. Thus, when a statistical expert in Department X retired, he or she was replaced by another statistical expert in Department X, not by a statistical expert, or even half a statistical expert, in the Department of Statistics. Not only that, if we argued too hard for such a replacement, we would be viewed by our colleagues in the other departments as dangerous revolutionaries to be suppressed, and our next campaign to retain a full slot after retirement would be just the time to send us a message. So the only way to expand our teaching was by cooperative action with the sister departments. An example would be to persuade the departments to allow their students to *choose* to take our introductory courses rather than to *require* them to take introductory courses in their own department.

Following the recommendations of the Committee on Statistics, we were given a half junior slot to be shared with another Social Science department; this seemed to be a sensible way to meet the desires of the departments for within house statistical expertise, and it has proved successful at Stanford … but nowhere else! For us, this experiment failed because we could never get a candidate who was regarded as acceptable by both departments. It is a rare junior faculty member who can meet the demands of two disciplines. Eventually, we were given another junior position.

Student Life

From the first days of the department, there has been a family atmosphere connecting students and faculty. Students were regularly invited to faculty homes for dinners and receptions for seminar speakers. We have two picnics each year. There were some pretty good Christmas parties, sometimes at the department, sometimes in the faculty homes, and usually with the kind of games favored by statisticians. At one of these, there was a prize for the person who guessed most closely the reading at midnight in kilowatt hours for energy consumption on our electric meter. As the witching hour approached, certain persons were to be found turning on all the lights in the upper stories and opening the windows, violating all norms of statistician's objectivity. It may have been the Bayesians. One year, the students produced a play, written by Stuart Newstead, "THE REAL TRAGEDY OF KING RICHARD II or HOW THE BAYESIAN WAS UNIFORMLY BEATEN", expressed in fine Shakespearean language skewering the faculty.

Another time, gamblin' Don Barry, who always tried to introduce a practical probability feature into the festivities, brought a set of videoed horse races, together with a program offering odds on each of the races. We would bet on the races according to the odds offered, watch the video, and win or lose fake money. Then a new race would be announced with different odds in the program with some of the same horses. The betting and paying off would continue. After a while we noticed uproarious behavior from the back of the room where Dave Pollard and some of his students were. Nope, they had noticed that if a horse won, his odds of winning increased in the next race; so, by looking at the future program they had a perfect wire on the results, and won every time!

Quoting from the Kuslan history, Jeff Simonoff writes:

> The way I always thought about the Department was that our being in Dana House couldn't have been more perfect, because it felt like a family (and you know who played the role of Mom – Barbara!). The faculty's office doors were always open, for professional or personal guidance. During my years at the Department, we went as a group for lunch to Commons almost every day – students, faculty and even Barbara. John and David were very regular participants, but Richard Savage often came, despite the logistical difficulties. Luncheon discussion could range from the fundamentals of Bayes methods to politics to the subtleties of NHL hockey. When I moved out the of graduate student dorm into an apartment, I had parties in two consecutive years, and they were attended not only by Stat students, but faculty as well. From everything I've heard from people who got their degrees at other stat departments, the warmth of the atmosphere at Yale was amazing, and absolutely unique. I consider myself very lucky to have been able to spend four years there.

Restructuring

In 1991, President Benno Schmidt embarked on a program to repair and refurbish Yale's physical plant, and decided that the Yale faculty should be reduced by 11% to help pay for the cost. A hiring freeze was instituted and a restructuring committee was set up to examine all of Yale's academic programs to decide where the cutting should take place. At this time, Frank Anscombe and Richard Savage had just retired, and we had just lost Stephan Morganthaler to ETH, so we were down to two tenured faculty (Pollard and Hartigan), and two non-tenured faculty (Chang and Roeder). We had a meeting of the faculty and graduate students so that everyone could learn what we knew of the situation, and I remember one of the students asking what was the chance that the department would still exist at the end of the process. I said about 95%, on the grounds that that was the chance that either Pollard or Hartigan would die in the next year. I am not sure that this number comforted either the faculty or the students.

However, we were approved to continue with our nominal faculty establishment, and we shortly appointed the eminent information theorist Andrew Barron to a tenured position, bringing substantive new directions in our research, and new energy and ideas for running the department.

Graduate Education

The standard course assignments in the department are for each faculty person to teach one undergraduate course, one introductory graduate course, and one advanced graduate course. We attempt to limit the number of regular courses taught every year to make it possible to preserve variety in the advanced courses, which could be on any topic that the instructor thought would gather an audience. In this way, we can teach material that we are doing current research on and give students ideas for thesis work. To name just a few, we have occasional courses on the Philosophy and History of statistics, Statistical Graphics, Probabilistic Convex Geometry, Probabilistic Network Algorithms, Weak Convergence and Empirical Processes. Stochastic Calculus, Counterexamples, Stochastic Models of Evolution, Computational Learning Theory, Classification, Resampling, Asymptotic Admissibility.

In addition to the regular visitor seminar, we had for a number of years a "surprise" lunchtime seminar in which the speaker was selected at random from whoever in the audience put their hands up. The students also ran a seminar for their own research in which the faculty were not particularly welcome! (I believe one of the faculty members was thought to ask too many questions.) At present we have in-house seminars on probabilistic networks, on genomic statistics, and also a statistical clinic oriented to applied problems and open to all students.

Undergraduate Teaching

In 1997, with David Pollard as Chairman, with substantial assistance from Donald Green, Chairman of the ubiquitous Committee on Statistics, we tried a new model for teaching introductory statistics courses. The teaching of the courses was shared with faculty members from Biology, Sociology, Political Science, and Psychology. In the beginning, a lecture on statistical methods was given by a teacher from the Statistics Department each week, and then a companion lecture was given by a teacher from one of the other participating departments illustrating the methods by examples taken from their own disciplines. Eventually, we switched to a new model in which the first half of the course was given by the statistician, and the second half by the teachers in the participating departments.

Amazingly, the cooperative model has worked, even though there are continuing complications in staffing and syllabus. In recent years, we have had the services of a brilliant teacher, Jonathan Reuning-Scherer, for the statistics part of the offering, and continued friendly cooperation with the departments. There are sections specializing in life sciences, political science, social sciences more generally, and data analysis, as well as a section on medicine led by David Salsburg, distinguished statistician from the pharmaceutical industry and author of *The Lady Tasting Tea: How Statistics Revolutionized Science in the Twentieth Century*.

Joseph Chang developed courses and seminars for a joint program in Stochastic Processes and Bioinformatics, and in 1998 received the William Clyde DeVane Medal—the oldest and highest award for distinguished scholarship and teaching in Yale College.

The Undergraduate Major

For many years, our undergraduate major was part of the Applied Math Program. We would have one or two students a year, who would take the advanced undergraduate courses in probability, theoretical statistics, and data analysis, some of our intermediate level graduate courses, and work on an undergraduate thesis with one of our faculty. Many of these students went on to advanced degrees in statistics or other disciplines.

We had been interested in starting our own major for some years but waited until the faculty had attained a sufficient size to offer a set of courses yearly that would support a high quality major. The major was proposed and approved in 2009. Prerequisites for the major are Multivariable Calculus, Linear Algebra, and a course in Computer Science; there are seven required courses in probability, statistical inference, and data analysis, including a senior project.

2010

In 2010, there are four tenured faculty (Barron, Chang, Pollard, and Zhou) and five non-tenured faculty (Chen, Emerson, Madiman, Reuning-Scherer, and Zhang). Harrison Zhou was promoted in this year, and also won the IMS Tweedie New Researcher Award. We teach 28 graduate courses and 12 undergraduate courses. There are 15 PhD students, 15 master's students, and four undergraduates in our just introduced undergraduate major. In other departments in the Faculty of Arts and Sciences, there are 15 graduate courses, and seven undergraduate courses where we do not share in the teaching. We have accomplished the goal of teaching most of the statistics courses in the Faculty of Arts and Sciences, and teach a good fraction of the introductory courses. We maintain good relations with the division of Biostatistics in Public Health, which is of about the same size and has been teaching statistics since 1918. When the department started in 1962, Colin White and Chester Bliss were joint appointments with us, and we have had a continuing series of joint appointments since then. We share seminars, students, faculty and data; there are a number of faculty in both departments with special interests in genetics.

We maintain a balance between research and teaching, and between theory and applications. At present, we have three open seminars each week, a general seminar, a seminar on probabilistic graph theory, and a seminar on Bioinformatics. We maintain a Friday consulting clinic to which Yale researchers present statistical problems, with two or three faculty and a number of graduate students attending.

Our Alumni

Former students have had senior faculty positions at Berkeley (Nolan), Stanford (Olshen), Columbia (DuMouchel), Duke (Berry), University of North Carolina (Carlstein), Minnesota (Oehlert), and at many other distinguished universities.

Don Barry is the president of Limerick University, Ireland. Antonio Possolo is the Director of the Statistical Engineering Division at the National Institute of Standards and Technology. Takashi Miyawaki is the Dean of the University of Tsukuba in Japan.

Faculty Accomplishments

Frank Anscombe, the founder of the department, was an eclectic Bayesian who did early fundamental work on sequential sampling, and on practical diagnostic methods in regression problems. Jimmy Savage was perhaps the most prominent American Bayesian in the post war period, especially after the production of his famous work "The Foundations of Statistics". John Hartigan wrote the first book "Clustering Algorithms" studying classification from a statistical point of view, and some of the algorithms, such as "k-means", are still widely used. He did some of the early work on resampling using subsamples. His Bayes theory book is the only Bayes book that attempts to formally deal with probabilistic reasoning in which some of the probabilities are infinite. David Pollard in his books on Convergence of Stochastic Processes and on Empirical Processes gave a rigorous and practical mathematical formulation for many of the standard asymptotic processes that arise in Statistics. Andrew Barron uses information theory to develop novel probabilistic limit theorems, studies risk analysis for Bayes methods, model selection problems, and optimum portfolio growth problems. Joseph Chang works in stochastic processes, statistical genetics, and bioinformatics; he developed a probabilistic model for evolution that permitted estimates of the time back to the most recent common ancestor of all present-day humans.

Selected Books and Papers, and Reference

John Hartigan, 1975, *Clustering Algorithms*, John Wiley

John Hartigan, 1983, *Bayes Theory*, Springer Verlag

David Pollard, 1984, *Convergence of Stochastic Processes,* Springer Verlag

David Pollard, 1990, *Empirical Processes: Theory and Applications.* Volume 2 NSF-CBMS Regional Conference Series in Probability and Statistics, IMS, Hayward CA.

Joseph Chang, 1999, Recent common ancestors of all present-day individuals. *Advances in Applied Probability*, 31: 1002–1038.

David Pollard, 2002, *A User's Guide to Measure Theoretic Probability,* Cambridge

Andrew Barron (1986) Entropy and the central limit theorem, *Annals of Probability*, 14: 336–342.

John Emerson, (with Miki Seltzer, and David Lin), 2009 Assessing Judging Bias: an Example from the 2000 Olympic Games. *The American Statistician* 63(2): 124–131.

Statistics Department 1968

First row: Susan Sumner, Barbara (Amato) Kuslan, Frank Anscombe, Marilyn Shields, Sue Ling. Second row: Paul Feder, Hilary Seal, Jimmie Savage, Chester Bliss, Colin White, Barry Margolin. Third row: Yuzo Hosoya, Bob Ling, Jack Alanen, John Mellman. Top row: Norman Johnson, Bill Chin, Pierre Pilon, Bill Winter, and Charles Patrick.

Harrison Zhou, (with G. Golubev, M. Nussbaum) (2010) Asymptotic Equivalence of Spectral Density Estimation and Gaussian White Noise *Annals of Statistics* 38: 181–214.

Mokshay Madiman (with S. Bobkov) (2011) Concentration of the information in data with log-concave distributions *Annals of Probability* 39:1528–1543

Stephen Stigler (1978) Mathematical Statistics in the Early States *Annals of Statistics* 6:239–265

Roster of Faculty, Visitors, and Staff 1963–2010

Prepared by Barbara Kuslan

Faculty

Francis J. Anscombe (1963–1988); Alan James (1963–1964); Leonard J. (Jimmie) Savage (1964–1971); John A. Hartigan (1969–2006); I. Richard Savage (1974–1991); David B. Pollard (1976–); Joseph (Joe) Chang (1989–); Andrew R. Barron (1992–); Huibin (Harrison) Zhou (2004–).

Statistics Department 2003

First row: Jay Emerson, Qi Yan, Jake Katz, Jonathan Reuning-Scherer. Second row: Andrew Barron, Hannes Leeb, Joe Chang, Hani Doss, John Ferguson. Third row: Gheorghe Doros, Summer Han, Daniela Doros, Dumitru Trucu, Marios Panayides. Fourth row: Kathryn Young, Susan Clapp (Perrone), Anna Kochetkova, Wei. Qiu, Dean Palejev, Xiao-Xian Luo, Jing Cao, Chandra Erdman, Andreas Glaser, John Hartigan.

Associate Professors

Barry H. Margolin (1967–1977); Robb J. Muirhead (1970–1978); Kathryn Roeder (1988–1994); Nicolas Hengartner (1993–1999); Hannes Leeb (2003–2007); John (Jay) Emerson (2003–).

Assistant Professors

Joseph (Jay) B. Kadane (1966–1969); Paul I. Feder (1967–1970); Mary Anne Maher (1977–1978); Kai Fun Yu (1978–1984); Stephan Morganthaler (1984–1988); Ross Ihaka (1985–1987); Offer Kella (1994–1995); Marten Wegkamp (1997–2002); Dragan Rudulov (2001–2002); Mokshay Madiman (2005–); Lisha Chen (2006–); Jing Zhang (2010–).

Lecturers and Research Staff

John A. Carpenter (1964–1965); Geoffrey F.W. Yeo (1964–1965); Chester I. Bliss (1964–1969); Hilary L. Seal (1964–1972); George A. Barnard (1965); Mary H.

Regier (1965–1966); James (Jim) Dickey (1965–1967); Richard Olshen (1966–1967); Edward J. Hannan (1973–1974); Willi Maurer (1972–1973); Beat Kleiner (1976); Jose J. Bernardo (1977–1978); Susan R. Wilson (1978–1979); Jon K. (Kim) Peck (1979–1983); Diccon R.E. Bancroft (1982–1985); Leticia Cuellar-Montoya (1993–1994); Jonna Kwiatkowski (2003); Marek Chawaski (2003); Jonathan Reuning-Scherer (2002–); David Salsburg (2004–); Fred Djang (2007); Balaji Raman (2009–2010); Michael Kane (2010).

Visiting Professors and Scholars

Robert J. Aumann (1964–1965); Harald Cramer (1965); Cedric A.B. Smith (1966); Joseph (Joe) Kruskal (1966–1967); Martin Schatzoff (1967–1968); Peter J. Huber (1971–1972); S. James Press (1972–73); Warren Hirsch (1972–1974); Leo Breiman (1973–1974); Laszlo Engelman (1976–1978); Doris C. Appleby (1980–1981); Adrian Baddeley (1981); Nell Sedransk (1984–1985); Benedikt Potscher (1986–1987); Grace Wahba (1987–1989); Daniel (Dan) G. Barry (1988); Gabor Szekely (1989); Ursula Gaither (1990); Bruce Lindsay (1990–1991); Evangelos Tabakis (1992); Bin Yu (1992–1993); Richard A. Vitale (1993–1994); Johannes Ledolter (1994–1995); Philip J. Everson (1995–1996); Junichi Takeuchi (1996–1997); Setanay Oner (1998); Philippe Barbe (1999–2000); Robert Muirhead (2000–2001); Hani Doss (2003–2004); Santiago Mavarro (2008); Antony Unwin (2009).

Referees

Ted Anderson
Colin Begg
Jim Berger
Roger Berger
James Booth
Hendricks Brown
Larry Brown
James Calvin
George Casella
Kathryn Chaloner
Herman Chernoff
Ted Colton
Richard Cornell
Dennis Cook
Noel Cressie
Peter Diggle
George Duncan
Mitchell Gail
Leon Gleser
Leo Goodman
Ronald Gulezian
William Harkness
John Hartigan*
J. Stuart Hunter
Peter Imrey
Jon Kettenring*
Tim Killeen
Uwe Koehn
Michael Kosorok
Larry Kupper
T. L. Lai
J. Richard Landis
Tom Louis
Jim Lynch
Robert Mason
Charles McCulloch
Joseph McKean
Michael Meyer
Mel Moeschberger
David Moore
Joseph Newton
Deborah Nolan
Ingram Olkin*
Joe Padgett
Juni Palmgren
Emanuel Parzen
Michael Perlman
Dawn Peters
Brett Presnell
Ronald Randles
Larry Ringer
Kathryn Roeder
Bob Riffenburgh
George Roussas
Richard Scheaffer
Robert Shumway
Robert Smythe
Terry Speed
Kenneth Stanley
Hal Stern
Stephen Stigler*
Maura Stokes
George Tiao
Luke Tierney
Robert Tsutakawa
Ray Waller
Marten Wegkamp
L. J. Wei
F. Tim Wright
Keying Ye
Marvin Zelen

* Reviewed more than one chapter

A. Agresti and X-L. Meng (eds.), *Strength in Numbers: The Rising of Academic Statistics Departments in the U.S.*, DOI: 10.1007/978-1-4614-3649-2,

Printed by Publishers' Graphics LLC
AMZ20130203.19.19.75